XVA Analysis

XVA Analysis: Probabilistic, Risk Measure, and Machine Learning Issues offers readers an up-to-date and comprehensive exploration of the X-Value Adjustment (XVA) universe and of the embedded risk measure issues inherent within it. The book tackles this subject through the triple lens of finance (wealth transfers), stochastic analysis (enlargement of filtration and BSDEs), and numerical computations.

The traditional credit valuation adjustment (CVA) desk compensates the trading desks for the cash flows that they lose in case of defaults of their counterparties. The Treasury of the bank funds the activity of the trading desks and of the CVA desk at the risk-free rate. The CVA desk and the Treasury charge their costs to the clients of the bank at a valuation level ensuring to the shareholders of the bank corresponding PnL processes that are martingales relative to a fininsurance probability measure calibrated to the market and consistent with the physical probability measure given the market. The management of the bank charges to the clients of the bank a capital valuation adjustment (KVA) risk premium, turning the overall dividend process of the bank shareholders into a submartingale in line with a target hurdle rate on their capital at risk within the bank.

This is the essence of the cost-of-capital XVA approach, which can also be used in reverse engineering mode, for determining the price range of a new deal that improves the implied hurdle rate of the bank shareholders. The advent of XVAs reflects a shift of paradigm regarding the pricing and risk management of financial derivatives, from hedging to balance sheet optimization.

Features

- A systematic coverage of the cost-of-capital XVA approach
- Unprecedented coverage of neural network regression methodologies
- Numerous illustrative figures and examples
- Suitable as supplementary reading for graduate students and as a practical reference for professional quantitative analysis and risk managers

Stéphane Crépey is a professor at the mathematics department of Université Paris Cité, in charge of the team mathematical finance and numerical probability at LPSM (Laboratoire de Probabilités, Statistique et Modélisation) and of the M2MO quantitative finance program.

Chapman & Hall/CRC Financial Mathematics Series

Recently Published Titles

Foundations of Quantitative Finance
Book IV: Distribution Functions and Expectations
Robert R. Reitano

Foundations of Quantitative Finance
Book V: General Measure and Integration Theory
Robert R. Reitano

Computational Methods in Finance, Second Edition
Ali Hirsa

Interest Rate Modeling
Theory and Practice, Third Edition
Lixin Wu

Data Science and Risk Analytics in Finance and Insurance
Tze Leung Lai and Haipeng Xing

Foundations of Quantitative Finance
Book VI: Densities, Transformed Distributions and Limit
Robert R. Reitano

Malliavin Calculus in Finance
Theory and Practice, Second Edition
Elisa Alòs and David Garcia Lorite

Arbitrage and Rational Decisions
Robert Nau

Equity Release Products
Radu S. Tunaru and Enoch B. Quaye

Quantitative Finance with Case Studies in Python
A Practical Guide to Investment Management, Trading and Financial Engineering
Chris Kelliher

Handbook of Quantitative Sustainable Finance
Edited by Peter Tankov, Ruixun Zhang

For more information about this series please visit: https://www.routledge.com/Chapman-and-HallCRC-Financial-Mathematics-Series/book-series/CHFINANCMTH

XVA Analysis

Probabilistic, Risk Measure, and Machine Learning Issues

Stéphane Crépey

CRC Press
Taylor & Francis Group
Boca Raton London New York

CRC Press is an imprint of the
Taylor & Francis Group, an **informa** business
A CHAPMAN & HALL BOOK

Front cover image: Stéphane Crépey

First edition published 2026
by CRC Press
2385 NW Executive Center Drive, Suite 320, Boca Raton FL 33431

and by CRC Press
4 Park Square, Milton Park, Abingdon, Oxon, OX14 4RN

CRC Press is an imprint of Taylor & Francis Group, LLC

ISBN: 978-1-041-01420-1 (hbk)
ISBN: 978-1-041-01475-1 (pbk)
ISBN: 978-1-003-61494-4 (ebk)

DOI: 10.1201/9781003614944

Typeset in Latin Modern font
by KnowledgeWorks Global Ltd.

Publisher's note: This book has been prepared from camera-ready copy provided by the authors.

To Claudio Albanese, for his invaluable source of inspiration, with friendship.

Foreword

When I talk with an academic colleague about XVAs for addressing the necessity to adopt a derivatives pricing approach at the bank portfolio level, he would suggest to proceed by utility maximization of the terminal wealth of the bank. That would be canonical and the ensuing price could perhaps be represented as an expectation under some risk-neutral measure. But a bank does not have the data regarding which utility should be used in this task and for which parameters. If it had the data, it would not have the computing power to solve the ensuing nonlinear optimization at the bank portfolio level. If it had the power, it would be unreasonably costly to do it. Even if it had all the resources of the world, a bank would never be willing to do it this way, because such an all-in-one, integrated approach is missing the expertise of its tens of thousands of traders and quants specialized on sub-tasks of the global problem, which should therefore be addressed in a "divide-and-conquer" manner: by determining, first, the trade-additive mark-to-market value of each deal, and, only in a second stage, a nonlinear correction valued at the portfolio level.

But on which theoretical basis to compute the latter? The answer is simple: we know it already. Since two centuries, bank shareholders expect a return on their capital at risk of an order of 10 to 15%.

Why bothering if this is the case? Well, because this is a bit vague. The book formalizes this idea in the case of the derivative portfolio of a bank. Note that for people who don't like the ideas of exogenous or target hurdle rates, the resulting cost-of-capital XVA approach can also be applied in the reverse engineering mode, for determining the price range of a new deal that improves the implied hurdle rate of the bank shareholders. Cost of capital proxies have always been used to estimate return on equity. Whether it is used in the direct or in the implied mode, the KVA is a refinement, dynamic and fine-tuned for derivative portfolios, but the base concept is far older than even the CVA.

All is not perfect. My academic colleague could object that the treatment of the "fininsurance" probability measure, advocated for XVA calculations in this book, is a bit cosmetic. Financial practitioners are bound to find the numerical calculations of this book too complex already. Hopefully, somebody will take over.

Stéphane Crépey
Paris, November 20, 2025

Contents

List of Figures

List of Tables

List of Algorithms

Preface

CONTEXT

The 2007–09 subprime and ensuing global financial crisis as well as the 2011 European sovereign debt crisis have highlighted counterparty credit risk. They have been the prelude of strongly reinforced regulatory collateral and capital requirements. Since then investment banks compute various cross-valuation adjustments (XVAs) to assess counterparty credit risk and its funding and capital implications. This having gone in hand with a simplification of the financial products themselves, it has been said that XVAs were the last generation of exotic options. Indeed XVAs deeply affect the derivative pricing task by making it global (portfolio-wide), nonlinear and entity dependent. A proper financial understanding of even the first-generation XVAs (CVA, DVA and FVA, where C sits for credit, D for debt and F for funding) refers to the distinction between firm and shareholder valuation, which mathematically goes to enlargement of filtration and absolutely continuous probability changes. Second-generation XVAs involve not only conditional expectations for prices but also conditional risk measures for MVA (margin valuation adjustment) and KVA (capital valuation adjustment) computations. Moreover, the fungibility of capital at risk with variation margins leads to an anticipated BSDE (backward stochastic differential equation) for the FVA, with a coefficient entailing a conditional risk measure of the one-year-ahead increment of the martingale part of the FVA itself. Finally, the trend of the regulation is to push participants to negotiate centrally via central counterparties. Extensive collateralization and central clearing trigger a shift from counterparty credit risk to liquidity and systemic risks.

CONTENTS OF THE BOOK

This book provides an up-to-date coverage of the XVA universe and of the embedded risk measure issues, from the triple angle of finance (wealth transfers), stochastic analysis (enlargement of filtration and BSDEs) and numerical computations. The XVA equations are addressed mathematically by projection on a reduced filtration in which the default of the bank only appears through its intensity (but the defaults of the bank counterparties are still present as binary events and not only probabilities). The ensuing all-inclusive

XVA add-on, to be sourced from clients incrementally at every new deal, reads (CVA + FVA + MVA + KVA), where C sits for credit, F for funding, M for (initial) margin and where the KVA is a cost of capital risk premium. This total amount can be interpreted dynamically as the cost of the possibility for the bank to go into run-off, while staying in line with shareholder interest, from any point in time onward if wished. From a computational viewpoint, a purely, multiply nested Monte Carlo treatment of the XVA equations would involve five compound layers of nested dependence (not to mention the ones induced by the time-discretization of the involved BSDEs). The approach favored in the book, instead, consists of neural net regressions and quantile regressions that are used for "cutting" the recursive levels of dependence of the XVA suite. These regressions are optimally implemented in pytorch with GPU acceleration. Importantly, they can be validated by companion twin Monte Carlo procedures. This is illustrated in the book by XVA computations on both toy and realistically large derivatives portfolios, in bilateral and centrally cleared trading setups. Beyond the XVA economico-mathematical foundations and numerical schemes that correspond to the first two parts of the book, the third part demonstrates that besides their primary pricing and accounting purposes, XVA metrics can also be extremely useful for risk, stress testing and reverse stress testing analysis.

Former XVA books by Brigo, Morini and Pallavicini (2013), Crépey, Bielecki and Brigo (2014), and Gregory (2015) were mainly about the aforementioned first-generation XVAs. Crépey, Bielecki and Brigo (2014, Part IV) also offer a thorough treatment of dynamic copula models of default times, which we capitalize on without returning to it in this book. But these books were all unaware of the wealth transfer issues that have since been understood to be crucial to the (even first-generation) XVA matter. Green (2015) is the first XVA book also encompassing MVA and KVA. The main theoretical differences between the XVA replication-based approach of his book and the cost-of-capital approach of this one, and their tangible consequences, are synthetized in II.§5. From a numerical viewpoint, the emphasis of this book is on neural network regression methodologies, unprecedented in book form, that allow capturing the XVA metrics not only at time 0 but at all nodes of a simulated grid of the risk factors. We also develop the applications of the XVA metrics to centrally cleared trading, in III.§4-§5 and Chapters VIII-IX, and to risk, in Chapters VIII to XI, the latter proposing a quantification of model risk in terms of hedging valuation adjustment (HVA).

PREREQUISITES

This book assumes some knowledge of stochastic analysis, mathematical finance and numerical finance at the MSc level (see e.g. Lamberton and Lapeyre

(1996) or Crépey (2013)). Some knowledge of corporate finance is also useful, in particular the Modigliani-Miller theory (see I.§3.E.2). By value-at-risk (VaR), we mean left quantile at a given confidence level $\alpha = 1 - \epsilon$, for some small and positive ϵ (from say 2.5% to 1‰); by expected shortfall (ES) of a loss ℓ we mean the expectation of ℓ given ℓ exceeds the corresponding value-at-risk of ℓ. ES is a sub-additive (and coherent) risk measure: unlike VaR, it acknowledges the risk benefit of diversification. The 2007–09 global financial crisis has prompted a regulatory shift from VaR to ES as the reference risk measure. As it results from Föllmer and Schied (2016, Proposition 4.51 page 234, Theorem 4.52 page 235, and Corollary 4.54 page 237), the expected shortfall operator is $(1-\alpha)^{-1}$ Lipschitz from the space of integrable losses ℓ with continuous distribution (endowed with the topology induced by the L_1 norm) into $\mathbb{R}$, and to $\mathbb{R}_+$ when restricted to centered losses ℓ.

NOTATION

A measurable space $(\Omega, \mathcal{A})$ is fixed throughout the book. By default, a random variable is real-valued and $\mathcal{A}$ measurable; we omit any dependence on $\omega \in \Omega$ in the notation. We write $\mathbf{1}_A$ and $\mathbb{1}_A$ for the indicator of a deterministic or random set A, and $x^{\pm} = \pm x \mathbf{1}_{\pm x > 0}$ for any real x. We denote by $\boldsymbol{\delta}_x$, a Dirac measure at x; by $X^{\tau} = X_{\cdot \wedge \tau}$, a process X stopped at some (possibly random) time τ; by $X^{\tau -} = \mathbb{1}_{[\![0,\tau[\![}X + \mathbb{1}_{[\![\tau,+\infty[\![}X_{\tau-}$ and ${}^{\tau-}X = X - X^{\tau-}$(assuming that the left-limit $X_{\tau-}$ exists), the processes X respectively stopped before and starting before τ.

Unless explicitly specified, an amount paid means effectively paid if positive, and received if negative. A similar convention applies to the notions of loss and gain or cost and benefit.

NUMBERING

Except for figures, tables and algorithms, which are numbered linearly throughout the book and recapitulated in the above lists, we use numbering per chapter. The numbering is local to the current environment, e.g. B means Subsection B in the current section, §3.B refers to Subsection B of Section 3 (other than the current one) in the current chapter, IV.§3.B to Subsection B of Section 3 in Chapter IV (other than the current chapter). Appendices are signaled by $(*)$.

INTRODUCTION

In counterparty credit risk complete markets, collateral and capital requirements would be indifferent to banks. The quantification by banks of market incompleteness based on the various XVA metrics has emerged as the unintended consequence of the FRTB banking reform (Bank for International Settlements (2019b)) and of more demanding regulatory capital requirements (European Parliament, 2013). The related risks are in fact reckoned today as the major risks for banks, well ahead of market risk (European Bank Authority, 2022, Figure 65 page 67).

The cost-of-capital XVA approach developed in this book provides a pricing, accounting, collateralization and dividend policy for a dealer bank, sustainable in the sense of ensuring to its shareholders a constant instantaneous return rate h (e.g. 10%) on their capital at risk, even in the limiting case of a portfolio that would be held on a run-off basis, i.e. without future deals. A contrario, Figure 1 shows the Ponzi scheme on financial derivatives preceding the 2007–09 global financial crisis, whereby, in the absence of such policy, the only way for banks to maintain their shareholders' profitability was to issue always more products and notionals. As detailed in Table 1, the cost-of-capital XVA approach can be seen as an investment banking, genuinely dynamic and continuous-time version of the Solvency II insurance methodology, driven by the same motivation for a sustainable (financial or insurance) system and economy.

Table 1: *Left*: Cost-of-capital XVA banking approach; *Right*: Solvency II insurance methodology (with SCR = $(\text{EC} - \text{KVA})^+$ for shareholder capital at risk everywhere).

contra-assets CA = **CVA** + **FVA**	**liabilities** best estimate, also called market consistent valuation
priced by conditional expectation of related future cash flows	priced by conditional expectation of related future cash flows
economic capital EC	solvency capital requirement
sized as a conditional expected shortfall of future losses over one year	sized as unconditional expected shortfall of future losses over each successive year
capital valuation adjustment **KVA**	market value margin or **risk margin**
sized as a supermartingale with drift coeff. hSCR and zero terminal condition	sized as summed future (deterministic) hSCR

The advent of XVAs reflects a shift of paradigm regarding the pricing and risk management of financial derivatives, from hedging to balance sheet optimization.

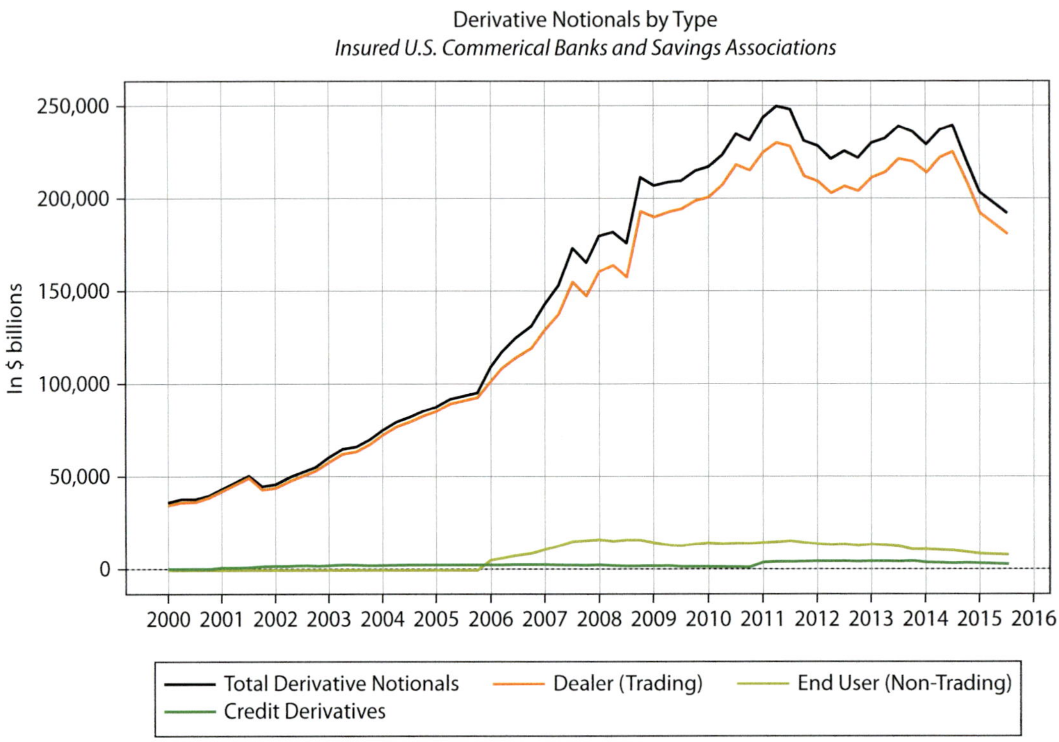

Figure 1: Ponzi scheme in the 2007–09 global financial crisis (*source*: Office of the Comptroller of the Currency, Q3 2015 quarterly bank trading revenue report).

The Sustainable Black-Scholes Equations

This chapter is a metaphorical introduction to our topic. The main driver of XVAs, default events, is not present, not even through some probabilities. We put it all down to liquidity! However the ideas of entity- and legacy-dependent pricing rules aligned to some hurdle rate are already here, as well as the semi-linear structure of the ensuing valuation equations or their anticipated feature in the case where the bank can use its capital at risk for funding its trading. A first glimpse into the associated model risk issue reveals that robust pricing may not be the most appropriate approach for such a purpose. We will indeed propose a rather different approach in the concluding Chapter XI of the book.

§1 INTRODUCTION

In incomplete markets, a basic Black-Scholes perspective has to be complemented by the valuation of market imperfections. Otherwise this results in Black-Scholes Ponzi schemes, such as the ones at the core of the 2007–09 global financial crisis (see Figure 1), where always more derivatives need to be issued for remunerating the capital attracted by the already opened positions. In this chapter we consider the sustainable Black-Scholes equations that arise for a portfolio of options if one adds to their trade additive Black-Scholes price, on top of a nonlinear funding cost, the cost of remunerating at a hurdle rate, and the residual risk left by imperfect hedging. Additionally, we assess the impact of model uncertainty in this setup.

§2 revisits the pricing of a book of options accounting for the cost of capital and cost of funding, which are material in incomplete markets. §3 specifies the pricing equations to a Markovian Black-Scholes setup. §4 assesses the impact of model risk in an uncertain volatility model (UVM) setup. §5 refines the ensuing model risk add-ons by accounting for calibrability constraints.

DOI: 10.1201/9781003614944-2

We consider a portfolio of options made of ω_i vanilla call options of maturity T_i and strike K_i on a stock S, with $0 < T_1 < \ldots < T_n = T$. Note that, if a corporate holds a bank payable, it typically has an appetite to close it, receive cash and restructure the hedge otherwise with a par contract (the bank would agree to close the deal as a market maker, charging fees for the new trade). Because of this natural selection, a bank is mostly in the receivables (i.e. "$\omega_i \geq 0$") in its derivative business with corporates.

§2 COST OF CAPITAL AND COST OF FUNDING

A Cost of Capital

In the presence of hedging imperfections resulting in a nonvanishing trading loss (and profit) process $\mathcal{L}$ of the bank, a conditional risk measure $\mathrm{EC} = \mathrm{EC}_t(\mathcal{L})$ must be dynamically computed and reserved by the bank as capital at risk, assimilated for simplicity in this chapter to an economic capital (EC), i.e. the level of capital at risk that a regulator would like to see on an economic, structural basis, defined as some to-be-specified risk measure (in a broad sense) of the trading losses of the bank.

Let r_t denote a risk-free short-term interest rate and $\beta_t = e^{-\int_0^t r_s ds}$ the corresponding risk-neutral discount factor. As first introduced in Albanese, Caenazzo and Crépey (2016, Section 5) and detailed in the next part of the book, the capital valuation adjustment (KVA) needed by the bank to remunerate its shareholders for their capital at risk at some average hurdle rate h (e.g. 10%) at any point in time in the future satisfies

$$\mathrm{KVA}_t = h\mathbb{E}_t \int_t^T \beta_t^{-1}\beta_s e^{-h(s-t)}\mathrm{EC}_s(\mathcal{L})ds, \tag{1}$$

where $\mathbb{E}_t$ stands for the conditional expectation with respect to some probability measure $\mathbb{Q}$ and model filtration $\mathfrak{G} = (\mathfrak{G}_t)_{0\leq t\leq T}$. In principle, the probability measure used in capital and cost of capital calculations should be the physical probability measure. However, in the context of optimization of a portfolio of derivatives, this physical measure is hard to estimate in a relevant way, especially for long maturities. As a consequence, we do all our price and risk computations under a risk-neutral measure $\mathbb{Q}$ calibrated to the market (or a family of such measures, in the context of model uncertainty later below), assuming no arbitrage. This basic view on $\mathbb{Q}$ will be refined in the introduction to the next part of the book.

B Cost of Funding

We assume that the bank can invest at the risk-free rate r but can only obtain unsecured funding at a shifted rate $r + \lambda > r$. This entails funding costs over risk-free rates and a related funding valuation adjustment (FVA) for the bank. Given our focus on capital and funding in this chapter, we ignore counterparty risk for simplicity, so that λ is interpreted as a pure funding liquidity basis. The bank "marks to the model" its derivative portfolio, assumed bought from a client at time 0, by means of an FVA-deducted value process Π. In order to exclude arbitrages in the primary market of hedging instruments, we assume that the vector gain process M of unit positions held in the hedging assets is a risk-neutral martingale (see e.g. Crépey, Bielecki and Brigo (2014, Sections 4.2-4.3)). The bank sets up a (possibly imperfect) hedge $(-\eta)$ in the hedging assets, for some predictable row-vector process η of the same dimension as M. We assume that the depreciation of Π, the funding expenditures, and the loss ηdM on the hedge, minus the option payoffs as they mature, are instantaneously realized into the trading loss process $\mathcal{L}$ of the bank. In particular, at any time t, the amount on the trading (funding) account of the bank is Π_t. Moreover, we assume that the capital at risk of the bank, proxied by its economic capital (EC) in this chapter, can be used by the trader for his funding purposes provided he pays to shareholders the risk-free rate on EC that they would make otherwise by depositing it (assuming it all cash for simplicity).

Note that the value process Π of the trade already includes the FVA as a deduction but ignores the KVA, which is considered as a risk adjustment computed in a second step. In other words, we assume that the trading account and the KVA account are kept separate from each other, the idea being that the KVA account is the business of the management of the bank, as opposed to the trading desks, in charge of Π. Rephrasing in mathematical terms the above description, the trading loss equation of the bank satisfies $\mathcal{L}_0 = y$, the accrued loss of the portfolio, and, for $t \in (0, T]$:

$$
\begin{aligned}
d\mathcal{L}_t = & -\underbrace{\sum_i \omega_i (S_{T_i} - K_i)^+ \boldsymbol{\delta}_{T_i}(dt)}_{\text{call payoffs}} \\
& + \underbrace{r_t \mathrm{EC}_t(\mathcal{L}) dt}_{\text{payment of internal lending of the EC funding source at the risk-free rate}} \\
& + \underbrace{\Big((r_t + \lambda_t)\big(\Pi_t - \mathrm{EC}_t(\mathcal{L})\big)^+ - r_t \big(\Pi_t - \mathrm{EC}_t(\mathcal{L})\big)^- \Big) dt}_{\text{portfolio funding costs/benefits}} \\
& + \underbrace{(-d\Pi_t)}_{\text{depreciation of }\Pi} + \underbrace{\eta_t dM_t}_{\text{loss on the hedge}} \\
= & -d\Pi_t - \sum_i \omega_i (S_{T_i} - K_i)^+ \boldsymbol{\delta}_{T_i}(dt) + \Big(\lambda_t \big(\Pi_t - \mathrm{EC}_t(\mathcal{L})\big)^+ + r_t \Pi_t \Big) dt + \eta_t dM_t.
\end{aligned} \tag{2}
$$

A martingale condition on the trading loss process $\mathcal{L}$ and the terminal condition $\Pi_T = 0$ lead to the following FVA-deducted valuation equation:

$$\begin{aligned}\Pi_t = & \underbrace{\mathbb{E}_t\Big[\sum_{t<T_i} \beta_t^{-1}\beta_{T_i}\omega_i(S_{T_i} - K_i)^+\Big]}_{=:P_t} \\ & - \underbrace{\mathbb{E}_t\Big[\int_t^T \beta_t^{-1}\beta_s\lambda_s\big(\Pi_s - \mathrm{EC}_s(\mathcal{L})\big)^+ ds\Big]}_{=:\mathrm{FVA}_t},\ t \in [0,T]\,.\end{aligned} \tag{3}$$

Observe that since we consider a portfolio of options with several maturities, we treat option payoffs as cash flows at their maturity times rather than a terminal condition in the equations, in particular $\Pi_T = 0$.

Also accounting for the KVA risk premium (1), the shareholder equity process is then a submartingale in line with a hurdle rate h on their capital at risk, which is the essence of a dynamic cost-of-capital approach (cf. Proposition II.1 page 59).

Note that the funding source provided by capital at risk creates a feedback loop from EC into FVA, which makes the FVA smaller. Also note that in the usual case of a risk measure EC that is only affected by the time fluctuations of $\mathcal{L}$, the equations (3) and in turn (1) are independent of the accrued loss y, which eventually does not affect Π nor the KVA.

If $\lambda = 0$, then, whatever the hedge η, Π reduces to P, which corresponds to the usual trade additive (linear) no-arbitrage pricing formula for a portfolio of options, with zero FVA, but with a KVA given by (1), depending on the hedge η. If $\lambda \neq 0$, we introduce the following backward stochastic differential equation (BSDE, see e.g. El Karoui, Peng and Quenez (1997) or Crépey (2013)):

$$\Pi_t^\star = \mathbb{E}_t\Big[\sum_{t<T_i} \beta_t^{-1}\beta_{T_i}\omega_i(S_{T_i} - K_i)^+ - \int_t^T \beta_t^{-1}\beta_s\lambda_s(\Pi_s^\star)^+ ds\Big],\ t \in [0,T]\,. \tag{4}$$

This is a BSDE with a monotone coefficient, admitting as such a unique square integrable solution $\Pi^\star$, provided λ is bounded from below and P is square integrable (see e.g. Kruse and Popier (2016, Section 4) and Theorem II.1(ii) page 64). If there exists a replicating hedge η, i.e. $\eta = \eta^\star$ such that the corresponding process $\mathcal{L}$ is constant in (2), i.e. such that $\eta_t^\star dM_t$ coincides with the martingale part of $\Pi^\star + \sum_{T_i \leq \cdot} \omega_i(S_{T_i} - K_i)^+$, then the resulting EC and KVA processes vanish (assuming $\mathrm{EC}(\mathcal{L} \equiv y) = 0$) and the ensuing FVA-deducted value process is given by $\Pi^\star$.

If $n = 1$ and $\omega_1 = 1$ (one long call position), then by application of the comparison theorem for BSDEs with a monotonic coefficient (see Kruse and

Popier (2016, Section 4)), we have $\Pi^\star \geq 0$, hence

$$\Pi^\star_t = \mathbb{E}_t\Big[\widetilde{\beta}_t^{-1}\widetilde{\beta}_{T_1}(S_{T_1} - K_1)^+\Big], \tag{5}$$

where $\widetilde{\beta}_t = e^{-\int_0^t (r+\lambda_s)ds}$. With respect to P, the value $\Pi^\star$ embeds an FVA rebate on the buying price by the bank (since we assumed a positive liquidity basis λ). If, instead, $n = \omega_1 = -1$ (one short call position), then we deduce likewise that $\Pi^\star \leq 0$, hence $\Pi^\star = P$.

But, apart from the above special cases where $\lambda = 0$ or $\eta = \eta^\star$, the equation (3) for Π is nonstandard due to the term $\mathrm{EC} = \mathrm{EC}_t(\mathcal{L})$ in the FVA, making (3) an anticipated BSDE (ABSDE) in the line of Peng and Yang (2009), with coefficient $\lambda_t\Big(\Pi_t - \mathrm{EC}_t(\mathcal{L})\Big)^+$.

§3 MARKOVIAN BLACK-SCHOLES SETUP

In this section we assume a constant risk-free rate r and a Black-Scholes stock S with volatility σ and constant dividend yield q. The risk-neutral martingale M is then taken as the gain process of a continuously rolled unit position on the stock S, assumed funded at the risk-free rate via a repo market, i.e. $dM_t = dS_t - (r-q)S_t dt$. We denote by $\mathcal{A}^{bs}_S = (r-q)S\partial_S + \frac{1}{2}\sigma^2 S^2 \partial^2_{S^2}$ the corresponding risk-neutral Black-Scholes generator.

Doing our modeling exercise in the context of the Black-Scholes model, where perfect replication, hence no KVA, is possible, may seem rather artificial. However, doing all the computations in a stylized Black-Scholes setup with a single risk factor S yields useful practical insights. In addition, this conveys the message that, in real-life incomplete markets, a basic Black-Scholes perspective has to be complemented by the valuation of market imperfections, otherwise this unavoidably results in Black-Scholes Ponzi schemes, such as the ones that have been involved in the global financial crisis, where always more derivatives are issued to remunerate the capital required by the already opened positions (see Figure 1, page 4).

In the Black-Scholes setup and assuming a stylized Markovian specification

$$\mathrm{EC}_t(\mathcal{L}) = f\sqrt{\frac{d\langle\mathcal{L}\rangle}{dt}} \tag{6}$$

(a stylized VaR proportional to the instantaneous volatility of the loss process $\mathcal{L}$) as well as $\lambda = \lambda(t, S_t)$, $\eta_t = \eta(t, S_t)$, then the above FVA and KVA equations can be reduced to the "sustainable Black-Scholes PDEs" (14), as follows.

We denote by u_{bs} the usual trade additive Black-Scholes value function of the option portfolio (except for option payoffs treated as dividends the way explained after (3)), by Δ_{bs} the corresponding Black-Scholes delta, by v the (admitted classical and such that $\beta_t S_t \partial_S v(t, S_t) dW_t$ is a martingale) solution to

$$v(T, S) = 0 \text{ on } (0, \infty) \text{ and, for } t < T,$$
$$\partial_t v + \mathcal{A}_S^{bs} v + \lambda\Big(u_{bs} - v - f\sigma S|\Delta_{bs} - \partial_S v - \eta|\Big)^+ - rv = 0 \text{ on } [0, T) \times (0, \infty), \tag{7}$$

and we set $u := u_{bs} - v$. Note that the function u coincides with u_i on each strip $[T_{i-1}, T_i) \times (0, \infty)$, where $(u_i)_{1 \le i \le n}$ is defined as the unique sequence of (admitted classical) solutions to the following PDE cascade, for i decreasing from n to 1, closing the system by setting $u_{n+1} = 0$ and $T_0 = 0$:

$$\begin{cases} u_i(T_i, S) = u_{i+1}(T_i, S) + \omega_i (S - K_i)^+ \text{ on } (0, \infty) \\ \partial_t u_i + \mathcal{A}_S^{bs} u_i - \lambda\Big(u_i - f\sigma S|\partial_S u_i - \eta|\Big)^+ - ru_i = 0 \text{ on } [T_{i-1}, T_i) \times (0, \infty). \end{cases} \tag{8}$$

Proposition. *$\Pi_t := u(t, S_t)$ and the associated continuous semimartingale*

$$\begin{aligned} d\mathcal{L}_t = & -d\Pi_t - \sum_i \omega_i (S_{T_i} - K_i)^+ \boldsymbol{\delta}_{T_i}(dt) \\ & + \Big(r_t \Pi_t + \lambda_t \Big(\Pi_t - f\sigma S_t \Big|\partial_S u(t, S_t) - \eta(t, S_t)\Big|\Big)^+\Big) dt + \eta_t dM_t \end{aligned} \tag{9}$$

solve (2)–(3) *and they satisfy*

$$\sqrt{\frac{d\langle \mathcal{L} \rangle}{dt}} = \sigma S_t \Big|\partial_S u(t, S_t) - \eta(t, S_t)\Big|. \tag{10}$$

Proof. By Itô calculus, we have

$$- dv(t, S_t) = -(\partial_t + \mathcal{A}_S^{bs}) v(t, S_t) dt - \sigma S \partial_S v(t, S_t) dW_t =$$
$$\lambda\Big(\Big(u_{bs} - v - f\sigma S|\Delta_{bs} - \partial_S v - \eta|\Big)^+ (t, S_t) - rv(t, S_t)\Big) dt - \sigma S \partial_S v(t, S_t) dW_t,$$

by (7). Hence $-d(\beta_t v(t, S_t)) = \beta_t \lambda\Big(\Big(u_{bs} - v - f\sigma S|\Delta_{bs} - \partial_S v - \eta|\Big)^+ (t, S_t) - rv(t, S_t)\Big) dt - \beta_t \sigma S \partial_S v(t, S_t) dW_t$ and

$$\Pi_t = P_t - \mathbb{E}_t\Big[\int_t^T \beta_t^{-1} \beta_s \lambda_s \Big(\Pi_s - f\sigma S|\partial_S u - \eta|(s, S_s)\Big)^+ ds\Big],\ t \in [0, T]\,. \tag{11}$$

Moreover, up to dt terms which do not contribute to the bracket of $\mathcal{L}$, $d\mathcal{L}_t$ boils down to

$$\begin{aligned} & - du_{bs}(t, S_t) + dv(t, S_t) - \sum_i \omega_i (S_{T_i} - K_i)^+ \boldsymbol{\delta}_{T_i}(dt) + \eta_t dM_t \\ & \quad = -\Big(du_{bs}(t, S_t) + \sum_i \omega_i (S_{T_i} - K_i)^+ \boldsymbol{\delta}_{T_i}(dt)\Big) + dv(t, S_t) + \eta_t dM_t, \end{aligned}$$

equal modulo dt (discount) terms to

$$-\sigma S(\Delta_{bs} - \partial_S v - \eta)(t, S_t)dW_t = -\sigma S(\partial_S u - \eta)(t, S_t)dW_t,$$

which proves (10). Hence (11) yields (3) and (9) yields (2). □

The ensuing FVA:= $P - \Pi$ and KVA processes are given as (cf. (3) and (1)):

$$\begin{aligned} \mathrm{FVA}_t(\mathcal{L}) &= \mathbb{E}_t\Big[\int_t^T e^{-r(s-t)}\lambda_s\Big(u(s,S_s) - f\sqrt{\frac{d\langle\mathcal{L}\rangle}{ds}}\Big)^+ ds\Big] \\ \mathrm{KVA}_t(\mathcal{L}) &= h\mathbb{E}_t\Big[\int_t^T e^{-(r+h)(s-t)} f\sqrt{\frac{d\langle\mathcal{L}\rangle}{ds}} ds\Big], \end{aligned} \tag{12}$$

with $\sqrt{\frac{d\langle\mathcal{L}\rangle}{dt}}$ given by (10). Setting $\eta = (1-\alpha)\partial_S u$, where α in $[0, 100\%]$ is a mis-hedge parameter, then $\sqrt{\frac{d\langle\mathcal{L}\rangle}{dt}}$ reduces to $\alpha\sigma S_t\big|\partial_S u(t, S_t)\big|$ and (12) yields

$$\begin{aligned} \mathrm{FVA}_t(\mathcal{L}) &= \mathbb{E}_t\Big[\int_t^T e^{-r(s-t)}\lambda_s\Big(u(s,S_s) - \alpha f\sigma S_s\big|\partial_S u(s,S_s)\big|\Big)^+ ds\Big] \\ &= v(t,S_t) = u_{bs}(t,S_t) - u(t,S_t), \\ \mathrm{KVA}_t(\mathcal{L}) &= h\mathbb{E}_t\Big[\int_t^T e^{-(r+h)(s-t)}\alpha f\sigma S_s\big|\partial_S u(s,S_s)\big| ds\Big] = w(t,S_t), \end{aligned} \tag{13}$$

where u_{bs} is the trade additive Black-Scholes portfolio value and where the FVA and KVA pricing functions v and w satisfy

$$\begin{cases} v(T,S) = w(T,S) = 0 \text{ on } (0,\infty) \\ \partial_t v + \mathcal{A}_S^{bs} v + \lambda\Big(u_{bs} - v - \alpha f\sigma S|\Delta_{bs} - \partial_S v|\Big)^+ - rv = 0 \text{ on } [0,T)\times(0,\infty) \\ \partial_t w + \mathcal{A}_S^{bs} w + \alpha h f\sigma S|\Delta_{bs} - \partial_S v| - (r+h)w = 0 \text{ on } [0,T)\times(0,\infty), \end{cases} \tag{14}$$

in which $\Delta_{bs} = \partial_S u_{bs}$.

These "sustainable Black-Scholes PDEs" (14) allow computing an FVA and KVA deducted price

$$u - w = u_{bs} - v - w$$

that would be sustainable for the bank even in the limit case of a portfolio held on a run-off basis, with no new trades ever entered in the future.

§4 VOLATILITY UNCERTAINTY

A relevant and important regulatory issue is the magnifying impact of model risk on the different XVA metrics. In this section, we assess model risk from the angle of Avellaneda, Levy and Paras (1995)'s uncertain volatility model

(UVM). Namely, we only assume positive bounds $\underline{\sigma}$ and $\overline{\sigma}$, but we do not assume any specific dynamics on the stock volatility process σ. Therefore, there is a model uncertainty about it. That is, we only consider $dM_t := \sigma_t S_t dW_t = dS_t - (r-q)S_t dt$ such that $\sigma_t \in [\underline{\sigma}, \overline{\sigma}]$ for every t.

We denote by $\mathcal{C}(\mathbb{R}_+)$ the space of continuous paths on $\mathbb{R}_+$, by C the canonical process on the space $\mathcal{C}(\mathbb{R}_+)$, by $\mathfrak{G} = (\mathfrak{G}_t)_{0\le t\le T}$ the canonical filtration generated by C, and by $\mathcal{Q}$ the set of $\mathfrak{G}$ local martingale probability measures for C. We recall from Soner, Touzi and Zhang (2012) that, for any probability measure $\mathbb{Q} \in \mathcal{Q}$, the process C satisfies $dC_t = a_t^{1/2} dW_t^{\mathbb{Q}}$, for some $\mathbb{Q}$ Brownian motion $W^{\mathbb{Q}}$, where a_t is the Lebesgue density of the quadratic variation of C. In the following we restrict attention to the probability measures $\mathbb{Q}$ such that $a_t^{1/2} \in [\underline{\sigma}, \overline{\sigma}]$ holds $dt \times \mathbb{Q}$ almost surely, still denoting by $\mathcal{Q}$ the (restricted) set of measures, and we model $dM_t = dS_t - (r-q)S_t dt$ as $S_t dC_t$.

Under each $\mathbb{Q} \in \mathcal{Q}$, similarly to (2), the trading loss equation of the bank is written, for $t \in (0,T]$, as:

$$\begin{aligned} d\mathcal{L}_t^{\mathbb{Q}} = -d\Pi_t^{\mathbb{Q}} - \sum_i \omega_i (S_{T_i} - K_i)^+ \boldsymbol{\delta}_{T_i}(dt) + \\ \Big(\lambda_t \Big(\Pi_t^{\mathbb{Q}} - \mathrm{EC}_t^{\mathbb{Q}}(\mathcal{L}^{\mathbb{Q}})\Big)^+ + r_t \Pi_t^{\mathbb{Q}}\Big) dt + \eta_t dM_t \end{aligned} \tag{15}$$

where $\mathrm{EC}^{\mathbb{Q}}$ is some conditional risk measure under $\mathbb{Q}$. The ensuing equation for the $\mathbb{Q}$ FVA-deducted value $\Pi^{\mathbb{Q}}$ appears as

$$\begin{aligned} \Pi_t^{\mathbb{Q}} = \mathbb{E}_t^{\mathbb{Q}} \Big[\sum_{t<T_i} \beta_t^{-1} \beta_{T_i} \omega_i (S_{T_i} - K_i)^+ \\ - \int_t^T \beta_t^{-1} \beta_s \lambda_s \Big(\Pi_s^{\mathbb{Q}} - \mathrm{EC}_s^{\mathbb{Q}}(\mathcal{L}^{\mathbb{Q}})\Big)^+ ds \Big], \, t \in [0,T]. \end{aligned} \tag{16}$$

Under each $\mathbb{Q}$, the trading desks of the bank should value the derivative portfolio $\Pi_0^{\mathbb{Q}}$ at time 0 (or $\Pi_t^{\mathbb{Q}}$ at time t). However, due to model uncertainty, they value it $\Pi_0 = \inf_{\mathbb{Q}\in\mathcal{Q}} \Pi_0^{\mathbb{Q}}$ (or at time t, $\Pi_t = \operatorname*{ess\,inf}_{\mathbb{Q}\in\mathcal{Q}} \Pi_t^{\mathbb{Q}}$).

At time t, $\mathrm{EC}_t^{\mathbb{Q}}(\mathcal{L}^{\mathbb{Q}})$ may depend on the whole future of the process $(\mathcal{L}_s^{\mathbb{Q}}$, $s \ge t)$. This makes (16) an anticipated BSDE under $\mathbb{Q}$ (see the last sentence of §2), with coefficient $\lambda_t \Big(\Pi_t^{\mathbb{Q}} - \mathrm{EC}_t^{\mathbb{Q}}(\mathcal{L}^{\mathbb{Q}})\Big)^+$.

For taking into consideration model risk, here meant as the impact of several $\mathbb{Q}$ (vs. the Darwinian model risk of Chapter XI, see Remark XI.1), we need the notion of second-order BSDE. Well-posedness results regarding second-order anticipated BSDEs are not yet available in the literature. Hence, we only give heuristic formulations in this regard. Namely, by analogy with the second-order BSDEs theory introduced by Soner, Touzi and Zhang (2012), we should have the following representation, where $\mathfrak{G}_+ = (\mathfrak{G}_{t+})_{0\le t\le T}$ is the right limit of $\mathfrak{G}$, i.e. $\mathfrak{G}_{t+} = \cap_{s>t} \mathfrak{G}_s$, $t \in [0,T)$, and $\mathfrak{G}_{T+} = \mathfrak{G}_T$:

There exists a process $\mathcal{L}$ such that, for each $\mathbb{Q} \in \mathcal{Q}$, $\mathcal{L}$ is a $\mathbb{Q}$ local martingale and it $\mathbb{Q}$ a.s. holds that

$$\begin{aligned} d\mathcal{L}_t = -d\Pi_t - \sum_i \omega_i (S_{T_i} - K_i)^+ \boldsymbol{\delta}_{T_i}(dt) \\ + \Big(\lambda_t \big(\Pi_t - \mathrm{EC}_t^{\mathbb{Q}}(\mathcal{L})\big)^+ + r_t \Pi_t\Big) dt + \eta_t dM_t + dA_t^{\mathbb{Q}}, \end{aligned} \tag{17}$$

where $\mathrm{EC}^{\mathbb{Q}}$ is some conditional risk measure and the family $\{A^{\mathbb{Q}}\}$ of non-decreasing processes satisfies the minimality condition

$$A_t^{\mathbb{Q}} = \operatorname*{ess\,inf^{\mathbb{Q}}}_{\mathbb{Q}' \in \mathcal{Q}(t,\mathbb{Q},\mathfrak{G}_+)} \mathbb{E}^{\mathbb{Q}'} \left[A_T^{\mathbb{Q}'} \middle| \mathfrak{G}_{t+} \right], \ 0 \le t \le T, \ \mathbb{Q} \text{ a.s.}, \ \forall \mathbb{Q} \in \mathcal{Q}, \tag{18}$$

where $\mathcal{Q}(t, \mathbb{Q}, \mathfrak{G}_+) := \big\{ \mathbb{Q}' \in \mathcal{Q}, \ \mathbb{Q}' = \mathbb{Q} \text{ on } \mathfrak{G}_{t+} \big\}$.

The corresponding equation for the FVA-deducted value Π would appear as

$$\begin{aligned} \Pi_t = \operatorname*{ess\,inf}_{\mathbb{Q}' \in \mathcal{Q}(t,\mathbb{Q},\mathfrak{G}_+)} \mathbb{E}_t^{\mathbb{Q}'} \Big[\sum_{t<T_i} \beta_t^{-1} \beta_{T_i} \omega_i (S_{T_i} - K_i)^+ \\ - \int_t^T \beta_t^{-1} \beta_s \lambda_s \big(\Pi_s - \mathrm{EC}_s^{\mathbb{Q}'}(\mathcal{L}) \big)^+ ds \Big], \ t \in [0, T], \ \mathbb{Q} \ a.s.. \end{aligned} \tag{19}$$

A Equations in the Robust Markovian Setting

In the setting of §3 (except for the now uncertain volatility), with VaR-like specification (6) of economic capital, $\mathrm{EC}_t^{\mathbb{Q}}(\mathcal{L}^{\mathbb{Q}})$ only depends on $(\mathcal{L}_t^{\mathbb{Q}})$ at time t, so that the ABSDE (19) reduces to a BSDE, amenable to rigorous statements.

According to the second-order BSDE theory of Soner et al. (2012), the PDE cascade (8) becomes $u_{n+1} = 0$ and, for i decreasing from n to 0,

$$\begin{cases} u_i(T_i, S) = u_{i+1}(T_i, S) + \omega_i (S - K_i)^+ \text{ on } (0, \infty) \text{ and, on } [T_{i-1}, T_i) \times (0, \infty), \\ \partial_t u_i + \min_{\sigma \in [\underline{\sigma}, \overline{\sigma}]} \Big[\mathcal{A}_S^{bs} u_i - \lambda \big(u_i - f \sigma S |\partial_S u_i - \eta| \big)^+ \Big] - r u_i = 0. \end{cases} \tag{20}$$

Let $\Pi_t := u(t, S_t)$ for u defined by $u_i(t, S)$ on each strip $[T_{i-1}, T_i) \times (0, \infty)$. The corresponding FVA can be defined as $\Pi^{\lambda=0} - \Pi$, and the ensuing KVA process satisfies, for each $\mathbb{Q} \in \mathcal{Q}$ (compare with (10) and (1)):

$$\mathrm{KVA}_t(\mathcal{L}) = h \operatorname*{esssup}_{\mathbb{Q}' \in \mathcal{Q}(t,\mathbb{Q},\mathfrak{G}_+)} \mathbb{E}_t^{\mathbb{Q}'} \Big[\int_t^T e^{-(r+h)(s-t)} f \sqrt{\frac{d\langle \mathcal{L} \rangle}{ds}} ds \Big], \ \mathbb{Q} \ a.s., \tag{21}$$

where $\sqrt{\frac{d\langle \mathcal{L} \rangle}{dt}} = a_t^{1/2} S_t \big| \partial_S u(t, S_t) - \eta(t, S_t) \big|$. In the case where $\eta = (1 - \alpha) \partial_S u$, we obtain

$$\mathrm{KVA}_t(\mathcal{L}) = w(t, S_t),$$

where

$$\begin{cases} w(T,S) = 0 \text{ on } (0,\infty) \\ \partial_t w + \max_{\sigma\in[\underline{\sigma},\overline{\sigma}]} \Big[\mathcal{A}_S^{bs} w + \alpha h f \sigma S|\partial_S u|\Big] - (r+h)w = 0 \text{ on } [0,T)\times(0,\infty), \end{cases} \quad (22)$$

with also $|\partial_S u_i - \eta| = \alpha|\partial_S u_i|$ in (20).

§5 OPTIMAL TRANSPORT APPROACH

Since vanilla call options are liquidly traded, their time 0 price components

$$\mathbb{E}^{\mathbb{Q}}\Big[\beta_{T_i}(S_{T_i} - K_i)^+\Big]$$

should not be seen as subject to model risk, but calibrated to the market. We need to refine our preliminary UVM assessment of model risk in order to account for these calibration constraints. For simplicity we consider a single call option (T,K), and we set $\lambda = 0$, focusing on the KVA in this section. Hence, the system (20) reduces to a single PDE with $\lambda = 0$, with solution denoted by u.

Tan and Touzi (2013) consider the optimal transport problem consisting of minimizing a cost among all continuous semimartingales with given initial and terminal distributions. They show an extension of the Kantorovitch duality to this context and suggest a finite differences scheme combined with the gradient projection algorithm to approximate the dual value. Their results can be applied to our setup as follows.

Let $\boldsymbol{\mu}_0 = \boldsymbol{\delta}_{S_0}$ denote the Dirac measure on the initial value of S_0, and let $\boldsymbol{\mu}_T$ denote the marginal distribution of S_T, inferred by calibration to the market prices of all European call options with maturity T (assuming quotations available for all strikes). Let

$$\mathcal{Q}(\boldsymbol{\mu}_0) = \{\mathbb{Q}\in\mathcal{Q} : \mathbb{Q}\circ S_0^{-1} = \boldsymbol{\mu}_0\},\ \mathcal{Q}(\boldsymbol{\mu}_0,\boldsymbol{\mu}_T) = \{\mathbb{Q}\in\mathcal{Q}(\boldsymbol{\mu}_0) : \mathbb{Q}\circ S_T^{-1} = \boldsymbol{\mu}_T\}.$$

From Remark 2.3 in Tan and Touzi (2013), $\mathcal{Q}(\boldsymbol{\mu}_0,\boldsymbol{\mu}_T)$ is not empty in our setting.

At time 0, the KVA with model uncertainty and terminal marginal constraint is defined as

$$\mathrm{KVA}_0(\mathcal{L}) = h \sup_{\mathbb{Q}\in\mathcal{Q}(\boldsymbol{\mu}_0,\boldsymbol{\mu}_T)} \mathbb{E}^{\mathbb{Q}}\Big[\int_0^T e^{-(r+h)(s)} f\sqrt{\frac{d\langle\mathcal{L}\rangle}{ds}}ds\Big], \quad (23)$$

where $\mathcal{L}$ represents the portfolio loss in the present setting, that is, the trading loss of the bank in a world with uncertain volatility subject to the law of S_T. However, it is not clear how to extrapolate the theory of Tan and Touzi (2013) to valuation at future time points when only the unconditional law of S_T is known. Hence for the sake of tractability we conservatively assume that $\mathcal{L}$ in (23) is the UVM one and we only apply the constraint to the outer expectation in (23) (and not to the conditional expectations that are hidden in $\mathcal{L}$).

With this understanding of (23), given any measure $\boldsymbol{\nu}$, we define

$$\boldsymbol{\nu}(\phi) = \int_{\mathbb{R}^d} \phi(x)\boldsymbol{\nu}(dx)$$

on the set $\mathcal{C}_b(\mathbb{R}^d)$ of all bounded continuous functions ϕ on $\mathbb{R}^d$. We can readily check that Assumptions 3.1–3.3 in Tan and Touzi (2013) are satisfied. By an application of their main duality result, we can thus rewrite the KVA as

$$\mathrm{KVA}_0(\mathcal{L}) = \inf_{\phi\in\mathcal{C}_b(\mathbb{R}^d)} \Big\{ \boldsymbol{\mu}_0(\Phi_0) - e^{-(r+h)T}\boldsymbol{\mu}_T(\phi)\Big\}, \tag{24}$$

where the "pseudo-payoff function" ϕ corresponds to a Lagrangian for the constrained optimization problem (23) and where

$$\Phi_0(x) = \sup_{\mathbb{Q}\in\mathcal{Q}(\boldsymbol{\delta}_x)} \mathbb{E}^{\mathbb{Q}}\Big[e^{-(r+h)T}\phi(S_T) + \int_0^T e^{-(r+h)s} hf\sqrt{\frac{d\langle\mathcal{L}\rangle}{ds}}ds\Big]. \tag{25}$$

Hence, the KVA in an optimal transport (OT) setting can be represented as an infimum of KVAs in a modified UVM setting.

A Equations in the Markovian Optimal Transport Setting

In the setting of §3 (but with now uncertain volatility subject to a known law of S_T), we consider the probability measures $\mathbb{Q}$ on the canonical space $(\Omega, \mathfrak{G}_T)$, under which the canonical process C is a local martingale on $[t, T]$. Define $\mathcal{Q}_t$ as the collection of all such martingale probability measures $\mathbb{Q}$ such that $a_s^{1/2} \in [\underline{\sigma}, \overline{\sigma}]$ holds $d\mathbb{Q}\times ds$ a.e. on $\Omega\times[t, T]$. Denote $\mathcal{Q}^{t,x} := \{\mathbb{Q}\in\mathcal{Q}_t : \mathbb{Q}[S_s = x, 0\le s\le t] = 1\}$. For any $\phi\in\mathcal{C}_b(\mathbb{R}^d)$, let

$$\Phi(t,x) = \sup_{\mathbb{Q}\in\mathcal{Q}^{t,x}} \mathbb{E}^{\mathbb{Q}}\Big[e^{-(r+h)(T-t)}\phi(S_T) + \int_t^T e^{-(r+h)(s-t)} hf\sqrt{\frac{d\langle\mathcal{L}\rangle}{ds}}ds\Big], \tag{26}$$

where $\sqrt{\frac{d\langle\mathcal{L}\rangle}{dt}} = a_t^{1/2}S_t\Big|\partial_S u(t,S_t) - \eta(t,S_t)\Big|$, in which u is the solution to (20) with $\lambda = 0$.

Then, in the case where $\eta = (1-\alpha)\partial_S u$, Φ is a viscosity solution to the dynamic programming equation

$$\begin{cases} \Phi(T,S) = \phi(S) \text{ on } (0,\infty) \\ \partial_t\Phi + \max_{\sigma\in[\underline{\sigma},\overline{\sigma}]}\Big[\mathcal{A}_S^{bs}\Phi + \alpha hf\sigma S|\partial_S u|\Big] - (r+h)\Phi = 0 \text{ on } [0,T)\times(0,\infty). \end{cases} \tag{27}$$

In view of (24), in the present OT setup, KVA_0 is obtained as the minimum of

$$\Phi(0, S_0) - e^{-(r+h)T} \int_{\mathbb{R}} \phi(x)\boldsymbol{\mu}_T(dx) \tag{28}$$

over all $\phi \in \mathcal{C}_b(\mathbb{R}^d)$. This minimization can be achieved numerically by the projected gradient descent of Tan and Touzi (2013, Section 5) or simply (as in our numerics below) by the Nelder-Mead simplex algorithm.

As a sanity check, observe that, if $\boldsymbol{\mu}_T$ is Black-Scholes σ and $\underline{\sigma} = \overline{\sigma} = \sigma$, then (28) is exactly the time-0 KVA of §3, independently of ϕ.

§6 NUMERICAL RESULTS

Figure 2 shows the results obtained by solving the related PDEs (and minimizing (28) in the OT setup) without model uncertainty as per §3.A (top panel), with UVM uncertainty as per §4.A (middle panel) and with OT uncertainty as per §5.A (bottom panel), for a level of the mis-hedge parameter α increasing from 0 to 100%. We used the following parameters:

$$S_0 = 100, \quad r = 2\%, \quad q = 0, \quad \sigma = 30\%,$$
$$\lambda = 200 \text{ bps}, \quad f = 1.2, \quad h = 10\%$$

and considered a single call option of maturity $T = 5$ years and strike $K = 107$.

The main observation from the top panel is that, unless the hedge is very good (of the order of 25% of mis-hedge or less), the KVA dominates the FVA, and becomes about ten times greater than the FVA in the absence of a hedge ($\alpha = 1$). This is logical given that EC has only an indirect reduction effect on the FVA, whereas it directly sizes the KVA.

Going to the middle panel, the FVA changes little, but both u and the KVA (unless the hedge is almost perfect) are tremendously impacted by the uncertainty on the volatility. Regarding the KVA this is in line with the fact that it is the cost of a risk measure, which nonlinearly amplifies the impact of perturbations to its input data.

In reality the time-0 price of a vanilla option such as the one considered in our numerics is given by the market, so there is no model risk on it, but only on the KVA. This is what is reflected by the OT bottom panel. The model risk on the KVA component, however, is essentially the same as in the UVM case, because it is conservatively assessed by using the UVM u in (27), fault of a developed theory of valuation at future time points under uncertain volatility subject to the unconditional law of S_T (see after (23)).

Conclusion

To assess the profitability of a deal, XVA desks are the first consulted desks in all major trades today. The above results in a toy model where all the

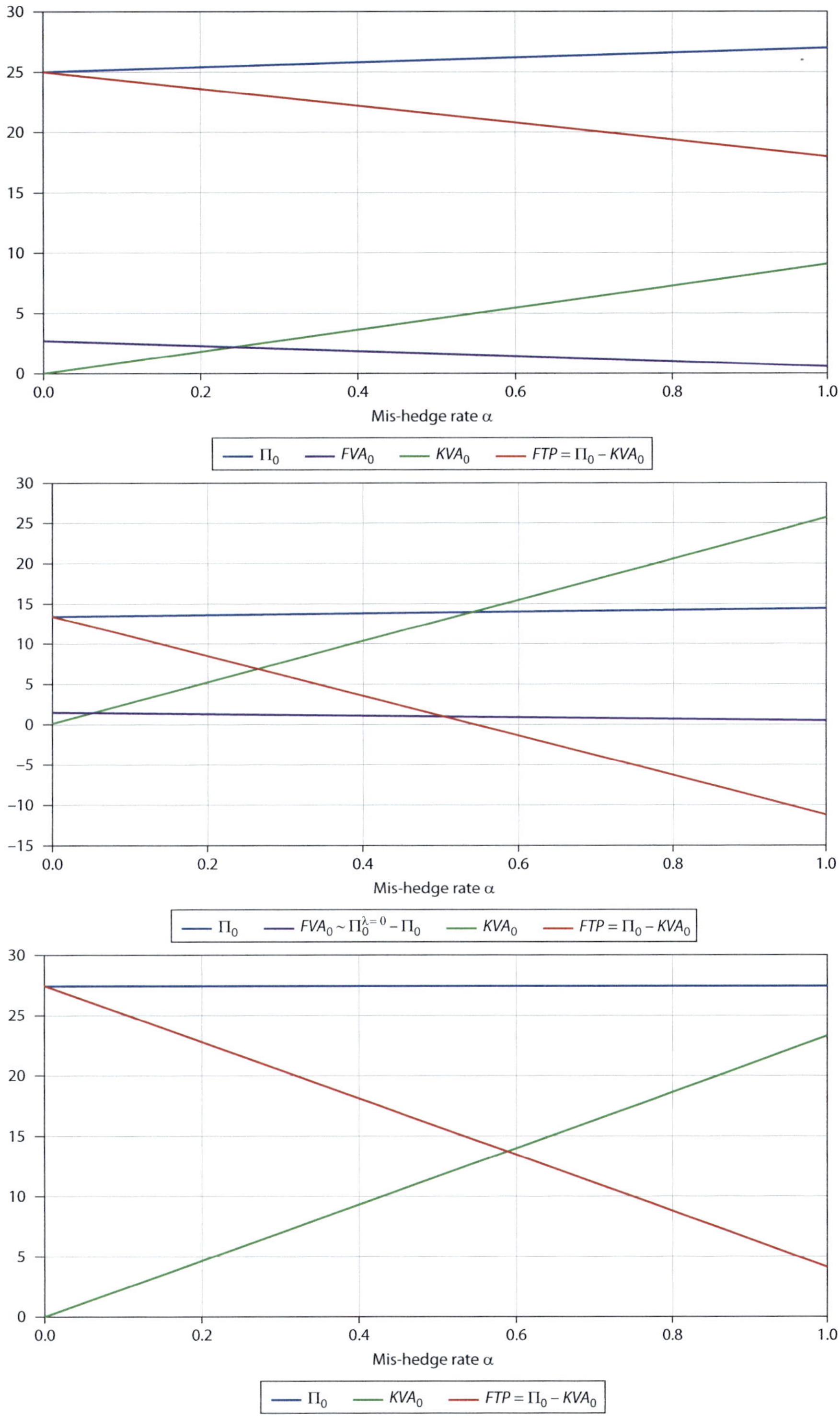

Figure 2: XVAs and FTP as a function of the mis-hedge parameter α. (*Top*) Without model uncertainty. (*Middle*) With UVM uncertainty ($\underline{\sigma} = 15\%, \overline{\sigma} = 60\%$). (*Bottom*) With OT uncertainty ($\underline{\sigma} = 15\%, \overline{\sigma} = 60\%, \sigma = 30\%$).

quantities of interest can be computed exactly (modulo the numerical error on one-dimensional PDE solutions) emphasize that, accounting for model risk, the relative importance of the KVA should become larger.

In order to transpose the results of this chapter to a counterparty credit risk setup, one can think of each option's payoff $(S_{T_i} - K_i)^+$ as the CVA exposure of the bank to the default at time of its counterparty i, at the (random) time T_i, with corresponding position of the bank $\omega_i S_{T_i}$ and margins received by the bank $\omega_i K_i$. However, in the counterparty credit risk setup, a relevant risk measure needs to be assessed at a one-year (as opposed to instantaneous in (6)) horizon, in order to leave time to credit events to develop. More generally, defaults (instead of liquidity) driven XVAs lead to developments of a different and more complex nature, which are the topic of the rest of the book.

PRICING

The cost-of-capital XVA approach has been introduced in Albanese, Caenazzo and Crépey (2016) and developed in various directions in a stream of papers, including Albanese, Crépey, Hoskinson and Saadeddine (2021) or Crépey (2022). The intent of this part of the book is to recap the economic and mathematical foundations of this approach, i.e. the cost-of-capital XVA conceptual framework. Chapter I provides an overview with all intuitions and insights. Chapters II and III deliver the underlying mathematics.

We use the risk-free asset (assumed to exist) as our numéraire. This choice of a numéraire simplifies equations (compare with Chapter 0) by removing all terms related to the (assumed risk-free) remuneration of all cash and collateral accounts. The funding issue is then refocused on the risky funding side of the problem, i.e. funding costs hereafter really means excess funding costs Chapter I is an overview providing all intuitions and insights, where the bank could equally borrow and lend at the risk-free rate.

In the incomplete market setup intrinsic to the XVA issue (see I.§3.E and II.§4), our reference probability measure $\mathbb{Q}^*$ is the following hybrid of pricing and physical probability measures:

Definition. Let there be given a σ-field $\mathcal{A}$, on which the physical probability measure is defined, and a financial sub-σ-field $\mathcal{B}$ of $\mathcal{A}$, on which a risk-neutral measure, equivalent to the restriction to $\mathcal{B}$ of the physical probability measure, is defined. The associated probability measure $\mathbb{Q}^*$ is the uniquely defined probability measure on $\mathcal{A}$, provided by Artzner, Eisele and Schmidt (2024, Proposition 4.1), such that *(i)* $\mathbb{Q}^*$ coincides with the underlying risk-neutral measure on $\mathcal{B}$ and *(ii)* $\mathbb{Q}^*$ and the physical probability measure coincide conditionally on $\mathcal{B}$.

Such a probability measure $\mathbb{Q}^*$ was introduced by Dybvig (1992) to deal with incomplete markets that are a mix of financial traded risk factors and unhedgeable ones (jumps to default, in our XVA setup). It was recently revisited in a finance and insurance context by Artzner et al. (2024), whence the qualifier of a “fininsurance” probability measure that we use for $\mathbb{Q}^*$ hereafter. The underlying risk-neutral pricing measure (hence, in view of *(i)*, $\mathbb{Q}^*$ itself) is calibrated to prices of fully collateralized transaction for which counterparty credit risk is negligible. The physical probability measure expresses user views on the unhedgeable risk factors. The uncertainty about $\mathbb{Q}^*$ can be dealt with by a Bayesian variation on our baseline XVA approach, whereby paths of alternative, co-calibrated models are combined in a global simulation (see X.§4.C).

We emphasize that, in general, the fininsurance probability measure does not coincide with the physical one, nor with the underlying risk-neutral measure. It is only in the unrealistic case of a bank that would not be hedged at all that $\mathbb{Q}^*$ would coincide with the physical measure. It is only in the other, equally unrealistic, extreme case of a perfectly hedged[1] and collateralized bank, that $\mathbb{Q}^*$ would coincide with the underlying risk-neutral measure. In such a perfectly hedged and collateralized case, the banks' trading loss, economic capital and KVA processes would vanish altogether, in line with the fact that the bank has no risk at all.

Given a model filtration $\mathfrak{G}$ on $\mathcal{A}$ with respect to which all our processes are adapted, with $(\mathfrak{G}_t, \mathbb{Q}^*)$ conditional expectation, denoted by $\mathbb{E}_t^*$, we also use throughout the book the bank survival probability measure $\mathbb{Q}$ associated with $\mathbb{Q}^*$ (equal to $\mathbb{Q}^*$ in the case of Chapter 0, where the bank was assumed default-free), defined in I.§4 (see also II.§3 for a practically equivalent reduction of filtration viewpoint). The $(\mathfrak{G}_t, \mathbb{Q})$ conditional expectation, variance, covariance, value-at-risk and expected shortfall (at some confidence level α specified by the context) are denoted by $\mathbb{E}_t$, $\mathbb{V}\mathrm{ar}_t$, $\mathbb{C}\mathrm{ov}_t$, $\mathbb{V}\mathrm{a}\mathbb{R}_t$ and $\mathbb{E}\mathbb{S}_t$ (and for $t = 0$, we drop all indices t).

Note that, for simplicity, we only consider European derivatives in this book.

[1]including against defaults, including its own, which is in fact unfeasible (see I.§3.E.1 and II.§4.A).

CHAPTER I

XVA Analysis From the Balance Sheet

XVAs denote various counterparty risk-related valuation adjustments that are applied to financial derivatives since the 2007–09 crisis. In this chapter we root a cost-of-capital XVA strategy in a balance sheet perspective which is key in identifying the economic meaning of the XVA terms. Our approach is first detailed in a static setup that is solved explicitly. It is then plugged in the dynamic and trade incremental context of a real derivative banking portfolio. The corresponding pricing, collateralization and dividends distribution policy ensures to bank shareholders a submartingale equity process corresponding to a target hurdle rate h on their capital at risk, consistently between and throughout deals. This is the essence of the cost-of-capital XVA approach.

§1 INTRODUCTION

XVAs, with X as C for credit, D for debt, F for funding, M for margin, or K for capital, are post-2007–09 crisis valuation adjustments for financial derivatives. In broad terms to be detailed later in this and the follow-up chapters (cf. Table 2 page 29), CVA is what the bank expects to lose due to counterparty defaults in the future; DVA (irrelevant for pricing but material to bank creditors as we will see) is what the bank expects to gain due to its own default; FVA is the expected cost for the bank of having to raise variation margin (rehypothecable collateral); MVA is the expected cost for the bank of having to raise initial margin (segregated collateral); KVA is the expected cost for the bank of having to remunerate its shareholders through dividends for their capital at risk.

Remark 1. To alleviate the notation, in the conceptual part of the chapter, i.e. hereafter in this section and in the next one, we only consider an FVA as the global cost of raising collateral for the bank, as opposed to a distinction, in the industry and in later sections, between an FVA, in the strict sense of

DOI: 10.1201/9781003614944-4

the cost of raising the variation margin, and an MVA for the cost of raising the initial margin.

XVAs deeply affect the derivative pricing task by making it global (portfolio-wide), nonlinear and entity dependent. However, before these technical implications, the fundamental point is to understand what really deserves to be priced and what does not, by rooting the pricing approach in a corresponding collateralization, accounting, and dividend policy of the bank.

Coming after several papers on the valuation of defaultable assets in the 90's, such as Duffie and Huang (1996), Bielecki and Rutkowski (2002, Eqn. (14.25) p. 448) obtained the formula

$$\mathrm{CVA} - \mathrm{DVA} \tag{1}$$

for the valuation of bilateral counterparty risk on a swap, assuming risk-free funding. This formula, rediscovered and generalized by others since the 2007–09 financial crisis (see e.g. Brigo and Capponi (2010)), is symmetrical, i.e., it is the negative of the analogous quantity considered from the point of view of the counterparty, consistent with the law of one price and the Modigliani and Miller (1958) theorem (see §3.E.2).

Around 2010, the materiality of the DVA windfall benefit of a bank at its own default time became the topic of intense debates in the quant and academic communities. At least, it seemed reasonable to admit that, if the own default risk of the bank was accounted for in the modeling, in the form of a DVA benefit, then the cost of funding (FVA) implication of this risk should be included as well, leading to the modified formula ($\mathrm{CVA} - \mathrm{DVA} + \mathrm{FVA}$). See for instance Burgard and Kjaer (2011, 2013, 2017), Crépey (2015), Brigo and Pallavicini (2014), or Bichuch, Capponi and Sturm (2018). See also Bielecki and Rutkowski (2015) for an abstract funding framework (without explicit reference to XVAs), generalizing Piterbarg (2010) to a nonlinear setup.

Then Hull and White (2012) objected that the FVA was only the compensator of another windfall benefit of the bank at its own default, corresponding to the non-reimbursement by the bank of its funding debt. Accounting for the corresponding "DVA2" (akin to the FDA in this book) brings back to the original valuation formula

$$\mathrm{CVA} - \mathrm{DVA} + \mathrm{FVA} - \mathrm{FDA} = \mathrm{CVA} - \mathrm{DVA}, \tag{2}$$

as $\mathrm{FVA} = \mathrm{FDA}$ (assuming risky funding fairly priced as we will see).

However, their argument implicitly assumes that the bank can perfectly hedge its own default: cf. Burgard and Kjaer (2013, end of Section 3.1) and see §3.E and II.§4. As a bank is an intrinsically leveraged entity, this is not the case in practice. One can mention the related corporate finance notion of debt overhang in Myers (1977), by which a project valuable for the firm as a whole may be rejected by shareholders because the project is mainly valuable to bondholders. But, until recently, such considerations were hardly considered in the field of derivative pricing.

The first ones to recast the XVA debate in the perspective of the balance sheet of the bank were Burgard and Kjaer (2011), to explain that an appropriately hedged derivative position has no impact on the dealer's funding costs. Also relying on balance sheet models of a dealer bank, Castagna (2014) and Andersen, Duffie and Song (2019) ended up with conflicting conclusions, namely that the FVA should, respectively should not be included in the valuation of financial derivatives. Adding a KVA, but in a replication framework, Green, Kenyon and Dennis (2014) concluded that both the FVA and the KVA should be included as add-ons in entry prices and as liabilities in the balance sheet.

A Contents of the Chapter

Our key premise is that counterparty risk entails two distinct but intertwined sources of market incompleteness:

- A bank cannot perfectly hedge counterparty default losses by lack of sufficiently liquid CDS markets;
- A bank can even less hedge its own jump-to-default exposure, because this would mean selling protection on its own default, which is impractical and, under certain juridictions, even legally forbidden.

In this chapter we specify the banking XVA metrics that align derivative entry prices to shareholder interest, given this impossibility for a bank to replicate the jump-to-default related cash flows. We develop a cost-of-capital XVA approach consistent with the accounting standards set out in IFRS 4 Phase II (see International Financial Reporting Standards (2013)), inspired from the Swiss solvency test and Solvency II insurance regulatory frameworks (see Swiss Federal Office of Private Insurance (2006) and Committee of European Insurance and Occupational Pensions Supervisors (2010)), which until recently had no analogue in the banking domain. Under this approach, the valuation (CL) of the so-called contra-liabilities and the cost of capital (KVA) are sourced from clients at trade inception, on top of the (CVA − DVA) complete market valuation of counterparty risk, in order to compensate bank shareholders for wealth transfer and risk on their capital.

The cost of the corresponding collateralization, accounting and dividend policy is, by contrast with the complete market valuation (CVA − DVA) of counterparty risk,

$$\mathrm{CVA} + \mathrm{FVA} + \mathrm{KVA}, \tag{3}$$

computed unilaterally in a certain sense (even though we do crucially include the default of the bank itself in our modeling), and charged to clients on an incremental run-off basis at every new deal.

All in one, our cost-of-capital XVA strategy makes shareholder equity a submartingale with drift corresponding to a hurdle rate h on shareholder capital at risk, consistently between and throughout deals. Thus we arrive at a sustainable strategy for profit retention, much like in the above-mentioned insurance regulation, but in a dynamically consistent, continuous-time and banking framework.

A.1 Outline and Contributions

§2 sets a financial stage where a bank is split across several trading desks and entails different stakeholders. §3 develops our cost-of-capital XVA approach in a one-period static setup. §4 revisits the approach at the dynamic and trade incremental levels.

Salient developments of this chapter are:

- The one-period static XVA model of §3, with explicit formulas for all the quantities at hand, offering a concrete grasp on the related wealth transfer and risk premium issues;
- Proposition 3, page 45, which establishes the connections between XVAs, the core equity tier 1 capital of the bank, and the bank shareholder equity;
- Proposition 4, page 48, which establishes that, under the XVA policy represented by the balance conditions (5) between deals and the counterparty risk add-on (44) throughout deals, bank shareholder equity is a submartingale with drift corresponding to a target hurdle rate h on shareholder capital at risk. This solves the puzzle according to which, on the one hand, XVA computations are performed on a run-off portfolio basis, while, on the other hand, they are used for computing pricing add-ons to new deals.

In particular (compare with the last paragraph before Subsection A above), our approach implies that the FVA should be included as an add-on in entry prices and as a liability in the balance sheet; the KVA should be included as an add-on in entry prices, but not as a liability in the balance sheet.

§2 BALANCE SHEET AND CAPITAL STRUCTURE MODEL OF THE BANK

We consider a dealer bank, which is a market maker involved in bilateral derivative portfolios. For simplicity, we only consider European derivatives. Collateral means cash or liquid assets that are posted to guarantee a netted set of transactions against defaults. It comes in two forms: ***variation***

margin, which is rehypothecable, i.e. fungible across netting sets, and ***initial margin***, which is segregated. We assume cash only collateral. Posted collateral is supposed to be remunerated at the risk-free rate.

The bank has two kinds of stakeholders, ***shareholders*** and ***bondholders***. The shareholders have control of the bank and are solely responsible for investment decisions before bank default. The bondholders represent the senior creditors of the bank, which have no power to make decisions until a bank default, but are protected by laws, of the pari-passu type, forbidding trades that would trigger value away from them to shareholders during the default resolution process of the bank. The bank also has junior creditors, represented in our framework by an ***external funder***, which can lend unsecured to the bank and is assumed to suffer an exogenously given loss-given-default in case of default of the bank.

We consider three kinds of business units within the bank:

- the ***trading desks***, who focus on the market risk of the contracts in their respective business lines;
- the ***CA desks***, i.e. the CVA desk and the funding desk (or Treasury) of the bank, in charge of what we call the contra-assets, i.e. special liabilities that arise from counterparty risk and its funding implications for the bank;
- the ***management*** of the bank, in charge of the dividend release policy of the bank.

See Figure 3, page 28, for the corresponding picture of the bank balance sheet, referring to Table 2, page 29 for a list of the main financial acronyms used across the book. Contra-liability valuation (CL) at the top of the figure is shown in dotted boxes because it is only of value to the bondholders (see §3.E and II.§4). Mark-to-market (MtM) or counterparty-risk-free valuation of the derivatives portfolio of the bank by the trading desks, as well as the corresponding collateral (trading account), are shown in dashed boxes at the bottom. Their role will essentially vanish in our setup, at least in our reference hedging case where we assume the trading desks are perfectly hedged. The dotted arrows in the left column represent trading losses of the CA desks in "normal years 1 to 39" and in an "exceptional year 40" with full depletion (i.e. refill via UC, under Assumption 1(ii) page 31) of RC, RM and SCR. The numberings yr1 to yr40 are fictitious yearly scenarios in line with a 97.5% expected shortfall of the one-year-ahead trading losses of the bank that we use for defining its economic capital. The dotted arrows in the right column symbolize the average depreciation in time of contra-assets between deals. The collateral between the bank and its counterparties is not shown to alleviate the picture.

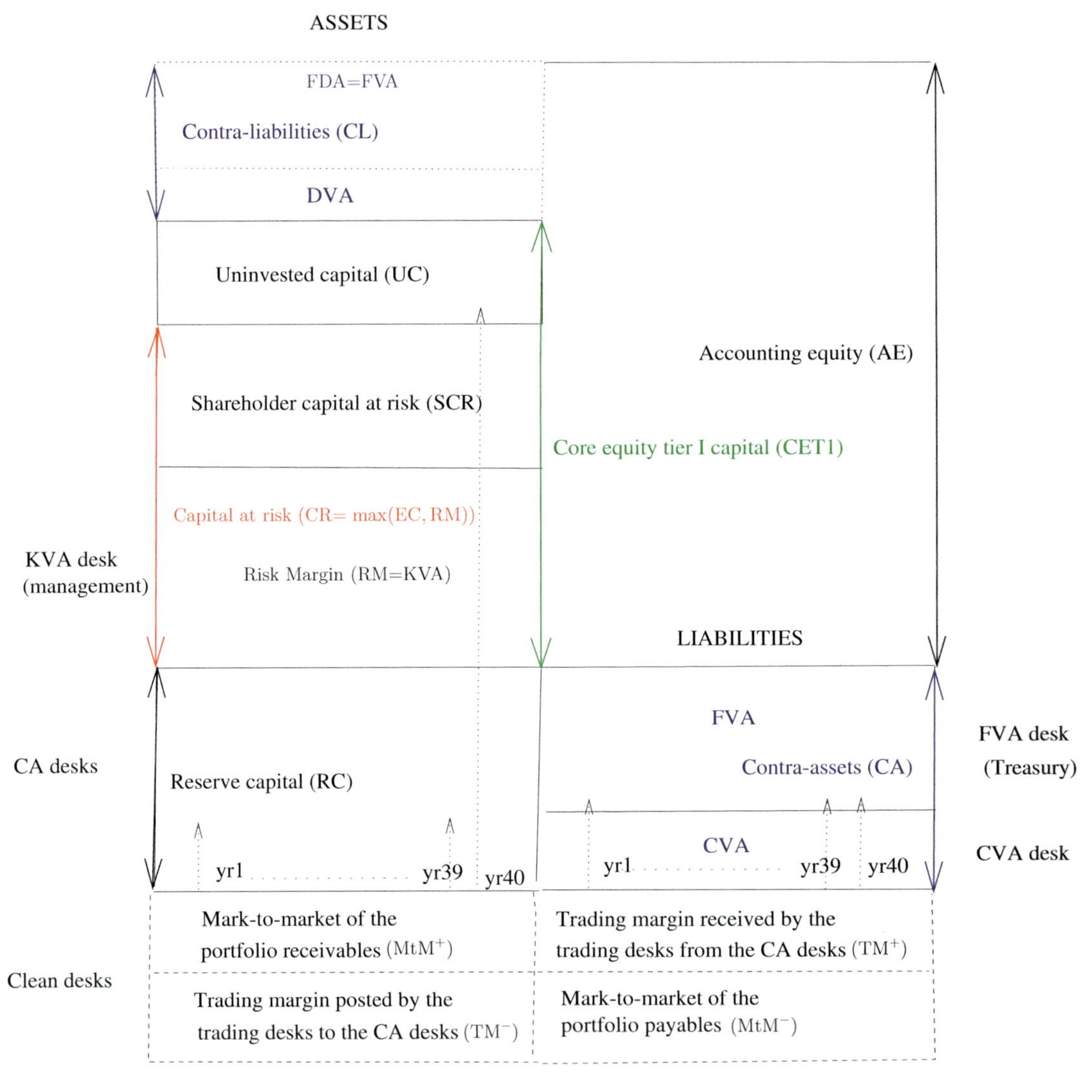

Figure 3: Balance sheet of a dealer bank (see the comments in the text).

Table 2: Main financial acronyms and place where they are introduced conceptually and/or specified mathematically within the book.

Amounts on dedicated cash accounts of the bank:

TM	Trading margin	Definition 1 and Assumption 1
RC	Reserve capital	Definition 1 and Assumption 1
RM	Risk margin	Definition 1 and Assumption 1
UC	Uninvested capital	Definition 1 and Assumption 1

Valuations:

CA	Contra-assets valuation	(4), (11), and VII.(1)
CL	Contra-liabilities valuation	Definition 1 and (13), (32), and (44)
CVA	Credit valuation adjustment	(12), (11), VII.(3), and VII.(10)-(11)
DVA	Debit valuation adjustment	(13) and (12)
FDA	Funding debit adjustment	(13) and (18)
FV	Firm valuation of counterparty risk	(16) and (18)
FVA	Funding valuation adjustment	Remark 1, (12), (11), and VII.(3)
KVA	Capital valuation adjustment	(5), (22), and (43)
MtM	Mark-to-market	(5)
MVA	Margin valuation adjustment	Remark 1, (29), VII.(3), and VII.(12)
XVA	Generic "X" valuation adjustment	First paragraph

Also:

CR	Capital at risk	(41)
CET1	Core equity tier I capital	(4) and (38)
EC	Economic capital	Definitions 3 and 5
FTP	Funds transfer price	(44)
SHC	Shareholder capital (or equity)	(4) and (39)
SCR	Shareholder capital at risk	Assumption 1 and (21)

The CA desks guarantee the business of the trading desks against counterparty defaults through a ***trading account*** storing (rehypothecable) collateral exchanged between the CA desks and the trading desks. The trading margin (TM) on the trading account also allows the trading desks to fund their trading at a risk-free rate that they pay to the Treasury of the bank (see the bottom, dashed boxes in Figure 3 page 28 and for a precise understanding of this mechanism, see III.§3, Assumption III.1 page 93 in particular).

In addition, the CA desks value the contra-assets (future counterparty default losses and funding expenditures), charge them to the corporate (non banking) clients at deal inception, deposit the corresponding payments in a ***reserve capital account***, and then are exposed to the corresponding payoffs. As time proceeds, contra-assets are realized and are covered by the CA desks with the reserve capital account.

On top of reserve capital, the so-called risk margin is sourced by the management of the bank from the clients at deal inception, deposited into a ***risk margin account***, and then gradually released as KVA payments into the shareholder dividend stream.

Another account contains the ***shareholder capital at risk*** earmarked by the bank to deal with exceptional trading losses (beyond the expected losses that are already accounted for by reserve capital).

Last, there is one more bank account with shareholder ***uninvested capital***.

All cash accounts are remunerated at the risk-free rate.

Definition 1. We write TM, RC, RM, SCR and UC for the respective (risk-free discounted) amounts on the trading, reserve capital, risk margin, shareholder capital at risk and uninvested capital accounts of the bank. We also define

$$\mathrm{SHC} = \mathrm{SCR} + \mathrm{UC}\,,\quad \mathrm{CET1} = \mathrm{RM} + \mathrm{SCR} + \mathrm{UC} = \mathrm{SHC} + \mathrm{RM}. \tag{4}$$

From a financial interpretation point of view, before bank default, SHC corresponds to ***shareholder capital*** (or ***equity***); CET1 is the ***core equity tier I capital*** of the bank, representing the financial strength of the bank assessed from a regulatory, structural solvency point of view, i.e. the sum between shareholder capital and the risk margin (which is also loss-absorbing), but excluding the value CL of the so-called contra-liabilities (see Figure 3 page 28). Indeed, the latter only benefits the bondholders (cf. §3.E and II.§4), hence it only enters accounting equity. Before the default of the bank, ***shareholders wealth*** and ***bondholders wealth*** are respectively given by $\mathrm{SHC}+\mathrm{RM}^{sh}$ and $\mathrm{CL}+\mathrm{RM}^{bh}$, for shareholder and bondholder components of RM to be detailed in Remark 10 page 39; shareholder and bondholder wealths sum up to the accounting equity $\mathrm{RM}+\mathrm{SCR}+\mathrm{UC}+\mathrm{CL}$, i.e. the wealth of the firm as a whole (see Figure 3).

Remark 2. The purpose of our capital structure model of the bank is *not* to model the default of the bank, like in a Merton (1974) model, as the point of negative equity (i.e. $\mathrm{CET1} < 0$). In the case of a bank, such a default model would be unrealistic. For instance, at the time of its collapse in April 2008, Bear Stearns had billions of capital. In fact, the legal definition of default is an unpaid coupon or cash flow, which is a liquidity (as opposed to solvency) issue. Eventually, we will model the default of the bank as a totally unpredictable event at some exogenous time τ calibrated to the credit default swap (CDS) curve referencing the bank. Indeed we view the latter as the most reliable and informative credit data regarding anticipations of market participants about future recapitalization, government intervention, bail-in and other bank failure resolution policies.

The aim of our capital structure model, instead, is to put in a balance sheet perspective the contra-assets and contra-liabilities of a dealer bank, items which are not present in the Merton model and play a key role in our XVA analysis.

The Volcker rule bans proprietary trading for a bank: see Paragraph 1851 in Section 619 from The United States Congress (2010). Accordingly, we typically (even though not always, cf. Remark II.4) assume a perfect market hedge of the derivative portfolio of the bank by the trading desks. By contrast, as jump-to-default exposures (own jump-to-default exposure, in particular) cannot be hedged by the bank (cf. §1.A), we typically (and conservatively) assume no XVA hedge.

Using the risk-free asset as our numéraire, we work on a measurable space $(\Omega, \mathcal{A})$ endowed with a fininsurance probability measure $\mathbb{Q}^*$ (see the introduction to this Pricing part of the book), with $\mathbb{Q}^*$ expectation denoted by $\mathbb{E}^*$.

A Run-Off Portfolio

Until §4.B, we consider the case of a portfolio held on a run-off basis, i.e. set up at time 0 and such that no new unplanned trades enter the portfolio in the future.

The cash flows of interest for us in this book (cumulative cash flow streams starting from 0 at time 0) then consist of the dynamic hedging cash flows $\mathcal{H}$ of the bank to the financial hedging markets, as well as the cash flows (other than those already contained in $\mathcal{H}$) $\mathcal{M}$ to the trading desks, $\mathcal{C}$ from the CVA desk and $\mathcal{F}$ from the funding desks of the bank (note that all cash flow differentials can be positive or negative). See e.g. §3.A and Proposition III.2 for concrete specifications in respective one-period and continuous-time setups.

Assumption 1. (i) (self-financing condition) RC+RM+SCR+UC−TM evolves like the received trading cash flows $\mathcal{M} - \mathcal{C} - \mathcal{F} - \mathcal{H}$.
(ii) (mark-to-model) The amounts on all the accounts but UC are marked-to-model (hence the last, residual amount, UC, plays the role of an adjustment variable). Specifically, we assume that the following ***shareholder balance conditions*** hold at all times:

$$\mathrm{TM} = \mathrm{MtM}\,,\ \ \mathrm{RC} = \mathrm{CA}\,,\ \ \mathrm{RM} = \mathrm{KVA}, \tag{5}$$

for theoretical target levels MtM, CA and KVA to be specified in later sections (along with the theoretical target level for SCR) in order to ensure to the shareholders a hurdle rate h on their capital at risk.

(iii) (agents) The initial amounts MtM_0, CA_0 and KVA_0 are provided by the clients at portfolio inception time 0. Between time 0 and the bank default time τ (excluded), the resets of the corresponding accounts are on bank shareholders. At the (assumed positive) bank default time τ, the property of the residual amount on the reserve capital and risk margin accounts is transferred from the shareholders to the bondholders of the bank (while the advent of the trading margin will be specified in Assumption III.1 page 93).

Remark 3. In an asymmetric setup with a price maker (such as a bank) and a price taker, the price maker passes his costs to the price taker. Accordingly, in our setup, the nonbanking (corporate) clients provide all the amounts to the trading, reserve capital and risk margin accounts of the bank which are required for setting the accounts to their theoretical target levels (5) corresponding to the portfolio at inception.

Principle 1. Under a cost-of-capital XVA approach:
(i) we define valuation so as to make shareholder trading losses (that include marked-to-model liability fluctuations) centered,
(ii) then we add a KVA risk premium (which is not a liability, hence is also loss-absorbing, i.e. part of capital at risk) in order to ensure to bank shareholders some positive hurdle rate h on their capital at risk.

In what follows, such an approach is developed, first, in a static setup, which can be solved explicitly, and then, in a dynamic and trade-incremental setup, as suitable for dealing with a real derivative banking portfolio.

§3 XVA ANALYSIS IN A STATIC SETUP

In this section, we present the cost-of-capital XVA approach for a portfolio made of a single deal, $\mathcal{P}$ (random variable promised to the bank), between a bank and a client, without prior endowment, in an elementary one-period setup. All the trading cash flows $\mathcal{P}$, $\mathcal{H}$, $\mathcal{M}$, $\mathcal{C}$ and $\mathcal{F}$ are then random variables (as opposed to processes in multi-period setups later). We first assume no collateral is exchanged between the bank and its client. Risky funding assets are assumed fairly priced by the market, in the sense that $\mathbb{E}^*\mathcal{F} = 0$.

The bank and client are both default-prone with zero recovery to each other. The bank also has zero recovery to its external funder. We denote by J and J_1 the survival indicators (random variables) of the bank and client at time 1, with default probability of the bank $\mathbb{Q}^*(J = 0) = \gamma$. Since prices and XVAs only matter at time 0 in a one-period setup; we identify all the XVA processes, as well as the counterparty-risk-free valuation P of the deal by the trading desks, with their values at time 0.

Remark 4. Nonzero, possibly random, recoveries could be introduced at no harm from a theoretical viewpoint and this will be done in the continuous-time setup later in the book. Nonzero recoveries are, of course, more realistic. But, from a qualitative viewpoint, that is our main objective here, they only soften the impact of the defaults.

The trader of the bank sells the deal to the client at the base price $P = \mathbb{E}^*\mathcal{P}$, to be adjusted in line with a cost of capital approach for counterparty credit risk and its implications in terms of capital and funding for the bank. We assume further that a back-to-back hedge is set up by the bank in the form of a deal with another bank, assumed default-free for simplicity, with price P, and variation margin P again, from the dealer bank to its hedge counterparty.

Remark 5. The hedge counterparty (and possibly the entity that funds the bank as well) is computing and charging likewise its own XVAs to its own corporate clients in the background, but this is of no concern to the dealer bank.

For any random variable $\mathcal{Y}$, we define

$$\mathcal{Y}^\circ = J\mathcal{Y} \text{ and } \mathcal{Y}^\bullet = -(1-J)\mathcal{Y}\,, \quad \text{hence } \mathcal{Y} = \mathcal{Y}^\circ - \mathcal{Y}^\bullet. \tag{6}$$

Let $\mathbb{E}$ denote the expectation with respect to the bank survival measure, say $\mathbb{Q}$, associated with $\mathbb{Q}^*$, i.e., for any random variable $\mathcal{Y}$,

$$\mathbb{E}\mathcal{Y} = (1-\gamma)^{-1}\mathbb{E}^*(\mathcal{Y}^\circ) \tag{7}$$

(which is also equal to $\mathbb{E}\mathcal{Y}^\circ$). The notion of bank survival measure was introduced in the continuous-time setup by Schönbucher (2004) (see §4). In the present static setup, (7) is nothing but the $\mathbb{Q}^*$ expectation of $\mathcal{Y}$ conditional on the survival of the bank (note that, whenever $\mathcal{Y}$ is independent from J, the right-hand side in (7) coincides with $\mathbb{E}^*\mathcal{Y}$).

Lemma 1. *For any random variable $\mathcal{Y}$ and constant Y, we have*

$$Y = \mathbb{E}^*(\mathcal{Y}^\circ + (1-J)Y) \Longleftrightarrow Y = \mathbb{E}\mathcal{Y}.$$

Proof. It holds that

$$Y = \mathbb{E}^*(J\mathcal{Y} + (1-J)Y) \Leftrightarrow \mathbb{E}^*(J(\mathcal{Y}-Y)) = 0 \Leftrightarrow \mathbb{E}(\mathcal{Y}-Y) = 0 \Leftrightarrow Y = \mathbb{E}\mathcal{Y},$$

where the middle equivalence is justified by (7). □

Remark 6. For simplicity in a first stage, we will ignore the possibility for the bank to use its capital at risk for its funding purposes, only considering in this respect reserve capital RC = CA (cf. (5)). The additional free funding source provided by capital at risk will be introduced later, as well as collateral between bank and clients, in Subsection D in the one-period setup and in VII.§1 in a continuous-time bilateral trading setup.

A Cash Flows

The general rule regarding the settlement of contracts of a defaulted netting set, to be instantiated in practical setups on a case-by-case basis, is as follows, assuming a zero recovery rate of the defaulted party for simplicity in this chapter:

Principle 2. If a counterparty in default is indebted toward the other beyond its posted margin, then this debt is only reimbursed at the level of this posted margin; otherwise, the debt between the two parties is fully settled.

Here debt is understood on a counterparty-risk-free basis.

One intuitively expects credit default cash flows of the form $\mathcal{C} = (D-M)^+$, where the "debt" D represents the pre-default value of the client derivative portfolio to the bank and M the margin posted by the client to the bank. We emphasize that such a counterparty credit default loss $\mathcal{C} = (D-M)^+$ (or $(1-R_1)(D-M)^+$ in the more general case of a recovery rate R_1 of the counterparty) should not be taken as an *assumption*, but arise as the result of a *computation* accounting for the cash flows of the portfolio *and its hedge*, derived in a specific market setup under the umbrella of the guiding principle 2 (or its nonzero recovery extension of Principle III.1 page 86). The exact outcome depends on the specification of the setup at hand: see e.g. Lemmas 2, III.2, III.4, VIII.2, or Proposition III.2. Such formulas cannot be safely *guessed*, they should be *derived* from first principles.

Lemma 2. *Given the (to-be-specified)* CA *amount (cf. Assumption 1(ii) page 31), the dynamic hedging, trading, credit and funding cash flows* $\mathcal{H}$, $\mathcal{M}$, $\mathcal{C}$ *and* $\mathcal{F}$ *of the bank and its trading loss (and profit)* $\mathcal{L}$ *are such that* $\mathcal{H} = \mathcal{M} = 0$ *and*

$$\begin{aligned}
&\mathcal{C} = (1-J_1)\mathcal{P}^+ - (1-J)\big(\mathcal{P}^- + (\mathcal{P}-P)^+\big)\\
&\mathcal{F} = \gamma(P-\mathrm{CA})^+ - (1-J)(P-\mathrm{CA})^+\\
&\mathcal{L} = \mathcal{C} + \mathcal{F} - \mathrm{CA},
\end{aligned} \tag{8}$$

hence

$$\begin{aligned}
&\mathcal{C}^\circ = J(1-J_1)\mathcal{P}^+\,,\ \ \mathcal{F}^\circ = J\gamma(P-\mathrm{CA})^+\\
&\mathcal{C}^\bullet = (1-J)\big(\mathcal{P}^- + (\mathcal{P}-P)^+ - (1-J_1)\mathcal{P}^+\big),\\
&\qquad\mathcal{F}^\bullet = (1-J)\big((P-\mathrm{CA})^+ - \gamma(P-\mathrm{CA})^+\big)\\
&\mathcal{L}^\circ = \mathcal{C}^\circ + \mathcal{F}^\circ - J\mathrm{CA}\,,\ \ \mathcal{L}^\bullet = \mathcal{C}^\bullet + \mathcal{F}^\bullet + (1-J)\mathrm{CA}.
\end{aligned} \tag{9}$$

Proof. On the claim side, accounting for the defaults, by application of Principle 2, the trading desk only gets from the client

$$J_1 J\mathcal{P} + (1-J_1)J(-\mathcal{P}^-) + J_1(1-J)\mathcal{P}^+ + (1-J_1)(1-J)\,0$$

instead of the promised cash flow $\mathcal{P}$. To restore the latter, the trading desk needs compensation by the CVA desk of

$$\begin{aligned}
&\mathcal{P} - J_1 J\mathcal{P} - (1-J_1)J(-\mathcal{P}^-) - J_1(1-J)\mathcal{P}^+ - (1-J_1)(1-J)\,0 \\
&= (1-J_1)J\mathcal{P}^+ - J_1(1-J)\mathcal{P}^- + (1-J_1)(1-J)\mathcal{P} \\
&= (1-J_1)J\mathcal{P}^+ - J_1(1-J)\mathcal{P}^- + (1-J_1)(1-J)(\mathcal{P}^+ - \mathcal{P}^-) \\
&= (1-J_1)\mathcal{P}^+ - (1-J)\mathcal{P}^-.
\end{aligned}$$

Similarly, to restore the cash flow $(-\mathcal{P})$ promised to the trading desk by the back-to-back hedge counterparty, the CVA needs to compensate the trading desk by the (negative) cash flow $-(1-J)(\mathcal{P}-P)^+$, whence $\mathcal{C}$ as in (8). These compensations by the CVA desk restore the perfect (market) hedge of the trading desk, i.e. $\mathcal{M} = 0$.

The Treasury of the bank incurs a funding cost $\gamma(P - \mathrm{CA})^+$ and enjoys a windfall funding benefit of $(1-J)(P-\mathrm{CA})^+$, whence $\mathcal{F}$ as per (8). The amount $(P-\mathrm{CA})^+$ corresponds to money borrowed at time 0 and kept at time 1 in the case where the bank defaults. Note that, if $P - \mathrm{CA} < 0$, then the bank is actually the lender (at the risk-free interest rate) at time 0 and it is reimbursed at time 1 whatever its default status at time 1 (assuming a risk-free lending third-party).

In total the bank indeed pays the trading cash flows $\mathcal{L}$ as per (8). □

Remark 7. The above derivation implicitly allows for negative equity (that arises whenever $\mathcal{L}^\circ > \mathrm{CET1}$, cf. (4)), which is interpreted as recapitalization. In a variant of the model excluding both recapitalization and negative equity, the default of the bank would be modeled in a structural fashion as the event $\{\widetilde{\mathcal{L}} = \mathrm{CET1}\}$, where

$$\widetilde{\mathcal{L}} = \Big((1-J_1)\mathcal{P}^+ + \gamma(P-\mathrm{CA})^+ - \mathrm{CA}\Big) \wedge \mathrm{CET1}, \tag{10}$$

and we would obtain, instead of $\mathcal{L}$ as in (8), the following trading loss for the bank:

$$\mathbb{1}_{\{\mathrm{CET1} > \widetilde{\mathcal{L}}\}} \widetilde{\mathcal{L}} + \mathbb{1}_{\{\mathrm{CET1} = \widetilde{\mathcal{L}}\}} \Big(\mathrm{CET1} - \mathcal{P}^- - (\mathcal{P}-P)^+ - (P-\mathrm{CA})^+\Big).$$

In this book we only consider models with recapitalization for the reasons explained in Remark 2.

Remark 8. Structural XVA approaches in a static setup have been proposed in Andersen, Duffie and Song (2019) (without KVA) and Kjaer (2019) (including the KVA). Their marginal, limiting results as a new deal size goes to zero are comparable to some of the results that we have here. But then, instead of developing a continuous time version of their corporate finance model and taking the small trade limit, these papers start the development of the continuous time model from the single period small trade limit model. By contrast, in our framework, we have end to end development in the present single period model and in the continuous time model of §4.

B Contra-assets and Contra-liabilities

To make shareholder trading losses centered (cf. the next-to-last paragraph of §2), the trading and the CA desks value by $\mathbb{Q}^*$ expectation their shareholder sensitive cash flows. These include, in case of default of the bank, the transfer from shareholders to bondholders of the residual value RC = CA on the reserve capital account (cf. Assumptions 1(ii)–(iii)). Accordingly:

Definition 2. We let

$$\mathrm{CA} = \mathrm{CVA} + \mathrm{FVA}, \tag{11}$$

where

$$\begin{aligned} \mathrm{CVA} &= \mathbb{E}^*\Big(\mathcal{C}^\circ + (1-J)\mathrm{CVA}\Big) \\ \mathrm{FVA} &= \mathbb{E}^*\Big(\mathcal{F}^\circ + (1-J)\mathrm{FVA}\Big), \end{aligned} \tag{12}$$

hence $\mathrm{CA} = \mathbb{E}^*\Big(\mathcal{C}^\circ + \mathcal{F}^\circ + (1-J)\mathrm{CA}\Big)$. We also define the contra-liabilities value

$$\mathrm{CL} = \mathrm{DVA} + \mathrm{FDA}, \tag{13}$$

where

$$\mathrm{DVA} = \mathbb{E}^*\Big(\mathcal{C}^\bullet + (1-J)\mathrm{CVA}\Big) \tag{14}$$

$$\mathrm{FDA} = \mathbb{E}^*\Big(\mathcal{F}^\bullet + (1-J)\mathrm{FVA}\Big). \tag{15}$$

Finally, we define the firm valuation of counterparty risk (i.e. its valuation from the point of view of the bank as a whole as we will see),

$$\mathrm{FV} = \mathbb{E}^*(\mathcal{C} + \mathcal{F}). \tag{16}$$

The terminal cash flows of the form $(1-J) \times \ldots$ in the CVA and FVA definitions (12), and in the KVA definition (22) later below, are consistent with the shareholder-centric perspective of Principle 1 that is relevant for the determination of entry prices. They can also be interpreted as the amounts of reserve capital and risk margin lost by the bank shareholders, hence valued as such in CA and KVA, as their property is transferred to the liquidator of the bank if the bank defaults. Due to these terminal cash flows, the definitions of the CVA and the FVA (and similarly so for the KVA later below) are in fact fix-point equations. However, the following result shows that these equations are well-posed and yield explicit formulas for all the quantities at hand.

Proposition 1. *We have*

$$\begin{aligned} \mathrm{CVA} &= \mathbb{E}\Big((1-J_1)\mathcal{P}^+\Big) \\ \mathrm{FVA} &= \gamma(P-\mathrm{CA})^+ = \frac{\gamma}{1+\gamma}(P-\mathrm{CVA})^+ \end{aligned} \tag{17}$$

and

$$\begin{aligned} &\mathbb{E}^*\mathcal{L}^\circ = \mathbb{E}\mathcal{L} = 0 \\ &\mathrm{FDA} = \mathrm{FVA} \\ &\mathrm{FV} = \mathbb{E}^*\mathcal{C} = \mathrm{CVA} - \mathrm{DVA} = \mathrm{CA} - \mathrm{CL}. \end{aligned} \tag{18}$$

Proof. The first identities in each line of (17) follow from Definition 2 by Lemma 1 and by definition of the involved cash flows in Lemma 2. Given (11), the formula $\mathrm{FVA} = \gamma(P-\mathrm{CA})^+$ in (17) is in fact a semi-linear equation

$$\mathrm{FVA} = \gamma(P - \mathrm{CVA} - \mathrm{FVA})^+. \tag{19}$$

But, as γ (a probability) is nonnegative, this equation has the unique solution given by the right-hand side in the third line of (17).

Regarding (18), we have

$$\mathbb{E}^*\mathcal{L}^\circ = (1-\gamma)\mathbb{E}\Big((1-J_1)\mathcal{P}^+ + \gamma(P-\mathrm{CA})^+ - \mathrm{CA}\Big) = 0,$$

by application of (7), the first line in (9), (17) and (11). Hence, using (7) again,

$$\mathbb{E}\mathcal{L} = (1-\gamma)^{-1}\mathbb{E}^*\mathcal{L}^\circ = 0.$$

This is the first line in (18), which implies the following ones by definition of the involved quantities and from the assumption that $\mathbb{E}^*\mathcal{F} = 0$. □

Remark 9. The (hedged) loss of the trading desks vanishes whatever the value of the counterparty-risk-free price P of the deal (since the latter cancels out with the hedge), but in view of Lemma 2, this value P is a key XVA input. The possibility for the trading desks to find hedge counterparties at the price P leads to assume that $P = \mathbb{E}^*\mathcal{P}$. From an engineering viewpoint, the above implies to convey the two probability measures $\mathbb{Q}^*$ and $\mathbb{Q}$ in the modeling. If, however, P is not only the value $\mathbb{E}^*\mathcal{P}$ of $\mathcal{P}$, but also its bank shareholder value $\mathbb{E}\mathcal{P}$, then $\mathbb{Q}$ modeling is sufficient.

By (7), $\mathbb{E}^*(J\mathcal{P}) = (1-\gamma)\mathbb{E}\mathcal{P} = (\mathbb{E}^*J)(\mathbb{E}\mathcal{P})$, by definition of γ. Having assumed that $\gamma \in (0,1)$, the additional assumption $\mathbb{E}\mathcal{P} = \mathbb{E}^*\mathcal{P}$ is therefore tantamount to

$$\mathbb{E}^*(J\mathcal{P}) = (\mathbb{E}^*J)(\mathbb{E}^*\mathcal{P}). \tag{20}$$

Note that $\mathcal{P} = \pm J$ violates (20) (having assumed $\gamma \in (0,1)$). We thus view a discrepancy between the valuation and the shareholder valuation of $\mathcal{P}$ as an indication of extreme dependence between the derivative portfolio and the default of the bank itself, such as the bank trading its own default risk, which should be considered with caution from a financial viewpoint. We conclude that the additional assumption $\mathbb{E}\mathcal{P} = \mathbb{E}^*\mathcal{P}$ is reasonable, and we advocate XVA modeling under $\mathbb{Q}$ on this basis.

C Capital Valuation Adjustment

Consistently with IFRS17, we define the KVA as a risk margin used for remunerating, at some hurdle rate h, the bank's shareholders for their capital at risk. Moreover, rather than relying on Pillar I regulatory capital models, we follow a more principled Pillar II economic-based approach. Indeed, regulatory capital models under Pillar I are consistent in intent with our assumptions, but they have numerous assumptions motivated by the desire to reduce the computational complexity.

Example 1. Regulatory capital is based on the notion of risk weighted assets (RWA), which is underlied by one of expected effective positive exposure (EEPE). In the case of an unsecured portfolio, the EEPE is defined as $\int_0^1 \left(\max_{s\in[0,t]} \mathbb{E}^*(P_s^+)\right)dt$, where P is the counterparty-risk free value of the portfolio. Hence regulatory capital takes a single counterparty-centric view; default clustering is neglected, as are complex tail dependencies; Gaussian copulas in the default RWA are highly stylized assumptions, and the possibility to create doom loops (e.g. between the CVA RWA and CDS hedges) is neglected. Pillar I approximations also unlink risk capital from the volatility of XVA metrics, albeit this feature has been identified by regulators as one of the primary drivers of the 2007–09 crisis.

Thus, we anchor the capital at risk of the bank at the level of an economic capital (EC) that a regulator would like to see on an economic, structural basis, defined as the 97.5% expected shortfall of the bank's losses over a one-year time horizon. Moreover, risk calculations are typically performed by banks "on a going concern", i.e. assuming that the bank itself does not default. Accordingly:

Definition 3. The economic capital (EC) of the bank is given by the 97.5% expected shortfall of the bank trading loss $\mathcal{L}$ under $\mathbb{Q}$, which we denote by[1] $\mathbb{ES}(\mathcal{L}^\circ)$.

CA corresponds to the expected costs for the shareholders of concluding the deal. In line with Principle 1(i) page 32, this CA component makes the

[1] note that, by definition of $\mathbb{Q}$, this expected shortfall does not depend on $\mathcal{L}^\bullet$.

shareholder trading loss $\mathcal{L}^\circ$ centered (cf. the first line in (18)). On top of expected shareholder costs, the bank charges the clients a risk margin (RM) as per Principle 1(ii). The risk margin (sized by the to-be-defined KVA in our setup) is not a liability. Therefore, it is loss-absorbing, hence part of capital at risk. Moreover, the KVA is originally sourced from the client (see Assumption 1(iii) page 31). As a consequence, *shareholder* capital at risk only consists of the *difference* between the (total) capital at risk and the KVA. Also accounting, regarding the $(1-J)$KVA term in (22), for the last part in Assumption 1(iii):

Definition 4. The capital at risk (CR) of the bank is given by max(EC, KVA) and the ensuing shareholder capital at risk (SCR) by

$$\mathrm{SCR} = \max(\mathrm{EC}, \mathrm{KVA}) - \mathrm{KVA} = (\mathrm{EC} - \mathrm{KVA})^+, \tag{21}$$

where, given some hurdle rate (target return-on-equity) h,

$$\mathrm{KVA} = \mathbb{E}^*\Big(h\mathrm{SCR}^\circ + (1-J)\mathrm{KVA}\Big). \tag{22}$$

Remark 10. In view of (22) and of the last balance condition in (5), we have (see after (4)):

$$\mathrm{RM}^{sh} = \mathbb{E}^*\Big(h\mathrm{SCR}^\circ\Big), \ \ \mathrm{RM}^{bh} = \mathbb{E}^*\Big((1-J)\mathrm{KVA}\Big). \tag{23}$$

We refer the reader to Lemma II.8(iii) page 74 for the analogous split of RM between shareholder and bondholder wealth in a dynamic, continuous-time setup.

Proposition 2. *We have*

$$\mathcal{L}^\circ = J(\mathcal{C} - \mathrm{CVA}), \ \ \mathrm{EC} = \mathbb{ES}(J(\mathcal{C} - \mathrm{CVA})) \tag{24}$$

$$\mathrm{KVA} = h\mathrm{SCR} = \frac{h}{1+h}\mathrm{EC} = \frac{h}{1+h}\mathbb{ES}(\mathcal{L}^\circ), \tag{25}$$

hence $0 \le \mathrm{KVA} \le \mathbb{ES}(\mathcal{L}^\circ)$.

Proof. (24) follows from the formula for $\mathcal{L}^\circ$ in (8) and the FVA formula in (17). The first identity in (25) follows from Lemma 1. The resulting KVA semi-linear equation (in view of (21)) is solved similarly to the FVA equation (19). The last statement follows from the nonnegativity of the expected shortfall of a centered random variable recalled in Prerequisites in the general introduction to the book. □

Remark 11. The reason why funding disappears from the bank trading loss on the survival event of the bank, i.e. $J(\mathcal{C} + \mathcal{F} - \mathrm{CA}) = J(\mathcal{C} - \mathrm{CVA})$, is because, in a one-period setup, the collateral borrowing requirements of the bank are simply constants. Hence, funding triggers no risk to the bank, but only a deterministic cost. In multi-period setups, as we will see, funding generates both costs and risk.

C.1 KVA Risk Premium and Indifference Pricing Interpretation

Assume the bank shareholders endowed with a utility function U on $\mathbb{R}$ such that $U(0) = 0$. In a shareholder indifference pricing framework, the risk margin would arise as per the equation

$$\mathbb{E}^* U(J(\mathrm{RM} - \mathcal{L})) = \mathbb{E}^* U(0) = 0 \tag{26}$$

(the expected utility of the bank shareholders without the deal), where

$$\mathbb{E}^* U(J(\mathrm{RM} - \mathcal{L}))) = \mathbb{E}^*\Big(JU(\mathrm{RM} - \mathcal{L}))\Big) = (1-\gamma)\mathbb{E}U(\mathrm{RM} - \mathcal{L})),$$

by (7). Hence

$$\mathbb{E}U(\mathrm{RM} - \mathcal{L})) = 0. \tag{27}$$

The corresponding RM is interpreted as the minimal admissible risk margin for the deal to occur, seen from bank shareholders' perspective.

Taking for concreteness $U(-\ell) = \frac{1-e^{\rho\ell}}{\rho}$, for some risk aversion parameter ρ, (27) yields $\mathrm{RM} = \rho^{-1} \ln \mathbb{E} e^{\rho\mathcal{L}} = \rho^{-1} \ln \mathbb{E} e^{\rho\mathcal{L}^\circ}$, by the observation following (7). In the limiting case where the shareholder risk aversion parameter $\rho \to 0$ and $\mathbb{E}U(-\mathcal{L})) \to -\mathbb{E}(\mathcal{L})) = 0$ (by the first line in (18)), then $\mathrm{RM} \to 0$.

In view of (5) and (25), the corresponding implied KVA and hurdle rate h are such that

$$\mathrm{KVA} = \rho^{-1} \ln \mathbb{E} e^{\rho\mathcal{L}^\circ}, \quad \frac{h}{1+h} = \frac{\rho^{-1} \ln \mathbb{E} e^{\rho\mathcal{L}^\circ}}{\mathbb{ES}(\mathcal{L}^\circ)}. \tag{28}$$

Hence, "for h and ρ small",

$$h \approx \frac{\mathbb{V}\mathrm{ar}(\mathcal{L}^\circ)}{2\mathbb{ES}(\mathcal{L}^\circ)} \rho$$

(as $\mathbb{E}(\mathcal{L}^\circ) = 0$), where $\mathbb{V}\mathrm{ar}$ is the $\mathbb{Q}$ variance operator. The hurdle rate h in our KVA setup thus plays the role of a risk aversion parameter, like ρ in the exponential utility framework.

An indifference price has a competitive interpretation. Assume that the bank is competing for the client with other banks. Then, in the limit of a continuum of competing banks with a continuum of indifference prices, whenever a bank makes a deal, this can only be at its indifference price. Our stylized indifference pricing model of a KVA defined by a constant hurdle rate h exogenizes (by comparison with the endogenous hurdle rate h in (28)) the impact on pricing of the competition between banks. It does so in a way that generalizes smoothly to a dynamic setup (see §4), as required to deal with a real derivative banking portfolio. It then provides a refined notion of return-on-equity for derivative portfolios, where a full-fledged optimization approach would be impractical (cf. the foreword to the book).

D Initial Margin and Fungibility of Capital at Risk as a Funding Source

In case of the variation margin (VM) that would be exchanged between the bank and its client, and of the initial margin that would be received (RIM) and posted (PIM) by the bank, at the level of, say for simplicity, some $\mathbb{Q}$ value-at-risk of $\pm(\mathcal{P} - \mathrm{VM})$, then

- $\mathcal{P}^+$ needs be replaced by $(\mathcal{P} - \mathrm{VM} - \mathrm{RIM})^+$ in $\mathcal{C}^\circ$ (cf. (9)) whence an accordingly modified (in principle: diminished) CVA,
- there would be an additional cash flow $J\gamma\mathrm{PIM}$ in F°, related to initial margin, triggering an additional adjustment MVA in CA, where
$$\mathrm{MVA} = \mathbb{E}^*\Big(J\gamma\mathrm{PIM} + (1-J)\mathrm{MVA}\Big) = \gamma\mathrm{PIM}; \tag{29}$$
- there would be an additional cash flow in $\mathcal{F}^\bullet$, also related to initial margin, triggering an additional adjustment MDA = MVA in CL;
- we would then have (cf. the corresponding lines in (9) and (25))
$$\begin{aligned}\mathcal{L}^\circ &= J(1-J_1)(\mathcal{P}{-}\mathrm{VM} - \mathrm{RIM})^+ - J\mathrm{CVA}\\ \mathrm{EC} &= \mathbb{ES}(\mathcal{L}^\circ)\,,\ \ \mathrm{KVA} = \frac{h}{1+h}\mathbb{ES}(\mathcal{L}^\circ) \le \mathbb{ES}(\mathcal{L}^\circ);\end{aligned} \tag{30}$$
- because of the additional MVA, the second FVA formula in (17) would become $\mathrm{FVA} = \frac{\gamma}{1+\gamma}(P - \mathrm{VM} - \mathrm{CVA} - \mathrm{MVA})^+$.

Moreover, capital at risk is available to the bank's Treasury as a free funding source for variation margin (cf. Remark 6). Accounting further for the latter, in view of the specification of capital at risk given in the first sentence of Definition 4, one obtains $(P - \mathrm{VM} - \mathrm{CA} - \max(\mathrm{EC}, \mathrm{KVA}))^+$ instead of $(P - \mathrm{CA})^+$ in (8) in $\mathcal{F}$. This results in the same CVA, MVA, $\mathcal{L}^\circ$, EC and KVA as in the bullet points above, but (cf. the second line in (17) and recall (11))

$$\begin{aligned}\mathrm{FVA} &= \gamma(P - \mathrm{VM} - \mathrm{CA} - \mathrm{EC})^+\\ &= \frac{\gamma}{1+\gamma}(P - \mathrm{VM} - \mathrm{CVA} - \mathrm{MVA} - \mathbb{ES}(\mathcal{L}^\circ))^+.\end{aligned} \tag{31}$$

Remark 12. The rationale for funding VM but not PIM from CA + max(EC, KVA) is as follows. The marginal cost of capital for using capital as a funding source for variation margin is nil, because when one posts cash against variation margin, the valuation of the collateralized hedge is reset to zero, and the total capital amount does not change. If, instead, the bank were to post capital as initial margin, then the bank would record a "margin receivable" entry on its balance sheet, which however cannot contribute to capital since this asset is too illiquid and impossible to unwind without unwinding all underlying derivatives. Hence, capital can only be used as VM, while PIM must be borrowed entirely.

E Funds Transfer Price

The funds transfer price (all-inclusive XVA rebate to P) aligning the deal to shareholder interest (in the sense of a given hurdle rate h, cf. the next-to-last paragraph of §2) is

$$\begin{aligned}\mathrm{FTP} &= \underbrace{\mathrm{CVA}+\mathrm{FVA}}_{\text{expected shareholder costs CA}} + \underbrace{\mathrm{KVA}}_{\text{shareholder risk premium}}\\ &= \underbrace{\mathrm{CVA}-\mathrm{DVA}}_{\text{firm valuation FV}} + \underbrace{\mathrm{DVA}+\mathrm{FDA}}_{\text{wealth transfer CL}} + \underbrace{\mathrm{KVA}}_{\text{shareholder risk premium}},\end{aligned} \tag{32}$$

where all terms are explicitly given in Propositions 1 and 2 (or the corresponding variants of Subsection D in the refined setup considered there).

E.1 Wealth Transfer Analysis

The above results implicitly assumed that the bank cannot hedge jump-to-default cash flows (cf. §1.A). To understand this, let us temporarily suppose, for the sake of the argument, that the bank would be able to hedge its own jump-to-default through a further deal, whereby the bank would deliver a payment $\mathcal{L}^\bullet$ at time 1 in exchange of a fee fairly valued as

$$\mathrm{CL} = \mathbb{E}^*\mathcal{L}^\bullet = \mathrm{DVA}+\mathrm{FDA}, \tag{33}$$

deposited in the reserve capital account of the bank at time 0.

We include this hedge and assume that the client would now contribute at the level of $\mathrm{FV} = \mathrm{CA} - \mathrm{CL}$ (cf. (18)), instead of CA before, to the reserve capital account of the bank at time 0. Then the amount that needs to be borrowed by the bank for implementing its strategy is still $\gamma(P-\mathrm{CA})^+$ as before (back to the baseline funding setup of Remark 6 page 33). But the trading loss of the bank becomes, instead of $\mathcal{L}$ before,

$$\mathcal{C}+\mathcal{F}-\mathrm{FV}+(\mathcal{L}^\bullet-\mathrm{CL}) = \mathcal{C}+\mathcal{F}-\mathrm{CA}+\mathcal{L}^\bullet = L+\mathcal{L}^\bullet = \mathcal{L}^\circ, \tag{34}$$

where the last line in (18) and the last identity in (9) were used in the first and second equality. By comparison with the setup of previous sections without own-default hedge by the bank:

- the shareholders are still indifferent to the deal in expected counterparty default and funding expenses terms,
- the recovery of the bondholders becomes zero,
- the client is better off by the amount $\mathrm{CA}-\mathrm{FV}=\mathrm{CL}$.

The CL originating cash flow $\mathcal{L}^{\bullet}$ has been hedged and monetized by the shareholders, who have passed the corresponding benefit to the client.

Under a cost-of-capital pricing approach, the bank would still charge its client a KVA add-on $\frac{h}{1+h}\mathbb{ES}(\mathcal{L}^{\circ})$, as risk compensation for the nonvanishing shareholder trading loss $\mathcal{L}^{\circ}$ still triggered by the deal. If, however, the bank could also hedge the (zero-valued, by the first line in (18)) loss $\mathcal{L}^{\circ}$, hence the totality of $\mathcal{L} = \mathcal{L}^{\circ} - \mathcal{L}^{\bullet}$ (instead of $\mathcal{L}^{\bullet}$ only in the above), then the trading loss and the KVA would vanish. As a result, the all-inclusive XVA add-on (rebate from the base price P) would boil down to

$$\mathrm{FV} = \mathrm{CVA} - \mathrm{DVA} \tag{35}$$

(cf. (1)), the value of counterparty risk and funding to the bank as a whole.

E.2 Connection with the Modigliani-Miller Theory

The Modigliani-Miller invariance result, with Modigliani and Miller (1958) as a seminal reference, consists of various facets of a broad statement that the funding and capital structure policies of a firm are irrelevant to the profitability of its investment decisions. Modigliani-Miller (MM) irrelevance, as we put it for brevity hereafter, was initially seen as a pure arbitrage result. However, it was later understood that there may be market incompleteness issues with it. So quoting Duffie and Sharer (1986, page 9),

> generically, shareholders find the span of incomplete markets a binding constraint [...] shareholders are not indifferent to the financial policy of the firm if it can change the span of markets (which is typically the case in incomplete markets)

or Gottardi (1995, page 197):

> When there are derivative securities and markets are incomplete the financial decisions of the firm have generally real effects.

A situation where shareholders may "find the span of incomplete markets a binding constraint" is when market completion is legally forbidden. This corresponds to the XVA case, which is also at the crossing between market incompleteness and the presence of derivatives pointed out above as the MM non-irrelevance case in Gottardi (1995). Specifically, the contra-assets and contra-liabilities that emerge endogenously from the impact of counterparty risk on the derivative portfolio of a bank cannot be "undone" by shareholders, because jump-to-default risk cannot be replicated by a bank.

As a consequence, MM irrelevance is expected to break down in the XVA setup. In fact, as visible on the FTP (counterparty risk pricing) formula (32) to be contrasted with the complete market FV as per (35) (cf. also (44) and

Proposition 4 in a dynamic and trade incremental setup below), cost of funding and cost of capital are material to banks and need be reflected in entry prices for ensuring shareholder indifference to the trades, i.e. preserving their hurdle rate throughout trades.

§4 XVA ANALYSIS IN A DYNAMIC SETUP

We now consider a dynamic, continuous-time setup, with model filtration $\mathfrak{G} = (\mathfrak{G}_t)_{t\in\mathbb{R}_+}$ and a (positive) bank default time τ endowed with an intensity γ. As introduced by Schönbucher (2004) and developed in Collin-Dufresne, Goldstein and Hugonnier (2004), the bank survival probability measure associated with the measure $\mathbb{Q}^*$ is then the probability measure $\mathbb{Q}$ with $(\mathfrak{G},\mathbb{Q}^*)$ density process $Je^{\int_0^{\cdot}\gamma_s ds}$ (assumed integrable), where $J = \mathbb{1}_{[\![0,\tau[\![}$ is the bank survival indicator process. In particular, writing $Y^{\circ} = JY + (1-J)Y_{\tau-}$ for any left-limited process Y, we have by application of the results of Crépey and Song (2017b) related to the condition (A) there (cf. also Collin-Dufresne, Goldstein and Hugonnier (2004, Lemma 1) and Crépey (2024, Section A)):

Lemma 3. *For every $\mathbb{Q}$ (resp. sub-) martingale Y, the process Y° is a $\mathbb{Q}^*$ (resp. sub-) martingale.* □

Remark 13. In the dynamic setup, the survival measure formulation is a light presentation, skipping the related integrability issues, of the reduction of filtration approach of Chapters II and III.

A Case of a Run-Off Portfolio

First, we consider the case of a portfolio held on a run-off basis (cf. §2.A). We denote by T the final maturity of the portfolio and we assume that all prices and XVAs vanish at time T if $T < \tau$. Then the mathematics of Chapters II and III show that all the qualitative insights provided by the one-period XVA analysis of §3 are still valid. Also accounting for CA dynamic hedges and possibly nonvanishing losses of the trading desks (see after Remark 2 page 30), the trading loss of the bank is now given by the process

$$\mathcal{L} = \mathcal{C} + \mathcal{F} + \mathrm{CA} - \mathrm{CA}_0 - (\mathcal{M} + \mathrm{MtM} - \mathrm{MtM}_0) + \mathcal{H} \tag{36}$$

and the bank *shareholder* trading loss by the $\mathbb{Q}$ (hence $\mathbb{Q}^*$, by Lemma 3) martingale

$$\mathcal{L}^{\circ} = \mathcal{C}^{\circ} + \mathcal{F}^{\circ} + \mathrm{CA}^{\circ} - \mathrm{CA}_0 - (\mathcal{M}^{\circ} + \mathrm{MtM}^{\circ} - \mathrm{MtM}_0) + \mathcal{H}^{\circ}. \tag{37}$$

In (36)–(37), we have CA = CVA + FVA as in (11); the processes $\mathcal{C}$, $\mathcal{F}$, CVA and FVA are continuous-time processes analogs, specified in concrete setups in chapters II-III and VII, of the eponymous quantities in §3 (which were constants or random variables there); $\mathcal{H}$, $\mathcal{M}$ and MtM, which were simply 0 in §3, do not need to be so in general (see after Remark 2), but they are always such that $(\mathcal{M}^\circ + \mathrm{MtM}^\circ - \mathrm{MtM}_0)$ and $\mathcal{H}^\circ$ are $\mathbb{Q}^*$ martingales (cf. Remark II.5 page 57).

Proposition 3. *The core equity tier 1 capital of the bank is given by*

$$\mathrm{CET1} = \mathrm{CET1}_0 - \mathcal{L}. \tag{38}$$

Shareholder equity is given by

$$\mathrm{SHC} = \mathrm{SHC}_0 - (\mathcal{L} + \mathrm{KVA} - \mathrm{KVA}_0). \tag{39}$$

Proof. In the continuous-time setup, Assumption 1(i) page 31 is written as

$$\mathrm{RC+RM+SCR+UC-TM-(RC+RM+SCR+UC-TM)_0} = \mathcal{M}-(\mathcal{C}+\mathcal{F}+\mathcal{H}).$$

Given the definition of CET1 in (4) and the balance conditions (5), this is equivalent to

$$\mathrm{CA} + \mathrm{CET1} - (\mathrm{CA} + \mathrm{CET1})_0 = -(\mathcal{C} + \mathcal{F}) + (\mathcal{M} + \mathrm{MtM} - \mathrm{MtM}_0) - \mathcal{H}.$$

In view of (36), we obtain (38).

As $\mathrm{SHC} = \mathrm{CET1} - \mathrm{RM}$ (cf. (4)), we have by (38):

$$\mathrm{SHC} = \mathrm{CET1}_0 - \mathcal{L} - \mathrm{RM} = \mathrm{CET1}_0 - \mathrm{RM}_0 - (\mathcal{L} + \mathrm{RM} - \mathrm{RM}_0),$$

which, by the third balance condition in (5), yields (39). □

Remark 14. It is only before τ that the right-hand-sides in (4) really deserve the respective interpretations of shareholder equity of the bank and core equity tier 1 capital. Hence, it is only the parts of (38) and (39) stopped before τ, i.e.

$$\mathrm{CET1}^\circ = \mathrm{CET1}_0 - \mathcal{L}^\circ\,, \quad \mathrm{SHC}^\circ = \mathrm{SHC}_0 - (\mathcal{L}^\circ + \mathrm{KVA}^\circ - \mathrm{KVA}_0), \tag{40}$$

which are meaningful financially.

As for economic capital and the KVA, by the same rationale as Definitions 3 and 4 in the static setup:

Definition 5. EC_t *is the time-t conditional 97.5% expected shortfall of* $(\mathcal{L}^\circ_{t+1} - \mathcal{L}^\circ_t)$ *under* $\mathbb{Q}$*, killed at the bank default time* τ.

Given a positive target hurdle rate h:

Definition 6. We set

$$\mathrm{CR} = \max(\mathrm{EC}, \mathrm{KVA}), \tag{41}$$

for a KVA process such that, for $t < \tau$,

$$\mathrm{KVA}_t = \mathbb{E}_t\Big[\int_t^T h\big(\mathrm{CR}_s - \mathrm{KVA}_s\big)ds\Big]. \tag{42}$$

Hence, for $t < \tau$,

$$\begin{aligned}\mathrm{KVA}_t &= \mathbb{E}_t\Big[\int_t^T he^{-h(s-t)}\mathrm{CR}_s ds\Big]\\ &= \mathbb{E}_t\Big[\int_t^T he^{-h(s-t)}\max(\mathrm{EC}_s, \mathrm{KVA}_s\big)ds\Big].\end{aligned} \tag{43}$$

The next-to-last identity is the continuous-time analog of the risk margin formula under the Swiss solvency test cost of capital methodology: see the beginning of the Introduction part of the book and see Swiss Federal Office of Private Insurance (2006, Section 6, middle of page 86 and top of page 88).

By Lemma 3, the continuous-time process KVA° that stems from (41)-(42) is a $\mathbb{Q}^*$ supermartingale with terminal condition $\mathrm{KVA}^\circ_T = 0$ on $\{T < \tau\}$ and drift coefficient $h\mathrm{SCR}$, where SCR is given as in (21), but for the dynamic EC process of Definition 5.

B Trade Incremental Cost-of-Capital XVA Strategy

In Subsection A above (as also in Chapters II-III), the derivative portfolio of the bank is assumed to be held on a run-off basis. But real-life banking derivative portfolios are incremental. In this subsection we specify the trade incremental XVA policy that allows one to extend the cost-of-capital XVA approach to incremental porfolios.

Assume a new deal shows up at time $\theta \in (0, \tau)$. We denote by $\Delta\cdot$, for any portfolio-related process, the difference between the time θ values of this process for the run-off versions of the portfolio with and without the new deal. In view of Remark 3 page 32, which is also relevant incrementally at every new deal:

Definition 7. We apply the following trade incremental pricing and accounting policy[2]:

- The trading desks pay ΔMtM to the client, and the CA desks add an amount ΔMtM to the trading account;

[2]with "add x" in the sense of "remove $(-x)$" if $x < 0$.

- The CA desks charge to the client an amount ΔCA and add it to the reserve capital account;
- The management of the bank charges the amount ΔKVA to the client and adds it to the risk margin account.

The funds transfer price of a deal is the all-inclusive XVA add-on charged by the bank to the client in the form of a rebate with respect to the mark-to-market ΔMtM of the deal. Under the above scheme, the overall price charged to the client for the deal is $\Delta\text{MtM} - \Delta\text{CA} - \Delta\text{KVA}$, i.e.

$$\begin{aligned}\text{FTP} &= \Delta\text{CA} + \Delta\text{KVA} = \Delta\text{CVA} + \Delta\text{FVA} + \Delta\text{KVA} \\ &= \Delta\text{FV} + \Delta\text{CL} + \Delta\text{KVA},\end{aligned} \tag{44}$$

by (11) and the last line in (18) (which still hold in continuous time, see Lemma II.8(i)-(ii) page 74) applied to the portfolios with and without the new deal.

Remark 15. As opposed to the ΔXVA terms, which entail portfolio-wide computations, ΔMtM reduces to trade-additive computations (as is intuitively clear and will be established mathematically by Eqn. III.(15)).

Obviously, the legacy portfolio of the bank has a key impact on the FTP. It may very well happen that the new deal is risk-reducing with respect to the portfolio, in which case $\text{FTP} < 0$, i.e. the overall, XVA-inclusive price charged by the bank to the client would be $\Delta\text{MtM} - \text{FTP} > \Delta\text{MtM}$ (subject, of course, to the commercial attitude adopted by the bank under such circumstances).

In order to exclude, for simplicity, jumps of our $\mathcal{L}$ and KVA processes at time θ (the ones related to the initial portfolio, but also those, starting at θ, corresponding to the augmented portfolio), we assume a quasi-left continuous model filtration $\mathfrak{G}$ and a $\mathfrak{G}$ predictable stopping time θ. The first assumption excludes that martingales can jump at predictable times. It is satisfied in all commonly used models and, in particular, in all jump-diffusion models with Lévy or Markov chain driven jumps. The second assumption is reasonable regarding the time at which a financial contract is concluded. Note that it was actually already assumed regarding the (fixed) time 0 at which the portfolio of the bank is supposed to have been set up in the first place.

Lemma 4. *Assuming the balance conditions* (5) *before θ and the new trade at time θ handled by the trade incremental policy of Definition 7, then shareholder equity* SHC$^\circ$ *(see Remark 14 page 45) is a $\mathbb{Q}^*$ submartingale on $[0,\theta] \cap \mathbb{R}_+$, with drift coefficient hSCR killed at τ.*

Proof. In the case of a trade incremental portfolio, a priori, the second identity in (40) is only guaranteed to hold *before θ*. However, in view of the observation made in Remark 3 and because, under our (harmless) technical assumptions, there can be no dividends arising from the portfolio expanded with the new deal (i.e. no jumps in the related processes $\mathcal{L}$ and KVA, defined on $[\theta, +\infty)$) at time θ itself, the process SHC does not jump at θ. The processes $\mathcal{L}$ and KVA related to the legacy portfolio cannot jump at θ either. As a result, the second identity in (40) still holds at θ. It is therefore valid on $[0, \theta] \cap \mathbb{R}_+$. The result then follows from the respective martingale and supermartingale properties of the (original) processes $\mathcal{L}^\circ$ and KVA$^\circ$ recalled before (37) and after Definition 6. □

The above XVA strategy can be iterated between and throughout every new trade. We call this approach the ***trade incremental cost-of-capital XVA strategy***. By an iterated application of Lemma 4 at every new trade, we obtain the following:

Proposition 4. *Under a dynamic and trade incremental cost-of-capital XVA strategy, shareholder equity* SHC$^\circ$ *is a* $\mathbb{Q}^*$ *submartingale on* $\mathbb{R}_+$, *with drift coefficient* hSCR *killed at* τ.

Thus, a trade incremental cost-of-capital XVA strategy results in a sustainable strategy for profits retention, both between and throughout deals, which was already the key principle behind Solvency II (see the Introduction to the book). Note that, without the KVA (i.e. for $h = 0$), the (risk-free discounted) shareholder equity process SHC$^\circ$ would only be a $\mathbb{Q}^*$ martingale, which could only be acceptable to shareholders without risk aversion (cf. §3.C.1 page 40).

In the current state of the market, even when they are computed, the KVA and even the MVA are not necessarily passed into entry prices. But they are strategically used for collateral and capital optimization purposes. This reflects a switch of paradigm in derivative management, from replication to balance sheet optimization.

CHAPTER II

The Cost-of-Capital XVA Approach in Continuous Time

Pedagogy is the art of repetition. This and the next chapter tell the same story as the previous one, but with a more axiomatic and mathematical focus here, versus intuitive and accounting there. Depending on the reader's background, these two chapters may be felt harder or easier to read than the previous one. In any case, they provide the rigorous treatment of the theory.

Since the 2007–09 crisis, derivative dealers charge their clients various add-ons, dubbed XVAs, meant to account for counterparty risk and its capital and funding implications. As banks cannot replicate the jump-to-default related cash flows, deals trigger wealth transfers and shareholders need to set capital at risk. In this chapter we detail mathematically the XVA policy of Chapter I, whereby so called contra-liabilities and cost of capital are sourced from bank clients at trade inceptions, on top of the firm valuation of counterparty risk, in order to guarantee to the shareholders a hurdle rate h on their capital at risk. The resulting all-inclusive XVA formula reads (CVA + FVA + KVA), where C stands for credit, F for funding, and where the KVA is a cost of capital risk premium. All these XVA metrics are portfolio-wide, nonnegative at the overall portfolio level and, despite the fact that we include the default of the bank itself in our modeling, they are ultimately unilateral. This makes them naturally in line with the requirement that capital at risk and reserve capital should not decrease simply because the credit risk of the bank has worsened.

§1 THE SUSTAINABLE PRICING AND DIVIDENDS PROBLEM

We devise a pricing and dividend policy for a dealer bank, **sustainable** in the sense of ensuring to its shareholders a constant instantaneous return rate h on

DOI: 10.1201/9781003614944-5

their capital at risk, even in the limiting case of a portfolio held on a run-off basis, i.e. without future deals.

Moreover, the corresponding policy of the bank should satisfy several **regulatory constraints**. Firstly, in order to comply with the Volcker rule that bans proprietary trading for a dealer bank, the market risk of the bank should be hedged as much as possible. As a result, mainly counterparty risk remains. Secondly, reserve capital should be maintained by the bank at the level of its *expected* counterparty credit losses, along two lines: the credit valuation adjustment (CVA) of the bank, meant to cope with the counterparty risk of the bank clients, i.e. with the expected losses of the bank due to client defaults; and the funding valuation adjustment (FVA), meant to cope with the counterparty risk of the bank itself, i.e. with its expected risky funding expenses. Thirdly, capital should be set at risk by the bank to deal with its *exceptional* (above expected) *losses*. The above return rate h is then meant as a hurdle rate for the bank shareholders, i.e. a risk premium for their capital at risk within the bank.

Reserve capital, like capital at risk, should obviously be ***nonnegative***. Furthermore, it should not decrease simply because the credit risk of the bank itself has worsened, a property which we refer to as ***monotonicity***: See Albanese and Andersen (2014, Section 3.1) for the relevant wordings from Basel Committee on Banking Supervision (2012) and Federal Register (2014).

Further requirements on a solution to the above sustainable pricing and dividend release policy problem are **economic interpretability** and **logical consistency** (for intellectual adhesion by market participants), **numerical feasibility** and **robustness** at the level of a realistic banking portfolio (for practicality), and **minimality** in the sense of being, all things equal, as cheap as possible (for competitiveness).

The mathematical foundation of a pricing and dividend policy satisfying all the above requirements is the main achievement of this chapter. Although we can not claim for uniqueness, we will see in §5.B that alternative XVA approaches in the literature breach several of the above requirements. For instance, it is not uncommon in the XVA literature to see some *possibly negative FVA metrics, or CVA and KVA metrics that tend to 0 when the default risk of the bank goes to infinity.*

The above is done in an abstract setup, which is then specified to bilateral and centrally cleared trading setups in the follow-up chapter. Related numerics are provided in the sequel of the book.

A Solution Setup

The starting point of our solution to the sustainable pricing and dividends problem is an organizational and accounting separation between three kinds of business units within the bank: the ***CA desks***, the ***trading desks*** and the ***management*** of the bank, see Figure 3 page 28.

The CA desks are themselves split between the CVA desk and the funding desk (or Treasury) of the bank, respectively in charge of the default risk triggered by clients and of the risky funding expenses of the bank. The corresponding cash flows are collectively called the contra-assets of the bank. In an informal sense made precise in this chapter, contra-assets (-liabilities) are bank liabilities (assets) that arise as a feedback effect of counterparty credit risk on the counterparty-risk-free assets (liabilities) of the bank.

The CA desks guarantee the trading desks against client and bank defaults, through an trading margin account, which also funds the trading of the trading desks at the risk-free rate. Thanks to this work accomplished by the CA desks, the trading desks can focus on the market risk of the contracts in their respective business lines, as if there was no counterparty risk: even if some of their positions are liquidated, this will occur at no loss from their perspective (if we ignore at least price impact effects, see Chapter IX). We denote by MtM the amount on the trading account of the bank (counted positively when posted by the CA desks) and we write

$$\mathrm{CA} = \mathrm{CVA} + \mathrm{FVA} \tag{1}$$

for the overall amount of reserve capital of the bank, which will correspond to the valuation of its contra-assets.

The management of the bank is in charge of its dividend release policy. We consider a level of capital at risk (CR) sufficient to make the bank resilient to a forty-year adverse event, i.e. greater than an economic capital (EC) defined as the expected shortfall of the losses of the bank over one year at the confidence level $\alpha = 97.5\% = 1 - \frac{1}{40}$. The implementation of a sustainable dividend remuneration policy requires a dedicated risk margin account, on which the bank profits are initially retained so that they can then be gradually released as dividends at a hurdle rate h on shareholder capital at risk (as opposed to being readily distributed as day-one profit). Counterparty default losses, as also funding payments, are materialities for default if not paid. By contrast, risk margin payments are at the discretion of the bank management, hence they do not represent an actual liability to the bank. As a consequence, the capital valuation adjustment (KVA) amount on the risk margin account is also loss-absorbing, hence part of capital at risk (CR). With minimality in view (see §1, Proposition 5 and §5.D), we thus set

$$\mathrm{CR} = \max(\mathrm{EC}, \mathrm{KVA}). \tag{2}$$

All bank accounts are marked-to-model, i.e. continuously and instantaneously readjusted to theoretical target levels, which will be defined in §2-§3 in view of yielding a solution to the sustainable pricing and dividends problem. All cash accounts of the bank, as well as all the collateral (assumed all cash for simplicity) posted as a guarantee for the trades, are remunerated at the risk-free rate.

In line with the sustainability requirement edicted in §1, the portfolio is supposed to be held on a run-off basis between the inception time 0 and its final maturity (the extension of the theory to trade incremental portfolios is in fact straightforward, the way already seen in I.§4.B). At the portfolio inception time 0, the trading desks pay MtM_0 to the clients; the CA desks add an amount MtM_0 on the trading account (if $\mathrm{MtM}_0 > 0$, whereas the trading desks put an amount $(-\mathrm{MtM}_0)$ on the trading account if $\mathrm{MtM}_0 < 0$); the CA desks charge the clients an amount CA_0 and add it to the reserve capital account; the management of the bank charges the amount KVA_0 to the clients and adds it to the risk margin account. Between time 0 and the bank default time τ (both excluded), mark-to-model readjustments of all bank accounts are on bank shareholders. If the bank defaults, any residual amount on the reserve capital and risk margin accounts, as well as any remaining trading cash flows, are transferred to the estate of the defaulted bank, dubbed the creditors of the bank hereafter, which is mandated to deal with the liquidation of the bank. These liquidation costs to be borne by the creditors are outside the scope of the model, as is also the primary business of the clients of the bank, which motivates their deals with the bank.

See Table 3 for a list of the main valuation acronyms used in this chapter.

Table 3: Main valuation acronyms and the place where they are introduced conceptually in this chapter and/or concretely specified in the follow-up ones.

CA	Contra-assets valuation	(1) and first row in Table 4
CL	Contra-liabilities valuation	Table 4
CR	Capital at risk	§1 and (2)
CVA	Credit valuation adjustment	§1, (14), (37) and III.(68)
DVA	Debit valuation adjustment	Table 4 and III.(72)
EC	Economic capital	§1 and Definition 6
FDA	Funding debit adjustment	Table 4 and III.(73)
FV	Fair valuation of counterparty risk	Table 4 and Lemma 8(ii)
FVA	Funding valuation adjustment	§1, (15), (38) and III.(69)
KVA	Capital valuation adjustment	§1.A and (19), (39), (53)
MtM	Mark-to-market	§1.A, (13), and III.(15)
SCR	Shareholder capital at risk	Definition 3

§2 THE COST-OF-CAPITAL XVA EQUATIONS

A Probabilistic Setup

Implicit to the above-sketched XVA framework is a probabilistic structure, i.e. a measurable space $(\Omega, \mathcal{A})$ endowed with a filtration $\mathfrak{G} = (\mathfrak{G}_t)_{t \in \mathbb{R}_+}$ satisfying the usual conditions of completeness and right-continuity. All the processes in the chapter are $\mathfrak{G}$ adapted and all the random times of interest are $\mathfrak{G}$ stopping times.

The XVA matter crucially entails nontraded assets (or trading constraints). The default risk of most of the clients of the bank is not liquidly priced in the market. Even if a liquid CDS market on the bank is available, this market is not accessible to the bank itself, which, in particular, cannot sell jump-to-default protection on itself. Given these hedging limitations of the bank, a fininsurance probability measure $\mathbb{Q}^*$, as defined in the introduction to this Pricing part of the book, is used for the linear valuation of cash flows of European derivatives (to which we restrict ourselves in this book), using the risk-free asset as our numéraire. The $\mathbb{Q}^*$ expectation and $(\mathfrak{G}_t, \mathbb{Q}^*)$ conditional expectation are denoted by $\mathbb{E}^*$ and $\mathbb{E}_t^*$.

We denote by T the sum between the final maturity of all claims in the portfolio (e.g. 50 years) and an upper bound $\delta \geq 0$ on the time of liquidating defaulted positions (typically considered in practice to be of the order of one to two weeks). All our cash flow and price processes are modeled as semimartingales, which are all taken in a càdlàg version. All our cash flow processes are cumulative starting from 0 at time 0, stopped at T and integrable.

For any left-limited process Y, we denote by $Y^{\tau-}$ and ${}^{\tau-}Y$ the respective processes Y stopped before the bank default time τ and starting before τ, i.e.

$$Y^{\tau-} = JY + (1 - J)Y_{\tau-}, \; {}^{\tau-}Y = Y - Y^{\tau-}, \tag{3}$$

where $J = \mathbb{1}_{[\![0,\tau[\![}$ is the survival indicator process of the bank.

Remark 1. So $Y^{\tau-}$ and ${}^{\tau-}Y$ correspond to Y° and $Y^{\bullet}$ in I.§4, where the latter nonstandard notations were used for emphasizing the connection between the continuous-time XVA approach of I.§4 and the one-period approach of I.§3.

Given counterparty risk and its funding and capital consequences, the pricing of a (assumed all European-style) derivative portfolio by the bank will depart from the $\mathbb{Q}^*$ valuation of its contractually promised cash flows. The following distinction between valuation and shareholder valuation will be important in this regard.

Definition 1. Given an optional, integrable process $\mathcal{Y}$ stopped at T (cumulative cash flow stream in the financial interpretation), we call:

(i) value process Z of $\mathcal{Y}$, the optional projection of $(\mathcal{Y}_T - \mathcal{Y})$, i.e.

$$Z_t = \mathbb{E}_t^*(\mathcal{Y}_T - \mathcal{Y}_t),\ t \le T, \tag{4}$$

and Z vanishes on $[T, +\infty)$;
(ii) shareholder value process Y of $\mathcal{Y}$, any process Y vanishing on $[T, +\infty)$ if $T < \tau$ and such that

$$Y_t = \mathbb{E}_t^*(\mathcal{Y}_{\tau-} - \mathcal{Y}_t + Y_{\tau-}),\ t < \tau. \tag{5}$$

Elementary manipulations show that the shareholder value equation (5), for a process Y vanishing on $[T, +\infty)$ if $T < \tau$, is equivalent to

$$Y_t^{\tau-} = \mathbb{E}_t^*(\mathcal{Y}_{\tau\wedge T}^{\tau-} - \mathcal{Y}_t^{\tau-} + \mathbb{1}_{\{\tau \le T\}} Y_\tau^{\tau-}),\ t \le \tau \wedge T. \tag{6}$$

In particular, $(\mathcal{Y} + Y)^{\tau-}$ is then a martingale (stopped before τ).

This makes apparent that the shareholder value equation (5) is actually an equation for $Y^{\tau-}$ and, in fact, a backward stochastic differential equation (BSDE) for $Y^{\tau-}$. This is a nonstandard BSDE (compare with El Karoui, Peng and Quenez (1997) and see also Crépey (2013)), stopped before the bank default time τ. Such a BSDE is tantamount to the notion of recursive valuation of defaultable securities in Collin-Dufresne, Goldstein and Hugonnier (2004, Section 3.2), in the special case where $R_t(x) = x$ there. This notion is shown to be well-posed in their Proposition 2, based on Schönbucher (2004)'s tool of the bank survival pricing measure (see around (25) below). In I.§4 we adopted the survival measure presentation to provide an intuitive overview of the cost-of-capital XVA approach, also well aligned with the one-period introduction of I.§3. In §3.B we will deal with shareholder valuation by a more comprehensive reduction of filtration methodology, which yields a more complete grasp on the related integrability issues.

B Abstract Trading Cash Flows

To avoid blurring the XVA conceptual picture by the combinatorial complexity of the financial network of the bank, in this chapter, we only work with abstract cash flows. We thus consider the dynamic hedging cash flows (inclusive of the cost of setting up the hedges)

$$\mathcal{H} = \mathcal{H}^{mtm} - \mathcal{H}^{cva} - \mathcal{H}^{fva}, \tag{7}$$

where $\mathcal{H}^{mtm}$ is the dynamic hedging loss of the trading desks, whilst $\mathcal{H}^{cva}$ and $\mathcal{H}^{fva}$ are the dynamic hedging gains of the CVA and funding desks. We also denote by $\mathcal{M}$, $\mathcal{C}$ and $\mathcal{F}$ the other (than dynamic hedging) trading cash flows to the trading desks and from the CVA and funding desks, respectively. Concrete specifications for these abstract cash flows $\mathcal{M}$, $\mathcal{C}$ and $\mathcal{F}$ will be provided in Proposition III.2. As for the dynamic hedging cash flows, $\mathcal{H}^{cva}$ and $\mathcal{H}^{fva}$ will typically be set to 0 (but not always, see Remark 4 page 56), while:

Example 1. Assuming a hedge of the trading desks implemented through a repo market on a Black-Scholes stock S with volatility σ, supposing no dividends and no repo basis on S, we have until the liquidation of the position that is hedged (with $\mathcal{H}^{mtm}$ and S in units of the risk-free asset that we use as our numéraire):

$$d\mathcal{H}_t^{mtm} = \zeta_t dS_t = \zeta_t \sigma S_t dW_t. \tag{8}$$

Here, W is a $(\mathfrak{G}, \mathbb{Q}^*)$ Brownian motion driving S, and ζ is the number of stocks shorted by the trading desks as their hedge. Given our choice of the risk-free asset as numéraire, the risk-free cost of funding the hedge is already included in (8).

Risky funding for the bank means borrowing at a nonnegative risky spread over the risk-free rate. We assume the funding debt of the bank is instantaneously liquidated at the bank default time τ. Accordingly:

Assumption 1. The process $\mathcal{F}$ is a martingale nondecreasing before τ and stopped at $\tau \wedge T$. Each of the processes $\Xi = \mathcal{H}^{mtm}, \mathcal{H}^{cva}, \mathcal{H}^{fva}$ (hence, the aggregated hedging loss $\mathcal{H}$) is such that the processes $\Xi^{\tau-}$ and ${}^{\tau-}\Xi$ are martingales.

Martingale assumptions on the hedging cash flows Ξ and on the risky funding cash flows $\mathcal{F}$ are in line with the definition of $\mathbb{Q}^*$ provided in the introduction to this Pricing part of the book. As bank shareholders only perceive bank pre-default cash flows, assuming that the $\Xi^{\tau-}$ processes are also martingales makes it consistent with a bank shareholder centric viewpoint of the different desks of the bank, in line with the fact that the shareholders have the control of the bank as long at it is nondefault (see the end of §1.A). A shareholder risk premium will be introduced in a second step, through KVA dividends.

Remark 2. Martingales with martingale $(\cdot)^{\tau-}$ component obviously include all martingales without jumps at τ, in particular all continuous martingales. By Crépey (2015, Lemma 2.1(ii)), they also include all the $\mathfrak{F}$ (càdlàg) martingales in an immersed reduction of filtration framework $\mathfrak{F} \subset \mathfrak{G}$, corresponding to the special case where $\mathbb{P} = \mathbb{Q}^*$ on $\mathfrak{F}_T$ in the setup of §3.A, provided the $\mathfrak{F}$ Azéma supermartingale of τ is continuous and nonincreasing.

Remark 3. In III.§3–§5, nonnegative losses triggered by client defaults accumulate before the default of the bank itself, so that $\mathcal{C}^{\tau-}$ is nondecreasing, if not for minor corrective terms related to a mismatch of temporality between some cash flows and their hedge during the liquidation periods that separate the defaults from their settlements (see Remark III.9 page 113). In particular, in the limiting case where defaults are instantaneously settled, $\mathcal{C}^{\tau-}$ is nondecreasing.

C Loss Processes

In our marked-to-model framework (see the end of §1.A), the gain process of the trading desks is given, inclusive of the corresponding hedging loss $\mathcal{H}^{mtm}$, by

$$\mathcal{M} + \mathrm{MtM} - \mathrm{MtM}_0 - \mathcal{H}^{mtm}, \tag{9}$$

for some theoretical target MtM level to be devised later in view of addressing the requirements of §1. Likewise, the loss processes of the CA desks are given, inclusive of their respective hedging gains, $\mathcal{H}^{cva}$ (for the CVA desk) and $\mathcal{H}^{fva}$ (for the funding desk), by

$$\begin{aligned} &\mathcal{C} + \mathrm{CVA} - \mathrm{CVA}_0 - \mathcal{H}^{cva} \\ &\mathcal{F} + \mathrm{FVA} - \mathrm{FVA}_0 - \mathcal{H}^{fva}, \end{aligned} \tag{10}$$

for some to-be-suitably-devised theoretical target CVA and FVA levels. Denoting CA=CVA+FVA (cf. (1)), the overall loss of the bank is (recalling (7))

$$\mathcal{L} = \mathcal{C} + \mathcal{F} + \mathrm{CA} - \mathrm{CA}_0 - (\mathcal{M} + \mathrm{MtM} - \mathrm{MtM}_0) + \mathcal{H}. \tag{11}$$

A dealer bank should not do proprietary trading (cf. §1). But the risk of financial loss as a consequence of client default is hard to hedge, because single name credit default swaps that could in principle be used for that purpose are illiquid. The possibility for the bank of hedging its own jump-to-default is even more questionable. Indeed, for the bank, hedging its default is tantamount to selling jump-to-default protection on itself, which is unfeasible (see §2.A). Accordingly, our *reference dynamic hedging case* is when the trading desks are perfectly hedged ($\mathcal{M} + \mathrm{MtM} - \mathrm{MtM}_0 - \mathcal{H}^{mtm} = 0$) whilst the CA desks are not hedged at all ($\mathcal{H}^{cva} = \mathcal{H}^{fva} = 0$), hence

$$\mathcal{H} = \mathcal{H}^{mtm} = \mathcal{M} + \mathrm{MtM} - \mathrm{MtM}_0 \text{ and } \mathcal{L} = \mathcal{C} + \mathcal{F} + \mathrm{CA} - \mathrm{CA}_0. \tag{12}$$

Example 2. Continuing with Example 1, the gain process of a trading desk long a delta-hedged option position with payoff $(S_T - K)^+$ (with maturity T, say) is given, until the liquidation of the position, by

$$\mathcal{M} + \mathrm{MtM} - \mathrm{MtM}_0 - \int_0^{\cdot} \zeta_t dS_t,$$

where $\mathcal{M} = (S_T - K)^+ \mathbb{1}_{[T,+\infty)}$ and MtM is the Black-Scholes price process of the option (vanishing from time T onward). This trading gain process is therefore 0 if the trader uses the Black-Scholes delta as his hedge, which he should do in order to conform to the Volcker rule.

Remark 4. Even if (12) is our reference dynamic hedging case, we do not assume it throughout the book. Indeed, on the one hand, the feasibility of the hedging loss $\mathcal{H}^{mtm} = \mathcal{M} + \mathrm{MtM} - \mathrm{MtM}_0$ by the trading desks is subject to the depth of the dynamic hedging markets (and also to model risk as discussed in Chapter XI), so supposing this hedging loss would be too restrictive. On the other hand, CVA traders do tentatively hedge their market (if not jump-to-default) risk (see VI.§6.A and Chapter XI, Conclusion), hence supposing $\mathcal{H}^{cva} = \mathcal{H}^{fva} = 0$ (i.e. $\mathcal{H} = \mathcal{H}^{mtm}$) would also be too restrictive.

D MtM, CVA and FVA

Definition 2. MtM, CVA and FVA are shareholder value processes of $\mathcal{M}$, $\mathcal{C}$, and $\mathcal{F}$.

That is, by Definition 1(ii), MtM, CVA and FVA are killed at T on $\{T < \tau\}$ and, for $t < \tau$,

$$\mathrm{MtM}_t = \mathbb{E}_t^*\Big(\mathcal{M}_{\tau-} - \mathcal{M}_t + \mathrm{MtM}_{\tau-}\Big), \tag{13}$$

$$\mathrm{CVA}_t = \mathbb{E}_t^*\Big(\mathcal{C}_{\tau-} - \mathcal{C}_t + \mathrm{CVA}_{\tau-}\Big), \tag{14}$$

$$\mathrm{FVA}_t = \mathbb{E}_t^*\Big(\mathcal{F}_{\tau-} - \mathcal{F}_t + \mathrm{FVA}_{\tau-}\Big). \tag{15}$$

Hence, by Assumption 1 and the sentence following (6), in line with Principle 1(i) page 32:

Remark 5. The processes MtM and CA = CVA + FVA (cf. (1)) are such that each of the trading gains and losses in (9)-(10), stopped before τ, is a martingale. So is therefore

$$\mathcal{L}^{\tau-} = \mathcal{C}^{\tau-} + \mathcal{F}^{\tau-} + \mathrm{CA}^{\tau-} - \mathrm{CA}_0 - (\mathcal{M}^{\tau-} + \mathrm{MtM}^{\tau-} - \mathrm{MtM}_0) + \mathcal{H}^{\tau-} \tag{16}$$

(cf. (11)), which represents the trading loss of the bank shareholders

Explicit assumptions ensuring the existence of the CVA and FVA shareholder value processes, and also their uniqueness within suitable spaces of square integrable solutions stopped before τ, will be provided in Lemma 4 and Propositions 2 and 3(ii). Regarding MtM, given Assumption 1 on $\mathcal{H}$, the reference dynamic hedging case (12) is only attainable provided not only $(\mathcal{M} + \mathrm{MtM})^{\tau-}$ (as dictated by (13), cf. again the line following (6)), but also ${}^{\tau-}(\mathcal{M} + \mathrm{MtM})$, hence $\mathcal{M} + \mathrm{MtM}$ as a whole, is a martingale. Even if we do not assume (12) throughout this chapter (cf. Remark 4), this motivates the following:

Assumption 2. MtM is the value process of $\mathcal{M}$ (in particular, $\mathrm{MtM}_T = 0$).

Remark 6. In further support of Assumption 2, note that even if the trading desks were able to find clients accepting to deal with the bank on the basis of an MtM process that would be the bank shareholder value of $\mathcal{M}$ but not its value process, the corresponding discrepancy between valuation and shareholder valuation of $\mathcal{M}$ would be an indication of extreme dependence between the market risk of the bank and its own default risk, such as the bank trading its own default risk, which should be considered with caution (cf. Remark I.9 page 37 and see also Remark 21, page 77).

The processes CVA and FVA are so far unconstrained on $[\![\tau, +\infty[\![\cap \big(\{\tau \leq T\} \times \mathbb{R}_+\big)$. We define them as zero there. As they already vanish on $[T, +\infty)$ if $T < \tau$, either of them, say Y, is in fact killed at $\tau \wedge T$, hence such that

$$
{}^{\tau-}Y = \mathbb{1}_{[\![\tau,+\infty[\![}(Y - Y_{\tau-}) = -\mathbb{1}_{[\![\tau,+\infty[\![}Y_{\tau-}. \tag{17}
$$

E Shareholder Capital at Risk and KVA

Since default risk can hardly be hedged, capital needs to be set at risk by shareholders, who therefore deserve, in the cost-of-capital pricing approach of this book, a further KVA add-on as a risk premium.

Economic capital (EC) is the level of capital at risk (CR) that a regulator would like to see on an economic basis. EC will be specified in Definition 6, page 68, once the necessary reduction of filtration setup is in place. On this basis, in view of (2), where KVA is provided by the clients in the first place (see §1.A):

Definition 3. We define the *shareholder* capital at risk (SCR), to be remunerated at a constant and nonnegative hurdle rate h, as

$$
\mathrm{SCR} = \mathrm{CR} - \mathrm{KVA} = \max(\mathrm{EC}, \mathrm{KVA}) - \mathrm{KVA} = (\mathrm{EC} - \mathrm{KVA})^+, \tag{18}
$$

where KVA is a shareholder value process of $\int_0^{\cdot} h\mathrm{SCR}_s ds$ killed at $\tau \wedge T$, i.e.

$$
\mathrm{KVA}_t = \mathbb{E}_t^*\Big[\int_t^{\tau\wedge T} h\big(\mathrm{EC}_s - \mathrm{KVA}_s\big)^+ ds + \mathrm{KVA}_{\tau-}\Big],\ t < \tau, \text{ and} \tag{19}
$$

$$
\text{KVA is killed at } \tau \wedge T. \tag{20}
$$

Explicit assumptions ensuring that such a KVA process is well defined will be provided by Proposition 4, page 72.

Remark 7. The process $\mathrm{KVA}^{\tau-}$ is a supermartingale with drift coefficient

$$
-h\mathrm{SCR} = -h\big(\mathrm{EC} - \mathrm{KVA}\big)^+. \tag{21}
$$

Note the following differential form of (19) (cf. (6)):

$$
\begin{aligned}
&\mathrm{KVA}_T^{\tau-} = 0 \text{ on } \{T < \tau\} \text{ and, for } t \leq \tau \wedge T,\\
&d\mathrm{KVA}_t^{\tau-} = -h\mathrm{SCR}_t dt + d\nu_t,\\
&\text{for some martingale } \nu.
\end{aligned} \tag{22}
$$

This formulation makes it apparent that the KVA corresponds to the amount to be maintained by the bank on its risk margin account in order to be in a position to deliver to its shareholders, dynamically into the future, a hurdle rate h on their capital at risk (SCR), in line with Principle 1(ii) page 32. In this sense our KVA addresses the sustainability requirement in §1. Moreover the amount on the risk margin account should land off at $\mathrm{KVA}_T = 0$ on $\{T < \tau\}$. Indeed, ending up in the negative at $T < \tau$ would mean an insufficient risk margin for ensuring the hurdle rate h to the shareholders. Conversely, ending up in the positive at $T < \tau$ would mean that the bank is unnecessarily expensive to its clients, which would contradict the minimality requirement in §1. See Proposition 5 for more on the minimality features of our KVA.

Proposition 1. *Shareholder dividends*

$$\mathcal{D} = -(\mathcal{L}^{\tau-} + \mathrm{KVA}^{\tau-} - \mathrm{KVA}_0) \tag{23}$$

are a submartingale stopped before τ, *with drift coefficient* $h\mathrm{SCR}$.

Proof. Shareholder trading gains $(-\mathcal{L}^{\tau-})$ and KVA risk margin payments result in a dividend stream (23) to shareholders. The stated properties of $\mathcal{D}$ follow from the observations made in Remarks 5 and 7. □

§3 WELL-POSEDNESS AND COMPARISON RESULTS

This section yields a technical specification of the setup of §2, in which the shareholder valuation equation (5) (in fact, a slight extension of it required for the XVA applications that follow) and the abstract XVA equations (14), (15) and (19), are shown to be well-posed.

A Reduced Stochastic Basis

We introduce a subfiltration $\mathfrak{F}$ of $\mathfrak{G}$, complete and right-continuous, quasi-left-continuous[1], and such that the bank default time τ is a $\mathfrak{G}$, but not an $\mathfrak{F}$, stopping time.

Remark 8. Regarding the relation between $\mathfrak{F}_\infty$ and the financial sub-σ-field $\mathcal{B}$ of $\mathcal{A}$ that appears in the definition of the fininsurance probability measure $\mathbb{Q}^*$ (see the introduction to this Pricing part of the book), as $\mathfrak{F}_\infty$ is meant to exclude the default of the bank, whereas $\mathcal{B}$ is meant to exclude nontraded risk factors starting with bank default, the big picture is (even if we do not explicitly use such an assumption): $\mathcal{B} \subset \mathfrak{F}_\infty \subset \mathcal{A}$.

Assumption 3. For any $\mathfrak{G}$ predictable, resp. optional process Y, there exists an $\mathfrak{F}$ predictable, resp. optional process Y', dubbed $\mathfrak{F}$ reduction of Y, that coincides with Y on $]\!]0, \tau]\!]$, resp. on $[\![0, \tau[\![$.

[1]needed in Subsection D below.

In particular, any $\mathfrak{G}$ stopping time θ admits an $\mathfrak{F}$ stopping time θ', dubbed $\mathfrak{F}$ reduction of θ, such that

$$\theta \wedge \tau = \theta' \wedge \tau. \tag{24}$$

Remark 9. As can be classically established by section theorem, for any $\mathfrak{G}$ progressive Lebesgue integrand X such that the $\mathfrak{G}$ predictable projection ${}^{\mathrm{p}}X$ exists,[2] the indistinguishable equality $\int_0^{\cdot} {}^{\mathrm{p}}X_s ds = \int_0^{\cdot} X_s ds$ holds. As a consequence, one can actually consider the $\mathfrak{F}$ predictable reduction X' of any $\mathfrak{G}$ progressive Lebesgue integrand X (even if this means replacing X by ${}^{\mathrm{p}}X$).

Definition 4. We call invariance probability measure, a probability measure $\mathbb{P}$ on $\mathfrak{F}_T$ equivalent to the restriction of $\mathbb{Q}^*$ to $\mathfrak{F}_T$, such that
(i) stopping before τ turns $(\mathfrak{F}, \mathbb{P})$ local martingales on $[0, T]$ into $(\mathfrak{G}, \mathbb{Q}^*)$ local martingales on $[\![0, \tau \wedge T]\!]$, and
(ii) the $\mathfrak{F}$ optional reductions of $(\mathfrak{G}, \mathbb{Q}^*)$ local martingales on $[\![0, \tau \wedge T]\!]$ without jump at τ are $(\mathfrak{F}, \mathbb{P})$ local martingales on $[0, T]$.

Remark 10. The basic "immersion" case where $\mathbb{P} = \mathbb{Q}^*$ on $\mathfrak{F}_T$ has already been mentioned in Remark 2, page 55. The "non-immersion" case corresponds to situations of extreme dependence ("hard wrong/right way risk") between the defaults of the bank and a client, or between the default of the bank and its portfolio exposure with a client (see Remark I.9, page 37, and Crépey and Song (2017a, 2016)).

We denote by $\mathsf{S} = \mathbb{Q}^*(\tau > \cdot \,|\, \mathfrak{F}_{\cdot})$ the $\mathfrak{F}$ Azéma supermartingale of τ, i.e. the $(\mathfrak{F}, \mathbb{Q}^*)$ optional projection of $J = \mathbb{1}_{[\![0,\tau[\![}$.

(Crépey and Song, 2017b, Lemma 2.3, Theorem 3.5 and Section 4.2) *Under Assumption 3, if* $\mathsf{S}_T > 0$ *a.s., then*

$$\textit{Two } \mathfrak{F} \textit{ optional processes that coincide before } \tau \textit{ coincide on } [0, T]. \tag{25}$$

Additionally assuming τ *endowed with a* $(\mathfrak{G}, \mathbb{Q}^*)$ *intensity process* $\gamma = \gamma J_-$ *such that* $e^{\int_0^{\tau \wedge T} \gamma_s ds}$ *is* $\mathbb{Q}^*$ *integrable, then there exists a unique invariance probability measure* $\mathbb{P}$ *on* $\mathfrak{F}_T$*. The measure* $\mathbb{P}$ *coincides with the restriction to* $\mathfrak{F}_T$ *of the bank survival probability measure associated with* $\mathbb{Q}^*$*, i.e. of the probability measure* $\mathbb{Q}$ *on* $(\Omega, \mathcal{A})$ *with* $(\mathfrak{G}, \mathbb{Q}^*)$ *density process* $\mathbb{1}_{\{\tau > \cdot \wedge T\}} e^{\int_0^{\cdot \wedge T} \gamma_s ds}$ *(see I.§4).*

Remark 11. $\mathbb{Q}$ and $\mathbb{P}$ should not be confused with the risk-neutral and physical measures that underlie $\mathbb{Q}^*$ (see the introduction to this Pricing part of the book).

---

[2] for which σ integrability of X valued at any stopping time, e.g. X bounded or càdlàg, is enough, see He, Wang and Yan (1992, Theorem V.5.2 page 136).

By reduction of filtration setup hereafter, we mean the setup of the above-recalled result, so that $\mathfrak{F}$ optional reductions of $\mathfrak{G}$ optional processes are uniquely defined on $[0, T]$, and $\mathfrak{F}$ reductions of $[0, T]$ valued $\mathfrak{G}$ stopping times are uniquely defined. Given our focus on the time interval $[0, T]$ in this chapter, we may and do assume that $\mathfrak{F}$ optional reductions are stopped at T (and are thus uniquely defined on $\mathbb{R}_+$), without loss of generality.

Lemma 1. *For any $\mathfrak{G}$ optional process Y, Y' is nondecreasing if $Y^{\tau-}$ is nondecreasing.*

Proof. If $Y^{\tau-}$ nondecreasing, then, for any constants $0 \le t \le s \le T$,

$$Y'_t \mathbb{1}_{\{s<\tau\}} = Y^{\tau-}_t \mathbb{1}_{\{s<\tau\}} \le Y^{\tau-}_s \mathbb{1}_{\{s<\tau\}} = Y'_s \mathbb{1}_{\{s<\tau\}},$$

hence $\mathbb{E}^*(Y'_t \mathbb{1}_{\{s<\tau\}} | \mathfrak{F}_s) \le \mathbb{E}^*(Y'_s \mathbb{1}_{\{s<\tau\}} | \mathfrak{F}_s)$, i.e. $Y'_t \mathsf{S}_s \le Y'_s \mathsf{S}_s$, where $\mathsf{S}_s > 0$—recalling from Dellacherie and Meyer (1980, Chapitre VI n° 17) that the set $\{\mathsf{S} > 0\}$ is a random interval starting from 0 included, which we assumed contains T. □

Unless explicitly mentioned, probabilistic statements hereafter still refer to the original stochastic basis $(\mathfrak{G}, \mathbb{Q}^*)$, with $\mathbb{Q}^*$ expectation and $(\mathfrak{G}_t, \mathbb{Q}^*)$ conditional expectation denoted by $\mathbb{E}^*$ and $\mathbb{E}^*_t$ as before. The $\mathbb{P}$ expectation and $(\mathfrak{F}_t, \mathbb{P})$ conditional expectation will be denoted by $\mathbb{E}'$ and $\mathbb{E}'_t$. We will work with the following spaces of processes:

- $\mathbb{L}'_2$, the space of $\mathfrak{F}$ progressive processes $\widetilde{X}$ such that

$$\left\|\widetilde{X}\right\|^2_{\mathbb{L}'_2} := \mathbb{E}'\Big[\int_0^T \widetilde{X}^2_s ds\Big] < +\infty;$$

- $\mathbb{L}_2$, the space of $\mathfrak{G}$ progressive processes X such that

$$\|X\|^2_{\mathbb{L}_2} := \mathbb{E}^*\Big[\int_0^{\tau\wedge T} e^{\int_0^s \gamma_u du} X^2_s ds\Big] = \left\|X'\right\|^2_{\mathbb{L}'_2} < +\infty, \tag{26}$$

  where the equality follows from the identity (9.4) in Crépey (2024); so $X \in \mathbb{L}_2 \Rightarrow X' \in \mathbb{L}'_2$;

- $\widetilde{\mathbb{S}}'_2$, resp. $\mathbb{S}'_2$, the spaces of $\mathfrak{F}$ adapted càdlàg processes $\widetilde{Y}$ such that

$$\left\|\widetilde{Y}\right\|^2_{\widetilde{\mathbb{S}}'_2} := \mathbb{E}'\Big[\sup_{t\in[0,T]} \widetilde{Y}^2_t\Big] < \infty,$$

  resp. and $\widetilde{Y}$ vanishes on $[T, +\infty)$;

- $\mathbb{S}_2$, the space of càdlàg $\mathfrak{G}$ adapted processes Y vanishing on $[\![\tau \wedge T, +\infty[\![$ and such that, denoting $Y_t^* = \sup_{s\in[0,t]} |Y_s|$:

$$\|Y\|^2_{\mathbb{S}_2} := \mathbb{E}^*\Big[Y_0^2 + \int_0^T J_s e^{\int_0^s \gamma_u du} d(Y^*)^2_s\Big] = \big\|Y'\big\|^2_{\widetilde{\mathbb{S}}'_2} < \infty, \quad (27)$$

where the equality is established as the formula (10.3) in Crépey (2024); so $Y \in \mathbb{S}_2 \Rightarrow Y' \in \mathbb{S}'_2$. Also note that, for $Y \in \mathbb{S}_2$,

$$\mathbb{E}^*\Big[\sup_{0\le t\le \tau\wedge T} Y_t^2\Big] \le \mathbb{E}^*\Big[Y_0^2 + \int_0^T J_s\, e^{\int_0^s \gamma_u du} d\big(Y^*\big)^2_s\Big] < \infty. \quad (28)$$

Lemma 2. *The $\mathfrak{F}$ optional reduction operator is an isometry from $\mathbb{S}_2$ onto $\mathbb{S}'_2$, with killing at τ as the reciprocal operator.*

***Proof*.** For $Y \in \mathbb{S}_2$,

$$Y'_T \mathsf{S}_T = \mathbb{E}^*(Y'_T \mathbb{1}_{\{\tau>T\}}|\mathfrak{F}_T) = \mathbb{E}^*(Y_T \mathbb{1}_{\{\tau>T\}}|\mathfrak{F}_T) = 0 \text{ a.e.},$$

hence $Y'_T = 0$ a.e. (having assumed $\mathsf{S}_T > 0$ a.s.), thus Y' vanishes on $[T, +\infty)$ (Y' being stopped at T as posited before Lemma 1). In view of (27), $\mathfrak{F}$ optional reduction thus applies $\mathbb{S}_2$ isometrically into $\mathbb{S}'_2$. Conversely, given $\widetilde{Y} \in \mathbb{S}'_2$, then $J\widetilde{Y} \in \mathbb{S}_2$, by (27) again, and $J\widetilde{Y} = \widetilde{Y}$ holds before τ, hence $\widetilde{Y} = (J\widetilde{Y})'$, by definition, existence and uniqueness of optional reductions. □

Remark 12. In contrast with the setup of Chapter VII, which is a continuous-time version of I.§3.D where more general anticipated XVA BSDEs are considered, in this and the following chapter no specific structure of the $(\mathfrak{F}, \mathbb{P})$ martingales is required. We do not even need to specify the risk drivers in the model. In practice (see e.g. IV.§4), those would typically consist of Brownian motions, driving the market risk factors, and of a continuous-time Markov chain (possibly modulated by the Brownian motions), used for driving the client default events.

B Shareholder Valuation Equation

Recall that the shareholder value equation (5), for a process Y vanishing on $[T, +\infty)$ if $T < \tau$, is nothing but the BSDE (6) for $Y^{\tau-}$. This applies to each of the MtM, CVA and FVA equations (13), (14) and (15). In the case of the KVA equation (19), the drift coefficient in the equation also depends on the KVA itself. To include this case, as well as the one of risky funding cash flows $\mathcal{F}$ encompassing the FVA process itself later in this chapter, we slightly extend the notion of shareholder valuation in Definition 1(ii), to cash flows including a component of the form

$$\int_0^{\cdot} J_t j_t(Y_t) dt, \quad (29)$$

for some random function $j = j_t(y)$ measurable with respect to the product of the $\mathfrak{F}$ predictable σ field by the Borel σ field on $\mathbb{R}$. We thus consider the following shareholder value equation, generalizing (5)–(6), for a $(\mathfrak{G}, \mathbb{Q}^*)$ semimartingale Y:

$$Y_t^{\tau-} = \mathbb{E}_t^*\Big(\mathcal{Y}_{\tau\wedge T}^{\tau-} - \mathcal{Y}_t^{\tau-} + \int_t^{\tau\wedge T} j_s(Y_s)ds + \mathbb{1}_{\{\tau\le T\}}Y_\tau^{\tau-}\Big),\ t \le \tau \wedge T, \text{ and} \quad (30)$$

$$Y = 0 \text{ on } [\![\tau\wedge T, +\infty[\![. \quad (31)$$

Likewise we consider the following reduced valuation equation for the reduction Y' of a $(\mathfrak{G}, \mathbb{Q}^*)$ semimartingale Y:

$$Y_t' = \mathbb{E}_t'\Big(\mathcal{Y}_T' - \mathcal{Y}_t' + \int_t^T j_s(Y_s')ds\Big), t \le T, \text{ and} \quad (32)$$

$$Y' = 0 \text{ on } [\![T, +\infty[\![. \quad (33)$$

Definition 5. We call $\mathbb{S}_2$ solution Y to the shareholder valuation equation (30), any $(\mathfrak{G}, \mathbb{Q}^*)$ semimartingale solution Y in $\mathbb{S}_2$ to (30) with $J\big(Y + \mathcal{Y} + \int_0^\cdot j_s(Y_s)ds\big)$ in $\mathbb{S}_2$. We call $\mathbb{S}_2'$ solution Y' to the reduced equation (32), any reduction of a $(\mathfrak{G}, \mathbb{Q}^*)$ semimartingale Y in $\mathbb{S}_2$ (hence $Y' \in \mathbb{S}_2'$ by Lemma 2), solving (32) with $(Y' + \mathcal{Y}' + \int_0^\cdot j_s(Y_s')ds)$ in $\mathbb{S}_2'$.

By well-posedness of an equation in a certain space of solutions, we mean existence and uniqueness of a solution in this class.

First we show an equivalence between the following differential variations on (30) and (32):

$$\begin{aligned} &Y_T^{\tau-} = 0 \text{ on } \{T < \tau\} \text{ and, for } t \le \tau\wedge T,\\ &dY_t^{\tau-} = -d\mathcal{Y}_t^{\tau-} - j_t(Y_t)dt + d\nu_t,\\ &\text{for some } (\mathfrak{G}, \mathbb{Q}^*) \text{ martingale } \nu \text{ such that } J\nu \in \mathbb{S}_2, \end{aligned} \quad (34)$$

respectively

$$\begin{aligned} &Y_T' = 0 \text{ and, for } t \le T,\\ &dY_t' = -d\mathcal{Y}_t' - j_t(Y_t')dt + d\mu_t,\\ &\text{for some } (\mathfrak{F}, \mathbb{P}) \text{ martingale } \mu \text{ in } \mathbb{S}_2'. \end{aligned} \quad (35)$$

Lemma 3. *For any* $(\mathfrak{G}, \mathbb{Q}^*)$ *semimartingale* Y*:*
(i) *If* $Y^{\tau-}$ *in* $\mathbb{S}_2$ *and* ν *solve* (34), *then* $Y', \mu = \nu'$ *in* $\mathbb{S}_2'$ *solve* (35)*;*
(ii) *Conversely, if* Y', μ *in* $\mathbb{S}_2'$ *solve* (35), *then* $Y^{\tau-} = JY', \nu = \mu^{\tau-}$ *solve* (34).

***Proof*.** By definition of $\mathfrak{F}$ optional reductions (stopped at T as posited before Lemma 1), the terminal condition in (35) implies the one in (34). Conversely, taking the $(\mathfrak{F}_T, \mathbb{Q}^*)$ conditional expectation of the terminal condition in (34) yields

$$0 = \mathbb{E}^*[Y_T^{\tau-}\mathbb{1}_{\{T<\tau\}}|\mathfrak{F}_T] = \mathbb{E}^*[Y_T'\mathbb{1}_{\{T<\tau\}}|\mathfrak{F}_T] = Y_T'\mathbb{Q}^*(\tau > T\,|\,\mathfrak{F}_T) = Y_T'\mathsf{S}_T,$$

hence $Y'_T = 0$ (as by assumption $\mathsf{S}_T > 0$, see above (25)), which is the terminal condition in (35).

The martingale condition in (35) implies the one in (34), by stopping before τ and application to $\nu = \mu^{\tau-}$ of Lemma 2 and of the condition (i) in Definition 4. Conversely, the martingale condition in (34) implies that $(Y', \mu = \nu')$ satisfies the second line in (35) on $[\![0, \tau[\![\cap [\![0, T]\!]$, hence on $[0, T]$, by (25). Moreover, by application of the condition (ii) in Definition 4 and of Lemma 2, $\mu = \nu'$ is an $(\mathfrak{F}, \mathbb{P})$ martingale in $\mathbb{S}'_2$. □

Theorem 1. **(i)** *The shareholder value equation* (30) *for* Y *in* $\mathbb{S}_2$ *is equivalent, through the bijection of Lemma 2, to the reduced valuation equation* (32) *for* $Y' \in \mathbb{S}'_2$.
(ii) *In the case where* $\mathcal{Y}'$ *is in* $\mathbb{S}'_2$, *if the random function* $v \mapsto j_t(v - \mathcal{Y}'_t)$ *is continuous and nonincreasing in* v, *if*

$$\mathbb{E}' \int_0^T \sup_{|v| \le r} |j_t(v - \mathcal{Y}'_t) - j_t(-\mathcal{Y}'_t)| dt < \infty \text{ for all } r > 0, \tag{36}$$

and if $j_\cdot(-\mathcal{Y}'_\cdot)$ *is in* $\mathbb{L}'_2$, *then the reduced valuation equation* (32) *for* Y' *in* $\mathbb{S}'_2$ *is well-posed, and so is the shareholder value equation* (30) *for* Y *in* $\mathbb{S}_2$.

Proof. **(i)** If Y is an $\mathbb{S}_2$ solution to (30), then Y in $\mathbb{S}_2$ and ν solve (34), for some ν. Hence Y' in $\mathbb{S}'_2$ and $\mu = \nu'$ solve (35), by Lemma 3. Therefore Y' is an $\mathbb{S}'_2$ solution to (32). Conversely, if Y' is an $\mathbb{S}'_2$ solution to (32), then Y' in $\mathbb{S}'_2$ and μ solve (35), for some μ. Hence $Y = JY'$ in $\mathbb{S}_2$ and $\nu = \mu^{\tau-}$ solve (34), by the converse in Lemma 3. Thus Y is an $\mathbb{S}_2$ solution to (30) (noting that $J\nu \in \mathbb{S}_2$ is $\mathbb{Q}^*$ square integrable over $[\![0, \tau \wedge T]\!]$, by (28), hence so is $\nu = \mu^{\tau-}$).
(ii) Under the additional assumptions made in the second part of the theorem, the well-posedness in $\mathbb{S}'_2$ of the reduced valuation equation (32) follows from Kruse and Popier (2016, Theorem 1) applied to the $(\mathfrak{F}, \mathbb{P})$ BSDE for $V' = Y' + \mathcal{Y}'$, i.e. to the $(\mathfrak{F}, \mathbb{P})$ BSDE with terminal condition $\mathcal{Y}'_T$ and monotone coefficient $v \mapsto j(v - \mathcal{Y}')$. The well-posedness of the shareholder value equation (30) for Y in $\mathbb{S}_2$ then follows from the first part of the theorem. □

Lemma 4. *The equations* (14), (15) *and* (19) *for* CVA, FVA *and* KVA *in* $\mathbb{S}_2$ *are respectively equivalent to the following equations for* CVA′, FVA′, *and* KVA′ *in* $\mathbb{S}'_2$: *For* $t \le T$,

$$\mathrm{CVA}'_t = \mathbb{E}'_t\Big(\mathcal{C}'_T - \mathcal{C}'_t\Big), \tag{37}$$

$$\mathrm{FVA}'_t = \mathbb{E}'_t\Big(\mathcal{F}'_T - \mathcal{F}'_t\Big), \tag{38}$$

$$\mathrm{KVA}'_t = \mathbb{E}'_t \int_t^T h(\mathrm{EC}'_s - \mathrm{KVA}'_s)^+ ds. \tag{39}$$

Assuming these reduced equations well-posed in $\mathbb{S}'_2$, the corresponding FVA *and* KVA *processes are nonnegative. So is also* CVA *provided $\mathcal{C}^{\tau-}$ is nondecreasing*[3]*.*

Proof. The equivalences hold by application of Theorem 1(i). The remaining statements are immediate consequences of Theorem 1(ii) and Eqns. (37)–(39), noting from Lemma 1 that $\mathcal{F}'$ is nondecreasing on $[0, T]$, by Assumption 1, as is also $\mathcal{C}'$ provided $\mathcal{C}^{\tau-}$ is nondecreasing. □

C CVA and FVA

We consider the CVA and FVA equations (14) and (15), as well as their reduced forms (37) and (38).

Regarding the risky funding cash flows, we postulate

$$d\mathcal{F}'_t = f_t(\mathrm{FVA}'_t)dt, \tag{40}$$

for some nonnegative random function $f = f_t(y)$ measurable with respect to the product of the $\mathfrak{F}$ predictable σ field by the Borel σ field on $\mathbb{R}$. The corresponding reduced FVA equation is

$$\mathrm{FVA}'_t = \mathbb{E}'_t \int_t^T f_s(\mathrm{FVA}'_s)ds,\ t \le T. \tag{41}$$

Proposition 2. (i) *If $\mathcal{C}' \in \mathbb{S}'_2$, then the formula (37) yields a well-defined* CVA$'$ *process in $\mathbb{S}'_2$, which is nonnegative provided $\mathcal{C}^{\tau-}$ is nondecreasing.*
(ii) *Assuming f continuous and nonincreasing in y, $\mathbb{E}' \int_0^T \sup_{|y| \le r} |f_t(y) - f_t(0)| dt < \infty$ for all $r > 0$, and $f_\cdot(0)$ in $\mathbb{L}'_2$, then the* FVA$'$ *equation (41) is well-posed in $\mathbb{S}'_2$ and* FVA *is nonnegative.*

Proof. By two successive applications of Theorem 1(ii), with $j = 0$ in the CVA case and $\mathcal{Y} = 0$ and $j = f$ in the FVA case, and by Lemma 4. □

The bank posts collateral to its counterparties in the form of rehypothecable collateral, which is fungible across netting sets, and segregated collateral, which is not. We denote by $\mathcal{R}$ and $\mathcal{S}$ two $\mathfrak{G}$ optional processes, with $\mathcal{S} \ge 0$, corresponding to the overall amounts of rehypothecable collateral and segregated collateral posted by the bank Treasury.

Assuming the bank default time τ endowed with an intensity $\gamma = J_-\gamma$, let $R \le \bar{R}$ denote two predictable, $[0, 1]$ valued recovery rates processes, corresponding to two bonds of different seniorities issued by the bank. We consider a market of risky funding assets defined by $U_0 = \bar{U}_0 = 1$ and, for $t \ge 0$,

$$dU_t = \lambda_t U_t dt + (1 - R_t)U_{t-} dJ_t = (1 - R_t)U_{t-}(dJ_t + \gamma_t dt)$$

[3]cf. Remark 3 page 55.

$$d\bar{U}_t = \widetilde{\lambda}_t \bar{U}_t dt + (1 - \bar{R}_t)\bar{U}_{t-} dJ_t = (1 - \bar{R}_t)\bar{U}_{t-}(dJ_t + \gamma_t dt), \tag{42}$$

where $dJ_t + \gamma_t dt$ is the compensated jump-to-default martingale of the bank and $\lambda = (1 - R)\gamma, \widetilde{\lambda} = (1 - \bar{R})\gamma$.

Assumption 4. $\bar{U}$ and U, where the recovery rate of the bank R is also the recovery rate of the bank toward its clients, are the martingale price processes of two risky funding assets used by the bank for its respective segregated collateral and residual[4] borrowing purposes. As the bank cannot sell default protection on itself, it can only take short positions in U and $\bar{U}$.

Remark 13. Assumption 4 implicitly supposes that the funders of the bank, i.e. the entities that fund the different segments of the debt of the bank (corresponding to its different funding sets), are default-free, and that the positions of the bank with its funders are instantaneously liquidated in case of default of the bank. This is all postulated for simplicity.

Remark 14. Random recoveries are in line with the uncertainty about the actual level of recovery rates that are only observed a posteriori and can reflect the possibility of liquidating various forms of collateral accounts for the output of liquidation procedures, legal resolutions, and other complex and unobservable features.

The motivation for Assumption 4 is that, as explained in Remark I.12 page 41, reserve capital (RC=CA) is a possible source of funding for rehypothecable collateral, whereas segregated collateral must be borrowed entirely. However, a recovery rate $\bar{R} \geq R$ may be available on the borrowing debt for segregated collateral, via margin lending schemes implemented through specialist lenders the way explained in III.§2.C.

Remark 15. For simplicity in this chapter, we assume that the bank does not use capital at risk (CR) for funding purposes. In fact, as already introduced in the one-period setup of I.§3.D, capital at risk can also be used by the bank, on top of reserve capital (CA), for its rehypothecable collateral funding purposes. This induces an intertwining of the FVA and the KVA, which will be considered in Chapter VII.

Proposition 3. *Regarding the risky*[5] *funding cash flows* $\mathcal{F}$*:*

[4]once everything else has already been accounted for, as prescribed by an overall self-financing condition.

[5]beyond the riskless rate that applies to all deposits, as already accounted for by our choice of the riskless asset as the numéraire.

(i) *We have*

$$\begin{aligned}\mathcal{F} &= \int_0^{\cdot} \lambda_t(\mathcal{R} - \mathrm{CA})_t^+ dt - (1 - R_\tau)(\mathcal{R} - \mathrm{CA})_{\tau-}^+ \mathbb{1}_{[\![\tau,\infty[\![} \\ &\quad + \int_0^{\cdot} \widetilde{\lambda}_t \mathcal{S}_t dt - (1 - \bar{R}_\tau)\mathcal{S}_{\tau-} \mathbb{1}_{[\![\tau,\infty[\![} \\ &= \int_0^{\cdot} \Big((1 - R_t)(\mathcal{R} - \mathrm{CA})_{t-}^+ + (1 - \bar{R}_t)\mathcal{S}_{t-}\Big)(dJ_t + \gamma_t dt),\end{aligned} \tag{43}$$

i.e.

$$\begin{aligned}\mathcal{F}^{\tau-} &= \int_0^{\cdot} \lambda_t(\mathcal{R} - \mathrm{CA})_t^+ dt + \int_0^{\cdot} \widetilde{\lambda}_t \mathcal{S}_t dt, \\ {}^{\tau-}(-\mathcal{F}) &= (1 - R_\tau)(\mathcal{R} - \mathrm{CA})_{\tau-}^+ \mathbb{1}_{[\![\tau,\infty[\![} + (1 - \bar{R}_\tau)\mathcal{S}_{\tau-} \mathbb{1}_{[\![\tau,\infty[\![}.\end{aligned} \tag{44}$$

(ii) *Assuming the reduction of filtration setup of §3.A and $\mathcal{C}' \in \widetilde{\mathbb{S}}_2'$, then*

$$\mathcal{F}' = \int_0^{\cdot} \lambda_t'(\mathcal{R}' - \mathrm{CVA}' - \mathrm{FVA}')_t^+ dt + \int_0^{\cdot} \widetilde{\lambda}_t' \mathcal{S}_t' dt, \tag{45}$$

which is of the form (40) *for*

$$f_t(y) = \lambda'(\mathcal{R}' - \mathrm{CVA}' - y)^+ + \widetilde{\lambda}'\mathcal{S}'; \tag{46}$$

if $\mathbb{E}' \int_0^T \lambda_t' dt < \infty$ *and* $\lambda'(\mathcal{R}' - \mathrm{CVA}')^+ + \widetilde{\lambda}'\mathcal{S}' \in \mathbb{L}_2'$, *then the* FVA *equation* (15) *is well-posed in* $\mathbb{S}_2$ *and we have, for* $t \le T$,

$$\mathrm{FVA}_t' = \mathbb{E}_t' \int_t^T \lambda_s'(\mathcal{R}' - \mathrm{CVA}' - \mathrm{FVA}')_s^+ ds + \mathbb{E}_t' \int_t^T \widetilde{\lambda}_s' \mathcal{S}_s' ds \tag{47}$$

as well as

$$\mathcal{L}' = \mathcal{C}' + \mathcal{F}' + \mathrm{CA}' - \mathrm{CA}_0 - (\mathcal{M}' + \mathrm{MtM}' - \mathrm{MtM}_0) + \mathcal{H}' \in \widetilde{\mathbb{S}}_2' \tag{48}$$

provided $\mathcal{M}' + \mathrm{MtM}' - \mathcal{H}' \in \widetilde{\mathbb{S}}_2'$[6].

Proof. **(i)** By Assumption 4, the risky funding strategy of the bank consists in maintaining short positions (consistent with the condition concluding Assumption 4) of $\frac{\mathcal{S}^{\tau-}}{\bar{U}^{\tau-}}$ units of the asset $\bar{U}$ and $\frac{((\mathcal{R}-\mathrm{CA})^+)^{\tau-}}{U^{\tau-}}$ units of U. Given our use of the risk-free asset as numéraire, the self-financing condition on the strategy of the funding desk is then written as

$$d\mathcal{F}_t = \frac{((\mathcal{R} - \mathrm{CA})^+)_t^{\tau-}}{U_t^{\tau-}} dU_t + \frac{\mathcal{S}_t^{\tau-}}{\bar{U}_t^{\tau-}} d\bar{U}_t. \tag{49}$$

This yields (43), which splits as (44).

---

[6]e.g. in the reference dynamic hedging case (12), where $\mathcal{M}' + \mathrm{MtM}' - \mathrm{MtM}_0 - \mathcal{H}' = 0$.

(ii) Assuming the reduction of filtration setup of §3.A, (45) follows from the first line in (44) by definition, existence and uniqueness of optional reductions viewed in §3.A (noting that $\mathrm{CA}' = \mathrm{CVA}' + \mathrm{FVA}'$, where CVA' is well defined under the stated hypotheses, by Proposition 2(i)). Hence the expression (45) is of the form (40), for f there given by (46). The remaining FVA statements follow by an application of Proposition 2(ii).

The process $\mathcal{L}'$ belongs to $\widetilde{\mathbb{S}}_2'$ as the sum (modulo a constant) between $\mathrm{CVA}' + \mathcal{C}'$, the $(\mathfrak{F}, \mathbb{P})$ martingale part of FVA', and $-(\mathcal{M}' + \mathrm{MtM}') + \mathcal{H}'$ (assumed in $\widetilde{\mathbb{S}}_2'$). □

Remark 16. The industry terminology distinguishes a strict FVA, in the sense of the cost of funding rehypothecable collateral, from an MVA, corresponding to the cost of funding segregated collateral. In this chapter, to spare one "VA" notation, we merge the two in an overall FVA meant in the broad sense of the cost of funding the derivative trading of the bank. The strict FVA and the MVA respectively correspond to the first and the second terms on the right-hand side of (47).

D Economic Capital

Capital requirements are focused on the solvency issue, because it is when a regulated firm becomes insolvent that the regulator may choose to intervene, take over, or restructure a firm. Specifically, Basel II Pillar II defines economic capital as the $\alpha = 99\%$ value-at-risk of the depletion of core equity tier I capital (CET1) over one year. Moreover, the fundamental review of the trading book (FRTB, cf. Basel Committee on Banking Supervision (2013)) required a shift from 99% value-at-risk to 97.5% expected shortfall as the reference risk measure in capital calculations.

In our setup, before bank default, CET1 depletions correspond to the shareholder trading loss process $\mathcal{L}^{\tau-}$ (see Proposition I.3 and Remark I.14 page 45). In addition, economic capital calculations are typically made by a bank "on a going concern", hence assuming that the bank itself does not default. Accordingly (cf. the last sentence in the result recalled after (25)), assuming the process $\mathcal{L}'$ is integrable under $\mathbb{P}$ (see Remark 18 below):

Definition 6. The economic capital (EC) of the bank is defined as $\mathrm{EC} = J\,\mathrm{ES}$, where, for $t \geq 0$, ES_t is the $(\mathfrak{F}_t, \mathbb{P})$ conditional expected shortfall of the random variable $\mathcal{L}'_{t+1} - \mathcal{L}'_t$ at the confidence level $\alpha = 97.5\%$, in the following $\mathfrak{F}$ predictable sense:

$$\mathrm{ES} = \inf_{\text{rationals } k} \Big(k + (1-\alpha)^{-1}\, {}^{p}\!\big[(\mathcal{L}'_{\cdot+1} - \mathcal{L}'_{\cdot} - k)^+\big] \Big). \tag{50}$$

Remark 17. In (50), ${}^{p}\cdot$ denotes the $(\mathfrak{F}, \mathbb{P})$ predictable projection operator that applies to any raw, non necessarily adapted, càdlàg process (see e.g. He, Wang and Yan (1992, Theorem V.5.2 page 136)), such as $(\mathcal{L}'_{\cdot+1} - \mathcal{L}'_{\cdot} - k)^+$ for any constant k. Regarding the formulation (50) of the expected shortfall process, see Crépey, Sabbagh and Song (2020, Lemma 2.1), which proves that, for every constant $t \geq 0$, $\mathrm{ES}_t = \mathbb{ES}_t(\mathcal{L}'_{t+1} - \mathcal{L}'_t)$ almost surely holds (under the assumed quasi-left continuity of $\mathfrak{F}$), where $\mathbb{ES}_t(\mathcal{L}'_{t+1} - \mathcal{L}'_t)$ is the $(\mathfrak{F}_t, \mathbb{P})$ conditional expected shortfall of $\mathcal{L}'_{t+1} - \mathcal{L}'_t$, with expected shortfall here in the usual sense (see Prerequisites in the general introduction to the book). The predictability property of the ES process makes $\mathrm{EC} = J\,\mathrm{ES}$ a suitable input to the BSDE (39).

Remark 18. Remark 5 and Definition 4(ii) imply that the process $\mathcal{L}'$ is an $(\mathfrak{F}, \mathbb{P})$ local martingale. Assuming its $\mathbb{P}$ integrability is not a practical restriction as, in concrete setups such as the one of Proposition 3(ii), $\mathcal{L}'$ is even in $\mathbb{S}'_2$.

In particular, the economic capital process is killed at $\tau \wedge T$ and, since the expected shortfall of a centered random variable is nonnegative (see Prerequisites in the general introduction of the book):

Remark 19. EC is nonnegative.

E KVA in the Case of a Default-Free Bank

In this section we temporarily suppose the bank default free, i.e., formally,

$$\text{“}\tau = +\infty, \ (\mathfrak{F}, \mathbb{P}) = (\mathfrak{G}, \mathbb{Q}^*).\text{”}$$

The results are then extended to the case of a defaultable bank in Subsection F.

So in this section we use the “$\cdot'$” notation, not in the sense of $\mathfrak{F}$ reduction (as $\mathfrak{F} = \mathfrak{G}$ here), but simply in order to distinguish the present equations from the ones in Subsection F, where $\mathfrak{F} \neq \mathfrak{G}$. In Subsection F the present data will then be interpreted as the $\mathfrak{F}$ reductions of the corresponding data there.

In particular, the process EC' is defined just like EC in Definition 6, except that $\tau = +\infty$ here. Given $C' \geq \mathrm{EC}' \geq 0$ representing a putative capital at risk process for the bank, we consider the auxiliary BSDE

$$K'_t = \mathbb{E}'_t \int_t^T h\left(C'_s - K'_s\right) ds, \ t \leq T, \tag{51}$$

with the same interpretation as the KVA (cf. the comment following (22)), but relative to any putative capital at risk process C', and simplified to the present setup of a risk-free bank.

Lemma 5. *If C' is in $\mathbb{L}'_2$, then the equation* (51) *for K' has for unique $\mathbb{S}'_2$ solution*

$$K'_t = \mathbb{E}'_t \int_t^T h e^{-h(s-t)} C'_s ds, \ 0 \leq t \leq T. \tag{52}$$

If $\mathcal{L}'$ is in $\widetilde{\mathbb{S}}_2'$, then EC' *is in* $\mathbb{L}_2'$ *and the* KVA' *equation* (39) *has a unique* $\mathbb{S}_2'$ *solution, such that*

$$\mathrm{KVA}_t' = \mathbb{E}_t' \int_t^T he^{-h(s-t)} \max(\mathrm{EC}_s', \mathrm{KVA}_s')ds,\ 0 \le t \le T. \tag{53}$$

Proof. If $\mathcal{L}'$ is in $\widetilde{\mathbb{S}}_2'$, then EC' is in $\mathbb{L}_2'$, by Definition 6 and $(1-\alpha)^{-1}$ Lipschitz property of the expected shortfall operator recalled in Prerequisites in the general introduction to the book. Moreover, the KVA' BSDE (39) has a nonincreasing coefficient

$$k_t(y) = h\Big(\mathrm{EC}_t' - y\Big)^+,\ y \in \mathbb{R}. \tag{54}$$

By Theorem 1(ii) applied with $\mathcal{Y} = 0$ and nonincreasing coefficient $j_t(y) = h(\mathrm{EC}_t' - y)^+$, the KVA' equation (39) has therefore a unique $\mathbb{S}_2'$ solution. This also holds for the linear BSDE (51), by even simpler considerations. Moreover, the $\mathbb{S}_2'$ solution K' to (52) solves (51).

The process KVA' is in $\mathbb{S}_2'$ with martingale part in $\mathbb{S}_2'$ and, by (39), we have, for $0 \le t \le T$,

$$\mathrm{KVA}_t' = \mathbb{E}_t' \int_t^T h\Big(\mathrm{EC}_s' - \mathrm{KVA}_s'\Big)^+ ds = \mathbb{E}_t' \int_t^T h\Big(\max(\mathrm{EC}_s', \mathrm{KVA}_s') - \mathrm{KVA}_s'\Big)ds. \tag{55}$$

Hence the process KVA' solves in $\mathbb{S}_2'$ the linear BSDE (51) corresponding to the implicit data $C' = \max(\mathrm{EC}', \mathrm{KVA}') \in \mathbb{L}_2'$. Equation (53) is the corresponding instantiation of (52). □

Assuming $\mathcal{L}'$ in $\widetilde{\mathbb{S}}_2'$, let

$$\mathrm{CR}' = \max(\mathrm{EC}', \mathrm{KVA}'), \tag{56}$$

where KVA' is the $\mathbb{S}_2'$ solution to (39). Note that CR' is nonnegative, as this is already the case for EC' as seen in Remark 19. To emphasize the dependence on C', we henceforth denote by $K' = K'(C')$ the solution (52) to the linear BSDE (51). In particular, (53) and (56) read as

$$\mathrm{KVA}' = K'(\mathrm{CR}'). \tag{57}$$

We define the set of admissible capital at risk processes as

$$\mathbb{A}\mathrm{dm}' = \{C' \in \mathbb{L}_2'; C' \ge \max\Big(\mathrm{EC}', K'(C')\Big)\}. \tag{58}$$

Here $C' \ge \mathrm{EC}'$ is the risk acceptability condition, while $C' \ge K'(C')$ expresses that the risk margin $K'(C')$, which would correspond through the hurdle rate h to the tentative capital at risk process C' (cf. the comment regarding the KVA made after (22)), is part of capital at risk (cf. the comment above (2)).

Lemma 6. *Assuming that* $\mathcal{L}'$ *is in* $\widetilde{\mathbb{S}}'_2$, *then*
(i) $\mathrm{CR}' = \min \mathbb{A}\mathrm{dm}', \mathrm{KVA}' = \min_{C' \in \mathbb{A}\mathrm{dm}'} K'(C')$;
(ii) *The process* KVA′ *is nondecreasing in the target hurdle rate* h.

Proof. **(i)** By (57),

$$\mathrm{CR}' = \max(\mathrm{EC}', \mathrm{KVA}') = \max\Big(\mathrm{EC}', K'(\mathrm{CR}')\Big).$$

Therefore $\mathrm{CR}' \in \mathbb{A}\mathrm{dm}'$. Moreover, for any $C' \in \mathbb{A}\mathrm{dm}'$, we have (cf. (54)):

$$k_t(K'_t(C')) = h\Big(\mathrm{EC}'_t - K'_t(C')\Big)^+ \le h(C'_t - K'_t(C')).$$

Hence the coefficient of the KVA′ BSDE (39) never exceeds the coefficient of the linear BSDE (51) when both coefficients are evaluated at the solution $K'_t(C')$ of (51). Since these are BSDEs with equal (null) terminal condition, the BSDE comparison principle of Proposition 4 in Kruse and Popier (2016)[7] applied to the BSDEs (51) and (39) yields $\mathrm{KVA}' \le K'(C')$. Consequently, $\mathrm{KVA}' = \min_{C' \in \mathbb{A}\mathrm{dm}'} K'(C')$ and, for any $C' \in \mathbb{A}\mathrm{dm}'$,

$$C' \ge \max(\mathrm{EC}', K'(C')) \ge \max(\mathrm{EC}', \mathrm{KVA}') = \mathrm{CR}'.$$

Thus $\mathrm{CR}' = \min \mathbb{A}\mathrm{dm}'$.
(ii) The coefficient (54) of the KVA′ BSDE (39) is nondecreasing in h. So is therefore the $\mathbb{S}'_2$ solution KVA′ to (39), by the BSDE comparison principle of Kruse and Popier (2016, Proposition 4) applied to the BSDE (39) for different h. □

F KVA in the Case of a Defaultable Bank

In the case of a defaultable bank, "·′" now denoting $\mathfrak{F}$ reduction (predictable or optional as relevant), we have by applications of Theorem 1(i) (with $\mathcal{Y} = 0$ there):

Lemma 7. *The equation*

$$K_t^{\tau-} = \mathbb{E}_t^*\Big[\int_t^{\tau\wedge T} h(C_s - K_s)ds + \mathbb{1}_{\{\tau \le T\}} K_\tau^{\tau-}\Big], \; t \le \tau \wedge T \qquad (59)$$

for a process $K = K(C)$ *in* $\mathbb{S}_2$ *is equivalent, through the bijection of Lemma 2, to the equation* (51) *for* K' *in* $\mathbb{S}'_2$.

Hence, given also Lemmas 4 and 5:

[7] note that jumps are not an issue for comparison in our setup, where the coefficient k "only depends on y"; cf. Kruse and Popier (2016, Assumption (H3')).

Proposition 4. *If $C' \in \mathbb{L}_2'$, then the equation* (59) *for K is well-posed in $\mathbb{S}_2$ and the $\mathfrak{F}$ optional reduction K' of its $\mathbb{S}_2$ solution K is the $\mathbb{S}_2'$ solution to* (51).

If $\mathcal{L}'$ is in $\widetilde{\mathbb{S}}_2'$ (e.g. in the setup of Proposition 3(ii)), then the KVA equation (19) *is well-posed in $\mathbb{S}_2$ and the $\mathfrak{F}$ optional reduction* KVA$'$ *of its $\mathbb{S}_2$ solution* KVA *is the $\mathbb{S}_2'$ solution to* (39) *or, equivalently,* (53).

In the case of a defaultable bank, writing $K = K(C)$ for the $\mathbb{S}_2$ solution to (59), the set of admissible capital at risk processes is defined by (cf. (58) and the following comments)

$$\mathbb{A}\mathrm{dm} = \{C \in \mathbb{L}_2; C \geq \max\big(\mathrm{EC}, K(C)\big)\}. \tag{60}$$

The following result shows that $\mathrm{CR} := \max(\mathrm{EC}, \mathrm{KVA})$ is in fact *the minimal and cheapest* capital at risk process C satisfying the risk admissibility condition $C \geq \mathrm{EC}$ and consistent with the target hurdle rate h on shareholder capital at risk.

Proposition 5. *Assuming that $\mathcal{L}'$ is in $\widetilde{\mathbb{S}}_2'$:*
(i) *We have* $\mathrm{CR} = \min \mathbb{A}\mathrm{dm}, \mathrm{KVA} = \min_{C\in\mathbb{A}\mathrm{dm}} K(C)$;
(ii) *The process* KVA *is nondecreasing in h.*

Proof. This follows by application of Lemma 6 and Proposition 4. □

The fact that KVA is continuous and nondecreasing in h as just seen (and in fact, unless the processes $\mathcal{L}'$ and EC would vanish, increasing) allows one to define the *implied hurdle rate* as the value of the target hurdle rate $h \in \mathbb{R}_+$ calibrated, through the KVA formula (53) valued at time $t = 0$, to the actual amount on the risk margin account of the bank (if this amount cannot be reached by any $h \in \mathbb{R}_+$, then the implied hurdle rate is deemed infinite).

See the concluding paragraph of I.§3.C for a simple example in a one-period XVA setup, where an implied hurdle rate can be characterized quite explicitly, shedding light on the impact thereon of the portfolio of the bank, the default intensities for the bank and its counterparty, and the risk aversion of the bank shareholders.

Cost of capital proxies have always been used to estimate return-on-equity. Whether it is used in the direct or in the implied mode, the KVA is a refinement, dynamic and fine-tuned for derivative portfolios, but the base concept is far older than even the CVA.

§4 WEALTH TRANSFER ANALYSIS

We now bring to light the symmetrical companions of the contra-assets, i.e. the contra-liabilities. Put together, contra-assets and contra-liabilities allow analyzing the wealth transfers triggered by the trading of the bank, which occur without giving rise to arbitrage opportunities to shareholders.

Using (17) that applies to $Y = \mathrm{CVA}$ and FVA, Figures 4–5 detail the split of the overall trading loss process of the bank, $\mathcal{L}$ in (11), as the difference between the pre-bank default loss process, i.e. the shareholder trading loss process $\mathcal{L}^{\tau-}$ as per (16), and the creditor gain process

$$
{}^{\tau-}(-\mathcal{L}) = {}^{\tau-}(-\mathcal{C}) + {}^{\tau-}(-\mathcal{F}) + \mathbb{1}_{[\![\tau,+\infty[\![}\mathrm{CA}_{\tau-} - {}^{\tau-}(-\mathcal{M} - \mathrm{MtM}) + {}^{\tau-}(-\mathcal{H}). \tag{61}
$$

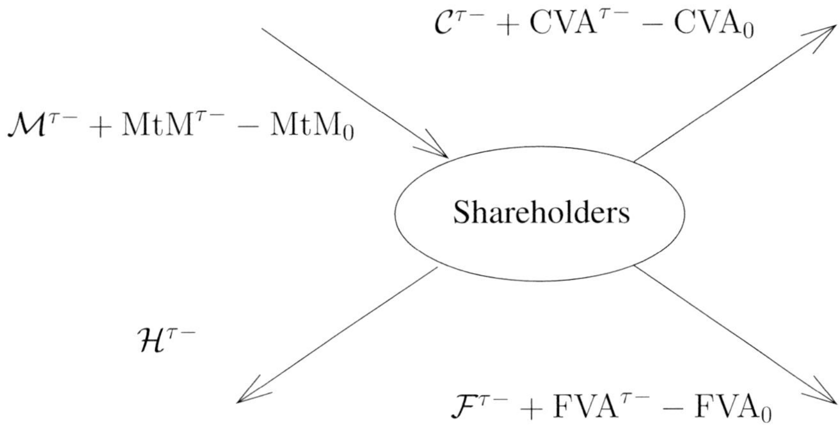

Figure 4: Pre-bank default trading cash flows.

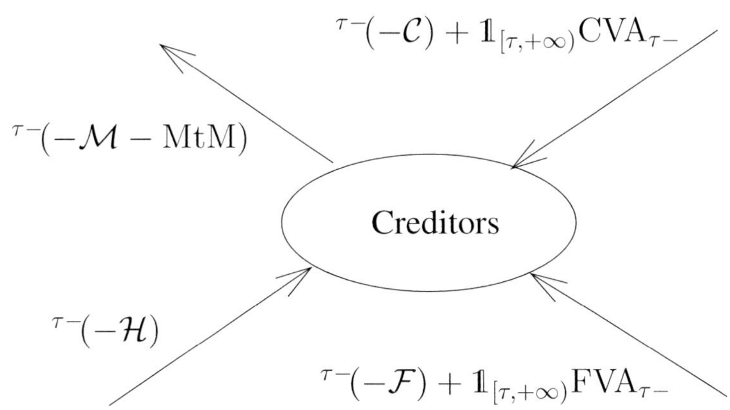

Figure 5: Contra-liabilities originating cash flows (i.e. bank trading cash flows from bank default onward).

Table 4 identifies various cash flows (column 3) and the related value processes in the sense of Definition 1(i) (columns 1 and 2) involved in the foregoing wealth transfer analysis. In particular:

Definition 7. We call debit valuation adjustment (DVA) and funding debit adjustment (FDA) the values of the cash flows ${}^{\tau-}(-\mathcal{C}) + \mathbb{1}_{[\![\tau,+\infty[\![}\mathrm{CVA}_{\tau-}$ and ${}^{\tau-}(-\mathcal{F}) + \mathbb{1}_{[\![\tau,+\infty[\![}\mathrm{FVA}_{\tau-}$.

A view on DVA and FDA as wealth transfers is consistent with the conclusions drawn in a structural default model of the bank by Andersen, Duffie and Song (2019) (who, however, do not deal with the KVA).

Table 4: Valuation acronym and name (columns 1 and 2) of various cash flows (column 3) involved in the XVA wealth transfer analysis.

acronym	name	cash flow valued
CA	Contra-assets valuation	$\mathcal{C}^{\tau-} + \mathcal{F}^{\tau-} + \mathbb{1}_{[\![\tau,+\infty[\![}\mathrm{CA}_{\tau-}$
CL	Contra-liabilities valuation	${}^{\tau-}(-\mathcal{L})$
DVA	Debit valuation adjustment	${}^{\tau-}(-\mathcal{C}) + \mathbb{1}_{[\![\tau,+\infty[\![}\mathrm{CVA}_{\tau-}$
FDA	Funding debit adjustment	${}^{\tau-}(-\mathcal{F}) + \mathbb{1}_{[\![\tau,+\infty[\![}\mathrm{FVA}_{\tau-}$
FV	Firm valuation of counterparty risk	$\mathcal{C} + \mathcal{F}$
KVA^{sh}	Shareholder KVA	$(\mathrm{KVA}_0 - \mathrm{KVA}^{\tau-})$
KVA^{cr}	Creditor KVA	$\mathbb{1}_{[\![\tau,+\infty[\![}\mathrm{KVA}_{\tau-}$

Lemma 8. *We have*
(i) $\mathrm{CL} = \mathrm{DVA} + \mathrm{FDA}$, *which is the value process of both* ${}^{\tau-}(-\mathcal{L})$ *and* $(-\mathcal{L})$*;*
(ii) $\mathrm{FVA} = \mathrm{FDA}$, $\mathrm{FV} = \mathrm{CA} - \mathrm{CL} = \mathrm{CVA} - \mathrm{DVA}$;
(iii) $\mathrm{KVA}^{sh}_t = J_t \mathbb{E}^*_t \int_t^\tau h\big(\mathrm{EC}_s - \mathrm{KVA}_s\big)^+ ds$ *and* $\mathrm{KVA}^{cr}_t = J_t \mathbb{E}^*_t \mathrm{KVA}_{\tau-}$, *for* $t \in \mathbb{R}_+$.

Proof. **(i)** holds by the definitions of Table 4, the formula (61) for ${}^{\tau-}(-\mathcal{L})$, and the martingale (hence, zero-value) properties of ${}^{\tau-}(\mathcal{M} + \mathrm{MtM})$, ${}^{\tau-}\mathcal{H}$ and $\mathcal{L}^{\tau-}$ (see Assumption 1 page 55 and Remark 5 page 57).
(ii) holds by the definitions of Table 4, by (14), (15), (1), and the fact that $\mathcal{F}$ is a martingale (hence, zero-valued) stopped at $\tau \wedge T$, by Assumption 1.
(iii) At $t < \tau$, the value processes of $(\mathrm{KVA}_0 - \mathrm{KVA}^{\tau-})$ and $\mathbb{1}_{[\![\tau,+\infty[\![}\mathrm{KVA}_{\tau-}$ are respectively worth $\mathbb{E}^*_t(\mathrm{KVA}_t - \mathrm{KVA}_{\tau-}) = \mathbb{E}^*_t \int_t^\tau h\big(\mathrm{EC}_s - \mathrm{KVA}_s\big)^+ ds$ (by (19)) and $\mathbb{E}^*_t \mathrm{KVA}_{\tau-}$. From τ onward these value processes vanish, as well as KVA^{sh} and KVA^{cr}. □

We assume that the shareholders have no other business than their involvement within the bank. The bank creditors, instead, have to face the liquidation costs of the bank, which are outside the scope of our model.

Definition 8. We call wealth transferred to the bank shareholders and creditors by the derivatives portfolio of the bank, denoted by $\mathcal{W}^{sh}$ and $\mathcal{W}^{cr}$, the sum between their respective (received) cash flows and the value process of their cash flows.

Proposition 6. The shareholders and creditors wealth transfer processes are the martingales

$$\begin{aligned} \mathcal{W}^{sh} &= -(\mathcal{L}^{\tau-} + \mathrm{KVA}^{\tau-} - \mathrm{KVA}_0) + \mathrm{KVA}^{sh} \text{ and} \qquad (62)\\ \mathcal{W}^{cr} &=^{\tau-} (-\mathcal{L}) + \mathrm{CL} + \mathbb{1}_{[\![\tau,+\infty[\![}\mathrm{KVA}_{\tau-} + \mathrm{KVA}^{cr}, \qquad (63) \end{aligned}$$

which start from the respective values KVA_0^{sh} and $\mathrm{CL}_0 + \mathrm{KVA}_0^{cr}$ at time 0.

Proof. The shareholder cash flows are given by $\mathcal{D}$ in (23), where $\mathcal{L}^{\tau-}$ is zero-valued (as a martingale) and $\mathrm{KVA}_0 - \mathrm{KVA}^{\tau-}$ is valued by KVA^{sh} (by definition of the latter in Table 4). The creditor cash flows are given by ${}^{\tau-}(-\mathcal{L}) + \mathbb{1}_{[\![\tau,+\infty[\![}\mathrm{KVA}_{\tau-}$, where ${}^{\tau-}(-\mathcal{L})$ is valued by CL, by Lemma 8(i), and $\mathbb{1}_{[\![\tau,+\infty[\![}\mathrm{KVA}_{\tau-}$ by KVA^{cr}, by definition of the latter in Table 4. □

Should the shareholders decide to put the bank in default at time 0 right after the portfolio has been set up, and they should not make any profit or loss, otherwise this would be an arbitrage for the shareholders. The fact that the shareholders wealth process $\mathcal{W}^{sh}$ starts from $\mathrm{KVA}_0^{sh} > 0$ (positive initial wealth transfer to shareholders, unless the KVA vanishes) suggests that our XVA framework may entail shareholder arbitrage. Yet, given the rules of default settlement stated in §1.A, upon bank default, the residual value on the (reserve capital and) risk margin account of the bank goes to creditors. So, even though KVA_0^{sh} is part of the wealth of the shareholders, the shareholders would not monetize KVA_0^{sh} by putting the bank in default at time 0 right after the portfolio has been set up. Hence the positive initial wealth transfer to shareholders does not entail such a shareholder arbitrage.

Likewise, the fact that the creditors wealth transfer martingale $\mathcal{W}^{cr}$ starts from $\mathrm{CL}_0 + \mathrm{KVA}_0^{cr}$ suggests that the derivative trading of the bank may entail a riskless profit to creditors. However, the scope of the model does not include the liquidation costs of the bank. For the creditors to monetize the wealth transfer triggered to them by the derivative portfolio of the bank, the bank has to default, and there is a substantial cost associated with that to the creditors.

A What-If Analysis

In this section we examine the consequences of an assumption that the bank could hedge out its risks completely, including default risk. As explained in §2.C, this assumption is counterfactual, for both practical and legal reasons. However, we endorse it here for the sake of the argument.

We recall from Lemma 8(i) that the value process of $(-\mathcal{L})$ is CL. To "complete the market", we now assume that the bank has access to a new deal with a client, insuring the payment of a cash flow stream $\mathcal{L}$ to the bank, along with a time 0 premium CL_0 (initiating the strategy at time 0 where the portfolio is settled). The deal would be fully collateralized, in the sense that the above cash flows would occur independently of the default status of the bank and its client.

Proposition 7. *Assume that the bank has access to the new deal as defined above, coming on top of the derivative portfolio of the bank and its dynamic hedge considered earlier in this chapter. Then*

$$\mathrm{MtM} - \mathrm{FV} = \mathrm{MtM} - (\mathrm{CVA} - \mathrm{DVA}) = \mathrm{MtM} - (\mathrm{CA} - \mathrm{CL})$$

(by Lemma 8(ii)) is a replication price process for the derivative portfolio of the bank, in the sense that the resulting loss process of the bank (impact of all the hedges included) vanishes, as does in turn the economic capital of the bank, its KVA, and the shareholder and creditor wealth transfer processes.

Proof. Here is the detail of a corresponding replication strategy, starting from time 0 where the portfolio is settled. The trading and CVA desks act exactly as before. The funding desk passes to the client (at time 0) and to the bank shareholders (on $(0, \tau)$) a diminished add-on FVA $-$ CL, instead of FVA before without the new deal. In addition, a new CL desk puts the upfront premium CL_0 of the new deal on a dedicated cash account, along with a matching liability of CL_0 on the bank balance sheet. This CL account, like all the other ones within the bank, is market-to-model, i.e. reset in continuous time to the value process of the corresponding liability (see §1.A), namely to the value CL of the cash flow $(-\mathcal{L})$ due by the bank under the terms of the new deal. Before bank default, the resets to the CL account, which accumulate to $\mathrm{CL}_0 - \mathrm{CL}$, are passed to bank shareholders, as is the $\mathcal{L}^{\tau-}$ component from the cash flows of the new deal (which thus do not stay on the balance sheet of the bank). Finally, from time τ onward, the ${}^{\tau-}\mathcal{L}$ component of the cash flows of the new deal hedges out the ${}^{\tau-}(-\mathcal{L})$ cash flows that were previously falling into the hands of the bank creditors.

As a result, the loss of the bank starting before τ (i.e. the process "${}^{\tau-}$(loss)") vanishes, whereas before τ the loss of the bank (i.e. the process "(loss)$^{\tau-}$") is given by

$$\underbrace{\mathcal{C}+\mathcal{F}+\mathrm{FV}-\mathrm{FV}_0-(\mathcal{M}+\mathrm{MtM}-\mathrm{MtM}_0)+\mathcal{H}}_{\text{modified loss of the CA desks}} \\ + \underbrace{-\mathcal{L}^{\tau-}+\mathrm{CL}-\mathrm{CL}_0}_{\text{new hedging loss components passed to shareholders}} \tag{64}$$
$$=\mathcal{C}+\mathcal{F}+\mathrm{CA}-\mathrm{CA}_0-(\mathcal{M}+\mathrm{MtM}-\mathrm{MtM}_0)+\mathcal{H}-\mathcal{L}=0, \tag{65}$$

where Lemma 8(ii) (and $\mathcal{L}=\mathcal{L}^{\tau-}$ before τ) was used in the first equality and (11) in the second one.

Shareholders bear no more trading risk, hence, economic capital and the KVA vanish, as do, in turn, the shareholder and creditor wealth transfer processes. □

Remark 20. Before τ, the amount available free of charge to the bank for its rehypothecable collateral funding purposes is $\mathrm{FV}+\mathrm{CL}=\mathrm{CA}$ as before. Hence, the risky funding cash flows for collateral, $\mathcal{F}$, are not modified by the new deal. The cash flows $\mathcal{C}$ are not affected by the new deal either.

Due to the new deal, the creditors are left without any resources to address the liquidation costs of the bank. The clients pay $(\mathrm{CA}_0-\mathrm{CL}_0)-\mathrm{MtM}_0$ instead of $(\mathrm{CA}_0+\mathrm{KVA}_0)-\mathrm{MtM}_0$ before, hence they are better off by the amount $\mathrm{CL}_0+\mathrm{KVA}_0$. As before, the bank shareholders are indifferent to the portfolio at the accordingly adjusted price paid by the bank. But this price is now $\mathrm{MtM}_0-(\mathrm{CA}_0-\mathrm{CL}_0)$, instead of $\mathrm{MtM}_0-(\mathrm{CA}_0+\mathrm{KVA}_0)$ before.

Remark 21. The reason why the new deal may alter the XVA picture so deeply is that this deal is different in nature from the contracts eligible for the bank derivative portfolio or its dynamic hedge earlier in this chapter. In fact, as $\mathcal{L}^{\tau-}$ is a martingale, the only case where the value process $\mathrm{CL}=\mathrm{DVA}+\mathrm{FVA}$ of $(-\mathcal{L})$ could also be its shareholder value (as required for eligibility to the derivative portfolio or its dynamic hedge earlier in this chapter, cf. the combined requirements of (13) and Assumption 2 page 73) would be when

$$\mathrm{CL}_t=\mathbb{E}_t^*\Big(-\mathcal{L}_{\tau-}^{\tau-}+\mathcal{L}_t^{\tau-}+\mathrm{CL}_{\tau-}\Big)=\mathbb{E}_t^*\mathrm{CL}_{\tau-},\ t<\tau, \tag{66}$$

e.g. if $\mathrm{CL}=\mathrm{DVA}+\mathrm{FVA}=0$ (and in this case only, under the reduction of filtration setup of §3.A, by Theorem 1(i) page 64). Hence the new deal cannot be seen as part of the derivative portfolio of the bank or its dynamic hedge earlier in this chapter.

Alternatively, one may try and interpret the trading loss

$$-\mathcal{L}+\mathrm{CL}-\mathrm{CL}_0$$

triggered by the new deal as an additional line of risky funding cash flows, coming on top of the risky funding cash flows $\mathcal{F}$ earlier in this chapter. However, as opposed to $\mathcal{F}$ as per (44), this new line of risky funding entails a loss of the bank at its own default, hence a benefit prior to τ, which is how the new deal allows the shareholders to monetize the default of the bank.

In any case, again, the new deal is unfeasible for the bank. In particular, a risky funding benefit of the bank at its own default means that the bank is effectively selling default protection on itself, which is even illegal (cf. the condition concluding Assumption 4 page 66).

Even if the new deal is unfeasible, the what-if analysis of Proposition 7 is enlightening on the nature of the XVAs: without the new deal, the wealth transfers to shareholders and creditors are given by (62) and (63), whereas Proposition 7 shows that these wealth transfers would vanish if the bank had access to the new deal. We conclude that the shareholder and creditor wealth transfers (62) and (63) are due to the inability of the bank to hedge, in particular, jump-to-default risk.

§5 DISCUSSION

We revisit, in the light of the requirements of §1, our cost-of-capital XVA solution to the sustainable pricing and dividend release policy problem.

A Regulatorily Admissible and Sustainable

As stated in Proposition 1 page 59, shareholder trading gains and KVA risk margin payments result in a $-(\mathcal{L}^{\tau-}+\mathrm{KVA}^{\tau-}-\mathrm{KVA}_0)$ dividend stream, turning the shareholder *equity* process (SHC) into a submartingale with growth rate h on their capital at risk (yet the shareholder *wealth* process is a martingale and the setup is non-arbitrable in the sense explained after Proposition 6 page 75). This holds even in the case of a portfolio held on a run-off basis, i.e. without the need to enter new deals for generating new profits. This feature addresses the sustainability requirement in §1. Moreover, I.§4.B shows that the sustainability property of Proposition 1 is still valid in the (realistic) case of a trade incremental portfolio, provided the corresponding trade incremental XVA policy is applied at every new deal.

Our setup crucially includes the default time τ of the bank itself, which is the essence of the contra-liabilities wealth transfer issue detailed in §4. However, accounting for all wealth transfers involved (cf. Proposition 6), we end up with portfolio-wide nonnegative[8] and ultimately unilateral CVA, FVA and KVA, which price the related (nondecreasing) cash flows until the final horizon T of the XVA problem (cf. (37)–(39)), as opposed to $\tau \wedge T$. This

[8] essentially (cf. Remark 3 page 55).

makes our approach naturally in line with the monotonicity requirement of §1 that capital at risk and reserve capital should not decrease simply because the credit risk of the bank has increased.

B Economically Credible and Logically Consistent

Whereas counterparty jump-to-default risk can fundamentally not be hedged, a large part of the XVA literature relies on a replication paradigm. However, as established in Proposition 7 page 76, in a theoretical, complete counterparty risk market, the all-inclusive XVA formula would simply be CVA − DVA, instead of CVA + FVA + KVA when market incompleteness is accounted for.

In particular, the Burgard and Kjaer (2011, 2013, 2017) FVA approach was pioneering, but it breaches several of the requirements stated in §1, namely: nonnegativity (with an FVA that may become negative in the limiting case of a deeply out-of-the-money portfolio), monotonicity (with CVA and KVA tending to 0 when the default risk of the bank goes to infinity), and economic realism (which is lacking in an "XVA replication paradigm"). Likewise, the Green, Kenyon and Dennis (2014) KVA approach was pioneering but it breaches monotonicity, economic realism, and (see Subsection D below) minimality. The XVA metrics of Bichuch, Capponi and Sturm (2018) also breach the nonnegativity and monotonicity requirements.

With respect to these XVA replication frameworks, our cost-of-capital XVA approach results in materially different XVA formulas and balance sheet implications. In particular:

- Despite the fact that we include the default of the bank itself in our modeling, our (portfolio-wide) XVA metrics are, ultimately, unilateral (hence do not tend to decrease simply because the credit risk of the bank has increased), and they are nonnegative (portfolio-wide);

- Our KVA is loss-absorbing: by contrast with the KVA of Green, Kenyon and Dennis (2014), it does not belong to the trading loss process $\mathcal{L}$ of the bank (it is not a liability like the CVA and the FVA, it would make no sense to try and replicate it);

- As a consequence of the previous point, our KVA discounts future capital at risk projections at the hurdle rate h (cf. (53)-(56)), which makes a big difference at the very long time horizon T (such as 50 years) of XVA computations (see the top right panel of Figure 36 page 206).

In addition, instead of working with economic capital, Green, Kenyon and Dennis (2014) use approximations in the form of scriptural regulatory capital specifications, which is less satisfying economically. It is also less self-consistent: under the cost-of-capital, economic capital-based, XVA approach, counterparty-risk-free valuations flow into CVA computations, which in turn flow into FVA computations, which all flow into KVA computations (see Figure 10 page 121). These connections make the counterparty-risk-free valuation, CA = CVA + FVA, and KVA equations, thus the derivative pricing problem as a whole, a self-contained and self-consistent problem.

C Numerically Feasible and Robust

These connections can also be exploited for numerical purposes. Chapter VII presents numerical applications on realistically large bilateral trade portfolios, mainly based on neural net regression computational strategies. See also Chapter X and Abbas-Turki, Diallo and Crépey (2018) for alternative, nested Monte Carlo computational strategies. All these works demonstrate the numerical feasibility and scalability of the cost-of-capital XVA approach.

The model risk inherent to economic capital-based KVA computations can be addressed by a Bayesian variant of our baseline cost-of-capital XVA approach. This is achieved by combining, in a global simulation, paths of the risk factors obtained in several "good" models, all econometrically realistic and calibrated to the market in counterparty-risk-free valuation terms (cf. X.§4.C). Drawing scenarios equally from each makes tails more leptokurtotic and risk measures greater as they are when one picks just a single (even good) model. The difference between the resulting enhanced KVA and a baseline, reference KVA, can be used as a reserve against model risk. For a more exhaustive development on model risk and XVAs, see Chapter XI.

D Minimal

An FVA for rehypothecable collateral computed at the level of a unique funding set, as in the first term of (47) (detailed further in the follow-up chapter as the first line of III.(69)), avoids the over-conservatism of FVAs for rehypothecable collateral sometimes calculated for simplicity by netting set and aggregated. Indeed, as developed in XI.§4.A, such simplification misses the FVA markdown corresponding to the rehypothecability of (eligible) collateral across the different netting sets of the bank. One should also account for the further FVA markdown due to the possibility for a bank to use its capital at risk as variation margin, which is done in Chapter VIII.

Our KVA is minimal in the sense of Proposition 5 page 72. An even cheaper KVA as in Albanese, Caenazzo and Crépey (2017, Proposition 4.2(v)) results from the following variation on our approach in this book: Upon the bank default, notwithstanding the bankruptcy rules recalled in the last paragraph of §1.A, the residual risk margin flows back into shareholder equity instead of going to creditors. Likewise, a cheaper FVA as in Albanese et al. (2017, Proposition 4.2(i)) (see after Eqn. X.(2) page 293) follows from asserting that, upon bank default, the residual reserve capital of the funding desk flows back into shareholder equity instead of going to creditors. However, these violations of the usual bankruptcy rules induce shareholder arbitrage, in the sense of a riskless profit strategy consisting for shareholders in letting the bank default instantaneously at time 0, right after the client portfolio has been setup and the corresponding reserve capital and risk margin amounts have been sourced from the clients. By contrast, as explained after Proposition 6 page 75, our approach in this book excludes such shareholder arbitrage opportunities.

Hence, "local departures" from our cost-of-capital XVA solution to the sustainable pricing and dividends problem of §1 may be a bit cheaper, but they are less self-consistent. As seen in Subsection B, more radically different approaches to the problem suffer from severe shortcomings with respect to the requirements of §1. In an intuitive formulation, we conclude that the cost of capital XVA solution to the sustainable pricing and dividends problem may not be the only solution, nor is it necessarily "globally minimal", but it has some "locally minimizing properties, at least in certain directions of the search space", and we are not aware of any other "distant solution".

CHAPTER III

Cash Flows Arithmetics

We apply the conceptual cost-of-capital XVA framework of Chapter II to the concrete portfolio of a dealer bank. One particular emphasis is on the understanding of the XVA implications of the hedging side of the portfolio: static back-to-back hedges of the cleared client netting sets of the bank, traded via central counterparties, versus dynamic hedge of the residual (non already statically offset) mark-to-market exposure of the bank, which can be achieved via futures-styles strategies implemented on exchanges. The XVA implications of the liquidation periods between the defaults and their settlements are also analyzed in detail.

§1 INTRODUCTION

A Executive Summary of Chapter II

MtM, CVA, and FVA are the bank shareholder value processes II.(5) of the cash flows impacting the eponymous trading desks of the bank, namely the dynamic hedging loss $\mathcal{H}$ of the bank and the other (than those already contained in $\mathcal{H}$) cash flow processes $\mathcal{M}$ to the trading desks, $\mathcal{C}$ from the CVA desk and $\mathcal{F}$ from the funding desk (or Treasury) of the bank. The ensuing loss process of the bank as a whole is given by

$$\mathcal{L} = \mathcal{C} + \mathcal{F} + \mathrm{CA} - \mathrm{CA}_0 - (\mathcal{M} + \mathrm{MtM} - \mathrm{MtM}_0) + \mathcal{H}, \tag{1}$$

where $\mathrm{CA} = \mathrm{CVA} + \mathrm{FVA}$. The trading loss process stopped before the default of the bank, $\mathcal{L}^{\tau-}$ (cf. II.(3)), is a martingale.

The KVA process is designed so as to turn the dividend process

$$-(\mathcal{L}^{\tau-} + \mathrm{KVA}^{\tau-} - \mathrm{KVA}_0)$$

of the bank shareholders into a submartingale with drift coefficient hSCR, where the nonnegative constant h is a target hurdle rate (e.g. 10%) set by the management of the bank, while the shareholder capital at risk process

DOI: 10.1201/9781003614944-6

(SCR) is sized dynamically based on the conditional expected shortfall of the increment of $\mathcal{L}^{\tau-}$ over the next year (cf. Definitions II.3 page 58 and II.6 page 68).

Conversely, relying on the above XVA setup, an implied hurdle rate can be reverse-engineered from the balance-sheet data of the bank. The performance of a new deal that would be concluded at a given price with a client can then be assessed by its incremental impact on the implied hurdle rate of the bank.

B Outline of the Chapter

The question addressed by the present chapter is:

What are the concrete cash flow processes $\mathcal{H}, \mathcal{M}, \mathcal{C}$, and $\mathcal{F}$?

This question is crucial for applications since, as visible from Subsection A, the processes $\mathcal{C}$ and $\mathcal{F}$ are the key input data to the XVA equations. As we will see in this chapter, the exact contents of $\mathcal{H}$ and $\mathcal{M}$ is key to a detailed understanding of the XVA implications of the derivative portfolio of the bank *and its hedge*: static back-to-back hedges of the cleared client netting sets of the bank, traded via central counterparties (§4), complemented by dynamic hedging of the residual (non-already statically offset) mark-to-market exposure of the bank (§3).

The primary contents of $\mathcal{H}$ is clear from its definition already illustrated by Examples II.1 and II.2. Note that the process $\mathcal{H}$ eventually disappears from the XVA equations in the reference dynamic hedging case where (see II.(12))

$$\mathcal{H} = \mathcal{M} + \mathrm{MtM} - \mathrm{MtM}_0, \text{ hence } \mathcal{L} = \mathcal{C} + \mathcal{F} + \mathrm{CA} - \mathrm{CA}_0. \tag{2}$$

But the process $\mathcal{H}$ becomes nontrivial and plays in fact a prominent role in Chapter XI, where market risk reintroduces itself into the XVA game, via model risk.

The concrete cash flows $\mathcal{M}$, $\mathcal{C}$ and $\mathcal{F}$ are derived in §3-§4 by application of the generic cash flow allocation principles of §2 to the actual financial network of the bank. Once the cash flows are specified, the XVA metrics are derived in §5 by application of the technical results of Chapter II.

C The Probabilistic Setup in a Nutshell

All our cash flow processes are cumulative, starting from 0 at time 0 and stopped at T, defined as the sum between the final maturity of the derivative portfolio of the bank (assessed on a run-off basis, e.g. 50 years) and an upper bound $\delta \geq 0$ on the time of liquidating defaulted positions (commonly considered to be of the order of one to two weeks). The default and liquidation times of the bank are denoted by τ and $\tau^\delta \geq \tau$, and we write $J = \mathbb{1}_{[\![0,\tau[\![}$.

All the cash flow and value processes are $\mathfrak{G}$ semimartingales in a càdlàg version, where the model filtration $\mathfrak{G} = (\mathfrak{G}_t)_{t\in\mathbb{R}_+}$ satisfies the usual conditions. All default and liquidation times (such as τ and τ^δ above) are assumed to be positive $\mathfrak{G}$ stopping times. We use the risk-free asset as a numéraire and we work under some fininsurance probability measure $\mathbb{Q}^*$. Unless explicitly mentioned, probabilistic statements refer to the full stochastic basis $(\mathfrak{G}, \mathbb{Q}^*)$, with time-$t$ conditional expectation denoted by $\mathbb{E}_t^*$, as opposed to the reduced stochastic basis $(\mathfrak{F}, \mathbb{P})$ specified in II.§3.A, such that $\mathfrak{F} \subset \mathfrak{G}$ on $[0, T]$ and $\mathbb{P} \sim \mathbb{Q}^*$ on $\mathfrak{F}_T$, with time-t conditional expectation denoted by $\mathbb{E}_t'$.

§2 CASH FLOWS ALLOCATION PRINCIPLES

A Financial Network of the Bank

As developed in the one-period setup of I.§3.E.1 and in continuous time in II.§4, banks can hardly hedge jump-to-default risk. Instead they mitigate it, by partitioning their derivative portfolio into netting sets of contracts which are jointly collateralized (guaranteed against default, to some extent) and are liquidated in case of default of the related counterparty or of the bank itself.

The financial network has been profoundly transformed over the last ten years with the increasingly dominant role assigned to central clearinghouse counterparties (CCPs), under the incentive of the G20. See Gregory (2014), Cont (2017) and Menkveld and Vuillemey (2021) for general CCP references. In September 2009 (G20 Research Group, 2009), the G20 imposed systematic settlement for over-the-counter (OTC) derivative contracts when they can be standardized through CCPs between two financial players, then recognized as "clearing members" of the CCPs acting as their intermediaries. In other words, the financial flows of a derivative contract between two counterparties pass through an intermediary CCP, the latter becoming the buyer of any selling counterparty and the seller of any buying counterparty of this contract. The CCPs also have the role of guaranteeing the continuity of payments of the contract cash flows even in the event of default of one of the counterparties. The contracts are guaranteed via the collection of collaterals (aka margin calls) from the two parties for which the CCPs act as intermediary, similar to the principles applied by financial exchanges. In the end, a few thousand of financial players having enough capital to face CCPs provide them with these collaterals collected from smaller players, financially reduced and unable to directly access the services of a CCP. But not all derivative products can be standardized and centrally cleared. Indeed, many products are specifically designed for the sole needs of clients wishing to protect themselves against certain market fluctuations or simply for speculative purposes.

Hence, the structure of financial markets has transformed itself from a structure of bilateral netting agreements operated between two players in the

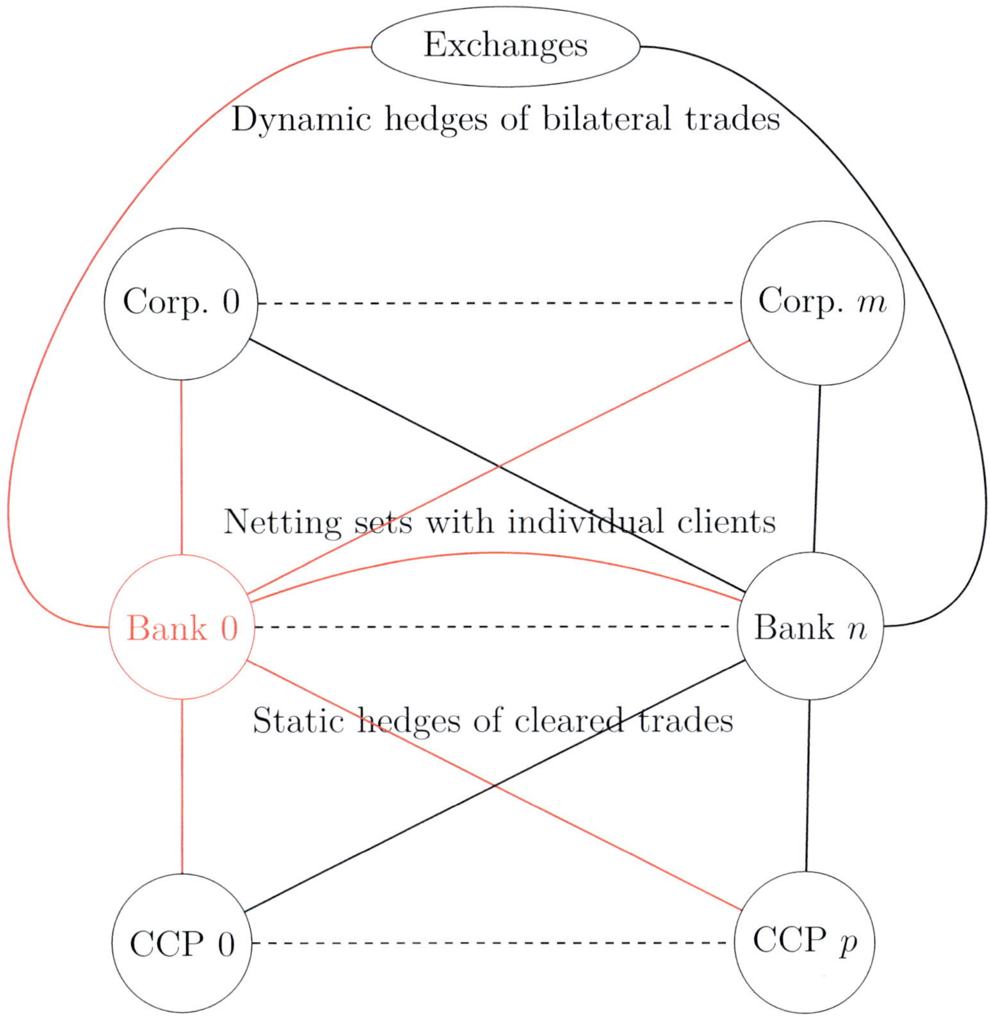

Figure 6: The network of (nonbanking) corporate clients, banks, CCPs and exchanges. The solid edges represent financial cash flows. Netting sets with individual clients, which can be corporate clients or other banks, correspond to the upper part of the picture (banks and above), and netting sets with CCPs to the lower part (below banks).

financial network, into a more complex structure. The latter is made up of tens of CCPs, each involving from a few units to a few hundred clearing members (with crossed participations and the ensuing joint membership issues), concentrating gigantic amounts of collaterals collected. The corresponding financial network is depicted in Figure 6, where the bank labeled by 0, with network connections emphasized in red, represents the reference bank, which XVA metrics are computed. The trades of the bank with CCPs and exchanges correspond to its respective static and dynamic hedges. More precisely, as will be detailed in §4, the trades of the bank with the CCPs, corresponding to the bottom part in Figure 6, are fully collateralized back-to-back mark-to-market hedges of client trades. The exchanges on top of Figure 6 denote platforms where futures-style positions can be used by the bank for hedging its residual market risk (not already statically offset between different deals) dynamically. Futures-style positions are tantamount to continuously rolled-over instantaneously maturing positions. As XVAs require a positive time horizon to develop, dynamic hedging by the bank is without XVA implications, if not for the model risk issue raised in Chapter XI. But dynamic hedging entail liquidity issues that are the topic of chapter IX (cf. Table 47 page 288).

B Liquidation and Funding Policies

As reasserted in the wake of the 2007–09 global financial crisis by the Volcker rule, a dealer bank should be hedged as much as possible, at least in terms of market risk. Jump-to-default risk, on the other hand, is hardly hedgeable in practice. Instead it is mitigated through netting and collateralization. Namely, designated netting sets of transactions between two given counterparties (two individual participants or a participant and the CCP) are jointly collateralized, i.e. guaranteed against the default of one or/and the other party. The collateral (or guarantee) comprises a variation margin, which tracks the mark-to-market (counterparty-risk-free value) of the netting set between the two parties, and nonnegative amounts of initial margin posted by each party to the other, which provide a defense against the risk of slippage of the value of the netting set away from its (frozen) variation margin during its liquidation period. In the case of transactions with a CCP, there is an additional layer of collateral in the form of a default fund meant as a defense against extreme and systemic risk. For each participant, variation margin is rehypothecable and fungible across all its netting sets. Initial margin is segregated at the netting set level. (Funded, see after Assumption III.2 page 97) default fund contributions (DFCs) are segregated at the clearing member level.

In words to be turned into equations in this chapter (see in particular the proofs of Lemmas 2 and 4), the standing rule regarding the liquidation of a defaulted portfolio is Principle I.2 page 23, extended to non-zero recoveries in the following form:

Principle 1. At the liquidation time of a netting set:

- if a counterparty in default is indebted toward the other beyond its posted margin[1], then this debt is only reimbursed at the level of this posted margin plus a fraction (recovery rate of the defaulted party) times the residual debt beyond the margin;

[1]which cannot occur for both jointly, by nonnegativity of initial margins, as we will see in Remark 5 page 95.

- otherwise the debt between the two parties is fully settled.

Here debt is understood as the sum between, on the one hand, the counterparty-risk-free value of the netting set at liquidation time and, on the other hand, the contractually promised cash flows of the netting set unpaid during the liquidation period that separates the default event from the settlement of the default.

Within the bank, the CVA desk is in charge of these netting set liquidation cash flows.

Regarding the risky funding cash flows, we assume the funding policy of Assumption II.4, so that $\mathcal{F}$ is given by II.(43), where the processes $\mathcal{R}$, resp. $\mathcal{S} \geq 0$, denote the overall amounts of rehypothecable, resp. segregated, collateral posted by the bank Treasury (see II.§3.C).

C Margin Lending

The unsecured borrowing spread process of a bank can be proxied by $\lambda = \gamma(1-R)$, where γ is its risk-neutral default intensity and R is its recovery rate. The best proxy for a term structure of λ is the CDS curve of the bank. As bank hurdle rates (e.g. 10%) are typically one order of magnitude greater than their CDS spreads, unsecured borrowing of initial margin is a cost-efficient strategy if compared to using equity capital (i.e. own assets) for this purpose. Accordingly, as of today, a common initial margin raising policy is unsecured borrowing. But this is still quite expensive. In this section we explain how a possibly blended funding spread $\widetilde{\lambda} \leq \lambda$ (i.e. $\bar{R} \geq R$ as postulated in Assumption II.4) may be accessible to the bank for funding its initial margin. This is done in a unified setup of the initial margin (IM) posted by a bank on a netting set with a bilateral counterparty or a CCP. The counterparty-risk-free value of the corresponding cash flows process $\mathcal{P}$ is denoted by P, and the associated variation margin by VM.

Under unsecured borrowing by the bank of its posted IM (PIM), if the bank defaults, then the external funder of the bank loses $(1-R)\mathrm{PIM}_{\tau-}$ at τ^{δ} (see the top panel of Figure 7), which the external funder charges $\lambda_s \mathrm{PIM}_s ds$ to the bank until its default. This is fair to the external funder (assumed default-free, cf. Remark II.13) in view of the identity

$$\mathbb{E}_t^*\Big[\mathbb{1}_{t<\tau<T}(1-R)\mathrm{PIM}_{\tau-}\Big] = \mathbb{E}_t^* \int_t^{\tau\wedge T} \lambda_s \mathrm{PIM}_s ds,\; 0 \leq t \leq \bar{\tau} \tag{3}$$

(under the integrability assumption that $\mathbb{1}_{[\![0,\tau[\![}\mathrm{PIM}_{\tau-} + \int_0^{\cdot\wedge\tau} \mathrm{PIM}_s \gamma_s ds$ is a $(\mathfrak{G}, \mathbb{Q}^*)$ martingale like $\mathbb{1}_{[\![0,\tau[\![} + \int_0^{\cdot\wedge\tau} \gamma_s ds$).

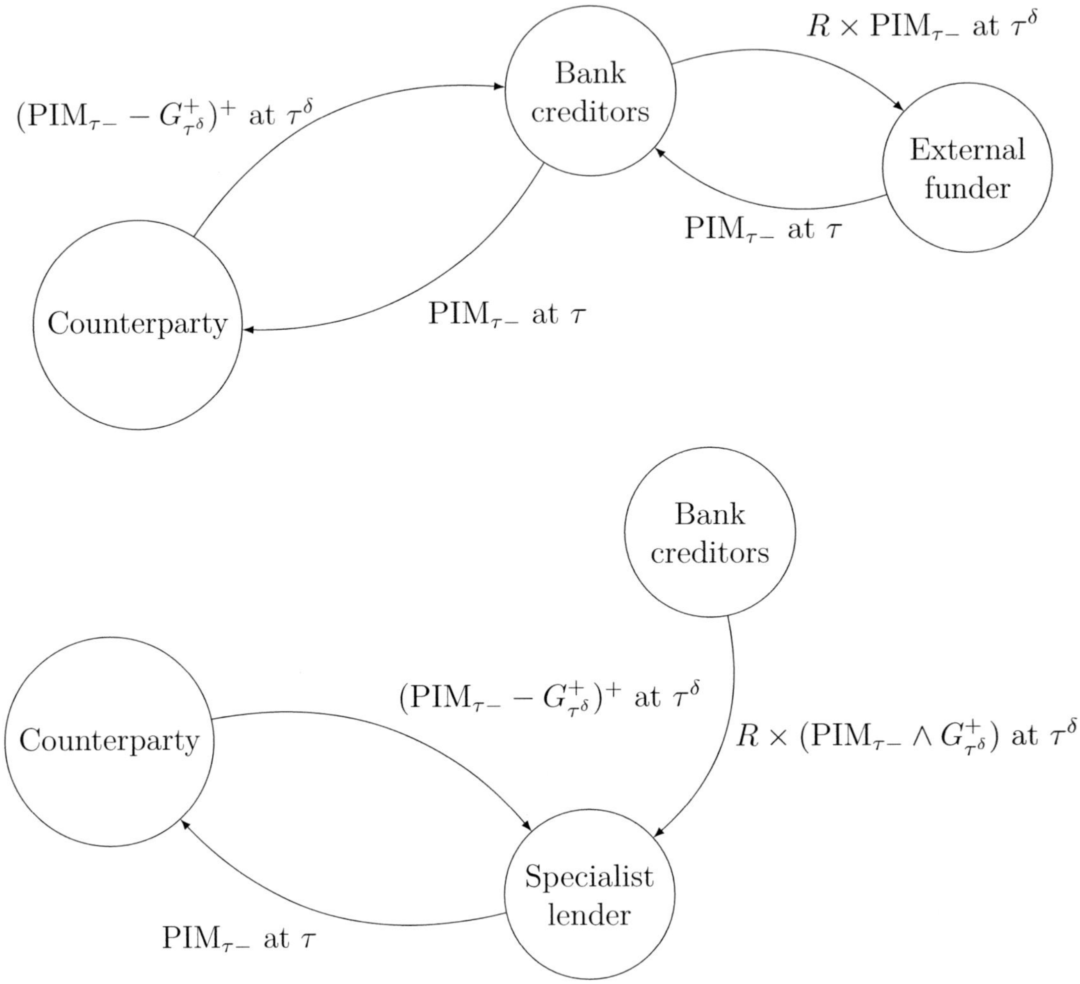

Figure 7: Reference bank own-default related funding cash-flows (using the risk-free asset as numéraire, i.e. ignoring remuneration of cash accounts at the risk-free rate, to alleviate the picture). The specialist lender on the bottom panel is lending the same IM amount to the counterparty (on behalf of the bank) than the external funder is lending to the bank (who is then lending it to the counterparty) on top. But, in case the bank defaults, the specialist lender receives back from the counterparty the portion of IM unused to cover losses. Hence, it is reimbursed at a much higher effective recovery rate than the nominal recovery rate R embedded in the bank credit spread.

However (see the bottom panel of Figure 7), we can consider an alternative initial margin lending scheme, whereby the IM is provided to the counterparty on behalf of the bank by a so called specialist lender, in exchange of some service fee. Under the terms of a legal agreement concluded between the counterparty and the specialist lender, in case of default of the bank, the specialist lender receives back from the counterparty the portion of IM unused to cover losses. As long as such margin lending is implemented off the balance sheet of the bank, is it not a violation of pari passu rules, just as repurchase agreements or mortgages are not: see Subsubsection C.1 below for more corporate details and implications. It is just a form of collateralised lending, which does not transfer wealth from senior creditors in the baseline case. In practice specialist lenders are private equity funds. For similar ideas regarding variation margin (VM), see Albanese, Brigo and Oertel (2013). However, such funding schemes are much more difficult to implement for VM, because VM is far larger and more volatile than IM.

We assume that the specialist lender funds are kept at the segregated account for initial margin. In case of default, this account is depleted by the amount $\mathrm{PIM}_{\tau-} - (\mathrm{PIM}_{\tau-} - G^+_{\tau^\delta})^+ = (G^+_{\tau^\delta} \wedge \mathrm{PIM}_{\tau-})$ (see the bottom panel of Figure 7), where the time-t gap G_t is given as

$$G_t = P_t + \mathcal{P}_t - \mathcal{P}_{t-\delta} - \mathrm{VM}_{(t-\delta)-}, \tag{4}$$

and the lender receives a claim against the bank estate. Hence (see the bottom panel of Figure 7), the exposure of the specialist lender to the default of the bank is

$$\mathrm{PIM}_{\tau-} - \Big((\mathrm{PIM}_{\tau-} - G^+_{\tau^\delta})^+ + R(G^+_{\tau^\delta} \wedge \mathrm{PIM}_{\tau-})\Big) = (1-R)(G^+_{\tau^\delta} \wedge \mathrm{PIM}_{\tau-}). \tag{5}$$

Let ξ be a $\mathfrak{G}$ predictable process, which exists by Corollary 3.23 2) in He, Wang and Yan (1992), such that

$$\xi_\tau = \mathbb{E}^*_{\tau-}\Big[(G^+_{\tau^\delta} \wedge \mathrm{PIM}_{\tau-})\Big] \leq \mathrm{PIM}_{\tau-}, \tag{6}$$

and let

$$\widetilde{\xi} := \xi \wedge \mathrm{PIM}_- \leq \mathrm{PIM}_-. \tag{7}$$

Hence $\widetilde{\xi}$ is another $\mathfrak{G}$ predictable process such that $\widetilde{\xi}_\tau = \xi_\tau \wedge \mathrm{PIM}_{\tau-} = \xi_\tau$, by (6). We assume that the bank, as long as it is non-default, pays continuous-time service fees $\lambda_s \widetilde{\xi}_s ds$ to the specialist lender.

Summing up, the specialist lender loses $(1-R)(\mathrm{PIM}_{\tau-} \wedge G^+_{\tau^\delta})$ at τ^δ, which the specialist lender charges $\lambda_s \widetilde{\xi}_s ds$ to the bank until the default of the latter. Such an arrangement can be deemed fair to the specialist lender (assumed default-risk-free), in view of the identities

$$\begin{aligned}
&\mathbb{E}^*_t\Big[\mathbb{1}_{t<\tau<T}(1-R)\Big(G^+_{\tau^\delta} \wedge \mathrm{PIM}_{\tau-}\Big)\Big] \\
&\quad = \mathbb{E}^*_t\Big[\mathbb{1}_{t<\tau<T}(1-R)\mathbb{E}^*_{\tau-}\Big[\Big(G^+_{\tau^\delta} \wedge \mathrm{PIM}_{\tau-}\Big)\Big]\Big] \\
&\quad = \mathbb{E}^*_t\Big[\mathbb{1}_{t<\tau<T}(1-R)\widetilde{\xi}_\tau\Big] = \mathbb{E}^*_t \int_t^{\tau\wedge T} \lambda_s \widetilde{\xi}_s ds,\ 0 \leq t \leq \bar{\tau}
\end{aligned} \tag{8}$$

(under the integrability assumption on $\widetilde{\xi}$ that $\mathbb{1}_{[\![0,\tau[\![}\widetilde{\xi}_\tau + \int_0^{\cdot\wedge\tau}\widetilde{\xi}_s\gamma_s ds$ is a $(\mathfrak{G},\mathbb{Q}^*)$ martingale like $\mathbb{1}_{[\![0,\tau[\![} + \int_0^{\cdot\wedge\tau}\gamma_s ds$).

Proposition 1. *The* MVA^{ub} *and* MVA^{sl} *of the bank corresponding to unsecured borrowing and specialist margin lending satisfy, for* $t<\tau$,

$$\mathrm{MVA}_t^{sl} = \mathbb{E}_t\int_t^T \lambda_s\widetilde{\xi}_s ds \le \mathbb{E}_t\int_t^T \lambda_s \mathrm{PIM}_s ds = \mathrm{MVA}_t^{ub}. \tag{9}$$

The blended funding spread $\widetilde{\lambda}$ *associated with margin lending, defined by* $\widetilde{\lambda}\mathrm{PIM}_- = \lambda\widetilde{\xi}$ *so that* $\mathrm{MVA}_t^{sl} = \mathbb{E}_t\int_t^T\widetilde{\lambda}_s\mathrm{PIM}_s ds$, *satisfies*

$$\frac{\widetilde{\lambda}}{\lambda} = \frac{\widetilde{\xi}}{\mathrm{PIM}_-} \le 1. \tag{10}$$

Proof. Under the respective unsecured borrowing and margin lending schemes for initial margin, the bank pays respective funding costs $\lambda_s\mathrm{PIM}_s ds$ and $\lambda_s\widetilde{\xi}_s ds$ until its default, which are cash flow streams with respective shareholder value $\mathbb{E}_t\int_t^T\lambda_s\mathrm{PIM}_s ds$ and $\mathbb{E}_t\int_t^T\lambda_s\widetilde{\xi}_s ds$. This shows the identities in (9), while the inequalities in (9)-(10) stem from the one in (7). □

Under a basic specification where PIM_s is set as a high quantile (time-s conditional value-at-risk) of G_{s^δ}[2], the blending ratio in (10) is typically much less than 1. Hence MVA^{sl} should be significantly less than MVA^{ub}. Equivalently, the corresponding blended recovery rate $\bar{R}$ in II.(42), i.e.

$$\bar{R}_t := (1-\frac{\widetilde{\lambda}_t}{\lambda_t}) + \frac{\widetilde{\lambda}_t}{\lambda_t}R_t$$

(as $\lambda=(1-R)\gamma$ and $\widetilde{\lambda}=(1-\bar{R})\gamma$), is typically much larger than the unsecured borrowing recovery rate R.

In the case of a bilateral netting set, given the very conservative levels of initial margin prescribed by the regulation since the emergence of ISDA's standard initial margin model (SIMM) for bilateral transactions, the blended spread $\widetilde{\lambda}_t$ is indeed much smaller than the unsecured funding spread λ: cf. Figure 53 page 299. In the case of a netting set with a CCP, if funded default fund contributions (DFCs, see above (22) below) are also usable for own default settlement as per current CCP designs, subordinating own DFC to IM would result in less IM consumption upon defaults, with $(G_{\tau^\delta}-\mathrm{DFC}_{\tau-})^+$ instead of $G^+_{\tau^\delta}$ in (5), hence even more efficient IM lending schemes: cf. Tables 23 page 235 and 38 page 281.

[2]cf. VII.(9) page 198, assuming there for simplicity continuous-time variation margining $\mathrm{VM}_t = P_t$ before time τ in (4).

We emphasize that the above results are independent of the specification of the PIM process, possibly inclusive of any (such as liquidity) add-ons, as long as the IM is provided by the specialist lender. Also note that, for the argument to be valid, the IM lender does not need to anticipate the nature of future trades, which in the case of a market maker, such as a bank, would be impossible. The argument is robust and independent of future dealings. The IM lender simply needs to know (which is public regulatory information) that the collateral posted by the bank is very conservative, no matter what trades are entered into the future.

From a "wrong-way risk" perspective, one may be worried that the margin lending service fee, $\lambda_s \widetilde{\xi}_s ds$ in the above mathematical setup, increases with the default risk of the bank (represented by $\lambda_s = (1 - R_s)\gamma_s$). But this is also the case of the instantaneous cost, $\lambda_s \mathrm{PIM}_s ds$, for unsecured borrowing of the IM. In fact, the ratio (10) between the two costs does not depend on λ.

C.1 Credit and Liquidity

Margin lending practicalities are an active investor (such as private equity fund) business, whereas the available liquidity is mostly with passive investors (such as insurance and pension funds). Hence, the implementation of margin lending naturally calls for a two-layered structure, whereby an active investor bridges to a passive one. Now, the above developments (whether they regard margin lending or unsecured borrowing) are only on the credit side of the problem, with short-term funding assumed to be continuously rolled over in time. As always with credit, there is another, liquidity side to it. This is the fact that the lender (the so-called external lender in the case of unsecured borrowing and specialist lender in the case of margin lending) may want to cease to roll-over its loan, not because of the credit risk of the borrower, but just because of liquidity squeeze in the market: the lender may be short of cash (or liquid assets), or want to keep the latter for other (own) purposes. It may then be argued that the liquidity issue is more stringent for margin lending, with its two-layered structuring, than for unsecured borrowing. This is also why margin lending is more difficult to implement for VM than for IM. In order for an IM margin lending business to obtain the blessing of a regulator, it would be better to address this liquidity issue in the legal structure of the setup. This could take the form of an option for the bank to force one roll-over by the specialist lender. Such a "liquidity option" would have to be priced into the structure. But unsecured borrowing is not exempt from liquidity issues either. A detailed comparison of unsecured borrowing and margin lending also accounting for these liquidity issues is left for further research.

§3 BILATERAL TRADING CASH FLOWS

Table 5 gathers the notation pertaining to a netting set ι of financial derivatives between the bank and an individual counterparty (nonbanking corporate client or other bank, as opposed to a CCP that is considered in the next section). In the case of a netting set between the bank and an individual counterparty, collateral may come in two layers: rehypothecable variation margins, which are flows designed to track the changes in the (counterparty-risk-free) value of the netting set, possibly complemented by segregated initial margins, meant to cover losses arising during the lapse of time that separates a default from its settlement. We refer the reader to VII.§1.B (in the continuous-time setup) or VIII.§5.A (in a one-period setup) for concrete specifications, not needed at this stage, regarding variation and initial margins. All the collateral processes VM^ι, PIM^ι and RIM^ι are assumed to be $\mathfrak{G}$ optional and killed at the final maturity of the bank portfolio. Regarding the first row in Table 5, we have by definition II.(4):

$$P_t^\iota = \mathbb{E}_t^*\left(\mathcal{P}_T^\iota - \mathcal{P}_t^\iota\right),\ t \le T, \tag{11}$$

and P^ι vanishes on $[T, +\infty)$.

Remark 1. By linearity, (11) is the sum over the netting set ι of the analogous quantity pertaining to each individual deal in ι (assuming all European-style derivatives). Hence, P^ι computations reduce to valuations at the individual trade level.

Also note that, by definition of the various stopping times involved and of T and δ in §1.C, we have, for all individual counterparty netting sets ι,

$$\tau_\iota \le \tau_\iota^\delta \le (\tau_\iota + \delta) \wedge T, \text{ whereas } \tau \le \tau^\delta \le \tau + \delta \tag{12}$$

holds regarding the bank itself.

In this section we restrict attention to bilateral netting sets and the dynamic hedge of their residual (non-already statically offset across different deals) mark-to-market exposure.

Remark 2. *Bilateral* netting sets are to be distinguished from *centrally cleared* (versus individual) netting sets, but also *cleared client* (but still individual) netting sets, both considered in §4.

The corresponding role of the CVA desk, its interaction with the trading desks, and the functioning of the trading account between the two, which were all informally introduced in I.§2 and II.§1, are detailed as follows.

Table 5: Notation pertaining to the individual counterparty netting set ι of contracts of the derivative portfolio of the bank.

$\mathcal{P}^\iota$ and P^ι	contractually promised cash flows and their value process
τ_ι	$T\wedge$ the default time of the related counterparty
τ_ι^δ	$T\wedge$ the end of the liquidation period of the related counterparty.
R^ι	predictable, $[0,1]$ valued recovery rate process of the related counterparty
$s_\iota = \tau_\iota \wedge \tau$	$T\wedge$ default time of the netting set ι
$t_\iota = \tau_\iota^\delta \wedge \tau^\delta$	$T\wedge$ liquidation time of the netting set ι
VM^ι	variation margin exchanged between the bank and the counterparty (positive when posted from the counter-party to the bank)
$\mathrm{RIM}^\iota \geq 0$	initial margin received by the the bank from the counterparty
$\mathrm{PIM}^\iota \geq 0$	initial margin posted by the bank to the counterparty
$\Gamma^\iota = \mathrm{VM}^\iota + \mathrm{RIM}^\iota$	overall collateral amount posted by the counterparty to the bank
$\Upsilon^\iota = (-\mathrm{VM}^\iota) + \mathrm{PIM}^\iota$	overall collateral amount posted by the bank to the counterparty

Assumption 1. For each (non already defaulted) bilateral netting set $\iota \equiv b$, the trading desks and the CVA desk maintain between them an amount P^b of rehypothecable collateral on the trading account. The process MtM corresponds to the total amount on the trading account (see I.(5)). At the time s_b where a bilateral netting set b defaults, the corresponding amount on the trading account is frozen at its level $P^b_{s_b-}$ in the wait of being transferred (property-wise) from the CVA desk to the trading desks at the liquidation time t_b. During the liquidation period $[\![s_b, t_b]\!]$, the CVA desk compensates the trading desks for their missed cash flows: contractual trading cash flows $d\mathcal{P}^b_t$ and mark-to-model fluctuations dP^b_t. In addition, at the liquidation time t_b, Principle 1, page 86, applies to the liquidation of the netting set (and the dynamic hedge of the trading desks related to the netting set b is unwound, i.e. the corresponding term in $\mathcal{H}$ is stopped at t_b).

A Cash Flows to the Trading Desks

Going by Assumption 1, the concrete processes $\mathcal{M}$ (cash flows to the trading desks other than their dynamic hedging cash flows) and MtM (amount on the trading account) corresponding to the present bilateral trading setup satisfy

$$\begin{aligned}
&\mathcal{M}_0 = 0,\ \mathrm{MtM}_0 = \textstyle\sum_b P^b_0 \text{ and, for } t \in (0,T],\\
&d\mathcal{M}_t = \sum_b \Big(\mathbb{1}_{\{t<s_b\}} d\mathcal{P}^b_t + \mathbb{1}_{\{s_b \le t \le t_b\}}(d\mathcal{P}^b_t + dP^b_t) + \boldsymbol{\delta}_{t_b}(dt) P^b_{s_b-}\Big), \qquad (13)\\
&d\mathrm{MtM}_t = \sum_b \Big(\mathbb{1}_{\{t<s_b\}} dP^b_t - \boldsymbol{\delta}_{t_b}(dt) P^b_{s_b-}\Big).
\end{aligned}$$

Lemma 1. *In a bilateral trading setup:*

(i) *we have*

$$\mathcal{M} = \sum_b \Big((\mathcal{P}^b)^{t_b} + \mathbb{1}_{[\![s_b, t_b [\![}(P^b - P^b_{s_b-}) + \mathbb{1}_{[\![t_b, +\infty [\![} P^b_{t_b} \Big), \tag{14}$$

$$\mathrm{MtM} = \sum_b (\mathbb{1}_{[\![0, s_b [\![} P^b + \mathbb{1}_{[\![s_b, t_b [\![} P^b_{s_b-}), \text{ and} \tag{15}$$

$$\mathcal{M} + \mathrm{MtM} = \sum_b (\mathcal{P}^b + P^b)^{t_b}; \tag{16}$$

(ii) MtM *is the value process of* $\mathcal{M}$ *(in particular,* $\mathcal{M}+\mathrm{MtM}$ *is a martingale).*

Proof. **(i)** All formulas follow from (13) by positivity of the $s_b = \tau_b \wedge \tau$ (having assumed all positive default times).
(ii) By (11), each of the $\mathcal{P}^b + P^b$ is a martingale. Hence so is the sum $(\mathcal{M} + \mathrm{MtM})$, by (16). Moreover, by (15) and definition of $s_b \leq t_b \leq T$ with $\iota \equiv b$ in Table 5, $\mathrm{MtM}_T = 0$. Hence the MtM process in (15) is the value process II.(4) of the process $\mathcal{M}$ in (14). □

Remark 3. For consistency with the additional (first stated in §1.A) requirement that MtM also be the shareholder value process II.(13) of $\mathcal{M}$, $\sum_b ((\mathcal{P}^b + P^b)^{t_b})^{\tau-}$ must also be a martingale. This is satisfied if each P^b is not only the value, but also the shareholder value process of $\mathcal{P}^b$ (so that $(\mathcal{P}^b + P^b)^{\tau-}$ is a martingale), which we assume hereafter. See Remarks II.2 page 55 and II.6 page 57 for a discussion of the ensuing setup.
Remark 4. Since $t_b = \tau^\delta \wedge \tau_b^\delta \leq \tau^\delta$, (14) and (15) imply that $\mathcal{M}$ and MtM are respectively stopped and killed at τ^δ.

B Collateral and Default Cash Flows

We recall from II.§3.C that the processes $\mathcal{R}$ and $\mathcal{S} \geq 0$ denote the overall amounts of rehypothecable collateral and segregated collateral posted by the bank Treasury.

Lemma 2. *In a bilateral trading setup, we have*

$$\mathcal{R} = (\mathrm{MtM} - \sum_b \mathrm{VM}^b),\ \mathcal{S} = \sum_b \mathrm{PIM}^b \tag{17}$$

and the trading cash flows $\mathcal{C}$ *from the* CVA *desk (other than their dynamic hedging cash flows) are*

$$\begin{aligned} \mathcal{C} = & \sum_{b;\tau_b \leq \tau^\delta} (1 - R^b_{t_b})\Big(P^b_{t_b} + \mathcal{P}^b_{t_b} - \mathcal{P}^b_{s_b-} - \Gamma^b_{s_b-}\Big)^+ \mathbb{1}_{[\![t_b, \infty [\![} \\ & - (1 - R_{t_b}) \sum_{b;\tau \leq \tau_b^\delta} \Big(P^b_{t_b} + \mathcal{P}^b_{t_b} - \mathcal{P}^b_{s_b-} + \Upsilon^b_{s_b-}\Big)^- \mathbb{1}_{[\![t_b, \infty [\![} \\ & + \sum_b \mathbb{1}_{[\![s_b, t_b [\![}\Big(\mathcal{P}^b + P^b - (\mathcal{P}^b + P^b)_{s_b-}\Big). \end{aligned} \tag{18}$$

Proof. In a bilateral trading setup, the rehypothecable collateral posted by the bank Treasury is the difference between the amount MtM on the trading account and the variation margin $\sum_b \mathrm{VM}^b$ posted to the bank by its counterparties. This yields the formula for $\mathcal{R}$ in (17), where the one for $\mathcal{S}$ corresponds to the initial margin posted by the Treasury of the bank.

By Assumption 1, during the liquidation period of the bilateral netting set b, the CVA desk delivers

$$\int_{s_b-}^{\cdot} \mathbb{1}_{t\le t_b}(d\mathcal{P}^b_t + dP^b_t) + P^b_{s_b-}\mathbb{1}_{[\![t_b,+\infty[\![} = \\ \mathbb{1}_{[\![s_b,+\infty[\![}((\mathcal{P}^b)^{t_b} - \mathcal{P}^b_{s_b-}) + \mathbb{1}_{[\![s_b,t_b[\![}(P^b - P^b_{s_b-}) + \mathbb{1}_{[\![t_b,+\infty[\![}P^b_{t_b}. \tag{19}$$

Moreover, by application of Principle 1 to the liquidation of the bilateral netting set b, the CVA desk receives the following amount from the counterparty at t_b, where $x_b = \mathcal{P}^b_{t_b} - \mathcal{P}^b_{s_b-} + P^b_{t_b}$:

$$\begin{aligned} &\mathbb{1}_{\{\tau_b\le t_b, x_b>\Gamma^b_{s_b-}\}}\Big(\Gamma^b_{s_b-} + R^b_{t_b}(x_b - \Gamma^b_{s_b-})\Big) \\ &\quad - \mathbb{1}_{\{\tau\le t_b,(-x_b)>\Upsilon^b_{s_b-}\}}\Big(\Upsilon^b_{s_b-} + R_{t_b}(-x_b - \Upsilon^b_{s_b-})\Big) \\ &\quad + \mathbb{1}_{\neg\Big\{\Big(\tau_b\le t_b, x_b>\Gamma^b_{s_b-}\Big)\vee\Big(\tau\le t_b,(-x_b)>\Upsilon^b_{s_b-}\Big)\Big\}} x_b \\ &= x_b - \mathbb{1}_{\{\tau_b\le t_b\}}(1-R^b_{t_b})(x_b - \Gamma^b_{s_b-})^+ + \mathbb{1}_{\{\tau\le t_b\}}(1-R_{t_b})(x_b + \Upsilon^b_{s_b-})^-. \end{aligned} \tag{20}$$

Subtracting $(\mathbb{1}_{[\![t_b,+\infty[\![}\times(20))$, where $x_b = \mathcal{P}^b_{t_b} - \mathcal{P}^b_{s_b-} + P^b_{t_b}$, from (19) yields the overall loss process of the CVA desk as

$$\begin{aligned} &\mathbb{1}_{\{\tau_b\le t_b\}}(1-R^b_{t_b})(x_b - \Gamma^b_{s_b-})^+\mathbb{1}_{[\![t_b,\infty[\![} - \mathbb{1}_{\{\tau\le t_b\}}(1-R_{t_b})(x_b + \Upsilon^b_{s_b-})^-\mathbb{1}_{[\![t_b,\infty[\![} \\ &\quad - \mathbb{1}_{[\![t_b,+\infty[\![}(\mathcal{P}^b_{t_b} - \mathcal{P}^b_{s_b-} + P^b_{t_b}) \\ &\quad + \mathbb{1}_{[\![s_b,+\infty[\![}((\mathcal{P}^b)^{t_b} - \mathcal{P}^b_{s_b-}) + \mathbb{1}_{[\![s_b,t_b[\![}(P^b - P^c_{s_b-}) + \mathbb{1}_{[\![t_b,+\infty[\![}P^b_{t_b} \\ &= (1-R^b_{t_b})(x_b - \Gamma^b_{s_b-})^+\mathbb{1}_{[\![t_b,\infty[\![} - \mathbb{1}_{\{\tau\le t_b\}}(1-R_{t_b})(x_b + \Upsilon^b_{s_b-})^-\mathbb{1}_{[\![t_b,\infty[\![} \\ &\quad + \mathbb{1}_{[\![s_b,t_b[\![}\Big(\mathcal{P}^b + P^b - (\mathcal{P}^b + P^b)_{s_b-}\Big), \end{aligned}$$

where, recalling $t_b = \tau^\delta_b \wedge \tau^\delta$,

$$\begin{aligned} &\{\tau_b \le t_b\} = \{\tau_b \le \tau^\delta_b\} \cap \{\tau_b \le \tau^\delta\} = \{\tau_b \le \tau^\delta\} \text{ and} \\ &\{\tau \le t_b\} = \{\tau \le \tau^\delta_b\} \cap \{\tau \le \tau^\delta\} = \{\tau \le \tau^\delta_b\}. \end{aligned} \tag{21}$$

By summation over the netting sets b, we obtain the formula (18) for $\mathcal{C}$. □

Remark 5. Since $\mathrm{RIM}^b, \mathrm{PIM}^b \ge 0$ in $\Gamma^b = \mathrm{VM}^b + \mathrm{RIM}^b$ and $\Upsilon^b = (-\mathrm{VM}^b) + \mathrm{PIM}^b$ (cf. Table 5), we have in (20):

$$\Big\{x_b > \Gamma^b_{t_b-}\Big\} \cap \Big\{(-x_b) > \Upsilon^b_{t_b-}\Big\} = \\ \Big\{x_b - \mathrm{VM}^b > \mathrm{RIM}^b_{t_b-}\Big\} \cap \Big\{(x_b - \mathrm{VM}^b) < -\mathrm{PIM}^b_{t_b-}\Big\} = \emptyset,$$

as announced in the footnote to Principle 1.

§4 CENTRALLY CLEARED TRADING CASH FLOWS

In this section we consider a financial network of the bank "orthogonal" to the one of §3, in the sense that the bank does all its trading with clients in the form of netting sets cleared with a central counterparty (CCP). In the case of a client netting set c of contracts cleared via the bank by a CCP (see §2.A and Chapters VIII-IX), there is a mirroring set of identical deals between the bank and the CCP. The trading margin and variation margin calls related to the mirroring trades are mirroring the ones on the originating client trade. The bank does not post any initial margin to the related client. In case the client of the netting set c defaults, not only the netting set c between the client and the bank is liquidated, but the mirroring trades between the bank and the CCP are also unwound.

We thus introduce a (single) CCP involving a finite number of clearing members, including the bank itself. The data relative to the different clearing members are indexed by a set of integers i, including $i = 0$ for the bank itself (and disjoint from any other index set in this chapter). We still use the notation of Table 5 relatively to the individual client netting sets (here, cleared netting sets) of the bank, which we index by $\iota \equiv c$ (for "cleared"), and we use the same notation, but with indices i instead of c, and from the point of view of the CCP (sign-wise), relatively to the portfolios of the clearing members toward the CCP. For instance, $\mathcal{P}^c$ denotes the cash flows contractually promised to the trading desks of the bank related to the cleared netting set c; $\mathcal{P}^i$ means the cash flows contractually promised by the trading desks of the clearing member i to the CCP, which, by the full collateralization mechanism provided by a counterparty-risk-free CCP as we will see (with unlimited unfunded default fund, see after Assumption 2 page 97), will coincide with the actual cash flows from the trading desks of the clearing member i to the CCP: see Figure 8, and see Figure 41 page 220 for a picture also encompassing bilateral trading.

Remark 6. In practice (see Figure 41 page 220), the trades of a clearing member with a CCP are partitioned between *(i)* so called proprietary trades, which are in effect hedges of its bilateral trades (as proprietary trading as such is forbidden for a dealer bank, which is not allowed to be globally directional in its trading) and *(ii)* mirroring trades of cleared client trades. Proprietary CCP trades and the corresponding bilateral trades can typically be rewired in the form of cleared client trades and CCP mirrorring trades, more efficient in XVA terms (see e.g. Albanese, Armenti and Crépey (2020, Figure 1) and note in particular that cleared nettings sets with clients do not attract any initial margin). Hence for simplicity we ignore proprietary CCP trades in this chapter; proprietary CCP trades will be reintroduced in the one-period model of Chapters VIII-IX.

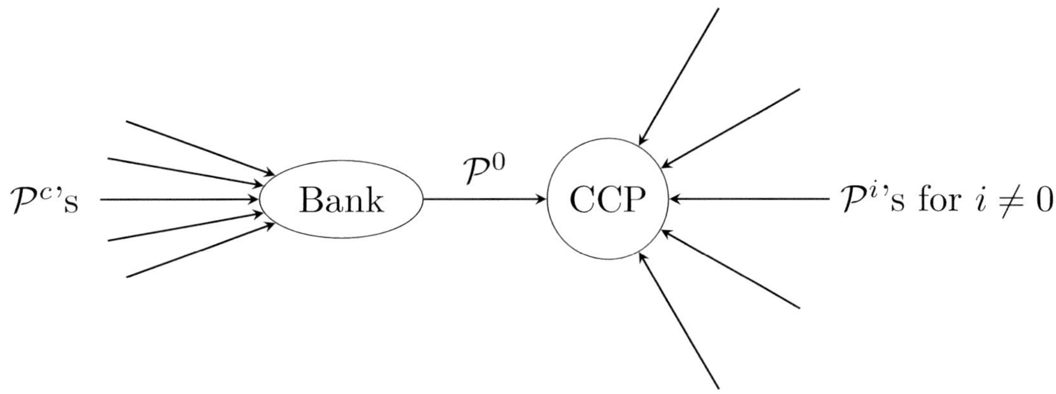

Figure 8: Cash flows $\mathcal{P}^c$ from the individual client (here cleared) netting sets c contractually promised to the trading desks of the bank, and cash flows $\mathcal{P}^i$ from the trading desks of the clearing member i to the CCP, in a trading network centrally cleared via a (single) CCP. The clearing members $i \neq 0$ (other than the bank) and their own clients (other than the CCP) are not shown to alleviate the picture.

All the deals between each clearing member i and the CCP are jointly collateralized in terms of variation margin VM^i and initial margin $\mathrm{PIM}^i \geq 0$ posted by the clearing member to the CCP, along with an additional layer of segregated collateral $\mathrm{DFC}^i \geq 0$, depending not only on the CCP portfolio of the clearing member i, but also on the CCP portfolios of the other members, dubbed (funded) default fund contribution of the clearing member i. We refer the reader to VIII.§5 for concrete specifications, not needed in this chapter, regarding the initial margin and default fund contributions posted by a clearing member of a CCP. All the collateral processes VM^i, PIM^i, DFC^i and VM^c are assumed to be $\mathfrak{G}$ optional. As the bank does not post any initial margin to its client in a cleared netting set c, we have

$$\mathrm{PIM}^c = 0 \text{ (hence } \Upsilon^c = -\mathrm{VM}^c). \tag{22}$$

The following hypothesis completes Principle 1 page 86 regarding the liquidation of a netting set in the case where the counterparty of the bank is not an individual counterparty, but a finite collection thereof, pooled in the form of a CCP, so that a rule is required to specify the allocation of the liquidation losses between the (surviving) clearing members.

Assumption 2. In case of a loss triggered by the liquidation of the CCP portfolio of a defaulted clearing member beyond its collateral posted to the CCP, then the loss in excess of its collateral is born by the surviving clearing members, proportionally to $\mathfrak{G}$ optional allocation weights

$$\mu^i \geq 0 \text{ such that } \mu^i = 0 \text{ on } [\![\tau_i, \infty[\![\text{ and } \textstyle\sum_i \mu^i = 1. \tag{23}$$

The ensuing losses for the clearing members are known as their unfunded default fund contributions (Ghamami, 2015).

Remark 7. The service closure, i.e. the closure of the activity of the clearing house on a given market or service, is usually specified in terms of events such as the unfunded default fund contribution of some (alive) member reaching a cap given as, e.g., twice the funded default fund. Given the high levels of initial margins that are used in practice, this is a very extreme tail event. Moreover, in case of service closure, the risk of a member is bounded above by the sum between its margins, three times its default fund contribution (assuming the above specification of service closure) and the cost of the liquidation of the service for this member. This cost is itself bounded by the notion of the member position, which would only be the actual cost in a scenario where all the assets of the CCP would jump to zero, also a very unlikely situation. In conclusion, the service closure event does not really matter regarding our present purpose of the XVA cost analysis of CCP membership. The default of the CCP as a whole (i.e. the closure of all its services) is an even more unlikely event, especially because a central bank would hardly allow it to occur in view of its systemic consequences. Hence, we may and do ignore the service closure and the default of the CCP in this chapter. So we assume the CCP is default-free, i.e. the existence of a default-free clearing member (other than the reference bank), referring the interested reader to Armakola and Laurent (2015) about CCP resilience and Duffie (2014) about alternative approaches to the design of insolvency and failure resolution regimes for CCPs.

Likewise, we ignore the skin-in-the-game of the CCP. This skin-in-the-game is the analog of a funded default fund contribution that would be posted by the CCP considered as an entity of its own, whereas, for simplicity in our setup, the CCP is nothing more than the collection of its clearing members. But the skin in the game is typically small and, as already demonstrated numerically Section 8.1 in Armenti and Crépey (2017, arXiv v1 version), negligible from a loss-absorbing, hence XVA, point of view.

A Back-to-back Hedging Cash Flows

Regarding the reference bank, similar to (14)–(15), but for cleared netting sets c here instead of bilateral netting sets b there:

Assumption 3. The cash flows $\mathcal{P}^0$ from the trading desks of the bank toward the CCP are given by

$$\mathcal{P}^0 = \sum_c \Big((\mathcal{P}^c)^{t_c} + \mathbb{1}_{[\![s_c, t_c [\![}(P^c - P^c_{s_c-}) + \mathbb{1}_{[\![t_c, +\infty [\![} P^c_{t_c} \Big). \tag{24}$$

As we will see, through the full collateralization mechanism provided by the CCP, these are both the promised and actual cash flows. Regarding now the globality of the cash flows to the trading desks:

Lemma 3. *In a trading setup centrally cleared via a unique CCP, we have*

$$\mathcal{M} = \text{MtM} = 0. \tag{25}$$

Proof. By the analysis having led to (13) here applied to the cleared netting sets c, the trading desks of the reference bank (clearing member indexed by 0) get the right-hand side in (24) on client deals. By (24), they deliver the same amount on CCP deals. Hence $\mathcal{M} = 0$. Besides, the trading margin calls on cleared trades of the bank being mirrored by identical trading margin calls on the mirroring trades between the bank and the CCP, we have $\text{MtM} = 0$. □

In this sense, in a centrally cleared trading setup, the deals of the bank with the CCP provide a fully collateralized back-to-back hedge to the deals with the clients. The bank is then statically hedged in terms of market risk, and there is no need for dynamic hedging (hence $\mathcal{H}^{mtm} = 0$, at least in the present setup of this chapter where we ignore model risk, which will be introduced in Chapter XI). But, contrary to a dynamic hedge with futures-style instruments, such a static hedge has XVA implications, much like in the one-period setup of I.§3, which are analysed hereafter.

In view of Lemma 1 and Remark 4 page 94, (24) implies that the value process P^0 of $\mathcal{P}^0$ is worth

$$P^0 = \sum_c (\mathbb{1}_{[\![0,s_c[\![} P^c + \mathbb{1}_{[\![s_c,t_c[\![} P^c_{s_c-}) \text{ (in particular, } P^0_0 = \textstyle\sum_c P^c_0), \tag{26}$$

and that

$$\mathcal{P}^0 + P^0 = \sum_c (\mathcal{P}^c + P^c)^{t_c}, \tag{27}$$

$$\mathcal{P}^0 \text{ is stopped at } \tau^\delta,\ P^0 \text{ is killed at } \tau^\delta\ . \tag{28}$$

We postulate a structure of the cash flows $\mathcal{P}^i$ of each clearing member i to the CCP analogous to (24) with respect to their own clients (not shown in Figure 8 to alleviate the picture) so that, in particular, P^i is killed at τ_i^δ.

B Collateral and Default Cash Flows

For any subset I of the clearing members, let τ_I^δ denote $T\wedge$ the (assumed $\mathfrak{G}$ stopping) time of occurrence of an instantaneous joint default of the clearing members I and only in I, let τ_I^δ denote $T\wedge$ the end of the corresponding liquidation period, and let the breach B^I denote the process

$$\begin{aligned} B^I &= \mathbb{1}_{[\![\tau_I^\delta,\infty[\![} \sum_{i\in I} (1 - R^i_{\tau_I^\delta})(P^i_{\tau_I^\delta} + \mathcal{P}^i_{\tau_I^\delta} - \mathcal{P}^i_{\tau_I-} - \text{VM}^i_{\tau_I-} - \text{PIM}^i_{\tau_I-} - \text{DFC}^i_{\tau_I-})^+ \\ &= \mathbb{1}_{[\![\tau_I^\delta,\infty[\![} \sum_{i\in I} (1 - R^i_{\tau_I^\delta})(\mathcal{P}^i_{\tau_I^\delta} - \mathcal{P}^i_{\tau_I-} - \text{VM}^i_{\tau_I-} - \text{PIM}^i_{\tau_I-} - \text{DFC}^i_{\tau_I-})^+ \end{aligned} \tag{29}$$

(as the $P^i_{\tau^\delta_I}$ for $i \in I$ vanish, cf. the last sentence of Subsection A). Hence

$$(B^I)^{\tau-} = \mathbb{1}_{\{\tau^\delta_I < \tau\}} B^I, \tag{30}$$

by definition II.(3).

Remark 8. The process $\sum_I B^I$ corresponds to the (cumulative) unfunded default fund of the CCP (see after Assumption 2).

Lemma 4. *In a trading setup centrally cleared via a unique CCP, we have*

$$\mathcal{R} = 0,\ \mathcal{S} = \mathrm{PIM}^0 + \mathrm{DFC}^0, \tag{31}$$

and the trading cash flows $\mathcal{C}$ from the CVA *desk (other than their dynamic hedging cash flows) are*

$$\begin{aligned}\mathcal{C} = &\sum_{c;\tau_c \le \tau^\delta} (1 - R^c_{t_c})\Big(P^c_{t_c} + \mathcal{P}^c_{t_c} - \mathcal{P}^c_{s_c-} - \Gamma^c_{s_c-}\Big)^+ \mathbb{1}_{[\![t_c,\infty[\![}\\ &\quad - \sum_{c;\tau \le \tau^\delta_c} (1 - R_{t_c})\Big(P^c_{t_c} + \mathcal{P}^c_{t_c} - \mathcal{P}^c_{s_c-} + \Upsilon^c_{s_c-}\Big)^- \mathbb{1}_{[\![t_c,\infty[\![}\\ &\qquad + \sum_c \mathbb{1}_{[\![s_c,t_c[\![}\Big(\mathcal{P}^c + P^c - (\mathcal{P}^c + P^c)_{s_c-}\Big)\\ &+ \sum_I \mu^0_{\tau^\delta_I} B^I - (1 - R^0_{\tau^\delta})(\mathcal{P}^0_{\tau^\delta} - \mathcal{P}^0_{\tau-} - \mathrm{VM}^0_{\tau-} - \mathrm{PIM}^0_{\tau-} - \mathrm{DFC}^0_{\tau-})^+ \mathbb{1}_{[\![\tau^\delta,\infty[\![}\\ &\qquad + \mathbb{1}_{[\![\tau,\tau^\delta[\![}(\mathcal{P}^0 - \mathcal{P}^0_{\tau-}).\end{aligned} \tag{32}$$

Proof. The variation margin calls on cleared trades of the bank are mirrored by identical variation margin calls on the mirroring trades between the bank and the CCP. Hence the cleared trades and their mirroring CCP trades have in aggregate no rehypothecable collateral implications for the bank Treasury, which, however, has the segregated collateral borrowing needs corresponding to the initial margins and funded default fund contribution to be posted by the bank (recalling from (22) that the bank does not post any initial margin to its clients). This yields (31).

The cash flows from the CVA desk are, on the one hand, the cash flows from the CVA desk related to the individual client (here cleared) netting sets c, i.e., by the same computation than the one having led to (18) (but for the netting set $\iota \equiv c$, here),

$$\begin{aligned}&\sum_{c;\tau_c \le \tau^\delta} (1 - R^c_{t_c})\Big(P^c_{t_c} + \mathcal{P}^c_{t_c} - \mathcal{P}^c_{s_c-} - \Gamma^c_{s_c-}\Big)^+ \mathbb{1}_{[\![t_c,\infty[\![}\\ &\quad - (1 - R_{t_c}) \sum_{c;\tau \le \tau^\delta_c} \Big(P^c_{t_c} + \mathcal{P}^c_{t_c} - \mathcal{P}^c_{s_c-} + \Upsilon^c_{s_c-}\Big)^- \mathbb{1}_{[\![t_c,\infty[\![}\\ &\qquad + \sum_c \mathbb{1}_{[\![s_c,t_c[\![}\Big(\mathcal{P}^c + P^c - (\mathcal{P}^c + P^c)_{s_c-}\Big).\end{aligned} \tag{33}$$

On the other hand, denoting by $\mathcal{U}^0$ the cumulative process of the unfunded default fund contributions of the bank (see after Assumption 2) and writing $x_0 := P^0_{\tau^\delta} + \mathcal{P}^0_{\tau^\delta} - \mathcal{P}^0_{\tau-} = \mathcal{P}^0_{\tau^\delta} - \mathcal{P}^0_{\tau-}$ (as $P^0_{\tau^\delta} = 0$), going by Principle 1 detailed as per Assumption 2 regarding the liquidation of clearing members, the cash flows needed from the CVA desk to restore the integrity of the promised cash flows of the trading desks are given by[3]

$$\begin{aligned}
&-\int_0^{\cdot} \mathbb{1}_{\tau\le t\le\tau^\delta} d\mathcal{P}^0_t + \mathcal{U}^0 + \mathbb{1}_{[\![\tau^\delta,\infty[\![}\times \\
&\Big(\mathbb{1}_{\{x_0\le \mathrm{VM}^0_{\tau-}+\mathrm{PIM}^0_{\tau-}+\mathrm{DFC}^0_{\tau-}\}}x_0 + \mathbb{1}_{\{x_0> \mathrm{VM}^0_{\tau-}+\mathrm{PIM}^0_{\tau-}+\mathrm{DFC}^0_{\tau-}\}} \\
&\qquad\Big(\mathrm{VM}^0_{\tau-} + \mathrm{PIM}^0_{\tau-} + \mathrm{DFC}^0_{\tau-} + R^0_{\tau^\delta}(x_0 - \mathrm{VM}^0_{\tau-} - \mathrm{PIM}^0_{\tau-} - \mathrm{DFC}^0_{\tau-})\Big)\Big) \\
&\quad = -\int_0^{\cdot} \mathbb{1}_{\tau\le t\le\tau^\delta} d\mathcal{P}^0_t + \mathcal{U}^0 + x_0\mathbb{1}_{[\![\tau^\delta,\infty[\![} \\
&\qquad - (1-R^0_{\tau^\delta})\Big(x_0 - \mathrm{VM}^0_{\tau-} - \mathrm{PIM}^0_{\tau-} - \mathrm{DFC}^0_{\tau-}\Big)^+ \mathbb{1}_{[\![\tau^\delta,\infty[\![},
\end{aligned} \tag{34}$$

where, as $x_0 = \mathcal{P}^0_{\tau^\delta} - \mathcal{P}^0_{\tau-}$,

$$\begin{aligned}
x_0\mathbb{1}_{[\![\tau^\delta,\infty[\![} - \int_0^{\cdot} \mathbb{1}_{\tau\le t\le\tau^\delta} d\mathcal{P}^0_t &= (\mathcal{P}^0_{\tau^\delta} - \mathcal{P}^0_{\tau-})\mathbb{1}_{[\![\tau^\delta,\infty[\![} - \int_0^{\cdot} \mathbb{1}_{\tau\le t\le\tau^\delta} d\mathcal{P}^0_t \\
&= \mathbb{1}_{[\![\tau,\tau^\delta[\![}(\mathcal{P}^0 - \mathcal{P}^0_{\tau-}).
\end{aligned} \tag{35}$$

Moreover, in view of the last line of (34) and by the symmetric cash flow analysis that applies to each clearing member of the CCP (cf. the last sentence of §4.A), at each time τ^δ_I, where I ranges over all the possible subsets of the clearing members, the loss to be taken over by the surviving clearing members through their unfunded default fund contributions is $B^I_{\tau^\delta_I}$ (cf. (29)). Given the allocation rule of Assumption 2 for these losses, we thus have in particular:

$$\mathcal{U}^0 = \sum_I \mu^0_{\tau^\delta_I} B^I. \tag{36}$$

Accounting for (35) and (36) into (34) and adding (33) yields (32). □

§5 SYNTHESIS

We now consider the realistic situation of a bank involved in an arbitrary combination of bilateral netting sets (as in §3) and of cleared netting sets and their mirroring CCP trades (as in §4), with possibly multiple CCPs (as

[3] like in the proof of (18), but simplified here by the fact that one of the counterparties in the netting set, namely the CCP, is default-free; the term $(-\int_0^{\cdot} \mathbb{1}_{\tau\le t\le\tau^\delta} d\mathcal{P}^0_t)$ is the analog of (19).

in Figure 41 page 220 but ignoring in this chapter proprietary trading, see Remark 6 page 96). The index ι now ranges over all the individual counterparty netting sets of the bank—individual as opposed to the netting sets between the bank and the CCPs: so an individual counterparty netting set ι below can be either a bilateral counterparty netting set b as in §3 or a cleared client netting set c as in §4. The bank can be a clearing member of several CCPs and we index all the CCP related quantities in §4 by an additional index ccp, ranging in an index set disjoint from any other index set in this chapter. For instance, for a given CCP ccp, the second index I in "ccp, I" ranges over all the possible subsets of clearing members of the CCP ccp; $\sum_{ccp,I}$ is a double sum over all CCPs and, for each of them, over all the corresponding subsets I.

A Cash Flows

Lemma 5. *Assuming the reduction of filtration setup of II.§3.A:*
(i) *For any $\mathfrak{G}$ optional process Y, if $Y^{\tau-}$ is nonnegative, then Y' is nonnegative.*
(ii) *For any $[0,T]$ valued $\mathfrak{G}$ stopping times θ and η such that $\theta \wedge \tau \leq \eta \wedge \tau$, we have $\theta' \leq \eta'$;*
(iii) *If, additionally, $\eta \leq \theta + c$ holds for some constant $c \geq 0$, then $\eta' \leq \theta' + c$.*

Proof. We recall from Dellacherie and Meyer (1980, Chapitre VI n° 17) that the set $\{\mathsf{S} > 0\}$ (where $\mathsf{S} = \mathbb{Q}^*(\tau > \cdot \,|\, \mathfrak{F}_\cdot))$ is a random interval starting from 0 included, which, in the reduction of filtration setup of II.§3.A, is assumed to contain T.
(i) If $Y^{\tau-}$ is nonnegative, then, for $t \leq T$,

$$Y'_t \mathbb{1}_{\{t<\tau\}} = Y^{\tau-}_t \mathbb{1}_{\{t<\tau\}} \geq 0,$$

hence $\mathbb{E}^*(Y'_t \mathbb{1}_{\{t<\tau\}} | \mathfrak{F}_t) \geq 0$, i.e. $Y'_t \mathsf{S}_t \geq 0$, where $\mathsf{S}_t > 0$.
(ii) If $\theta \wedge \tau \leq \eta \wedge \tau$, then $\mathbb{1}_{[\![\theta\wedge\tau,+\infty[\![} - \mathbb{1}_{[\![\eta\wedge\tau,+\infty[\![} \geq 0$, hence part (i) implies that $\mathbb{1}_{[\![\theta',+\infty[\![} - \mathbb{1}_{[\![\eta',+\infty[\![} \geq 0$, i.e. $\theta' \leq \eta'$.
(iii) If, additionally, $\eta \leq \theta + c$ holds for some constant $c \geq 0$, then part (i) implies that $\theta' \leq \eta'$ and we have

$$\mathbb{1}_{[\![0,\tau[\![}(\eta' - \theta')\mathbb{1}_{[\![\eta',+\infty[\![} = \mathbb{1}_{[\![0,\tau[\![}(\eta - \theta)\mathbb{1}_{[\![\eta,+\infty[\![} \leq c\mathbb{1}_{[\![0,\tau[\![}\mathbb{1}_{[\![\eta,+\infty[\![}, \tag{37}$$

where the left equality means that, for $t \geq 0$,

$$\mathbb{1}_{\{\eta' \leq t < \tau\}}(\eta' - \theta') = \mathbb{1}_{\{\eta \leq t < \tau\}}(\eta - \theta'). \tag{38}$$

To justify (38), note that the indicators in (38) must coincide; otherwise, this would imply that either $\eta' \le t < \tau$ holds but not $\eta \le t < \tau$, in which case $\eta > t$ and $\eta' \wedge \tau \le t < \eta \wedge \tau$, or that $\eta \le t < \tau$ holds but not $\eta' \le t < \tau$, in which case $\eta' > t$ and $\eta \wedge \tau \le t < \eta' \wedge \tau$, violating in both cases the identity $\eta \wedge \tau = \eta' \wedge \tau$ (cf. II.(24)). Moreover, as we already know from part (ii) that $\theta' \le \eta'$ and given the assumption $\theta \wedge \tau \le \eta \wedge \tau$, if the indicators in (38) are both equal to one, then the identities $\eta \wedge \tau = \eta' \wedge \tau$ and $\theta \wedge \tau = \theta' \wedge \tau$ imply that $\theta = \theta'$ and $\eta = \eta'$, so that $\eta' - \theta' = \eta - \theta'$, and the identity (38) holds (which is also obviously the case if the indicators in (38) are both equal to zero).

By valuation at time η' of the left and right processes in (37) and by taking the $(\mathfrak{F}_{\eta'}, \mathbb{Q}^*)$ conditional expectation of the resulting expressions, we obtain

$$\mathbb{E}^*((\eta' - \theta')\mathbb{1}_{\{\eta'<\tau\}}|\mathfrak{F}_{\eta'}) \le c\mathbb{E}^*(\mathbb{1}_{\{\eta'<\tau\}}\mathbb{1}_{\{\eta\le\eta'\}}|\mathfrak{F}_{\eta'}) \le c\mathbb{E}^*(\mathbb{1}_{\{\eta'<\tau\}}|\mathfrak{F}_{\eta'}),$$

i.e. $(\eta' - \theta')\mathsf{S}_{\eta'} \le c\mathsf{S}_{\eta'}$, where $\mathsf{S}_{\eta'} > 0$ a.s.. □

Lemma 6. *For each individual counterparty netting set ι, we have*

$$t_\iota - s_\iota \le \delta \tag{39}$$

and, assuming the reduction of filtration setup of II.§3.A,

$$0 \le \tau_\iota' \le (\tau_\iota^\delta)' \le T,\ (\tau_\iota^\delta)' - \tau_\iota' \le \delta. \tag{40}$$

Proof. We compute

$$\begin{aligned} t_\iota - s_\iota &= \tau_\iota^\delta \wedge \tau^\delta - s_\iota = (\tau_\iota^\delta - s_\iota) \wedge (\tau^\delta - s_\iota) = (\tau_\iota^\delta - \tau_\iota \wedge \tau) \wedge (\tau^\delta - \tau_\iota \wedge \tau) \\ &= \Big((\tau_\iota^\delta - \tau_\iota) \vee (\tau_\iota^\delta - \tau)\Big) \wedge \Big((\tau^\delta - \tau_\iota) \vee (\tau^\delta - \tau)\Big) \\ &\le \Big(\delta \vee (\tau_\iota^\delta - \tau)\Big) \wedge \Big(\delta \vee (\tau^\delta - \tau_\iota)\Big), \end{aligned}$$

by (12), where, by (12) again: If $\tau \vee \tau_\iota = \tau$, then $\tau_\iota^\delta - \tau \le \tau_\iota^\delta - \tau_\iota \le \delta$, otherwise $\tau \vee \tau_\iota = \tau_\iota$ and $\tau^\delta - \tau_\iota \le \tau^\delta - \tau \le \delta$. This yieds (39), from which (40) follows, in the reduction of filtration setup of II.§3.A, by application of Lemma 5. □

Writing $M^\iota = \mathcal{P}^\iota + P^\iota$, we denote

$$\begin{aligned} \mathcal{Y}^{ccp} &= \mathbb{1}_{[\![\tau,\tau^\delta[\![}(\mathcal{P}^{ccp,0} - \mathcal{P}^{ccp,0}_{\tau-}) \\ \mathcal{Y}^\iota &= \mathbb{1}_{[\![s_\iota,t_\iota[\![}\Big(M^\iota - M^\iota_{s_\iota-}\Big) \end{aligned} \tag{41}$$

and, in the reduction of filtration setup of II.§3.A,

$$\mathcal{X}^{\iota} = \mathbb{1}_{[\![\tau'_{\iota},(\tau^{\delta}_{\iota})'[\![}\Big((M^{\iota})' - (M^{\iota})'_{\tau'_{\iota}-}\Big). \tag{42}$$

As we will see these are corrective liquidation period CVA and DVA related cash flows. Note that they all vanish identically in the instantaneous liquidation case where $\delta = 0$ (so that the τ^{δ}_{ι} = the τ_{ι} and $\tau^{\delta} = \tau$, by (12), and the $(\tau^{\delta}_{\iota})'$ = the τ'_{ι}, by (40)).

Proposition 2. *The cash flows of a bank involved in an arbitrary combination of bilateral netting sets and of netting sets cleared with one among several available CCPs satisfy*

$$\mathcal{M} = \sum_b \Big((\mathcal{P}^b)^{t_b} + \mathbb{1}_{[\![s_b,t_b[\![}(P^b - P^b_{s_b-}) + \mathbb{1}_{[\![t_b,+\infty[\![}P^b_{t_b}\Big), \text{ hence} \tag{43}$$

$$\mathrm{MtM} = \sum_b (\mathbb{1}_{[\![0,s_b[\![}P^b + \mathbb{1}_{[\![s_b,t_b[\![}P^b_{s_b-}), \tag{44}$$

$$\mathcal{M} + \mathrm{MtM} = \sum_b (\mathcal{P}^b + P^b)^{t_b}, \tag{45}$$

$$\mathcal{R} = \mathrm{MtM} - \sum_b \mathrm{VM}^b,\ \mathcal{S} = \sum_b \mathrm{PIM}^b + \sum_{ccp}(\mathrm{PIM}^{ccp,0} + \mathrm{DFC}^{ccp,0}), \tag{46}$$

$$\begin{aligned}
\mathcal{C} - \sum_{\iota}\mathcal{Y}^{\iota} - \sum_{ccp}\mathcal{Y}^{ccp} = &\sum_{\iota;\tau_{\iota}\le\tau^{\delta}} (1 - R^{\iota}_{t_{\iota}})\Big(P^{\iota}_{t_{\iota}} + \mathcal{P}^{\iota}_{t_{\iota}} - \mathcal{P}^{\iota}_{s_{\iota}-} - \Gamma^{\iota}_{s_{\iota}-}\Big)^{+}\mathbb{1}_{[\![t_{\iota},\infty[\![}\\
&- \sum_{\iota;\tau\le\tau^{\delta}_{\iota}} (1 - R_{t_{\iota}})\Big(P^{\iota}_{t_{\iota}} + \mathcal{P}^{\iota}_{t_{\iota}} - \mathcal{P}^{\iota}_{s_{\iota}-} + \Upsilon^{\iota}_{s_{\iota}-}\Big)^{-}\mathbb{1}_{[\![t_{\iota},\infty[\![}\\
&+ \sum_{ccp,I} \mu^{ccp,0}_{\tau^{\delta}_I} B^{ccp,I} \\
&- \sum_{ccp}(1 - R^{ccp,0}_{\tau^{\delta}})(\mathcal{P}^{ccp,0}_{\tau^{\delta}} - \mathcal{P}^{ccp,0}_{\tau-} - \mathrm{VM}^{ccp,0}_{\tau-} - \mathrm{PIM}^{ccp,0}_{\tau-} - \mathrm{DFC}^{ccp,0}_{\tau-})^{+}\mathbb{1}_{[\![\tau^{\delta},\infty[\![},
\end{aligned} \tag{47}$$

$$\begin{aligned}
\mathcal{F} = &\int_0^{\cdot} \lambda_t\Big(\mathrm{MtM} - \sum_b \mathrm{VM}^b - \mathrm{CA}\Big)^{+}_t dt\\
&-(1 - R_{\tau})\Big(\mathrm{MtM} - \sum_b \mathrm{VM}^b - \mathrm{CA}\Big)^{+}_{\tau-}\mathbb{1}_{[\![\tau,\infty[\![}\\
&+ \int_0^{\cdot} \tilde{\lambda}_t\Big(\sum_b \mathrm{PIM}^b + \sum_{ccp}(\mathrm{PIM}^{ccp,0} + \mathrm{DFC}^{ccp,0})\Big)_t dt\\
&-(1 - \bar{R}_{\tau})\Big(\sum_b \mathrm{PIM}^b + \sum_{ccp}(\mathrm{PIM}^{ccp,0} + \mathrm{DFC}^{ccp,0})\Big)_{\tau-}\mathbb{1}_{[\![\tau,\infty[\![},
\end{aligned} \tag{48}$$

i.e.[4]

$$
\begin{aligned}
&\mathcal{C}^{\tau-} - \sum_\iota (\mathcal{Y}^\iota)^{\tau-} = \sum_{\iota;\tau_\iota^\delta<\tau} (1-R^\iota_{t_\iota})\Big(P^\iota_{t_\iota} + \mathcal{P}^\iota_{t_\iota} - \mathcal{P}^\iota_{s_\iota-} - \Gamma^\iota_{s_\iota-}\Big)^+ \mathbb{1}_{[\![t_\iota,\infty[\![}\\
&\quad + \sum_{ccp,I} \mu^{ccp,0}_{\tau_I^\delta} B^{ccp,I}\\
&{}^{\tau-}(-\mathcal{C}) + \sum_\iota {}^{\tau-}(\mathcal{Y}^\iota) + \sum_{ccp} \mathcal{Y}^{ccp} =\\
&\quad \sum_{\iota;\tau\le\tau_\iota^\delta} (1-R_{t_\iota})(P^\iota_{t_\iota} + \mathcal{P}^\iota_{t_\iota} - \mathcal{P}^\iota_{s_\iota-} + \Upsilon^\iota_{s_\iota-})^- \mathbb{1}_{[\![t_\iota,\infty[\![}\\
&\qquad - \sum_{\iota;\tau_\iota\le\tau^\delta,\tau_\iota^\delta\ge\tau} (1-R^\iota_{t_\iota})\Big(P^\iota_{t_\iota} + \mathcal{P}^\iota_{t_\iota} - \mathcal{P}^\iota_{s_\iota-} - \Gamma^\iota_{s_\iota-}\Big)^+ \mathbb{1}_{[\![t_\iota,\infty[\![}\\
&\quad + \sum_{ccp}(1-R^{ccp,0}_{\tau^\delta})(\mathcal{P}^{ccp,0}_{\tau^\delta} - \mathcal{P}^{ccp,0}_{\tau-} - \mathrm{VM}^{ccp,0}_{\tau-} - \mathrm{PIM}^{ccp,0}_{\tau-} - \mathrm{DFC}^{ccp,0}_{\tau-})^+ \mathbb{1}_{[\![\tau^\delta,\infty[\![}
\end{aligned}
\tag{49}
$$

and

$$
\begin{aligned}
\mathcal{F}^{\tau-} &= \int_0^\cdot \lambda_t(\mathrm{MtM} - \sum_b \mathrm{VM}^b - \mathrm{CA})^+_t dt +\\
&\quad \int_0^\cdot \widetilde{\lambda}_t\Big(\sum_b \mathrm{PIM}^b + \sum_{ccp}(\mathrm{PIM}^{ccp,0} + \mathrm{DFC}^{ccp,0})\Big)_t dt\\
{}^{\tau-}(-\mathcal{F}) &= (1-R_\tau)\Big(\mathrm{MtM} - \sum_b \mathrm{VM}^b - \mathrm{CA}\Big)^+_{\tau-} \mathbb{1}_{[\![\tau,\infty[\![}\\
&\quad + (1-\bar{R}_\tau)\Big(\sum_b \mathrm{PIM}^b + \sum_{ccp}(\mathrm{PIM}^{ccp,0} + \mathrm{DFC}^{ccp,0})\Big)_{\tau-} \mathbb{1}_{[\![\tau,\infty[\![}.
\end{aligned}
\tag{50}
$$

Assuming the reduction of filtration setup of II.§3.A, we have

$$
\mathcal{L}' = \mathcal{C}' + \mathcal{F}' + \mathrm{CA}' - \mathrm{CA}_0 - (\mathcal{M}' + \mathrm{MtM}' - \mathrm{MtM}_0) + \mathcal{H}', \tag{51}
$$

where

$$
\mathrm{MtM}' = \sum_b \Big((P^b)'\mathbb{1}_{[\![0,(\tau_b)'[\![} + \mathbb{1}_{[\![(\tau_b)',(\tau_b^\delta)'[\![}(P^b)'_{\tau_b'-}\Big), \tag{52}
$$

$$
\mathcal{M}' + \mathrm{MtM}' = \sum_b ((\mathcal{P}^b + P^b)')^{(\tau_b^\delta)'} \tag{53}
$$

$$
\mathcal{C}' - \sum_\iota \mathcal{X}^\iota = \sum_\iota (1 - (R^\iota)'_{(\tau_\iota^\delta)'}) \tag{54}
$$

$$
\Big((P^\iota)'_{(\tau_\iota^\delta)'} + (\mathcal{P}^\iota)'_{(\tau_\iota^\delta)'} - (\mathcal{P}^\iota)'_{\tau_\iota'-} - (\Gamma^\iota)'_{\tau_\iota'-}\Big)^+ \mathbb{1}_{[\![(\tau_\iota^\delta)',\infty[\![} \tag{55}
$$

$$
+ \sum_{ccp,I} (\mu^{ccp,0})'_{(\tau_I^\delta)'} (B^{ccp,I})', \tag{56}
$$

[4] recalling II.(3).

$$\mathcal{F}' = \int_0^{\cdot} \lambda'_t\Big(\mathrm{MtM}' - \sum_b (\mathrm{VM}^b)' - \mathrm{CA}'\Big)^+_t dt +$$
$$\int_0^{\cdot} \widetilde{\lambda}'_t\Big(\sum_b (\mathrm{PIM}^b)' + \sum_{ccp}\Big(\mathrm{PIM}^{ccp,0} + \mathrm{DFC}^{ccp,0}\Big)'\Big)_t dt, \qquad (57)$$

and

$$(B^{ccp,I})' = \sum_{i\in I}\Big(1 - (R^i)'_{(\tau_I^\delta)'}\Big) \qquad (58)$$
$$\Big(\mathcal{P}^{ccp,i}_{(\tau_I^\delta)'} - \mathcal{P}^{ccp,i}_{\tau_I'-} - \mathrm{VM}^{ccp,i}_{\tau_I'-} - \mathrm{PIM}^{ccp,i}_{\tau_I'-} - \mathrm{DFC}^{ccp,i}_{\tau_I'-}\Big)\mathbb{1}_{[\![(\tau_I^\delta)',\infty[\![}. \qquad (59)$$

Proof. We partition the portfolio of the reference bank into the collections of its cleared netting sets ccp, c and their netting set of mirroring deals $ccp, 0$, CCP by CCP, complemented by the collection of the bilateral netting sets b of the bank and their dynamic hedge.

We then apply the analysis of §4, CCP by CCP, and the analysis of §3 to the collection of the bilateral netting sets and their dynamic hedge. As detailed below, this results in respective cash flows $\mathcal{M}^{ccp} = \mathcal{R}^{ccp} = 0, \mathcal{S}^{ccp}, \mathcal{C}^{ccp}$ stemming from the cleared netting sets ccp, c and their back-to-back hedge $ccp, 0$, for each CCP ccp, and $\mathcal{M}^{bil}, \mathcal{R}^{bil}, \mathcal{S}^{bil}, \mathcal{C}^{bil}$ stemming from the collection bil of all the bilateral netting sets of the bank and their dynamic hedges, such that

$$\begin{aligned}
\mathcal{S}^{ccp} &= \mathrm{PIM}^{ccp,0} + \mathrm{DFC}^{ccp,0},\\
\mathcal{C}^{ccp} &= \sum_{c;\tau_c\le\tau^\delta}(1 - R^{ccp,c}_{t_c})\Big(P^{ccp,c}_{t_c} + \mathcal{P}^{ccp,c}_{t_c} - \mathcal{P}^{ccp,c}_{s_c-} - \Gamma^{ccp,c}_{s_c-}\Big)^+\mathbb{1}_{[\![t_c,\infty[\![}\\
&\quad - \sum_{c;\tau\le\tau_c^\delta}(1 - R_{t_c})\Big(P^{ccp,c}_{t_c} + \mathcal{P}^{ccp,c}_{t_c} - \mathcal{P}^{ccp,c}_{s_c-} + \Upsilon^{ccp,c}_{s_c-}\Big)^-\mathbb{1}_{[\![t_c,\infty[\![}\\
&\qquad + \sum_c \mathbb{1}_{[\![s_c,t_c[\![}\Big(\mathcal{P}^{ccp,c} + P^{ccp,c} - (\mathcal{P}^{ccp,c} + P^{ccp,c})_{s_c-}\Big)\\
&\quad + \sum_I \mu^{ccp,0}_{\tau_I^\delta} B^{ccp,I}\\
&\quad -(1 - R^{ccp,0}_{\tau^\delta})(\mathcal{P}^{ccp,0}_{\tau^\delta} - \mathcal{P}^{ccp,0}_{\tau-} - \mathrm{VM}^{ccp,0}_{\tau-} - \mathrm{PIM}^{ccp,0}_{\tau-} - \mathrm{DFC}^{ccp,0}_{\tau-})^+\mathbb{1}_{[\![\tau^\delta,\infty[\![}\\
&\qquad + \mathbb{1}_{[\![\tau,\tau^\delta[\![}(\mathcal{P}^{ccp,0} - \mathcal{P}^{ccp,0}_{\tau-}),\\
\mathcal{M}^{bil} &= \sum_b\Big((\mathcal{P}^b)^{t_b} + \mathbb{1}_{[\![s_b,t_b[\![}(P^b - P^b_{s_b-}) + \mathbb{1}_{[\![t_b,+\infty[\![}P^b_{t_b}\Big),\\
\mathrm{MtM}^{bil} &= \sum_b(\mathbb{1}_{[\![0,s_b[\![}P^b + \mathbb{1}_{[\![s_b,t_b[\![}P^b_{s_b-}),\\
\mathcal{R}^{bil} &= \mathrm{MtM}^{bil} - \sum_b \mathrm{VM}^b,\ \mathcal{S}^{bil} = \sum_b \mathrm{PIM}^b,
\end{aligned}$$

$$\mathcal{C}^{bil} = \sum_{b;\tau_b \le \tau^\delta} (1 - R^b_{t_b})\Big(P^b_{t_b} + \mathcal{P}^b_{t_b} - \mathcal{P}^b_{s_b-} - \Gamma^b_{s_b-}\Big)^+ \mathbb{1}_{[\![t_b,\infty[\![}$$
$$-(1 - R_{t_b}) \sum_{b;\tau \le \tau^\delta_b} \Big(P^b_{t_b} + \mathcal{P}^b_{t_b} - \mathcal{P}^b_{s_b-} + \Upsilon^b_{s_b-}\Big)^- \mathbb{1}_{[\![t_b,\infty[\![}$$
$$+\sum_b \mathbb{1}_{[\![s_b,t_b[\![}\Big(\mathcal{P}^b + P^b - (\mathcal{P}^b + P^b)_{s_b-}\Big).$$

For each CCP ccp, the formulas for $\mathcal{S}^{ccp}$ and $\mathcal{C}^{ccp}$ in the above follow by application of the respective formulas (31) and (32) to the collection of the cleared netting sets ccp, c and their netting set of mirroring deals $ccp, 0$. The formulas for $\mathcal{M}^{bil}$ and MtM^{bil} are obtained by application of Lemma 1 to the collection of the bilateral netting sets b and their dynamic hedge. The formulas for $\mathcal{R}^{bil}$, $\mathcal{S}^{bil}$ and $\mathcal{C}^{bil}$ follow by applications of the respective formulas (17) and (18) to the collection of the bilateral netting sets b and their hedge.

The formulas (43) through (47) then follow from the above ones by summation through the identity $\mathcal{Y} = (\sum_{ccp} \mathcal{Y}^{ccp}) + \mathcal{Y}^{bil}$ that applies for each cash flow $\mathcal{Y} = \mathcal{M}, \mathcal{R}, \mathcal{S}, \mathcal{C}$, reflecting the partition of the portfolio of the bank introduced in the beginning of this proof, noting that the set of all the individual client netting sets ι of the bank is itself partitioned into the subsets of the cleared netting sets ccp, c, for each CCP ccp, complemented by the subset of the bilateral netting sets b.

The formula (47) for $\mathcal{C}$ splits as (noting that $(\mathcal{Y}^{ccp})^{\tau-} = 0$)

$$(\mathcal{C} - \sum_\iota \mathcal{Y}^\iota)^{\tau-} = \sum_{\iota;\tau_\iota \le \tau^\delta, t_\iota < \tau} (1 - R^\iota_{t_\iota})\Big(P^\iota_{t_\iota} + \mathcal{P}^\iota_{t_\iota} - \mathcal{P}^\iota_{s_\iota-} - \Gamma^\iota_{s_\iota-}\Big)^+ \mathbb{1}_{[\![t_\iota,\infty[\![}$$
$$- \sum_{\iota;\tau \le \tau^\delta_\iota, t_\iota < \tau} (1 - R_{t_\iota})\Big(P^\iota_{t_\iota} + \mathcal{P}^\iota_{t_\iota} - \mathcal{P}^\iota_{s_\iota-} + \Upsilon^\iota_{s_\iota-}\Big)^- \mathbb{1}_{[\![t_\iota,\infty[\![}$$
$$+ \sum_{ccp,I} \mu^{ccp,0}_{\tau^\delta_I} B^{ccp,I}$$

$$^{\tau-}(-\mathcal{C} + \sum_\iota \mathcal{Y}^\iota) + \sum_{ccp} \mathcal{Y}^{ccp} = \tag{60}$$
$$\sum_{\iota;\tau \le \tau^\delta_\iota, t_\iota \ge \tau} (1 - R_{t_\iota})(P^\iota_{t_\iota} + \mathcal{P}^\iota_{t_\iota} - \mathcal{P}^\iota_{s_\iota-} + \Upsilon^\iota_{s_\iota-})^- \mathbb{1}_{[\![t_\iota,\infty[\![}$$
$$- \sum_{\iota;\tau_\iota \le \tau^\delta, t_\iota \ge \tau} (1 - R^\iota_{t_\iota})\Big(P^\iota_{t_\iota} + \mathcal{P}^\iota_{t_\iota} - \mathcal{P}^\iota_{s_\iota-} - \Gamma^\iota_{s_\iota-}\Big)^+ \mathbb{1}_{[\![t_\iota,\infty[\![}$$
$$+ \sum_{ccp}(1 - R^{ccp,0}_{\tau^\delta})(\mathcal{P}^{ccp,0}_{\tau^\delta} - \mathcal{P}^{ccp,0}_{\tau-} - \mathrm{VM}^{ccp,0}_{\tau-} - \mathrm{PIM}^{ccp,0}_{\tau-} - \mathrm{DFC}^{ccp,0}_{\tau-})^+ \mathbb{1}_{[\![\tau^\delta,\infty[\![}.$$

Regarding the index sets over which the sums are taken in (60) versus (49), note that, as $t_\iota = \tau^\delta_\iota \wedge \tau^\delta$: (i) $t_\iota < \tau \Leftrightarrow \tau^\delta_\iota < \tau$, hence $\{\iota; \tau_\iota \le \tau^\delta, t_\iota < \tau\} = \{\iota; \tau^\delta_\iota < \tau\}$, (ii) $\{\tau \le \tau^\delta_\iota, t_\iota < \tau\} = \emptyset$, (iii) $\tau \le \tau^\delta_\iota \Rightarrow t_\iota \ge \tau$, hence $\{\iota; \tau \le \tau^\delta_\iota, t_\iota \ge \tau\} = \{\iota; \tau \le \tau^\delta_\iota\}$; (iv) $\{c; \tau_\iota \le \tau^\delta, t_\iota \ge \tau\} = \{c; \tau_\iota \le \tau^\delta, \tau^\delta_\iota \ge \tau\}$. Hence (60) reduces to (49).

The formulas in (48) and (50) for $\mathcal{F}$ and its $\mathcal{F}^{\tau-}$ and ${}^{\tau-}(-\mathcal{F})$ components are obtained by feeding II.(43)-(44) with $\mathcal{R}$ and $\mathcal{S}$ from (46).

Assuming the reduction of filtration setup of II.§3.A, (51) follows from (1) by definition, existence and uniqueness of optional reductions (cf. II.(25)). Moreover, we have, for each individual client netting set ι, $s_\iota \wedge \tau = (\tau_\iota \wedge \tau) \wedge \tau = \tau_\iota \wedge \tau$ and $t_\iota \wedge \tau = (\tau_\iota^\delta \wedge \tau^\delta) \wedge \tau = \tau_\iota^\delta \wedge \tau$, hence $s'_\iota = \tau'_\iota$ and $t'_\iota = (\tau_\iota^\delta)'$, by definition, existence and uniqueness of reductions of $[0,T]$ valued stopping times. Then (52) and (53) follow from the expressions for MtM and ($\mathcal{M}$ + MtM) in (43)-(45), while (56), (57) and (58) respectively follow from the expressions of $(\mathcal{C})^{\tau-}$ in (49), where $(\mathcal{Y}^\iota)' = \mathcal{X}^\iota$, of $(\mathcal{F})^{\tau-}$ in (50) and of $(B^I)^{\tau-}$ in (29)-(30). □

In particular, we have the following cash flow additivity properties.

Corollary 1. *$\mathcal{M}$ is additive over bilateral trades, $\mathcal{C}$ is additive over individual client netting sets ι and CCPs. The different terms in $\mathcal{F}$ are additive over the corresponding funding sets: bilateral client and CCP netting sets b and $ccp, 0$ for the segregated collateral $\mathcal{S}$ related components of $\mathcal{F}$ (last two lines in* (48)*), while the rehypothecable collateral $\mathcal{R}$ related component of $\mathcal{F}$ (first two lines in* (48)*) can only be assessed at the level of the overall derivatives portfolio of the bank.*

***Proof*.** The assertion regarding $\mathcal{M}$ follows from (43) by trade additivity of each of the terms $\mathcal{P}^b$ and P^b noted in Remark 1. The statements regarding $\mathcal{C}$ and $\mathcal{F}$ follow by inspection of the related processes in (47)-(48). □

B Valuation Adjustments

Lemma 7. *Given stopping times $\alpha \le \beta$ such that $0 \le \beta - \alpha \le \delta$ and a process $Y = \mathbb{1}_{[\![\alpha,\beta[\![}X$, for a (nonnecessarily adapted) càdlàg process X stopped at $(T-\delta)$ and such that $X_{\alpha-} = 0$:*
(i) *The process $Y = \mathbb{1}_{[\![\alpha,\beta[\![}X$ is killed at T;*
(ii) *If Y is adapted, then, for $t \le T$, $\mathbb{E}_t^*(Y_T - Y_t) = -Y_t$.*

***Proof*.** **(i)** $T \le t$ and $\alpha \le t < \beta \implies \alpha \ge \beta - \delta > t - \delta \ge T - \delta$, so that ($X$ being stopped at $T-\delta$) $X_t = X_{\alpha-} = 0$ and $Y_t = 0$, which also holds in case $T \le t$ and $\neg(\alpha \le t < \beta)$.
(ii) As $Y_T = 0$, if Y is adapted, then $\mathbb{E}_t^*(Y_T - Y_t) = -Y_t$. □

On top of the processes $\mathcal{Y}^{ccp}, \mathcal{Y}^\iota$ and $\mathcal{X}^\iota$ already defined in (41)-(42), let further

$$\begin{aligned} \mathcal{Z}^\iota &= \mathbb{1}_{[\![\tau,t_\iota[\![}\mathbb{1}_{\{\tau_\iota<\tau\le t_\iota\}}(M^\iota - M^\iota_{\tau-}) = \mathbb{1}_{[\![\tau,\tau\vee t_\iota[\![}\mathbb{1}_{\{\tau_\iota<\tau\le t_\iota\}}(M^\iota - M^\iota_{\tau-}) \\ \mathcal{V}^\iota &= -\mathbb{1}_{[\![t_\iota,+\infty[\![}\mathbb{1}_{\{\tau_\iota<\tau\le t_\iota\}}(M^\iota_{\tau-} - M^\iota_{s_\iota-}). \end{aligned} \quad (61)$$

Note that all the processes $\mathcal{Y}^{ccp}, \mathcal{Y}^{\iota}, \mathbb{1}_{\{\tau\leq\tau_\iota\}}\mathcal{Y}^{\iota}, \mathcal{Z}^{\iota}$ and $\mathcal{V}^{\iota}$ are $\mathfrak{G}$ adapted and start from 0 at time 0; assuming the reduction of filtration setup of II.§3.A, the process $\mathcal{X}^{\iota}$ is $\mathfrak{F}$ adapted and starts from 0 at time 0.

Lemma 8. **(i)** *Each process $\mathcal{Y}^{ccp}$ is killed at T and, for $t \leq T$, the time-t value of $\mathcal{Y}^{ccp}$ is $-\mathcal{Y}^{ccp}_t = -\mathbb{1}_{\tau\leq t<\tau^\delta}(\mathcal{P}^{ccp,0}_t - \mathcal{P}^{ccp,0}_{\tau-})$.*
(ii) *Each process $\mathbb{1}_{\{\tau\leq\tau_\iota\}}\mathcal{Y}^{\iota}$ is killed at T and, for $t \leq T$, the time-t value of $\mathbb{1}_{\{\tau\leq\tau_\iota\}}\mathcal{Y}^{\iota}$ is $-\mathbb{1}_{\{\tau\leq\tau_\iota\}}\mathcal{Y}^{\iota}_t = -\mathbb{1}_{\{s_\iota\leq t<t_\iota\}}\mathbb{1}_{\{\tau\leq\tau_\iota\}}\big(M^\iota_t - M^\iota_{s_\iota-}\big)$.*
(iii) *Each process $\mathcal{Z}^{\iota}$ is killed at T and, for $t \leq T$, the time-t value of $\mathcal{Z}^{\iota}$ is $-\mathcal{Z}^{\iota}_t = -\mathbb{1}_{\{\tau\leq t<t_\iota\}}\mathbb{1}_{\{\tau_\iota<\tau\leq t_\iota\}}(M^\iota_t - M^\iota_{\tau-})$.*
(iv) *Assuming the reduction of filtration setup of II.§3.A, each process $\mathcal{X}^{\iota}$ as per (42) (cf. (54)) is killed at T and, for $t \leq T$, $\mathbb{E}'_t(\mathcal{X}^\iota_T - \mathcal{X}^\iota_t) = -\mathcal{X}^\iota_t = -\mathbb{1}_{\{\tau'_\iota\leq t<(\tau^\delta_\iota)'\}}\big((M^\iota)'_t - (M^\iota)'_{\tau'_\iota-}\big)$.*

Proof. Recall from §1.C that T exceeds by δ the final maturity of all claims in all considered portfolios. Hence all the processes $\mathcal{P}^{\cdot}$, as well as their value processes $P^{\cdot}$, are stopped not only at T, but even at $T-\delta$: in particular (cf. (24)),

$$d\mathcal{P}^{ccp,0}_t = \sum_c \Big(\mathbb{1}_{\{t\leq t_c\}}d\mathcal{P}^{ccp,c}_t + \mathbb{1}_{\{s_c\leq t<t_c\}}dP^{ccp,c}_t + P^{ccp,c}_{t_c}\boldsymbol{\delta}_{t_c}(dt)\Big) = \mathbb{1}_{t\leq T-\delta}d\mathcal{P}^{ccp,0}_t.$$

Assuming the reduction of filtration setup of II.§3.A (with optional reductions stopped at T), let us now show that

$$(M^\iota)' = (((M^\iota)^{T-\delta})')^T = ((M^\iota)')^{T-\delta}, \tag{62}$$

so that $(M^\iota)'$ is also stopped at $T-\delta$. To justify (the right equality in) (62), note that $((M^\iota)^{T-\delta})'$ and $(M^\iota)')^{T-\delta}$ are two $\mathfrak{F}$ optional processes coinciding with $(M^\iota)^{T-\delta}$ before τ. This holds by definition of an optional reduction, regarding the former. Regarding the latter, we have $(M^\iota)'\mathbb{1}_{[\![0,\tau[\![} = M^\iota\mathbb{1}_{[\![0,\tau[\![}$, which implies by stopping at $T-\delta$ that $((M^\iota)')^{T-\delta}(\mathbb{1}_{[\![0,\tau[\![})^{T-\delta} = (M^\iota)^{T-\delta}(\mathbb{1}_{[\![0,\tau[\![})^{T-\delta}$. But $(\mathbb{1}_{[\![0,\tau[\![})^{T-\delta} = \mathbb{1}_{\{T-\delta\geq\tau\}}\mathbb{1}_{[\![0,\tau[\![} + \mathbb{1}_{\{T-\delta<\tau\}}$, hence $\mathbb{1}_{[\![0,\tau[\![}(\mathbb{1}_{[\![0,\tau[\![})^{T-\delta} = \mathbb{1}_{[\![0,\tau[\![}$, therefore multiplying the above identity by $\mathbb{1}_{[\![0,\tau[\![}$ yields $((M^\iota)')^{T-\delta}\mathbb{1}_{[\![0,\tau[\![} = (M^\iota)^{T-\delta}\mathbb{1}_{[\![0,\tau[\![}$, so that $((M^\iota)')^{T-\delta}$ indeed coincides with $(M^\iota)^{T-\delta}$ before τ, like$((M^\iota)^{T-\delta})'$. By uniqueness II.(25) of optional reductions, the processes $((M^\iota)^{T-\delta})'$ and $(M^\iota)')^{T-\delta}$ therefore coincide.

All results then follow by application of Lemma 7, with
(i) $\alpha = \tau, \beta = \tau^\delta, X = \mathcal{P}^{ccp,0} - \mathcal{P}^{ccp,0}_{\tau-}$.
(ii) $\alpha = s_\iota, \beta = t_\iota, X = \mathbb{1}_{\{\tau\leq\tau_\iota\}}(M^\iota - M^\iota_{s_\iota-})$
(iii) $\alpha = \tau, \beta = \tau\vee t_\iota, X = \mathbb{1}_{\{\tau_\iota<\tau\leq t_\iota\}}(M^\iota - M^\iota_{\tau-})$
(iv) $\alpha = \tau'_\iota, \beta = (\tau^\delta_\iota)', X = (M^\iota)' - (M^\iota)'_{\tau'_\iota-}$. □

Lemma 9. **(i)** *The process $\mathcal{V}$ is stopped at T.*

(ii) *For $t \le T$, we have*

$$\mathbb{E}_t^* \mathcal{V}_T^\iota = \mathcal{V}_t^\iota + W_t^\iota, \tag{63}$$

where

$$W_t^\iota = -\mathbb{1}_{\{t<t_\iota\}} \mathbb{E}_t^* \Big[\mathbb{1}_{\{\tau_\iota < \tau \le t_\iota \le T\}} (M_{\tau-}^\iota - M_{\tau_\iota -}^\iota) \Big]$$

Proof. **(i)** On $\{T \ge t_\iota\}$, $\mathcal{V}$ is worth $-\mathbb{1}_{\{\tau_\iota<\tau\le t_\iota\}}(M_{\tau-}^\iota - M_{s_\iota -}^\iota)$ on $[T,+\infty[$. On $\{T < t_\iota\} \setminus \{\tau_\iota < \tau \le t_\iota\}$, $\mathcal{V}$ vanishes identically, which also holds on $\{T < t_\iota\} \cap \{\tau_\iota < \tau \le t_\iota\}$, where $\tau > \tau_\iota = s_\iota \ge t_\iota - \delta > T - \delta$ (using (39)), so that $M_{\tau-}^\iota = M_{s_\iota -}^\iota$.
(ii) For $t \le T$,

$$\begin{aligned}
\mathbb{E}_t^* \mathcal{V}_T^\iota &= -\mathbb{E}_t^* \Big[\mathbb{1}_{\{\tau_\iota<\tau\le t_\iota \le T\}} (M_{\tau-}^\iota - M_{s_\iota-}^\iota)\Big)\Big] \\
&= -\mathbb{E}_t^* \Big[\mathbb{1}_{\{t \ge t_\iota\}} \mathbb{1}_{\{\tau_\iota<\tau\le t_\iota\}} (M_{\tau-}^\iota - M_{s_\iota-}^\iota)\Big)\Big] \\
&\qquad - \mathbb{E}_t^* \Big[\mathbb{1}_{\{t < t_\iota\}} \mathbb{1}_{\{\tau_\iota<\tau\le t_\iota \le T\}} (M_{\tau-}^\iota - M_{s_\iota-}^\iota)\Big)\Big] \\
&= \mathcal{V}_t^\iota - \mathbb{E}_t^* \Big[\mathbb{1}_{\{t < t_\iota\}} \mathbb{1}_{\{\tau_\iota<\tau\le t_\iota \le T\}} (M_{\tau-}^\iota - M_{s_\iota-}^\iota)\Big)\Big] \\
&= \mathcal{V}_t^\iota + W_t^\iota. \ \square
\end{aligned}$$

Lemma 10. (i) *We have*

$$^{\tau-}(\mathcal{Y}^\iota) = \mathbb{1}_{\{\tau \le \tau_\iota\}} \mathcal{Y}^\iota + \mathcal{Z}^\iota + \mathcal{V}^\iota. \tag{64}$$

(ii) *The time-t value of $^{\tau-}(\mathcal{Y}^\iota)$ is $-\mathbb{1}_{\{\tau\le\tau_\iota\}}\mathcal{Y}_t^\iota - \mathcal{Z}_t^\iota + W_t^\iota$.*

Proof. **(i)** By a repeated use of the definitions II.(3), we have, also using in the third line the identity $(XY)^{\tau-} = X^{\tau-}Y^{\tau-}$ that holds for any left-limited processes (as both sides are stopped before τ and coincide with XY before τ),

$$\begin{aligned}
&^{\tau-}\Big(\mathbb{1}_{[\![s_\iota,t_\iota[\![}\big(M^\iota - M_{s_\iota-}^\iota\big)\Big) \\
&\quad = \mathbb{1}_{[\![s_\iota,t_\iota[\![}\big(M^\iota - M_{s_\iota-}^\iota\big) - \Big(\mathbb{1}_{[\![s_\iota,t_\iota[\![}\big(M^\iota - M_{s_\iota-}^\iota\big)\Big)^{\tau-} \\
&\quad = \mathbb{1}_{[\![s_\iota,t_\iota[\![}\big(M^\iota - M_{s_\iota-}^\iota\big) - \Big(\pm M^\iota + (M^\iota)^{\tau-} - M_{s_\iota-}^\iota\Big)(\mathbb{1}_{[\![s_\iota,t_\iota[\![})^{\tau-} \\
&\quad =^{\tau-}(\mathbb{1}_{[\![s_\iota,t_\iota[\![})\big(M^\iota - M_{s_\iota-}^\iota\big) +^{\tau-}(M^\iota)(\mathbb{1}_{[\![s_\iota,t_\iota[\![})^{\tau-},
\end{aligned}$$

where

$$\begin{aligned}
(\mathbb{1}_{[\![s_\iota,t_\iota[\![})^{\tau-} &= \mathbb{1}_{\{\tau>t_\iota\}}\mathbb{1}_{[\![s_\iota,t_\iota[\![} + \mathbb{1}_{\{\tau_\iota<\tau\le t_\iota\}}\mathbb{1}_{[\![s_\iota,+\infty[\![} \\
^{\tau-}(\mathbb{1}_{[\![s_\iota,t_\iota[\![}) &= \mathbb{1}_{[\![s_\iota,t_\iota[\![} - (\mathbb{1}_{[\![s_\iota,t_\iota[\![})^{\tau-} = \mathbb{1}_{\{\tau\le\tau_\iota\}}\mathbb{1}_{[\![s_\iota,t_\iota[\![} - \mathbb{1}_{\{\tau_\iota<\tau\le t_\iota\}}\mathbb{1}_{[\![t_\iota,+\infty[\![}
\end{aligned}$$

$$^{\tau-}(M^\iota) = \mathbb{1}_{[\![\tau,+\infty[\![}(M^\iota - M^\iota_{\tau-}).$$

Hence

$$\begin{aligned}
^{\tau-}(\mathcal{Y}^\iota) &= \Big(\mathbb{1}_{\{\tau\le\tau_\iota\}}\mathbb{1}_{[\![s_\iota,t_\iota[\![} - \mathbb{1}_{\{\tau_\iota<\tau\le t_\iota\}}\mathbb{1}_{[\![t_\iota,+\infty[\![}\Big)\Big(M^\iota - M^\iota_{s_\iota-}\Big)\\
&\quad + \Big(\mathbb{1}_{\{\tau>t_\iota\}}\mathbb{1}_{[\![s_\iota,t_\iota[\![} + \mathbb{1}_{\{\tau_\iota<\tau\le t_\iota\}}\mathbb{1}_{[\![s_\iota,+\infty[\![}\Big)\mathbb{1}_{[\![\tau,+\infty[\![}(M^\iota - M^\iota_{\tau-})\\
&= \Big(\mathbb{1}_{\{\tau\le\tau_\iota\}}\mathbb{1}_{[\![s_\iota,t_\iota[\![}\Big(M^\iota - M^\iota_{s_\iota-}\Big)\\
&\quad + \mathbb{1}_{\{\tau>t_\iota\}}\mathbb{1}_{[\![s_\iota,t_\iota[\![}\mathbb{1}_{[\![\tau,+\infty[\![}(M^\iota - M^\iota_{\tau-})\\
&\quad - \mathbb{1}_{\{\tau_\iota<\tau\le t_\iota\}}\mathbb{1}_{[\![t_\iota,+\infty[\![}\Big(M^\iota - M^\iota_{s_\iota-}\Big) + \mathbb{1}_{\{\tau_\iota<\tau\le t_\iota\}}\mathbb{1}_{[\![s_\iota,+\infty[\![}\mathbb{1}_{[\![\tau,+\infty[\![}(M^\iota - M^\iota_{\tau-}),
\end{aligned}$$

where the first two lines respectively equal $\mathbb{1}_{\{\tau\le\tau_\iota\}}\mathcal{Y}^\iota$ and 0, whereas the last line equals

$$\begin{aligned}
&- \mathbb{1}_{\{\tau_\iota<\tau\le t_\iota\}}\mathbb{1}_{[\![t_\iota,+\infty[\![}\Big(M^\iota - M^\iota_{s_\iota-}\Big) + \mathbb{1}_{\{\tau_\iota<\tau\le t_\iota\}}\mathbb{1}_{[\![t_\iota,+\infty[\![}(M^\iota - M^\iota_{\tau-})\\
&\qquad + \mathbb{1}_{\{\tau_\iota<\tau\le t_\iota\}}\mathbb{1}_{[\![s_\iota,t_\iota[\![}\mathbb{1}_{[\![\tau,+\infty[\![}(M^\iota - M^\iota_{\tau-})\\
&= \mathcal{V}^\iota + \mathbb{1}_{\{\tau_\iota<\tau\le t_\iota\}}\mathbb{1}_{[\![\tau,t_\iota[\![}(M^\iota - M^\iota_{\tau-}) = \mathcal{V}^\iota + \mathcal{Z}^\iota.
\end{aligned}$$

Hence we obtain (64).
(ii) follows from part (i) and Lemmas 8(ii)-(iii) and 9(ii). □

Recall the spaces of processes $\mathbb{L}_2$, $\mathbb{S}_2$,$\mathbb{L}'_2$, $\widetilde{\mathbb{S}}'_2$ and $\mathbb{S}'_2$ of II.§3.A.

Theorem 1. *In the reduction of filtration setup of II.§3.A for a bank involved in an arbitrary combination of bilateral netting sets and of netting sets cleared with one among several available CCPs, if*

$$\sum_\iota (1-(R^\iota)'_{(\tau_\iota^\delta)'})\Big((P^\iota)'_{(\tau_\iota^\delta)'} + (\mathcal{P}^\iota)'_{(\tau_\iota^\delta)'} - (\mathcal{P}^\iota)'_{\tau'_\iota-} - (\Gamma^\iota)'_{\tau'_\iota-}\Big)^+ \mathbb{1}_{\{(\tau_\iota^\delta)'<T\}} \quad (65)$$

$$+ \sum_{ccp,I} (\mu^{ccp,0})'_{(\tau_I^\delta)'}(B^{ccp,I})'_{(\tau_I^\delta)'}\mathbb{1}_{\{(\tau_I^\delta)'<T\}} \quad (66)$$

is $\mathbb{P}$ square integrable, that $\sum_\iota \mathcal{X}^\iota \in \widetilde{\mathbb{S}}'_2$, $\mathbb{E}'\int_0^T \lambda'_t dt < \infty$, and that

$$\lambda'\Big(\mathrm{MtM}' - \sum_b (\mathrm{VM}^b)'\Big)^+ + \widetilde{\lambda}'\Big(\sum_\iota (\mathrm{PIM}^\iota)' + \sum_{ccp}(\mathrm{PIM}^{ccp,0} + \mathrm{DFC}^{ccp,0})'\Big) \text{ and } \lambda'\sum_\iota \mathcal{X}^\iota \quad (67)$$

are in $\mathbb{L}'_2$, then the CVA *and* FVA *equations II.(14)-(15) are well-posed in $\mathbb{S}_2$*

and we have, for $t \le T$,

$$\begin{aligned}
&\mathrm{CVA}'_t + \sum_\iota \mathcal{X}^\iota_t = \\
&\sum_\iota \mathbb{E}'_t\Big[(1-(R^\iota)'_{(\tau^\delta_\iota)'})\Big((P^\iota)'_{(\tau^\delta_\iota)'} + (\mathcal{P}^\iota)'_{(\tau^\delta_\iota)'} - (\mathcal{P}^\iota)'_{\tau'_\iota -} - (\Gamma^\iota)'_{\tau'_\iota -}\Big)^+ \mathbb{1}_{\{t<(\tau^\delta_\iota)'\le T\}}\Big] \\
&\quad + \sum_{ccp,I} \mathbb{E}'_t[(\mu^{ccp,0})'_{(\tau^\delta_I)'}(B^{ccp,I})'_{(\tau^\delta_I)'}\mathbb{1}_{\{t<(\tau^\delta_I)'\le T\}}],
\end{aligned} \tag{68}$$

$$\begin{aligned}
\mathrm{FVA}'_t = \mathbb{E}'_t \int_t^T \lambda'_s \Big(\mathrm{MtM}' - \sum_b (\mathrm{VM}^b)' - \mathrm{CVA}' - \mathrm{FVA}'\Big)^+_s ds + \\
\mathbb{E}'_t \int_t^T \widetilde{\lambda}'_s \Big(\sum_b (\mathrm{PIM}^b)' + \sum_{ccp}\Big(\mathrm{PIM}^{ccp,0} + \mathrm{DFC}^{ccp,0}\Big)'\Big)_s ds,
\end{aligned} \tag{69}$$

$$\begin{aligned}
d\mathcal{L}'_t = &\sum_\iota (1-(R^\iota)'_{(\tau^\delta_\iota)'})\Big((P^\iota)'_{(\tau^\delta_\iota)'} + (\mathcal{P}^\iota)'_{(\tau^\delta_\iota)'} - (\mathcal{P}^\iota)'_{\tau'_\iota -} - (\Gamma^\iota)'_{\tau'_\iota -}\Big)^+ \boldsymbol{\delta}_{(\tau^\delta_\iota)'}(dt) \\
&+ \sum_{ccp,I} (\mu^{ccp,0})'_{(\tau^\delta_I)'}(B^{ccp,I})'_{(\tau^\delta_I)'}\boldsymbol{\delta}_{(\tau^\delta_I)'}(dt) \\
&+ d(\mathrm{CVA}' + \sum_\iota \mathcal{X}^\iota)_t \\
&+ \lambda'_t\Big(\mathrm{MtM}' - \sum_b (\mathrm{VM}^b)' - \mathrm{CVA}' - \mathrm{FVA}'\Big)^+_t dt \\
&+ \widetilde{\lambda}'_t\Big(\sum_b (\mathrm{PIM}^b)' + \sum_{ccp}(\mathrm{PIM}^{ccp,0} + \mathrm{DFC}^{ccp,0})'\Big)_t dt \\
&+ d\mathrm{FVA}'_t \\
&- \sum_b \mathbb{1}_{\{t\le(\tau^\delta_b)'\}} d(\mathcal{P}^b + P^b)'_t + d\mathcal{H}'_t.
\end{aligned} \tag{70}$$

Moreover, if $\sum_b((\mathcal{P}^b + P^b)')^{(\tau^\delta_b)'} - \mathcal{H}'$ *is in* $\widetilde{\mathbb{S}}'^{\,5}_2$, *then the* KVA *equation II.*(19) *is well-posed in* $\mathbb{S}_2$ *and we have, for* $t \in [0,T]$,

$$\mathrm{KVA}'_t = \mathbb{E}'_t \int_t^T h e^{-h(s-t)} \max(\mathrm{EC}'_s, \mathrm{KVA}'_s) ds. \tag{71}$$

Regarding the contra-liabilities of Definition II.7 page 74, we have

$$\begin{aligned}
&\mathrm{DVA}_t + \sum_{ccp} \mathcal{Y}^{ccp}_t - \sum_\iota \mathbb{1}_{\{\tau\le\tau_\iota\}}\mathcal{Y}^\iota_t - \sum_\iota \mathcal{Z}^\iota_t + \sum_\iota W^\iota_t = \\
&\mathbb{E}^*_t\Big[\sum_{\iota;\tau\le\tau^\delta_\iota} \mathbb{1}_{\{t<t_\iota\le T\}}(1-R_{t_\iota})(P^\iota_{t_\iota} + \mathcal{P}^\iota_{t_\iota} - \mathcal{P}^\iota_{s_\iota -} + \Upsilon^\iota_{s_\iota -})^-\Big] \\
&-\mathbb{E}^*_t\Big[\sum_{\iota;\tau_\iota\le\tau^\delta,\tau^\delta_\iota\ge\tau} \mathbb{1}_{\{t<t_\iota\le T\}}(1-R^\iota_{t_\iota})\Big(P^\iota_{t_\iota} + \mathcal{P}^\iota_{t_\iota} - \mathcal{P}^\iota_{s_\iota -} - \Gamma^\iota_{s_\iota -}\Big)^+\Big]
\end{aligned} \tag{72}$$

[5] e.g. in the reference dynamic hedging case (2), where $\sum_b((\mathcal{P}^b + P^b)')^{(\tau^\delta_b)'} - \mathrm{MtM}_0 - \mathcal{H}' = 0$.

$$+\mathbb{E}_t^*\Big[\sum_{ccp}(1-R^{ccp,0}_{\tau^\delta})(\mathcal{P}^{ccp,0}_{\tau^\delta}-\mathcal{P}^{ccp,0}_{\tau-}-\mathrm{VM}^{ccp,0}_{\tau-}-\mathrm{PIM}^{ccp,0}_{\tau-}-\mathrm{DFC}^{ccp,0}_{\tau-})^+\Big]$$
$$+\mathbb{E}_t^*\Big[\mathbb{1}_{\{t<\tau\le T\}}\mathrm{CVA}_{\tau-}\Big],$$
$$\mathrm{FDA}_t=\mathbb{E}_t^*\Big[\mathbb{1}_{\{t<\tau\le T\}}(1-R_\tau)(\mathrm{MtM}-\sum_b\mathrm{VM}^b-\mathrm{CA})^+_{\tau-}\Big]$$
$$+\mathbb{E}_t^*\Big[\mathbb{1}_{\{t<\tau\le T\}}(1-\bar{R}_\tau)\Big(\sum_b\mathrm{PIM}^b+\sum_{ccp}(\mathrm{PIM}^{ccp,0}+\mathrm{DFC}^{ccp,0})\Big)_{\tau-}\Big]$$
$$+\mathbb{E}_t^*\Big[\mathbb{1}_{\{t<\tau\le T\}}\mathrm{FVA}_{\tau-}\Big]=\mathrm{FVA}_t. \tag{73}$$

Proof. The CVA related statement follows by application of Proposition 2(i) with $\mathcal{C}'-\sum_\iota\mathcal{X}^\iota$ given by (56), which is nondecreasing and worth (65) at T, also using Lemma 8(iv) to obtain the liquidation period corrective term $\sum_\iota\mathcal{X}^\iota_t$ in (68).

Note that $\mathrm{CVA}'+\sum_\iota\mathcal{X}^\iota\ge 0$, by (68). Hence, also using the 1-Lipschitz property of $x\mapsto(a-x)^+$ in the third line, we have:

$$\Big(\mathrm{MtM}'-\sum_b(\mathrm{VM}^b)'-\mathrm{CVA}'\Big)^+$$
$$=\Big(\mathrm{MtM}'-\sum_b(\mathrm{VM}^b)'-(\mathrm{CVA}'+\sum_\iota\mathcal{X}^\iota)+\sum_\iota\mathcal{X}^\iota\Big)^+ \tag{74}$$
$$\le\Big(\mathrm{MtM}'-\sum_b(\mathrm{VM}^b)'-(\mathrm{CVA}'+\sum_\iota\mathcal{X}^\iota)\Big)^++|\sum_\iota\mathcal{X}^\iota|$$
$$\le\Big(\mathrm{MtM}'-\sum_b(\mathrm{VM}^b)'\Big)^++|\sum_\iota\mathcal{X}^\iota|. \tag{75}$$

The FVA-related statement then follows by application of Proposition II.2(ii), with $\mathcal{R}$ and $\mathcal{S}$ there given by (46) so that, by (75), $\lambda'(\mathcal{R}'-\mathrm{CVA}')^++\widetilde{\lambda}'\mathcal{S}'$ is in $\mathbb{L}_2'$ provided the two terms in (67) are in $\mathbb{L}_2'$.

The dynamics (70) for $\mathcal{L}'$ are obtained by plugging (53)–(57) into (51). This process $\mathcal{L}'$ belongs to $\widetilde{\mathbb{S}}_2'$ as the sum (modulo a constant) between $\mathrm{CVA}'+\mathcal{C}'$, the $(\mathfrak{F},\mathbb{P})$ martingale part of FVA', and $-\sum_b((\mathcal{P}^b+P^b)')^{(\tau_b^\delta)'}+\mathcal{H}'$, assumed in $\widetilde{\mathbb{S}}_2'$. By an application of Proposition II.2(iii) (given (53)), the KVA equation II.(19) is then well-posed in $\mathbb{S}_2$ and the KVA' formula (71) holds.

In view of Definition II.7, the DVA and FDA formulas respectively follow from the expressions of ${}^{\tau-}(-\mathcal{C})$ in (49), also using Lemmas 8(i) and 10(ii), and ${}^{\tau-}(-\mathcal{F})$ in (50). The last equality in (73) is a general consequence, established in Lemma II.8(ii), of the martingale property of $\mathcal{F}$. □

Remark 9. The CVA with liquidation period δ is given by two dominant nonnegative terms (the two terms on the right-hand side in (68)) and a corrective explicit term $\sum_\iota\mathcal{X}^\iota$ only nonnull during the liquidation periods (which

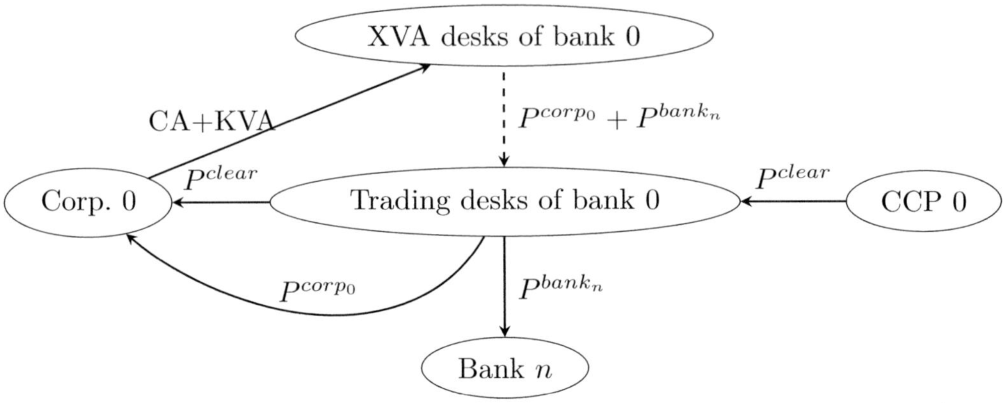

Figure 9: Zoom on the reference bank labeled by 0 in Figure 6, focusing on its netting sets with Corp. 0, Bank n and CCP 0. The solid arrows represent the direction of deal entry payments between the bank, its clients and the CCP. The reference bank 0 buys assets from its (nonbanking) corporate clients, at a (CA + KVA)-deducted price with respect to their counterparty-risk-free valuation, P^{ι} (cf. Table 5), at which transactions between banks are concluded. The dashed arrow refers to the trading margin amount $P^{corp_0} + P^{bank_n}$ posted by the Treasury to the trading desks of Bank 0, so that these can setup bilateral netting sets valued P^{corp_0} and P^{bank_n} (on a counterparty-risk-free basis) with Corp. 0 and Bank n, the way detailed in §3. The aggregate $P^{corp_0} + P^{bank_n}$ is dynamically hedged on exchanges (not shown to alleviate the picture). In addition, the bank has a netting set of cleared deals with Corp. 0, with counterparty-risk-free value P^{clear}, which are statically replicated by identical deals between the bank and CCP 0, in the way detailed in §4.

therefore vanishes identically in the $\delta = 0$ case where all liquidations are instantaneous). Similar comments apply to the DVA.

CONCLUSION

The portfolio-wide, all XVA-inclusive, cost-of-capital pricing formula is

$$\text{MtM} - (\text{CVA} + \text{FVA} + \text{KVA}), \tag{76}$$

charged to the clients of the deals at the portfolio inception time 0. The trading margin MtM corresponds essentially[6] to the counterparty-risk-free value of the portfolio. The reserve capital amount CA = CVA + FVA is used by the CVA desk and the bank Treasury for coping with the expected client default losses and risky funding expenses of the bank. The KVA risk margin amount is used by the management of the bank for gradually releasing a dividend risk

[6] and exactly so at time 0, see (44).

premium to the shareholders of the bank, at a hurdle rate h on their capital at risk. Figure 9 provides a focus on a subset of the (red) financial network of the reference bank labeled by 0 in Figure 6, detailing further the respective roles of the different desks within the bank.

NUMERICAL METHODS

Following the increase in regulation since the global financial crisis of 2007–09, banks are subject to a huge number of risk calculations. Banks are also required to calculate various XVA metrics, i.e. valuation adjustments to account for counterparty risk and its consequences in terms of capital and collateral costs. These calculations must be carried out at different levels of aggregation: the bank netting sets (client portfolios) for CVA calculations, the broader level of funding sets for financing cost calculations, and even the bank's balance sheet as a whole for economic capital and cost of capital calculations.

The numerical toolbox of market finance traditionally rests on three pillars: explicit (exact or approximate) formulas, deterministic numerical schemes and simulation schemes. The first pillar includes Fourier transform formulas in affine jump diffusion models, or implied volatility asymptotics in the SABR model. The second one comprises the finite differences or elements techniques for the derivatives pricing PDEs. The third one is made of Monte Carlo methods based on the Feynman-Kac representation of the solutions of linear PDEs, or on the probabilistic formulation of the solutions of nonlinear PDEs as backward stochastic differential equations (BDSEs).

From the numerical viewpoint, this book emphasizes statistical learning methods on simulated data, dubbed probabilistic machine learning hereafter. Learning is then conceived, not as a way of modelling from data (since the data are simulated within a predefined financial model or class of models), but as a fourth term in the previous toolbox. Quantitative finance indeed offers a vast field of application for statistical learning techniques implemented on simulated data. This is again a well-established tradition, originating in the numerical simulation/regression schemes for pricing Bermudan options à la Tsitsiklis and Van Roy (2001) and Longstaff and Schwartz (2001), which have since been considerably extended to BSDEs (Gobet, Lemor and Warin, 2005; Gobet, 2016). Nevertheless, we can speak of a recent breakthrough in this field, which has been disrupted by the flood of learning techniques (E, Han and Jentzen, 2017; Huré, Pham and Warin, 2020).

This book mostly reflects our second-generation XVA GPU implementation. The first generation, still represented by Chapter X, consisted of nested Monte Carlo techniques specialized to XVA applications as per Albanese, Caenazzo and Crépey (2017) and Abbas-Turki, Diallo and Crépey (2018). The second generation takes advantage of machine learning techniques and packages that are used for XVA-related regression and quantile regression tasks. Compared to a regulatory capital-based KVA approach, an economic capital based KVA approach is then not only economically more satisfying (see Example I.1 and II.§5.B-C), but also simpler to implement.

CHAPTER IV

Generalities

§1 THE XVA COMPUTATIONAL CHALLENGE

Figure 10 yields a representation, in the form of an XVA dependence tree, of the continuous-time XVA equations of Theorem III.1 page 111 (see also VII.§1.A when capital at risk is deemed fungible with variation margin).

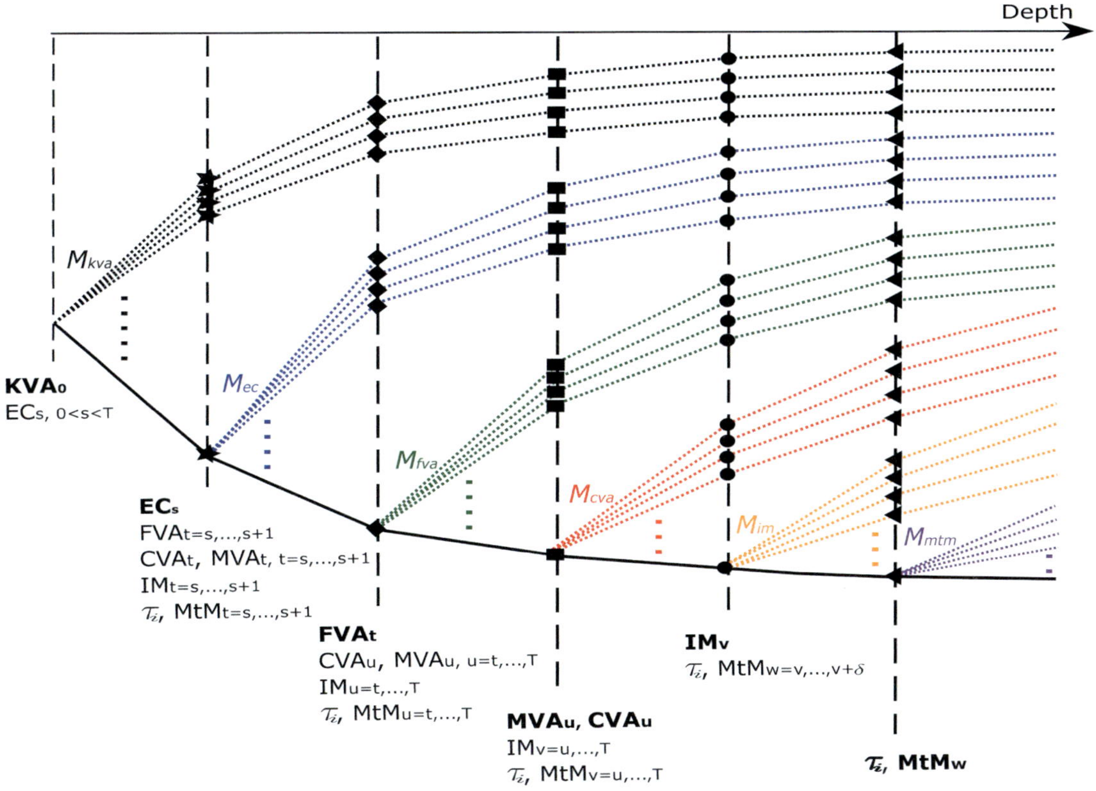

Figure 10: The XVA equations dependence tree (*source*: Abbas-Turki, Diallo and Crépey (2018)).

DOI: 10.1201/9781003614944-8

As visible from the corresponding equations in Chapters III and VII (see Corollary III.1 in particular), the CVA and the MVA of the bank are defined at the netting set level (or counterparty i of the bank with default time denoted by τ_i in Figure 10). The FVA and the KVA, instead, are defined in terms of semilinear equations that can only be formulated at the level of the overall derivative portfolio of the bank. Moreover, as the FVA and KVA equations are only semilinear, their solution necessitates some backwardation, i.e. pricing at each date of a pricing grid, going backward in time. In addition, the fungibility of capital at risk with variation margin (cf. Remark I.12 page 41 and Chapter VII) induces a coupling between, on the one hand, the FVA and KVA backward SDEs and, on the other hand, the forward SDE for the trading loss process $\mathcal{L}$ of the bank.

On top of being holistic computations encompassing all the derivative contracts of the bank, our XVA computations involve not only conditional expectations but also conditional value-at-risks and expected shortfall for the dynamic initial margin and economic capital calculations. Moreover, they are driven, to a large extent, by default scenarios, which are discrete in nature, in the tails of the loss distributions, and appear nonlinearly in the equations so that they have to be simulated. And with all this, XVA computations require a certain accuracy so that trade incremental XVA calculations, which are required as XVA add-ons to derivative entry prices (cf. I.§4.B), are not in the numerical noise of the machinery.

A XVA Computational Strategies

Our XVA computational strategies always involve an outer Monte Carlo simulation of risk factor trajectories and the pricing of the mark-to-market (MtM) cube, i.e. the pathwise counterparty-risk-free valuation of each netting set. This outer Monte Carlo simulation is combined with inner simulations, interpolations and/or regressions (or enhanced machine learning techniques, regarding the two latter) for deriving the associated XVA metrics. Backwardation in time is also used for the FVA and EC/KVA computations. In the case of the anticipated XVA BSDEs accounting for the fungibility of capital at risk with variation margin, the forward-backward system for the bank trading loss $\mathcal{L}$, the FVA and the KVA can be decoupled by the Picard iteration of Algorithm 7 page 200. An alternative without Picard iterations is the explicit scheme of Abbas-Turki, Crépey, Li and Saadeddine (2025).

In view of Figure 10, a purely, multiply nested Monte Carlo XVA computational strategy would involve five nested layers of simulation (with respective numbers of paths $M_{xva} \sim \sqrt{M_{mtm}}$, see Abbas-Turki, Diallo and Crépey (2018, Section 3.3)), plus the ones required by the time backwardations for the FVA and KVA BSDEs. Such computations are of course, unfeasible at the scale of the derivatives portfolio of a bank. In this book, with the exception of Chapter X where a term structure approximation of economic capital makes simply nested Monte Carlo computations sufficient to estimate the whole XVA suite, we use no nested Monte Carlo other than for benchmarking purposes: beyond the base MtM layer in the XVA dependence tree, each successive layer (from right to left in Figure 10, and at each Picard iteration if any) is "learned" (or regressed) instead.

A.1 Related literature

Supervised learning tasks can be subdivided into two main categories (Murphy, 2012). The first one is classification problems, for which training a neural net allows learning a (conditional) probability that a discrete random variable takes its different possible values (such as 'cat or dog' for an image), by empirical minimization of an entropic criterion[1], based on a number of labeled observations (images and their correct classification, e.g. 'cat or dog'). The second category consists of regression tasks, for which training a neural net allows learning a conditional expectation $\mathbb{E}(\xi|\mathcal{X})$, by empirical minimization of a least squares criterion based on observations of the pair of random variables $(\mathcal{X}, \xi)$. Our XVA applications fall in the second category, in the special case where the $(\mathcal{X}, \xi)$ are simulated, assuming the data-generating process known. For validation purposes, we rely on out-of-sample twin Monte Carlo procedures exposed in §2.A and §3.A (and on nested Monte Carlo procedures whenever available).

In the last decade, such neural net regression-based simulation techniques have rapidly imposed themselves as a worthy player in the field of numerical methods for PDEs and BSDEs: see e.g. E, Han and Jentzen (2017) or Huré, Pham and Warin (2020). There are good reasons for this evolution, starting with universal approximation theorems, or density results justifying the use of neural network as a versatile parameterization: see Kidger and Lyons (2020) for the fixed-width case, and Hornik (1991), Cybenko (1989) for the fixed-depth case. Other incentives for using neural networks are the ability to automatically infer a linear regression basis provided by the trained hidden layers (see Figure 13 page 140), or the ease of transfer learning (Pan and Yang, 2009; Bozinovski, 2020). However, the control of the error arising from the numerical optimizations for the embedded training tasks (a priori error bounds for numerical solutions to (3) in our setup) is a largely open issue: in the nonconvex numerical optimization case typical of neural net training tasks, only idealized versions of stochastic gradient descents with over-parameterized neural nets can be shown to converge to the global minimum of the empirical loss, and only in-sample (Chizat and Bach, 2018; Du, Lee, Li, Wang and Zhai,

[1] or Kullback-Leibler divergence, or maximization of a likelihood.

2019); otherwise, only local minima can be guaranteed, under assumptions such as the ones in Lei, Hu, Li and Tang (2019). These local minima are often close to global minima (Choromanska, Henaff, Mathieu, Arous and LeCun, 2015), so that this local versus global minimization issue is commonly considered as a false problem in practice. Still, given the regulatory requirements to which banks are subject, the twin Monte Carlo validation procedures of IV.§2.A and IV.§3.A are important tools in the XVA context.

On the XVA side, the use of regression-based Monte Carlo simulations is not a novelty by itself. It was already presented in Cesari, Aquilina and Charpillon (2010) as a key CVA computational paradigm, intended to avoid nested Monte Carlo. However, from Cesari et al. (2010) to the neural net regressions of Huge and Savine (2020) or Gaussian process regressions as per Crépey and Dixon (2020), regressions are only used for computing the mark-to-market cube of the bank. The CVA of the bank at time 0 (only) is then obtained by integration of the so-called expected positive exposure relative to each netting set against the credit curve of the corresponding counterparty (see (Crépey, Bielecki and Brigo, 2014, Eqn. (3.25) page 76)), and summation over netting sets. Similar approaches are also used in the industry for computing simplistic FVA metrics (see XI.§4.A).

By contrast, in this book, we aim at learning the XVA metrics as processes, i.e. at every node of a simulation for all risk factors, based on a mark-to-market cube computed by model analytics at the forward simulation stage. Regressions for the mark-to-market of derivatives à la Cesari et al. (2010) are typically multiple parametric regressions in diffusive and low-dimensional setups, as opposed to high-dimensional neural net regressions in hybrid diffusive / Markov chain setups in our pathwise XVA case. Gnoatto, Picarelli and Reisinger (2023) deep-hedge and learn the CVA and the FVA, but this is in a purely diffusive setup, once the default of the bank and its (assumed single) counterparty eliminated from the equations by the reduction of filtration technique of Crépey and Song (2015). As clearly visible on Theorem III.1, in the realistic case of a bank involved in transactions with several (in practice, many, e.g. several thousands of) counterparties, this technique of reduction is useful for handling the default of the bank itself, but it cannot eliminate the defaults of the counterparties. Indeed these enter the ensuing FVA and KVA equations in a nonlinear fashion, so that, unless oversimplified XVA metrics are considered, there is no other choice than simulating these defaults and including them in the training as we do.

So, as opposed to the industry standard expected exposure based XVA computational approaches that are simpler but only applicable for deriving rather rudimentary XVA metrics at time 0, our algorithms require the simulation of the defaults of the counterparties of the bank. We refer the reader to Crépey, Bielecki and Brigo (2014, Part IV) regarding the simulation of the defaults in dynamic copula models with nontrivial credit/credit and credit/-market dependence, which should also be compatible with the property required in II.§3.A that the bank default time τ be an invariance time. Defaults simulation also raises a numerical variance issue requiring special care, which is the topic of Chapter V.

§2 NEURAL NET REGRESSION

For $(\mathcal{X}, \xi)$ random in $\mathbb{R}^d \times \mathbb{R}$ with ξ square integrable and d possibly "large", a neural net regression aims at approximating the function

$$\mathbb{R}^d \ni x \mapsto u(x) := \mathbb{E}(\xi | \mathcal{X} = x) \in \mathbb{R}. \tag{1}$$

We first recall the equivalence of minimizing, over any convex subset $\mathcal{S}$ of the set $\mathcal{B}(\mathbb{R}^d)$ of the Borel measurable functions $h : \mathbb{R}^d \to \mathbb{R}$ (cf. Eqn. (4) below), the mean quadratic error between $h(\mathcal{X})$ and

(i) cash flows ξ, i.e. minimizing the regression error $\mathbb{E}\left[(\xi - h(\mathcal{X}))^2\right]^{\frac{1}{2}}$, and

(ii) the conditional expectation $\mathbb{E}[\xi | \mathcal{X}]$, i.e. minimizing the "pricing error" $\mathbb{E}\left[\left(\mathbb{E}[\xi | \mathcal{X}] - h(\mathcal{X})\right)^2\right]^{\frac{1}{2}}$ (interpreting $\mathbb{E}[\xi | \mathcal{X}]$ as a price).

This equivalence, which can be seen geometrically as a consequence of Pythagora's theorem (see Figure 11), is established algebraically as follows:

Lemma 1. *Assuming ξ square integrable:*
(i) *For any function h in $\mathcal{B}(\mathbb{R}^d)$ such that $h(\mathcal{X})$ is square integrable,*

$$\mathbb{E}\left[(h(\mathcal{X}) - \xi)^2\right] = \mathbb{E}\left[\left(h(\mathcal{X}) - \mathbb{E}[\xi | \mathcal{X}]\right)^2\right] + \mathbb{E}\left[\mathbb{V}\mathrm{ar}(\xi | \mathcal{X})\right] \tag{2}$$

holds in $\mathbb{R}_+$;
(ii) *$u(\mathcal{X})$ (cf. (1)) is square integrable and*

$$u(\cdot) \in \underset{h \in \mathcal{B}(\mathbb{R}^d)}{\operatorname{argmin}} \mathbb{E}\left[\left(\xi - h(\mathcal{X})\right)^2\right]. \tag{3}$$

Proof. **(i)** For any $h \in \mathcal{B}(\mathbb{R}^d)$ such that $h(\mathcal{X})$ is square integrable, we have in $\mathbb{R}_+$:

$$\begin{aligned}\mathbb{E}\left[(h(\mathcal{X}) - \xi)^2\right] = \mathbb{E}\left[\left(h(\mathcal{X}) - \mathbb{E}[\xi | \mathcal{X}]\right)^2\right] + \mathbb{E}\left[\left(\mathbb{E}[\xi | \mathcal{X}] - \xi\right)^2\right] \\ + 2\mathbb{E}\Big[\left(h(\mathcal{X}) - \mathbb{E}[\xi | \mathcal{X}]\right)\left(\mathbb{E}[\xi | \mathcal{X}] - \xi\right)\Big],\end{aligned}$$

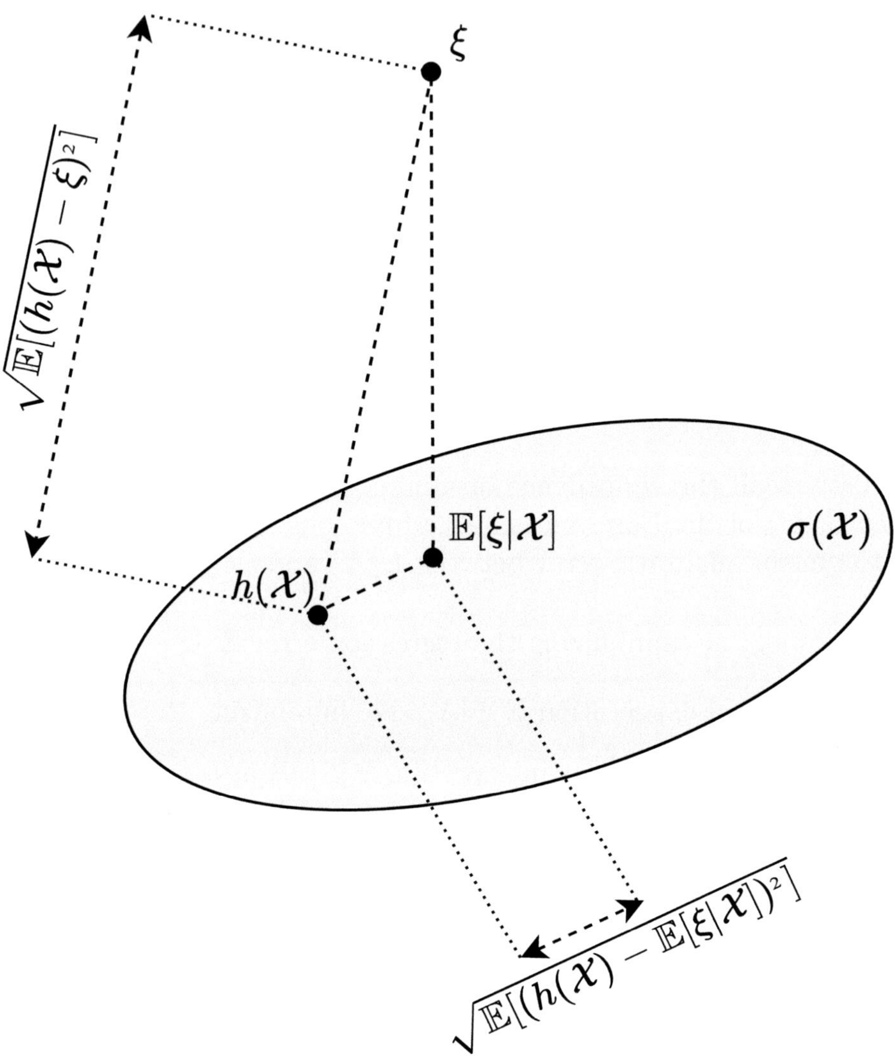

Figure 11: Regression error $\mathbb{E}\left[(\xi - h(\mathcal{X}))^2\right]^{\frac{1}{2}}$ vs. pricing error $\mathbb{E}\left[\left(\mathbb{E}[\xi\,|\mathcal{X}] - h(\mathcal{X})\right)^2\right]^{\frac{1}{2}}$.

where

$$\begin{aligned}
&\mathbb{E}\Big[\left(h(\mathcal{X}) - \mathbb{E}[\xi\,|\mathcal{X}]\right)\left(\mathbb{E}[\xi\,|\mathcal{X}] - \xi\right)\Big] \\
&\qquad = \mathbb{E}\Big[\mathbb{E}\Big[\left(h(\mathcal{X}) - \mathbb{E}[\xi\,|\mathcal{X}]\right)\left(\mathbb{E}[\xi\,|\mathcal{X}] - \xi\right)\,|\mathcal{X}\Big]\Big] = 0,
\end{aligned}$$

while

$$\mathbb{E}\Big[\Big(\mathbb{E}\left[\xi\,|\mathcal{X}\right]-\xi\Big)^{2}\Big]=\mathbb{E}(\xi^{2})-\mathbb{E}\Big[\Big(\mathbb{E}\left[\xi\,|\mathcal{X}\right]\Big)^{2}\Big]$$
$$=\mathbb{E}\Big[\mathbb{E}\left[\xi^{2}\,|\mathcal{X}\right]-\Big(\mathbb{E}\left[\xi\,|\mathcal{X}\right]\Big)^{2}\Big]=\mathbb{E}\left[\mathbb{V}\mathrm{ar}\left(\xi\,|\mathcal{X}\right)\right].$$

(ii) The square integrability of $u(\mathcal{X})$ follows from the conditional Jensen inequality. Moreover, by (2), where $\mathbb{E}\left[\mathbb{V}\mathrm{ar}\left(\xi\,|\mathcal{X}\right)\right]$, does not depend on h, we have for any subset $\mathcal{S}$ of $\mathcal{B}(\mathbb{R}^{d})$:

$$\underset{h\in\mathcal{S}}{\operatorname{argmin}}\,\mathbb{E}\Big[\Big(\xi-h(\mathcal{X})\Big)^{2}\Big]=\underset{h\in\mathcal{S}}{\operatorname{argmin}}\,\mathbb{E}\Big[\Big(\mathbb{E}\left[\xi\,|\mathcal{X}\right]-h\left(\mathcal{X}\right)\Big)^{2}\Big]. \tag{4}$$

For $\mathcal{S}=\mathcal{B}(\mathbb{R}^{d})$, this implies (3). □

A learning algorithm for the function u in (1) follows by replacing in the representation (3):

- $\mathbb{E}$, by a sample mean $\widehat{\mathbb{E}}$ over i.i.d. samples $(\mathcal{X}_j,\xi_j)$ of $(\mathcal{X},\xi)$;
- $\mathcal{B}(\mathbb{R}^{d})$, by an hypothesis space $\mathcal{H}$ of parameter functions $h=h_\theta(\mathrm{x})$, e.g. neural networks with a suitable architecture;
- global minimization by numerical minimization, e.g. Adam stochastic gradient descent (SGD) for neural networks.

A A Posteriori Twin Monte Carlo Validation Procedure

Switching from minimization w.r.t. functions $h(x)$ to minimization w.r.t. the weights and biases θ of neural network parameterizations $h_\theta(x)$ for $h(x)$ makes the resulting minimization problem finite dimensional but also nonconvex. As detailed in §1.A.1, there is no known learning algorithm solving neural network nonconvex minimization training problems with an a priori error bound. Moreover, in regression pricing setups, the quantity of interest is the pricing error, not the regression error (see Figure 11). However, when the learning terminates, the training error corresponds to the regression error and we do not know the pricing error.

But a key asset of probabilistic machine learning procedures for any conditional expectation is the availability of a companion "twin Monte Carlo validation procedure" allowing one to validate any predictor. Namely, assuming one has access to the generative process of the data (i.e. in probabilistic machine learning applications), one can estimate the distance of any guess $h(\mathcal{X})$ to the groundtruth (unknown) $u(\mathcal{X})=\mathbb{E}(\xi|\mathcal{X})$ using an a posteriori companion Monte Carlo procedure, dubbed "twin Monte Carlo" below. Let thus $\xi^{(1)}$ and $\xi^{(2)}$ denote two independent copies of ξ conditional on $\mathcal{X}$, i.e. such that $\mathbb{E}[\varphi_1(\xi^{(1)})\varphi_2(\xi^{(2)})|\mathcal{X}] = \mathbb{E}[\varphi_1(\xi^{(1)})|\mathcal{X}]\mathbb{E}[\varphi_2(\xi^{(2)})|\mathcal{X}]$ holds for any Borel bounded functions φ_1 and φ_2.

Proposition 1. *For any Borel function $h = h(x)$ (e.g. a neural net estimate of $\mathbb{E}[\xi|\mathcal{X}]$) such that $h(\mathcal{X})$ is square integrable, we have*

$$\mathbb{E}\Big[(h(\mathcal{X}) - \mathbb{E}[\xi|\mathcal{X}])^2\Big] = \mathbb{E}\Big[h(\mathcal{X})^2 - (\xi^{(1)} + \xi^{(2)})h(\mathcal{X}) + \xi^{(1)}\xi^{(2)}\Big]. \qquad (5)$$

***Proof*.** We compute

$$(h(\mathcal{X}) - \mathbb{E}[\xi|\mathcal{X}])^2 = (h(\mathcal{X}))^2 - 2h(\mathcal{X})\mathbb{E}[\xi|\mathcal{X}] + (\mathbb{E}[\xi|\mathcal{X}])^2,$$

where $2h(\mathcal{X})\mathbb{E}[\xi|\mathcal{X}] = \mathbb{E}[(\xi^{(1)} + \xi^{(2)})h(\mathcal{X})|\mathcal{X}]$ while, by conditional independence,

$$(\mathbb{E}[\xi|\mathcal{X}])^2 = \mathbb{E}[\xi^{(1)}|\mathcal{X}]\mathbb{E}[\xi^{(2)}|\mathcal{X}] = \mathbb{E}[\xi^{(1)}\xi^{(2)}|\mathcal{X}].$$

Taking outer expectations in the above yields (5). □

The twin Monte Carlo validation procedure for a predictor $h(\mathcal{X})$ of $u(\mathcal{X})$ is then based on a Monte Carlo estimate *twin-stat* for the right-hand side in (5): see Algorithm 1. But such a Monte Carlo estimate *twin-stat* can be negative. To remedy (at least partially) this limitation, the corresponding 95% upper bound, which is found to be positive most of the times, is calculated alongside the sample mean *twin-stat*.

Thus one can approximate the L_2 error of any predictor of a conditional expectation, without any knowledge on the latter, using only two inner paths per outer one. Moreover the accuracy of the ensuing twin Monte-Carlo estimates is controlled by the corresponding confidence interval. In the case where the twin Monte Carlo validation estimates, after having been confirmed to be accurate by drawing enough samples, are not good enough, one can improve the predictor h by playing with the three degrees of freedom provided by the three bullet points following Lemma 1.

A.1 Twin Monte Carlo Validation for BSDEs

Likewise, in the BSDE discretization setups of Chapter VII or Abbas-Turki, Crépey, Li and Saadeddine (2025) for FVA and KVA calculations, one can, iteratively in time, Monte Carlo estimate the L_2 norm of $\epsilon_{t_i} = |\widetilde{Y}_{t_i} - \widehat{Y}_{t_i}|$, where $\widetilde{Y}_{t_i}$ and $\widehat{Y}_{t_i}$ denote the theoretical (with true conditional expectation) and empirical (with trained conditional expectation) dynamic programming estimates of Y_{t_i} based on an estimate $\widehat{Y}_{t_{i+1}}$ for the solution $Y_{t_{i+1}}$ of the BSDE at the next time step t_{i+1}: see Abbas-Turki et al. (2025, Section 4).

§3 NEURAL NET QUANTILE REGRESSION

Most commonly used financial risk measures are value-at-risk (VaR) and expected shortfall (ES) at a given confidence (quantile) level $\alpha = 1-\epsilon$, for some small and positive ϵ (see Prerequisites in the preliminaries of the book). We just presented neural net regression for conditional expectations in a simulation environment, along with the corresponding a posteriori twin Monte Carlo validation procedure. Here we explain how this can be extended to conditional value-at-risk and expected shortfall computations.

Let ξ represent a future loss random variable, integrable with continuous and increasing cdf F, and let $\epsilon \in (0,1)$. Then, at the level of confidence α,

$$\begin{aligned} &\mathbb{V}\mathrm{a}\mathbb{R}(\xi) = F^{-1}(\alpha), \\ &\mathbb{ES}(\xi) = \mathbb{E}[\xi|\xi > \mathbb{V}\mathrm{a}\mathbb{R}(\xi)] = \epsilon^{-1}\mathbb{E}[\xi \mathbb{1}_{\{\xi > \mathbb{V}\mathrm{a}\mathbb{R}(\xi)\}}]. \end{aligned}$$

For $(\mathcal{X}, \xi)$ random in $\mathbb{R}^d \times \mathbb{R}$ with ξ integrable and d possibly "large", quantile regression aims at approximating the functions

$$\mathbb{R}^d \ni x \mapsto (q(x), s(x)) := (\mathbb{V}\mathrm{a}\mathbb{R}(\xi|\mathcal{X} = x), \mathbb{ES}(\xi|\mathcal{X} = x)) \in \mathbb{R}^2,$$

using spaces of hypotheses (e.g. families of functions represented in terms of

---

Algorithm 1: Twin Monte Carlo validation for a predictor $h(\mathcal{X})$ of $u(\mathcal{X}) = \mathbb{E}(\xi|\mathcal{X})$ (a nontrivial normalization *norm* $\neq 1$ is used in Table 11 page 184).

---

input : out-of-sample $\{(\mathcal{X}_j, \xi_j^{(1)}, \xi_j^{(2)})\}_{j=1}^m$ with independent draws $\mathcal{X}_j$ of $\mathcal{X}$ and $\xi_j^{(1)}, \xi_j^{(2)}$ independent copies of ξ given $\mathcal{X} = \mathcal{X}_j$; *norm* (set to 1 by default)

output: Estimation of the normalized root mean square relative error and its 95%-confidence upper bound for a predictor $h(\mathcal{X})$ of $\mathbb{E}(\xi|\mathcal{X})$

Compute $\textit{twin-stat} = \frac{1}{m}\sum_{j=1}^m \left[(h(\mathcal{X}_j))^2 - (\xi_j^{(1)} + \xi_j^{(2)})h(\mathcal{X}_j) + \xi_j^{(1)}\xi_j^{(2)}\right]$

Compute $\textit{twin-stdev} =$
$\sqrt{\frac{1}{m}\sum_{j=1}^m \left[(h(\mathcal{X}_j))^2 - (\xi_j^{(1)} + \xi_j^{(2)})h(\mathcal{X}_j) + \xi_j^{(1)}\xi_j^{(2)} - \textit{twin-stat}\right]^2}$

Compute $\textit{twin-up} = \sqrt{\textit{twin-stat} + \frac{2}{\sqrt{m}}\textit{twin-stdev}}$

if $\textit{twin-stat} > 0$ **then**

  Compute $\textit{twin-err} = \sqrt{\textit{twin-stat}}/\textit{norm}$

else

  Set *twin-err* to N/A

end

Output *twin-err* and $\textit{twin-ub} := \textit{twin-up}/\textit{norm}$

---

neural nets) $\mathcal{F}$ and $\mathcal{G} \subset \mathcal{B}(\mathbb{R}^d)$ for $q(\cdot)$ and $s(\cdot)$, respectively, and i.i.d. samples $(\mathcal{X}_j, \xi_j)$ of $(\mathcal{X}, \xi)$.

This is usually implemented by means of the pinball loss function $p(y, q) = \epsilon^{-1}(y - q)^+ + q$, based on the following representation of the (unconditional) VaR and ES.

Theorem 1 in Rockafellar and Uryasev (2000). *In the above setup, we have*

$$\mathbb{VaR}(\xi) = \underset{q \in \mathbb{R}}{\operatorname{argmin}}\, \mathbb{E}[p(\xi, q)] \tag{6}$$

$$\mathbb{ES}(\xi) = \mathbb{E}p(\xi, \mathbb{VaR}(\xi)) = \underset{s \in \mathbb{R}}{\operatorname{argmin}}\, \mathbb{E}\Big[\big(p(\xi, \mathbb{VaR}(\xi)) - s\big)^2\Big]. \tag{7}$$

Proof. $\mathbb{E}[p(\xi, q)]$ is convex in q and

$$\partial_q \mathbb{E}[p(\xi, q)] = -\epsilon^{-1}\mathbb{Q}[\xi > q] + 1,$$

which vanishes for $\mathbb{Q}[\xi > q] = \epsilon$, i.e. for $\mathbb{Q}[\xi \le q] = \alpha$, that is, for $q = \mathbb{VaR}(\xi)$. This proves (6). Moreover,

$$\mathbb{ES}(\xi) = \epsilon^{-1}\mathbb{E}[\xi \mathbb{1}_{\{\xi > \mathbb{VaR}(\xi)\}}] = \mathbb{E}[p(\xi, \mathbb{VaR}(\xi))].$$

This is the first identity in (7), from which the second one follows by the least-squares characterization of expectation (cf. (4)). $\square$

In the conditional case, we have the corresponding functional representations:

Lemma 2. *We have* $q(\cdot) \in \operatorname{argmin}_{f \in \mathcal{B}(\mathbb{R}^d)} \mathbb{E}p(\xi, f(\mathcal{X}))$. *If* ξ *is square integrable, then so are* $q(\mathcal{X})$ *and* $s(\mathcal{X})$ *and*

$$s(\cdot) \in \underset{g \in \mathcal{B}(\mathbb{R}^d)}{\operatorname{argmin}}\, \mathbb{E}\Big[\big(p(\xi, q(\mathcal{X})) - g(\mathcal{X})\big)^2\Big].$$

Proof. For any constant $x \in \mathbb{R}$ and function $f \in \mathcal{B}(\mathbb{R}^d)$, (6) yields

$$\epsilon^{-1}\mathbb{E}\left((\xi - q(x))^+ | \mathcal{X} = x\right) + q(x) \le \epsilon^{-1}\mathbb{E}\left((\xi - f(x))^+ | \mathcal{X} = x\right) + f(x).$$

Hence

$$\epsilon^{-1}\mathbb{E}\left((\xi - q(x))^+ | \mathcal{X}\right) + q(\mathcal{X}) \le \epsilon^{-1}\mathbb{E}\left((\xi - f(\mathcal{X}))^+ | \mathcal{X}\right) + f(\mathcal{X})$$

and, by taking expectation,

$$\mathbb{E}[\epsilon^{-1}(\xi - q(x))^+ + q(\mathcal{X})] \le \mathbb{E}[\epsilon^{-1}(\xi - f(\mathcal{X}))^+ + f(\mathcal{X})].$$

This proves the stated argmin property of $q(\cdot)$. Assuming ξ square integrable, the square integrability of $q(\mathcal{X})$ and $s(\mathcal{X})$ follows from the inequalities

$-\alpha^{-1}\mathbb{E}(\xi^{-}|\mathcal{X}) \leq q(\mathcal{X}) \leq \epsilon^{-1}\mathbb{E}(\xi^{+}|\mathcal{X})$ and $|\mathbb{E}(\xi|\mathcal{X})| \leq \epsilon^{-1}\mathbb{E}(|\xi|\,|\,\mathcal{X})$ that hold with probability one (Barrera, Crépey, Gobet, Nguyen and Saadeddine, 2024, Lemma 2.2). Since

$$s(\mathcal{X}) = \epsilon^{-1}\mathbb{E}[(\xi - q(\mathcal{X}))^{+}\,|\mathcal{X}] + q(\mathcal{X}) = \mathbb{E}[\epsilon^{-1}\,(\xi - q(\mathcal{X}))^{+} + q(\mathcal{X})|\mathcal{X}],$$

the argmin representation stated for $s(\cdot)$ follows from (3). □
The (conditional) VaR and ES learning algorithm follows by replacing in the argmin properties of $q(\cdot)$ and $s(\cdot)$ provided by Lemma 2:

- $\mathbb{E}$, by sample means $\widehat{\mathbb{E}}$ over i.i.d. samples $(\mathcal{X}_j, \xi_j)$ of $(\mathcal{X}, \xi)$,
- $\mathcal{B}(\mathbb{R}^d)$, by suitable hypothesis spaces $\mathcal{F}$ and $\mathcal{G}$ (e.g. neural networks) for $q(\cdot)$ and $s(\cdot)$,
- global minimization by numerical minimization, e.g. Adam SGD for neural networks.

Neural nets often show particularly good generalization and scalability performances (in the XVA setup, see VII.§3.E). In the case of conditional value-at-risk and expected shortfall computations, neural quantile regression is also easier to implement than more naive methods, such as the resimulation and sort-based scheme of Barrera, Crépey, Diallo, Fort, Gobet and Stazhynski (2019) for the value-at-risk and expected shortfall at each outer node of a nested Monte Carlo simulation. In particular, when neural networks are used as hypothesis spaces, for estimating $s(\cdot)$, retraining with a least squares loss function the last layer of the neural network, already obtained for $q(\cdot)$ is typically enough (Barrera et al., 2024, Section 4.2).

A A Posteriori Twin Monte Carlo Validation Procedure

What we showed for a conditional expectation $u(\cdot) = \mathbb{E}[\xi|\mathcal{X} = \cdot]$ in §2.A is extendible to conditional VaR $q(\cdot)$ and ES $s(\cdot)$ as follows. The related comments apply mutatis mutandis.

Proposition 2. *Let f and g in $\mathcal{B}(\mathbb{R}^d)$ be two tentative approximations of q and s. Introducing two conditionally independent copies $\xi^{(1)}$ and $\xi^{(2)}$ of ξ given $\mathcal{X}$, we have*

$$\mathbb{E}\Big[\Big(\mathbb{Q}[\xi \geq f(\mathcal{X})|\mathcal{X}] - \epsilon\Big)^2\Big] = \epsilon^2 - \epsilon(\mathbb{Q}(\xi^{(1)} > f(\mathcal{X})) + \mathbb{Q}(\xi^{(2)} > f(\mathcal{X}))) + \mathbb{Q}[\xi^{(1)} \wedge \xi^{(2)} > f(\mathcal{X})] \quad (8)$$

and a control bound detailed in Barrera et al. (2024, Proposition 3.4(ii)), similarly amenable to twin Monte Carlo simulation (see after Proposition 1), holds for $\mathbb{E}\Big[\Big(g(\mathcal{X}) - s(\mathcal{X})\Big)^2\Big]$.

Proof. We have

$$\mathbb{E}\Big[\big(\mathbb{Q}[\xi \geq f(\mathcal{X})|\mathcal{X}] - \epsilon\big)^2\Big] = \mathbb{E}[\big(\mathbb{Q}[\xi \geq f(\mathcal{X})|\mathcal{X}]\big)^2] + \epsilon^2 - 2\epsilon\mathbb{Q}[\xi \geq f(\mathcal{X})],$$

where

$$\begin{aligned}\big(\mathbb{Q}[\xi \geq f(\mathcal{X})|\mathcal{X}]\big)^2 &= \mathbb{Q}[\xi^{(1)} \geq f(\mathcal{X})|\mathcal{X}]\mathbb{Q}[\xi^{(2)} \geq f(\mathcal{X})|\mathcal{X}] \\ &= \mathbb{Q}[\xi^{(1)} \wedge \xi^{(2)} \geq f(\mathcal{X})|\mathcal{X}]. \ \Box\end{aligned}$$

§4 MARKET AND CREDIT MODEL IN CONTINUOUS TIME

For our XVA numerical experiments in this part of the book, we consider a bank trading derivative contracts in different economies $\langle e \rangle$ with various bilateral counterparties b. The currency corresponding to the economy labeled by $\langle 0 \rangle$ is taken as the reference currency, namely the one of the reference bank, indexed by 0, which XVA metrics are computed. Let there be given the short rate process $r^{\langle e \rangle}$ in each economy $\langle e \rangle$, as well as the exchange rate process $\chi^{\langle e \rangle}$ from the currency of each economy $\langle e \rangle$ to the reference currency $\langle 0 \rangle$ (including, for notational convenience, $\chi^{\langle 0 \rangle} \equiv 1$). The risk-neutral measure corresponding to each numeraire $\exp(\int_0^{\cdot} r_s^{\langle e \rangle} ds)$ is denoted by $\mathbb{Q}^{\langle e \rangle}$. Our standing probability measures $\mathbb{Q}^*$ and $\mathbb{Q}$ in this book should then be taken as $\mathbb{Q}^{\langle 0 \rangle}$ and the associated survival probability measure of the reference bank. In such a multi-currency setup, we recall the no-arbitrage consistency condition

$$\frac{d\mathbb{Q}^{\langle e \rangle}}{d\mathbb{Q}^{\langle 0 \rangle}}\bigg|_{\mathfrak{G}_t} = \exp\Big(\int_0^t (r_s^{\langle e \rangle} - r_s^{\langle 0 \rangle})ds\Big)\frac{\chi_t^{\langle e \rangle}}{\chi_0^{\langle e \rangle}} \tag{9}$$

stemming from the fact that, for any non-dividend-paying (European) asset Z priced in a currency $e \geq 1$, $\exp(-\int_0^{\cdot} r_s^{\langle e \rangle} ds)Z$ and $\exp(-\int_0^{\cdot} r_s^{\langle 0 \rangle} ds)\chi^{\langle e \rangle} Z$ should be martingales with respect to $\mathbb{Q}^{\langle e \rangle}$ and $\mathbb{Q}^{\langle 0 \rangle}$ respectively, hence

$$\mathbb{E}^{\mathbb{Q}^{\langle e \rangle}}[e^{-\int_0^t r_s^{\langle e \rangle} ds} Z_t] = Z_0 = \frac{1}{\chi_0^{\langle e \rangle}} \mathbb{E}^{\mathbb{Q}^{\langle 0 \rangle}}[e^{-\int_0^t r_s^{\langle 0 \rangle} ds} \chi_t^{\langle e \rangle} Z_t].$$

Each counterparty b of the bank is supposed to have a stochastic default time τ^b with default intensity process γ^b and we denote by $\gamma^0 \equiv \gamma$ the default intensity of the reference bank.

A Dynamics

For every economy $\langle e \rangle$ and counterparty b (or the bank itself "$b = 0$"), let $B^{r,\langle e \rangle}$, $B^{\chi,\langle e \rangle}$ and $B^{\gamma,b}$ denote $\mathbb{Q}^{\langle 0 \rangle}$ Brownian motions. For every economy $\langle e \rangle$, the short-rate $r^{\langle e\rangle}$ and the exchange rate $\chi^{\langle e\rangle}$ against the reference currency respectively follow Vasicek (cf. (12)) and log-normal dynamics

$$\begin{aligned} dr_t^{\langle e\rangle} &= \left(a^{\langle e\rangle}(b^{\langle e\rangle} - r_t^{\langle e\rangle}) - \rho^{\langle e\rangle}\sigma^{r,\langle e\rangle}\sigma^{\chi,\langle e\rangle}\right)dt + \sigma^{r,\langle e\rangle}dB_t^{r,\langle e\rangle} \\ d\log\chi_t^{\langle e\rangle} &= (r_t^{\langle 0\rangle} - r_t^{\langle e\rangle} - \frac{1}{2}|\sigma^{\chi,\langle e\rangle}|^2)dt + \sigma^{\chi,\langle e\rangle}dB_t^{\chi,\langle e\rangle}. \end{aligned} \tag{10}$$

For both the bank ("$b = 0$") and every bilateral counterparty b ($\neq 0$), the default intensity process γ^b follows CIR dynamics

$$d\gamma_t^b = \alpha^b(\delta^b - \gamma_t^b)dt + \nu^b\sqrt{\gamma_t^b}dB_t^{\gamma,b}. \tag{11}$$

In the above, $a^{\langle\cdot\rangle}$, $b^{\langle\cdot\rangle}$, $\sigma^{\cdot,\langle\cdot\rangle}$, $\alpha^{\cdot}$, $\delta^{\cdot}$, $\nu^{\cdot}$ are model parameters meant to be calibrated using liquid market instruments (simply set to "arbitrary but reasonable values" in our numerics), and so is the correlation $\rho^{\langle e\rangle}$ between $B^{r,\langle e\rangle}$ and $B^{\chi,\langle e\rangle}$. By (9),

$$\left.\frac{d\mathbb{Q}^{\langle e\rangle}}{d\mathbb{Q}^{\langle 0\rangle}}\right|_{\mathfrak{G}_t} = \exp\left(\sigma^{\chi,\langle e\rangle}B_t^{\chi,\langle e\rangle} - \frac{1}{2}(\sigma^{\chi,\langle e\rangle})^2 t\right).$$

Hence, by Girsanov's theorem,

$$\widetilde{B}_t^{r,\langle e\rangle} = B_t^{r,\langle e\rangle} - \sigma^{\chi,\langle e\rangle}\rho^{\langle e\rangle}t$$

is a $\mathbb{Q}^{\langle e\rangle}$ standard Brownian motion and we have

$$dr_t^{\langle e\rangle} = a^{\langle e\rangle}(b^{\langle e\rangle} - r_t^{\langle e\rangle})dt + \sigma^{r,\langle e\rangle}d\widetilde{B}_t^{r,\langle e\rangle}. \tag{12}$$

For every counterparty b, its default time τ^b is modeled as a the stopping time $\inf\{t > 0; \int_0^t \gamma_s^b ds \geq \epsilon^b\}$, where ϵ^b is an independent standard exponential random variable. That is, for every $t \geq 0$,

$$\mathbb{1}_{\{\tau^b \leq t\}} = 1 \Leftrightarrow \tau^b \leq t \Leftrightarrow \int_0^t \gamma_s^b ds \geq \epsilon^b. \tag{13}$$

Soft market/credit and/or credit/credit dependence can be conveyed via the default intensities. Hard credit/credit dependence can be implemented by common shock extensions of this basic default setup along the lines of Crépey, Bielecki and Brigo (2014, Chapters 8–10) (see also Crépey and Song (2016) and Elouerkhaoui (2007, 2017)).

B Portfolio

For the instruments, we assume a book comprised of interest rate swaps at par at inception. For each swap, we denote the set of its reset dates by $reset$ and by t_- and t_+ the reset dates respectively immediately preceding and following

t. Note that the price of a zero-coupon bond is given in closed form in the above model. We assume that successive reset dates are regularly spaced by δ, that the swap is spot starting, i.e. $0 \in reset$, and that the swap is paying fixed $\delta\Sigma$, where Σ is the swap rate, and receiving floating $\frac{1}{\mathrm{ZC}_{t_-}(t)} - 1$, where $\mathrm{ZC}_t(t')$ is the price of a zero-coupon bond at time t with maturity t', at each reset date $t \in reset \setminus \{0\}$. Denoting by P_t^{sw} the price of the swap at time t in units of the underlying currency, we have for all $t \leq t_{max} := \max reset$:

$$P_t^{sw} = \begin{cases} \dfrac{\mathrm{ZC}_t(t_+)}{\mathrm{ZC}_{t_-}(t_+)} - \mathrm{ZC}_t(t_{max}) - \delta\Sigma \displaystyle\sum_{t' \in reset, t' > t} \mathrm{ZC}_t(t') & \text{if } t \notin reset \setminus \{0\} \\ \dfrac{1}{\mathrm{ZC}_{t_-}(t)} - \mathrm{ZC}_t(t_{max}) - \delta\Sigma(1 + \displaystyle\sum_{t' \in reset, t' > t} \mathrm{ZC}_t(t')) & \text{if } t \in reset \setminus \{0\} \\ 1 - \mathrm{ZC}_0(t_{max}) - \delta\Sigma \displaystyle\sum_{t' \in reset \setminus \{0\}} \mathrm{ZC}_0(t') & \text{if } t = 0. \end{cases}$$

These prices are then to be multiplied by the cross-currency exchange rate processes to have all prices in the reference currency.

Remark 1. The path-dependence induced by the previous reset date can be resorbed by including the short rates of that date among the risk factors in the model.

CHAPTER V

Pathwise CVA Regressions With Oversimulated Defaults

In this chapter, with XVAs in view, we consider the computation by simulation and neural net regression of conditional expectations, or more general elicitable statistics, of functionals of processes (X, Y). Here an exogenous component Y (Markov by itself, e.g. market risk factors) is time-consuming to simulate, while the endogenous component X (jointly Markov with Y, e.g. client defaults) is quick to simulate given Y, but is responsible for most of the variance of the simulated payoff. To address the related variance issue, we introduce a conditionally independent, hierarchical simulation scheme, where several paths of X are simulated for each simulated path of Y. We derive heuristics on the number of paths of Y and, for each of them, of X, that should be simulated. The resulting algorithm is implemented on a graphics processing unit (GPU) combining Python/CUDA and learning with PyTorch. A CVA case study with a nested Monte Carlo benchmark shows that the hierarchical simulation technique is key to the success of the learning approach on hybrid diffusive and default features.

§1 INTRODUCTION

A Financial Motivation

Greensill defaulted on March 8, 2021, a collapse estimated by British parliamentarians to trigger a cost for UK taxpayers of up to £5bn: cf. https://www.theguardian.com/business/2021/apr/28/greensill-collapse-could-cost-uk-taxpayer-up-to-5bn-mps-told, accessed on May 16, 2022. Greensill fell short of capital because they lent to Gupta against future invoices which then did not materialize. A projection of Greensill's capital requirements including the tail risk related to Gupta's default, would have highlighted a sizable concentrated and unsecured credit risk to a junk-rated counterparty. This example emphasizes the importance of performing proper default risk simulations, as

DOI: 10.1201/9781003614944-9

opposed to credit spread simulations simply, as typically done in the industry for the sake of simplicity. Another example (in fact, our main motivation for the purposes of this book) is the pathwise XVA regression approach of Chapter VII, which also requires a hybrid market and credit setup, where the actual defaults of the clients of the bank are simulated.

However, to obtain the required level of accuracy in a simulation setup with defaults, one needs a very large number of simulations—100 times more, say, than in a purely diffusive model, where 100 is in reference to credit spreads that would be of the order of 1%. A factor of 100 is not necessarily a sizable amount as far as the simulation of the risk factors is involved. But it also means that the mark-to-market cube of pathwise prices of all trades of the bank becomes 100 times bigger. When applied at the level of a realistic banking portfolio, this becomes prohibitive in terms of both computation time and memory occupancy.

B Contribution and Outline of the Chapter

To overcome this problem, we introduce an acceleration technique for the computation by simulation and regression of conditional expectations of functions Y of Markov pairs (X, Y), where Y is an exogenous component, Markov by itself, whose simulation is time-consuming, while the endogenous component X (jointly Markov with Y) is quick to simulate given Y, but also responsible for most of the variance of the simulated payoff. The idea, which we call hierarchical simulation, is then to draw an optimized number of realizations of X conditional on each simulation of Y. For example, in the above-mentioned XVA regression framework, we simulate a few hundred paths of client defaults X conditionally on each path of the market risk factors Y. Proceeding in this way, the computational burden of the mark-to-market cube is not amplified by the simulation of the client defaults.

In §2, we introduce a neural net learning framework for conditional expectations, iterated in time as they appear naturally in dynamic pricing problems, taking into account the dynamics of the problem by means of a backward pricing algorithm. In §3, we identify the variance issue raised by the coexistence of risk factors evolving at different paces in the problem (e.g. market risk versus default indicator processes) and we propose a hierarchical simulation approach to address it. §4 illustrates our approach with a CVA numerical case study.

Remark 1. Although our CVA case study only covers quadratic risk minimization (for benchmarking reasons), the approach of this chapter is valid for more general loss functions and applies to the learning of any elicitable statistics. In particular, via the Rockafellar and Uryasev (2000) representation of value-at-risk and expected shortfall of a given loss (random variable) in terms of "far out-of-the-money call options" on that loss (see IV.(6)-(7)), our hierarchical simulation approach is also relevant for learning conditional value-at-risk and expected shortfall as per IV.§3 in hybrid market and default simulation setups. Such an approach is even particularly relevant in these cases, where the fact that defaults are responsible for most of the variance of the payoff is then intrinsic to the far out-of-the-money feature of the corresponding "option". The oversimulation technique of the present chapter is therefore also crucial for the computations of Chapter VII.

§2 NEURAL REGRESSION SETUP

Neural regressions were addressed at a general level in IV.§2. Here we provide more details in the case of interest in this chapter. A reference probability space, with corresponding probability measure and expectation denoted by $\mathbb{Q}$ and $\mathbb{E}$, is fixed throughout the chapter. The state spaces of X and Y are taken as $\mathbb{R}^p$ and $\mathbb{R}^q$, for some positive integers p and q. We identify $\mathbb{R}^{p+q}$ with $\mathbb{R}^p \times \mathbb{R}^q$ and write $\phi(z)$ or $\phi(x, y)$ interchangeably, where z is the concatenation of x and y, for every function ϕ defined over $\mathbb{R}^{p+q}$ or $\mathbb{R}^p \times \mathbb{R}^q$, and for every $x \in \mathbb{R}^p$ and $y \in \mathbb{R}^q$.

In the (default risk) case of a Markov chain-like component X, referred to hereafter as the Markov chain X case (but with transition intensities modulated by Y), we assume, without loss of generality in this case, that X evolves on the vertices $\{0, 1\}^p$ of the unit cube in $\mathbb{R}^p$. We take the problem after discretisation of time (if the latter was continuous in the first place), for a time step set to one year for ease of notation.

We then consider $(X_i)_{0 \leq i \leq n}$ and $(Y_i)_{0 \leq i \leq n}$ as discrete-time processes on the time grid. Our goal is to estimate, for every i, conditional expectations of the form

$$\Pi_i = \mathbb{E}[\xi_{i,n} \,|X_i, Y_i], \tag{1}$$

where

$$\xi_{i,n} = f_i(X_i, \ldots, X_n, Y_i, \ldots, Y_n). \tag{2}$$

Here f_i is a measurable real function such that $\xi_{i,n}$ is a square-integrable random variable.

Conditional expectations such as (1) can be estimated via linear regression using a finite sample. This is ubiquitous in quantitative finance since the Bermudan Monte Carlo papers of Tsitsiklis and Van Roy (2001) and Longstaff and Schwartz (2001). In order to estimate the conditional expectation in (1), one draws i.i.d. samples $\{(X_i^\imath, Y_i^\imath, \xi_{i,n}^\imath)\}_{\imath \in I}$ of $(X_i, Y_i, \xi_{i,n})$, where I is a finite set of indices. Then, given a feature map $\phi : \mathbb{R}^{p+q} \to \mathbb{R}^m$ (for some positive integer m), one linearly regresses $\{\xi_{i,n}^\imath\}_{\imath \in I}$ against $\{\phi(X_i^\imath, Y_i^\imath)\}_{\imath \in I}$, solving for

$$\widehat{w}_i \in \operatorname*{argmin}_{w_i \in \mathbb{R}^m} \sum_{\imath \in I} (\xi_{i,n}^\imath - w_i^\top \phi(X_i^\imath, Y_i^\imath))^2. \tag{3}$$

One then uses $\widehat{w}_i^\top \phi(X_i, Y_i)$ as an approximation for Π_i.

The above procedure is justified by the characterization, in the square integrable case, of conditional expectations as orthogonal projections (cf. Lemma IV.1(ii)), i.e.

$$\mathbb{E}[\xi_{i,n} \,|X_i, Y_i] = \Phi_i^\star(X_i, Y_i) \quad \text{a.s.},$$

where, denoting by $\mathcal{B}(E)$ the set of Borel measurable real functions on a metric space E,

$$\Phi_i^\star \in \underset{\Phi \in \mathcal{B}(\mathbb{R}^p \times \mathbb{R}^q)}{\operatorname{argmin}} \; \mathbb{E}[(\xi_{i,n} - \Phi(X_i, Y_i))^2]. \tag{4}$$

One recovers the linear regression formulation (3) by approximating the expectation by an empirical mean and restricting the search space to the functions of the form $\mathbb{R}^{p+q} \ni (x, y) \mapsto w_i^\top \phi(x, y)$, where $w_i \in \mathbb{R}^m$.

A Neural Net Parameterization

Linear regression by means of a priori, explicit factors has a reasonable chance of success when $\Phi_i^\star$ is simple enough and the feature mapping ϕ can be judiciously chosen, usually from expert knowledge. This is, however, not always the case, e.g. when considering portfolio-wide XVA metrics, which exhibit non-trivial dependencies on the many risk factors being regressed against. It is then impossible to manually devise a satisfactory feature mapping ϕ. Figure 12 shows how a linear regression with the raw risk factors as features fails in learning the conditional CVA of an elementary portfolio made of a single call option, while the neural net estimator almost matches with the nested Monte Carlo estimator (see §4 for more numerical details). In the Markov

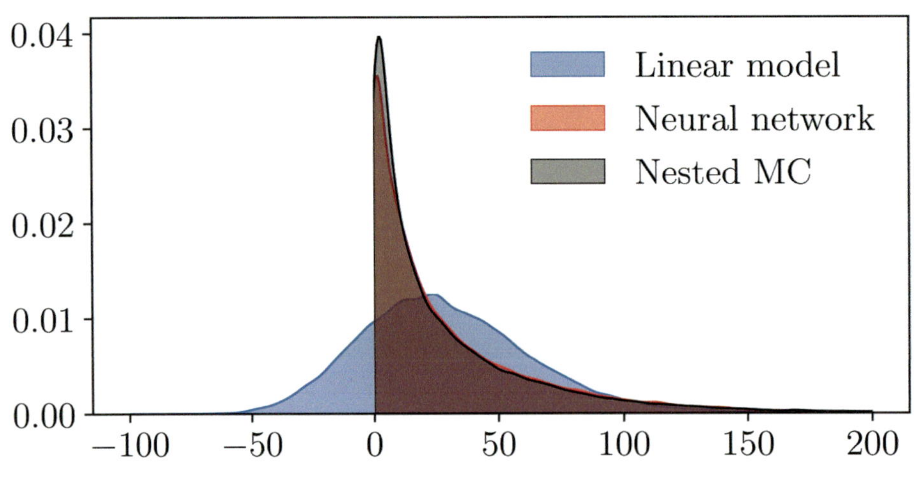

Figure 12: Density plot of the CVA of a vanilla call, at mid-life of the option.

chain X case, we face the additional peculiarity of a hybrid regression setting, with discrete and continuous natures of the X and Y model components.

As explained in IV.§2, neural networks propose an alternative way to parameterize and learn the feature map. Let $\mathcal{NN}_{p+q,h,u,\varsigma}$ denote the set of functions of the form

$$\mathbb{R}^{p+q} \ni z \mapsto \zeta(z; W^{[h+1]}, \ldots, W^{[1]}, b^{[h+1]}, \ldots, b^{[1]}) = \zeta^{[h+1]}(z; W, b),$$

where

$$W^{[h+1]} \in \mathbb{R}^{1\times u}, \ldots, W^{[\ell]} \in \mathbb{R}^{u\times u}, \ldots, W^{[1]} \in \mathbb{R}^{u\times(p+q)}$$

are weight matrices,

$$b^{[h+1]} \in \mathbb{R}, \ldots, b^{[\ell]} \in \mathbb{R}^{u}, \ldots, b^{[1]} \in \mathbb{R}^{u}$$

are bias offsets, W and b are the respective concatenations of the $W^{[\ell]}$ and of the $b^{[\ell]}$, ς is a scalar nonlinearity applied element-wise and, for every $z \in \mathbb{R}^{p+q}$,

$$\begin{aligned}
&\zeta^{[0]}(z; W, b) = z \\
&\zeta^{[\ell]}(z; W, b) = \varsigma(W^{[\ell]}\zeta^{[\ell-1]}(z; W, b) + b^{[\ell]}), \quad \ell = 1, \ldots, h \\
&\zeta^{[h+1]}(z; W, b) = W^{[h+1]}\zeta^{[h]}(z; W, b) + b^{[h+1]}.
\end{aligned}$$

The function $z \mapsto \zeta^{[h]}(z; W, b)$ can be seen as a nonlinear feature mapping from $\mathbb{R}^{p+q}$ to $\mathbb{R}^{u}$, parameterized by $W^{[h]}, \ldots, W^{[1]}, b^{[h]}, \ldots, b^{[1]}$ (for a given activation function ς): see Figure 13 and Goodfellow, Bengio and Courville (2016).

On top of the set $\mathcal{NN}_{p+q,h,u,\varsigma}$ of real-valued neural networks taking inputs from $\mathbb{R}^{p+q}$, with h hidden layers, u units per hidden layer (and one output neuron), and ς as the activation function, we also define

$$\mathcal{NN}^{+}_{p+q,h,u,\varsigma} = \{\mathbb{R}^{p+q} \ni z \mapsto (f(z))^{+} + \mu, f \in \mathcal{NN}_{p+q,h,u,\varsigma}, \mu \in \mathbb{R}\}. \quad (5)$$

This specification ensures positivity of the output when the additive constant μ is nonnegative and is useful for learning positive (e.g. XVA) functions. The additive constant μ is introduced in order to improve the fit of the first moment of the target function.

B Local Training Algorithm

Learning the conditional expectation (1) in a positive neural net search space consists in applying the same empirical risk minimization (3) approximation as in linear regression, using this time $\mathcal{NN}^{+}_{p+q,h,u,\varsigma}$ as the search space, i.e. solving for

$$\widehat{\Phi}_i \in \underset{\Phi\in\mathcal{NN}^{+}_{p+q,h,u,\varsigma}}{\operatorname{argmin}} \sum_{\imath\in I}(\xi^{\imath}_{i,n} - \Phi(X^{\imath}_{i}, Y^{\imath}_{i}))^2. \quad (6)$$

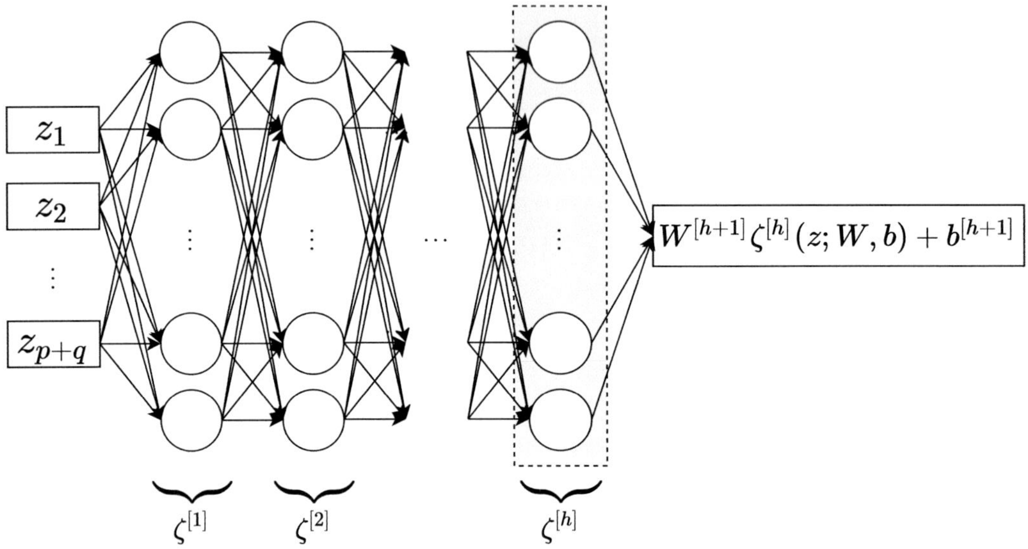

Figure 13: A feedforward neural network. The last hidden layer of our network, $\zeta^{[h]}$, can be seen as a parameterization of an endogenous, implicit linear regression basis.

This is achieved by mini-batch stochastic gradient descent.

For learning a positive output (e.g. an XVA function at some given positive pricing time), the addition of a ReLU activation $\cdot^+$ at the output layer in (5) can jeopardize the learning as the gradient may vanish at a certain SGD iteration and the parameters are then frozen irrespective of the number of subsequent iterations. Thus, for more stability of the learning procedure, we first perform the first half of SGD steps on the network without the ReLU at the output layer. Then, still without the ReLU, we fine-tune the weights of the output layer by optimizing with respect to those weights only (freezing the weights of the hidden layers), which can be done in closed form in the case of quadratic risk minimization.

Remark 2. This fine-tuning step is not achievable in closed-form in the case of quantile regression (cf. IV.§3 and Remark 1 page 136). However, the corresponding output optimization problem is still convex and, as such, easier to solve numerically.

Finally, we restore the ReLU at the output layer to proceed with the second half of the SGD iterations.

We also chose to retain the best set of parameters among those explored during the SGD iterations. Figure 14 shows the corresponding improvement in generalization when applied in the context of the CVA case study of §4. The ensuing learning scheme is detailed in Algorithm 2. Note that we presented vanilla SGD iterations only for the sake of simplicity. In practice, accelerated SGD methods like Adam (Kingma and Ba (2014)) are used instead.

Algorithm 2: Baseline training scheme for learning a conditional expectation as per (6) at a given time-step i

---

name : BaseAlg

input : $\{(X_i^\imath, Y_i^\imath, \xi_{i,n}^\imath), \imath \in I\}$, a partition B of I into batches, a number of epochs $E \in \mathbb{N}^\star$, a learning rate $\eta > 0$, initial values for the network parameters W, b and μ

output: Trained parameters W_{best}, b_{best} and μ_{best}

define $\mathcal{L}(W, b, \mu, \text{batch}, \text{pos}) =$

$$\begin{cases} \dfrac{1}{|\text{batch}|} \displaystyle\sum_{\imath \in \text{batch}} (\zeta^{[h+1]}(X_i^\imath, Y_i^\imath; W, b) + \mu - \xi_{i,n}^\imath)^2 & \text{if pos} = 0 \\ \dfrac{1}{|\text{batch}|} \displaystyle\sum_{\imath \in \text{batch}} ((\zeta^{[h+1]}(X_i^\imath, Y_i^\imath; W, b))^+ + \mu - \xi_{i,n}^\imath)^2 & \text{if pos} = 1 \end{cases}$$

$\mathcal{L}_{\text{best}} \leftarrow \infty$, pos $\leftarrow 0$

for $epoch = 1, \ldots, E$ **do** `// loop over epochs`

  for $batch \in B$ **do** `// loop over batches`

    for $\ell = 1, \ldots, h+1$ **do**

      $W^{[\ell]} \leftarrow W^{[\ell]} - \eta \nabla_{W^{[\ell]}} \mathcal{L}(W, b, \mu, \text{batch}, \text{pos})$

      $b^{[\ell]} \leftarrow b^{[\ell]} - \eta \nabla_{b^{[\ell]}} \mathcal{L}(W, b, \mu, \text{batch}, \text{pos})$

    end

    $\mu \leftarrow \mu - \eta \partial_\mu \mathcal{L}(W, b, \mu, \text{batch}, \text{pos})$

  end

  if $epoch = \left\lfloor \frac{E}{2} \right\rfloor$ **then** `// tune weights of last layer`

    $(W^{[h+1]}, b^{[h+1]}) \leftarrow$

    $\underset{\widetilde{W}^{[h+1]}, \widetilde{b}^{[h+1]}}{\operatorname{argmin}} \mathcal{L}(\{W^{[0]}, \ldots, W^{[h]}, \widetilde{W}^{[h+1]}\}, \{b^{[0]}, \ldots, b^{[h]}, \widetilde{b}^{[h+1]}\}, \mu, obs, 0)$

    pos $\leftarrow 1$

  end

  if $\mathcal{L}(W, b, \mu, I, 1) < \mathcal{L}_{best}$ **then** `// keep track of best parameters`

    $\mathcal{L}_{\text{best}} \leftarrow \mathcal{L}(W, b, \mu, obs, 1)$

    $W_{\text{best}} \leftarrow W$

    $b_{\text{best}} \leftarrow b$

    $\mu_{\text{best}} \leftarrow \mu$

  end

end

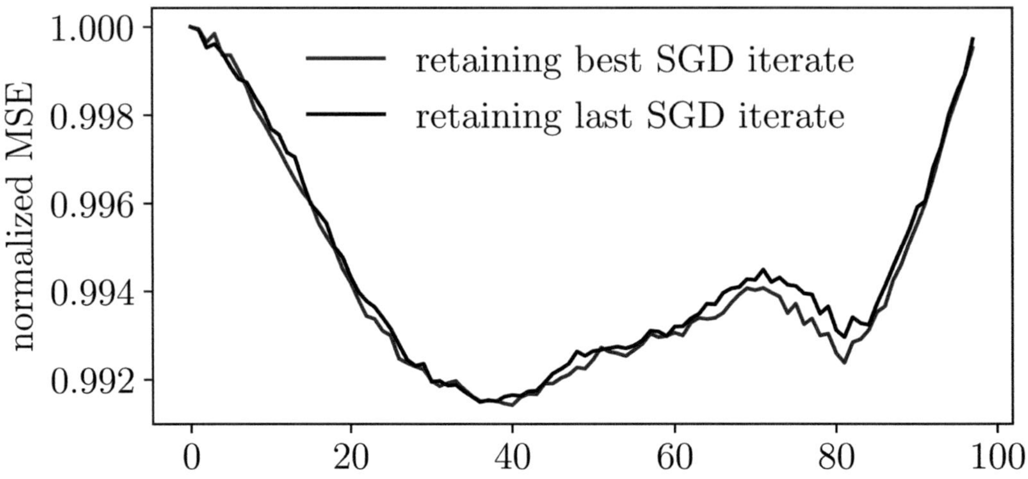

Figure 14: Out-of-sample MSEs against labels $\xi_{i,n} \equiv \tilde{\xi}^{cva}_{i,n}$ as per (12) at different time-steps divided by the variance of the labels in the context of the CVA case study of §4.

C Backward Learning

In the setup of the pathwise pricing problem (1), at each pricing time i, a separate learning problem is solved by Algorithm 2 page 141. Since the algorithm returns for each problem a local minimum, it is possible to end up with an approximation of the pricing function $\mathbb{E}[\xi_{i,n} | X_i = x, Y_i = y]$ (cf. (1)) with noisy paths (i.e. with respect to time i) if the local minima are not close to each other, even for fixed x and y. Yet, for two consecutive time-steps i and $i+1$, the learning problems are similar. One possible refinement is, after having learned Π_{i+1}, to initialize the parameters of the network at time i with the parameters of the network trained at time $i+1$. This not only smoothes the results across regression times, but also accelerates convergence. Early stopping could be used to help reduce training time further while improving regularization.

We obtain an algorithm which starts the learnings at the last time step n and, proceeding backward in time until time step 1, reuses each time the previously trained weights and biases as an initialization for the next learning. The ensuing backward learning scheme is detailed in Algorithm 3. This process of reusing knowledge from a different but related learning task can be seen as a form of transfer learning (see IV.§1.A.1).

Remark 3. A variation on the above would be forward learning. We favor the backward learning scheme because it is the only one that is amenable to more general backward stochastic differential equations, such as the FVA and the KVA equations of Chapters III and VII. In addition, in these XVA

Algorithm 3: Baseline training scheme for learning a conditional expectation as per (6) at all times i

input : $\{(X_i^\imath, Y_i^\imath, \xi_{i,n}^\imath), \imath \in I, 1 \le i \le n\}$, a partition B of I into batches, a number of epochs $E \in \mathbb{N}^\star$, a learning rate $\eta > 0$
output: $\widehat{\Phi}_1, \ldots, \widehat{\Phi}_n$
initialize parameters W_{n+1}, b_{n+1} and μ_{n+1} of the network at terminal time-step n
for $i = n \ldots 1$ **do**
  $W_i, b_i, \mu_i \leftarrow \text{BaseAlg}(\{(X_i^\imath, Y_i^\imath, \xi_{i,n}^\imath), \imath \in I\}, B, E, \eta, W_{i+1}, b_{i+1}, \mu_{i+1})$
  $\widehat{\Phi}_i \leftarrow \{x \mapsto \zeta^{[h+1]}(x, y; W_i, b_i) + \mu_i\}$
end

applications, the labels (cash flows)/features (risk factors) corresponding to times i closer to the final maturity n of the portfolio have a lower/higher variance. Hence the training task corresponding to a lower time step i is harder. In order to see this in a simplified setup, consider linear regression instead of neural networks. When close to the time step $i = 0$, the variances of the regressors tend to 0, which leads to ill-conditioned covariance matrices, while the variance of the regressands increases, which makes the regression even more unstable.

Proceeding backward in time, we thus solve successively harder and harder problems, but problems which benefit from the knowledge acquired solving the previous ones: a virtuous divide-and-conquer strategy.

§3 HIERARCHICAL SIMULATION

Learning tasks involving defaults such as (6) may be challenging, even with optimized training schemes. As should always be the first scrutiny with machine learning, a difficulty can come from the data, i.e. from the simulation part in our case. Specifically, a large variance of the estimated population loss function can jeopardize the learning approach, which we address in what follows by a suitable hierarchical simulation approach.

A Identification of the Variance Contributions Using Automatic Relevance Determination

In this part we show how to hierarchize the variance impact of explanatory variables using automatic relevance determination (ARD). As detailed in Rasmussen and Williams (2006, Sections 5.1, 5.4.3, 6.6 and 8.3.7), ARD is a Bayesian procedure for feature selection consists of estimating the relevance of the features by maximizing a marginal likelihood. In our case we apply a

Gaussian process regression based ARD to quantify empirically the impact of the variances of X and Y on that of Y.

Toward this aim, we treat the vector of the parameters of the data-generating process (DGP, e.g. a financial model), denoted by ρ, as a latent variable endowed with some instrumental distribution.

Example 1. In the CVA setup of IV.§4.A, ρ is the collection of the drift and diffusion parameters of the SDEs IV.(10)-(11) for all $\langle e \rangle$ and b.

Given ρ, we sample the following time-averages of the variances of $X_1, \dots, X_n$, $Y_1, \dots, Y_n$ and $\xi^1, \dots, \xi^n$:

$$\begin{aligned} V(X|\rho) &= \frac{1}{n+1} \sum_{i=0}^{n} \widehat{\mathbb{E}}^{\rho}[(X_i - \widehat{\mathbb{E}}^{\rho}[X_i])^2] \\ V(Y|\rho) &= \frac{1}{n+1} \sum_{i=0}^{n} \widehat{\mathbb{E}}^{\rho}[(Y_i - \widehat{\mathbb{E}}^{\rho}[Y_i])^2] \\ V(\xi|\rho) &= \frac{1}{n+1} \sum_{i=0}^{n} \widehat{\mathbb{E}}^{\rho}[(\xi_{i,n} - \widehat{\mathbb{E}}^{\rho}[\xi_{i,n}])^2], \end{aligned}$$

meant componentwise in the vector cases of X and Y, where $\widehat{\mathbb{E}}^{\rho}$ is an empirical average over paths sampled for a given realization ρ of the DGP parameters. Then, based on a finite sample of ρ and on the corresponding realizations of the triple $(V(X|\rho), V(Y|\rho), V(\xi|\rho))$, we perform a Gaussian process regression (Rasmussen and Williams, 2006) of $V(\xi|\rho)$ against $V(X|\rho)$ and $V(Y|\rho)$. In this procedure we use an anisotropic kernel $\exp(-\sum_{j=1}^{p} \frac{(v_j - v'_j)^2}{2\lambda_{x,j}^2} - \sum_{j=1}^{q} \frac{(w_j - w'_j)^2}{2\lambda_{y,j}^2})$, where the hyperparameters $\lambda_{x,1}, \dots, \lambda_{x,p}$ and $\lambda_{y,1}, \dots, \lambda_{y,q}$ are characteristic length-scales for the corresponding components of $V(X|\rho)$ and $V(Y|\rho)$. Maximizing the marginal likelihood on the dataset allows to recover those length-scales and these can then be interpreted as relevance estimates for the different input variables. The higher the inverse length-scale gets, the more the corresponding variable influences the output (payoff variance, in our case).

This procedure is then itself randomized, i.e. run multiple times on the restricted datasets corresponding to different sub-samplings of ρ. This provides a distribution of the fitted hyper-parameters λ in the above, while being also less prone to over-fitting and local minima issues. A similar analysis was used in Bergstra and Bengio (2012) to study the relevance of different neural network hyper-parameters with respect to the validation loss.

Example 2. Figure 15 reveals the dominance of the impact of the variance of the X_i on that of the $\xi_{i,n} \equiv \mathrm{DL}_n - \mathrm{DL}_i$ as per (9)–(11) in the context of the CVA case study of §4 (in Figure 15 for a single-client portfolio of the bank), where Y is the vector of the market risk factor processes and X is the default indicator process of the client.

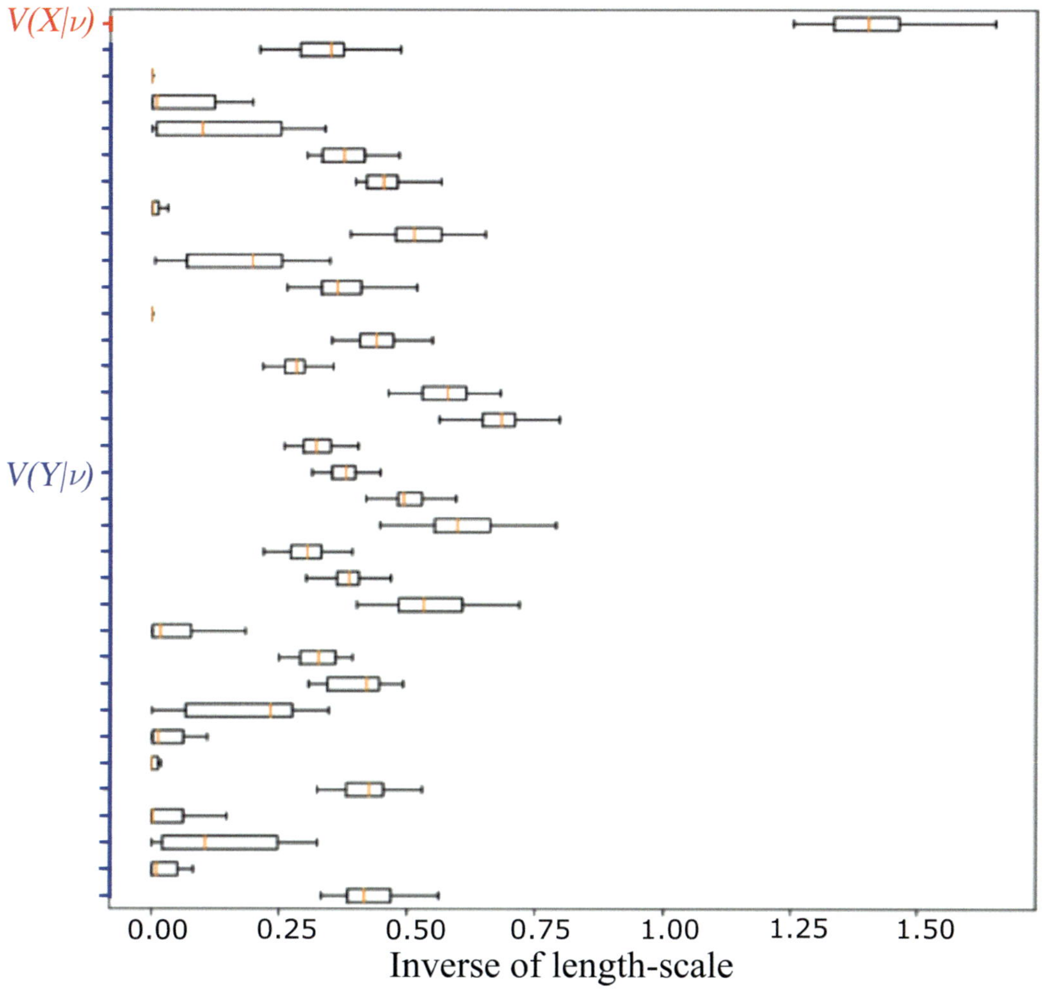

Figure 15: CVA case with a single client in §4.B: Box-plot of the inverse length-scales obtained by randomized Gaussian process regressions of the conditional variances of the cash flows $\xi_{i,n} \equiv \mathrm{DL}_n - \mathrm{DL}_i$ as per (9)-(11) against the conditional variances of the risk factors X_i (default indicator process of the client) and Y_i, where conditional here is in reference to the parameters of the financial model IV.(10)-(11) for all $\langle e \rangle$ and b, treated as a random vector with a postulated distribution.

B Learning on Hierarchically Simulated Paths

If X contributes more to the variance of ξ than Y, then, in order to deal with the resulting variance issue regarding the associated simulation/learning scheme, an idea is to simulate more realizations of X than Y, even if this means giving up the independence of the simulation setup. More precisely, we simulate M i.i.d paths $Y^1, \ldots, Y^M$ of Y and, for every $k \in \{1, \ldots, M\}$ and $i \in \{1, \ldots n\}$, we simulate N i.i.d realizations

$X_i^{k,1}, \dots, X_i^{k,N}$ of X_i conditional on Y^k. For every i, this yields a sample $(X_i^{k,l}, Y_i^k, \xi_{i,n}^{k,l})$, $k \in \{1, \dots, M\}, l \in \{1, \dots, N\}$ of $(X_i, Y_i, \xi_{i,n})$ of size MN, where, within each block k, independence between the $X^{k,l}$ only holds conditionally on Y^k.

Algorithm 3 page 143 is then run on the resulting hierarchically simulated dataset by taking $I = \{1, \dots, M\} \times \{1, \dots, N\}$, with $Y^{k,l} := Y^k$ for all l. For implementation efficiency reasons pertaining to memory contiguity, the set of indices of the $\imath$th batch, with $1 \leq \imath \leq |B|$, is chosen to be $\{(k,l) \in \{1, \dots, M\} \times \{1, \dots, N\} : (\imath - 1)\frac{|I|}{|B|} \leq (l-1)N + (k-1) + 1 < \imath \frac{|I|}{|B|}\}$.

Hierarchical simulation in the above sense can be thought of as a form of data augmentation procedure (Shorten and Khoshgoftaar, 2019), but in a simulation setup where one knows how to generate the data perfectly. The main question is then to which extent one should augment the data, i.e. the choice of the hierarchical simulation parameters M and N, which is the focus of the sequel of this section.

Remark 4. Hierarchical simulation is different in nature from importance sampling that favors particular events, e.g., in the portfolio credit risk setup of Carmona and Crépey (2010), default versus survival. In an XVA setup, some metrics, like the CVA, need default events for being properly estimated, whereas others, like the FVA, require survival events. Hence what one needs is richness regarding both default and survival events, which is what hierarchical simulation provides.

C Choosing the Hierarchical Simulation Factor

Assume that simulating Y_i costs P times more than it takes for simulating X_i given a path $\{Y_j\}_{j \leq i}$ in terms of computation time. The hierarchical simulation factor N can be chosen so as to minimize the variance of the loss $\frac{1}{MN}\sum_{k=1}^M \sum_{l=1}^N f_i(\theta, X_i^{k,l}, \dots, X_n^{k,l}, Y_i^k, \dots, Y_n^k)$ with respect to N, under a budget constraint $M(N+P) = B$, where f_i is the point-wise loss of our learning task at time-step i, e.g. $f_i(\theta, X_i, \dots, X_n, Y_i, \dots, Y_n) = (\xi_{i,n} - \Phi_i^\theta(X_i, Y_i))^2$, where Φ_i^θ is a neural net (element of $\mathcal{NN}^+_{p+q,h,u,\varsigma}$) with parameters collectively denoted by θ (cf. (6)).

For ease of notation in this subsection, we write $f_i(\theta, X^{k,l}, Y^k)$ and $f_i(\theta, X, Y)$ instead of $f_i(\theta, X_i^{k,l}, \dots, X_n^{k,l}, Y_i^k, \dots, Y_n^k)$ and $f_i(\theta, X_i, \dots, X_n, Y_i, \dots, Y_n)$ (it is then implied that X and Y play formally the role of vectors containing their path from time-step i up to n).

Proposition 1. *The hierarchical simulation factor that minimizes the variance of the loss* $\frac{1}{MN}\sum_{k=1}^M \sum_{l=1}^N f_i(\theta, X^{k,l}, Y^k)$ *with respect to* N*, subject to the* *budget constraint* $M(N+P)=B$, *is*

$$N_i^\theta = \sqrt{\frac{Q_i^\theta P}{R_i^\theta}}, \tag{7}$$

where

$$R_i^\theta = \mathrm{Cov}\Big(f_i(\theta, X^{1,1}, Y^1), f_i(\theta, X^{1,2}, Y^1)\Big) = \mathbb{V}\mathrm{ar}\Big(\mathbb{E}(f_i(\theta, X^{1,1}, Y^1)|Y^1)\Big)$$
$$Q_i^\theta = \mathbb{E}\Big(\mathbb{V}\mathrm{ar}(f_i(\theta, X^{1,1}, Y^1)|Y^1)\Big) = \mathbb{V}\mathrm{ar}(f_i(\theta, X^{1,1}, Y^1)) - R_i^\theta.$$

***Proof*.** After rearranging terms, we obtain

$$\mathbb{V}\mathrm{ar}(\frac{1}{MN}\sum_{k=1}^{M}\sum_{l=1}^{N} f_i(\theta, X^{k,l}, Y^k)) = \frac{R_i^\theta}{B}(\frac{1}{N}(N - \sqrt{\frac{Q_i^\theta P}{R_i^\theta}})^2 + (\sqrt{\frac{Q_i^\theta}{R_i^\theta}} + \sqrt{P})^2),$$

where

$$Q_i^\theta = \mathbb{E}[(f_i(\theta, X^{1,1}, Y^1))^2] - \mathbb{E}[f_i(\theta, X^{1,1}, Y^1) f_i(\theta, X^{1,2}, Y^1)]$$
$$R_i^\theta = \mathbb{E}[f_i(\theta, X^{1,1}, Y^1) g(\theta, X^{1,2}, Y^1)] - (\mathbb{E}[f_i(\theta, X^{1,1}, Y^1)])^2. \ \square$$

Note that $Q_i^\theta + R_i^\theta = \mathbb{V}\mathrm{ar}(f_i(\theta, X^{1,1}, Y^1))$. The quotient

$$\frac{Q_i^\theta}{R_i^\theta} = \frac{\mathbb{E}\Big(\mathbb{V}\mathrm{ar}(f_i(\theta, X^{1,1}, Y^1)|Y^1)\Big)}{\mathbb{V}\mathrm{ar}\Big(\mathbb{E}(f_i(\theta, X^{1,1}, Y^1)|Y^1)\Big)}$$

in (7) measures the relative contributions of X and Y to the variance of the loss estimator. To estimate the values of Q_i^θ and of R_i^θ therein, one only needs to simulate $(X^{1,1}, X^{1,2}, Y^1)$, i.e., with respect to the bare simulation of (X, Y), one extra simulation of X conditional on each realization of Y.

As a fixed value of N has to be chosen throughout the simulation and training task, for the above result to be of practical use, N_i^θ as per (7) has to be reasonably stable with respect to both pricing time steps i and SGD iterations (the transfer learning scheme of §2.C is also advantageous in this respect in that it stabilizes the learning). If so, this leads to the following:

Heuristic 1. Choose for N the average of the values N_i^θ obtained during the SGD iterations and the time steps. Make for M the corresponding choice deduced from the budget constraint, i.e. $M = \frac{B}{N+P}$.

Note that N depends only on P, and M on P and B. If P is not analytically known, it can be deduced from simulation times of experiments corresponding to the same M but different N. Namely, let B and B' the budgets corresponding to configurations (M, N) and (M, N'). We have

$$\frac{B}{B'} = \frac{P+N}{P+N'}. \tag{8}$$

One can deduce P by identifying the ratio in (8) to that of the execution times of (M, N) and (M, N'). For doing so, it is preferable to choose M large enough to avoid time measurement noise that may be due to caching or parallelization of the simulations.

§4 CVA CASE STUDY

We illustrate our approach numerically by a CVA case study. In this context the probability measure $\mathbb{Q}$ corresponds to the bank survival measure associated with a reference fininsurance measure $\mathbb{Q}^*$ calibrated to the market (see the introduction to the Pricing part of this book).

A Market and Credit Model

We consider an Euler-Marayama time-discretization of the model of IV.§4.A. We use the same notation for the continuous-time processes and their discrete-time approximations, with time-step equal to one year to alleviate the notation. We thus consider a bank trading derivative contracts in different economies $\langle e \rangle$ with various bilateral counterparties b. The currency corresponding to the economy labeled by $\langle 0 \rangle$ is taken as the reference currency. Each counterparty b of the bank has a default-time τ^b with a stochastic intensity process γ^b.

Remark 5. In practice, the time-discretizations are stepping through a refined simulation time grid. This simulation grid is also used when integrating numerically some of the above diffusions, e.g. the default intensities in IV.(13), or for defining risk-neutral discount factors $\beta_i^{\langle 0 \rangle}$ associated with the reference currency by approximating $(-\ln \beta_i^{\langle 0 \rangle})$ using numerical integration[1] of the reference short-term interest rate $r^{\langle 0 \rangle}$ on $[0, i]$. Learning, pricing and checking for default events, instead, are only done at the coarser pricing time steps. Hence, although we step through the fine time grid in our discretized diffusions, we only need to store the values of the processes at the pricing time steps.

We define X as the collection of all the default indicator processes of the counterparties b and Y as the collection of all the short interest rate, exchange-rate and default intensity processes r, χ and γ, endowed with the filtration generated by the innovation in the model, i.e. the collection of all the Gaussian and exponential variables involved at the increasing time steps $i \in 1 \mathinner{..} n$. Note that both Y (by itself) and (X, Y) (jointly) are Markov processes with respect to this filtration.

[1] it is also possible to jointly simulate exactly $r^{\langle 0 \rangle}$ and its integral without the need for numerical integration, see for example Glasserman (2003).

B Learning the CVA

We denote by P_i^b the mark-to-market at time i, from the point of view of the reference bank and in units of the reference currency, of all the contracts with the counterparty b (assuming a unique netting set per counterparty). By mark-to-market we mean trade additive counterparty-risk-free valuation, i.e. the risk-neutral conditional expectation of the future contractually promised cash flows, expressed in units of the reference currency and discounted at the risk-free rate $r^{\langle 0\rangle}$. We restrict ourselves to interest-rate derivatives for which mark-to-market valuation at time i is a function of Y_i, by the nature of the cash-flows and the Markov property of Y (see Remark IV.1 page 134). In terms of the default loss process

$$\mathrm{DL}_i = \sum_b \sum_{j=0}^{i-1} (P_j^b)^+ \mathbb{1}_{\{j<\tau^b\le(j+1)\}}, \tag{9}$$

the time-discretized CVA of the bank at the time step i is given by

$$\mathrm{CVA}_i = \mathbb{E}\Big[\underbrace{\sum_b \sum_{j=i}^{n-1} (P_j^b)^+ \mathbb{1}_{\{j<\tau^b\le j+1\}}}_{\mathrm{DL}_n - \mathrm{DL}_i}\Big| X_i, Y_i\Big]. \tag{10}$$

Hence $\mathrm{CVA}_i = \Phi_i^\star(X_i, Y_i)$, where

$$\Phi_i^\star \in \underset{\Phi\in\mathcal{B}(\mathbb{R}^{p+q})}{\operatorname{argmin}} \ \mathbb{E}\Big[\Big(\mathrm{DL}_n - \mathrm{DL}_i - \Phi(X_i, Y_i)\Big)^2\Big]. \tag{11}$$

We also mention the following default intensity-based formula for the CVA of the bank (cf. Eqn. VII.(10)):

$$\widetilde{\mathrm{CVA}}_i = \mathbb{E}\Big[\underbrace{\sum_b \sum_{j=i}^{n-1} (P_j^b)^+ (e^{-\sum_{\iota=i}^{j}\gamma_\iota^b} - e^{-\sum_{\iota=i}^{j+1}\gamma_\iota^b}) \mathbb{1}_{\{i<\tau^b\}}}_{=:\widetilde{\xi}_{i,n}^{cva}}\Big| X_i, Y_i\Big], \tag{12}$$

which converges to the same continuous-time limit as CVA_i when the time discretisation step[2] goes to zero. Hence $\widetilde{\mathrm{CVA}}_i = \widetilde{\Phi}_i^\star(X_i, Y_i)$, where (cf. (4))

$$\widetilde{\Phi}_i^\star \in \underset{\Phi\in\mathcal{B}(\mathbb{R}^{p+q})}{\operatorname{argmin}} \ \mathbb{E}\Big[\Big(\widetilde{\xi}_{i,n}^{cva} - \Phi(X_i, Y_i)\Big)^2\Big]. \tag{13}$$

We reiterate that Algorithm 3 page 143, with hierarchical simulation of

[2]Conventionally set to one in this chapter; the discrepancy between CVA_i and $\widetilde{\mathrm{CVA}}_i$ is just a time-discretization error due to the replacement of integrals by sums under the exponentials in (12), cf. VI.(9).

(X, Y), is generically applicable to all the XVA metrics. The focus on the CVA in this chapter is for benchmarking purposes. Were it for the CVA only, the regression learning scheme with minimal variance is obviously the one based on (12) (which is the one on which we exclusively rely on Chapter VI). Also note that learning the CVA counterparty by counterparty, exploiting the linearity of the conditional expectation for this purpose would lead to as many regressions as there are counterparties of the bank, which for realistic banking portfolios would be extremely inefficient.

On top of the learning schemes (13) and (11) associated with the formulations (12) and (10), another computational alternative is nested Monte Carlo (NMC) as detailed in Abbas-Turki, Diallo and Crépey (2018), where the CVA at the outer nodes of the simulation is computed by inner Monte Carlo simulations for (10) or (12). This variety of approaches is useful for benchmarking purposes. Specifically, we implemented the learning procedure of Algorithm 3 page 143 in PyTorch with custom CUDA kernels for label generation during the backward iterations, the way detailed in Subsection F below (see also the accompanying github repository). In addition we implemented an optimized CUDA benchmark involving nested simulations, using the intensity-based formulation (12) for the inner CVA computations. Due to the heavy calculation, this nested Monte Carlo CVA is only computed at a few pricing times and, unless otherwise stated, using 128 inner paths.

In the following experiments, we assume that the bank is trading derivatives in 10 economies with 8 bilateral counterparties. Implementing the discretized market and default model, we get a total of 10 interest rates, 9 cross-currency rates and 8 default intensities. This yields 27 diffusive market risk factors and 8 default indicator processes. For time-stepping, we use $n = 100$ pricing time steps and 25 simulation sub-steps per pricing time step (see Remark 5 page 148). We consider a portfolio of 500 interest rate swaps with random characteristics (notional, currency, counterparty and maturity between 0 and $T = 10$ years); the P^b are thus analytic. All swaps are priced at par at inception. There is no initial margin nor variation margin. For all the runs of the simulations in this section, whether for training or testing, we use $M = 16384$ paths for the market risk factors Y.

C CVA Learning Results From Default Intensities

The validation of our CVA neural regression methodology is first done in the simpler intensity-based setup (12)-(13). The neural regressions are validated by nested Monte Carlo as explained above. Figures 16, 17 and 18 show that the CVA thus learned from the default intensities is consistent with that obtained from the corresponding nested Monte Carlo simulation. The CVA error profile on Figure 16 reveals slightly more difficulty in learning the earlier CVAs. This is because of a higher variance of the corresponding cash flows (integrated over longer time frames) in conjunction with a lower variance of the features (risk factors diffused over shorter time horizons, see §2.C).

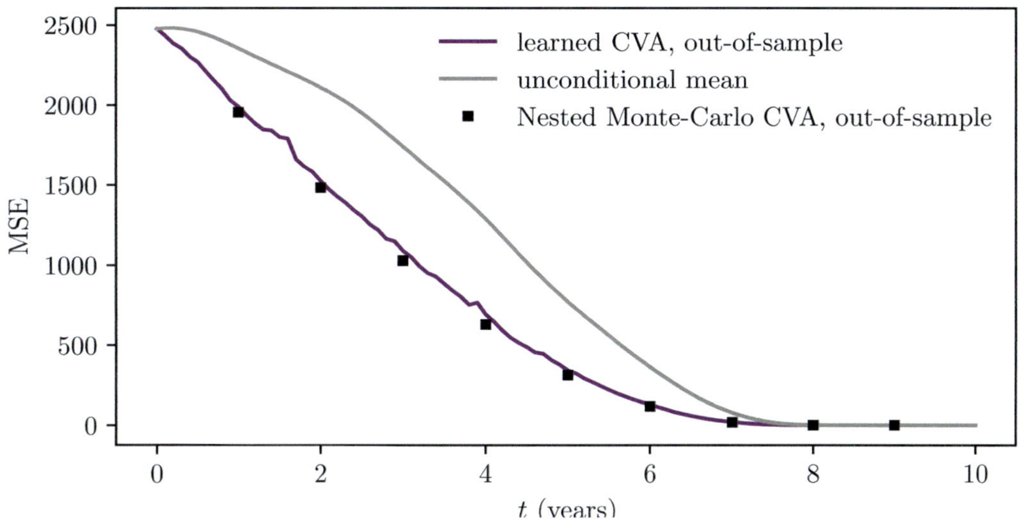

Figure 16: Empirical quadratic loss of various CVA predictors at each pricing time-step. The lower, the closer to the true conditional expectation (cf. IV.(2)). At each pricing time step the unconditional mean is the regression estimate against constants. Since the nested Monte Carlo method is computationally expensive, NMC was carried out only once every 10 pricing time-steps.

Table 6 shows the computational cost and accuracy of the nested Monte Carlo method for different numbers of inner paths, using 32,768 outer paths. The convergence is already achieved for approximately 128 inner paths, in line with the nested Monte Carlo square root rule recalled in an XVA setup in Abbas-Turki, Diallo and Crépey (2018, Section 3.3). Table 7 and Figure 19 show that good accuracy can be achieved through learning at a lower computational cost than through nested Monte Carlo, while also enjoying the advantages of the approach being (semi-)parametric. Indeed, once the CVA is learned, one only pays the cost of inference later on, which is generally negligible compared to training time. By contrast, a nested Monte Carlo approach would require to relaunch the nested simulations every time the CVA estimator is needed on new paths.

More generally, in the presence of multiple number of XVA layers, a purely nested Monte Carlo approach would require multiple layers of nested simulations (cf. Figure 10 page 121), which would amount to a computational time that is exponential in the number of XVA layers, while the computational complexity for the learning approach is linear.

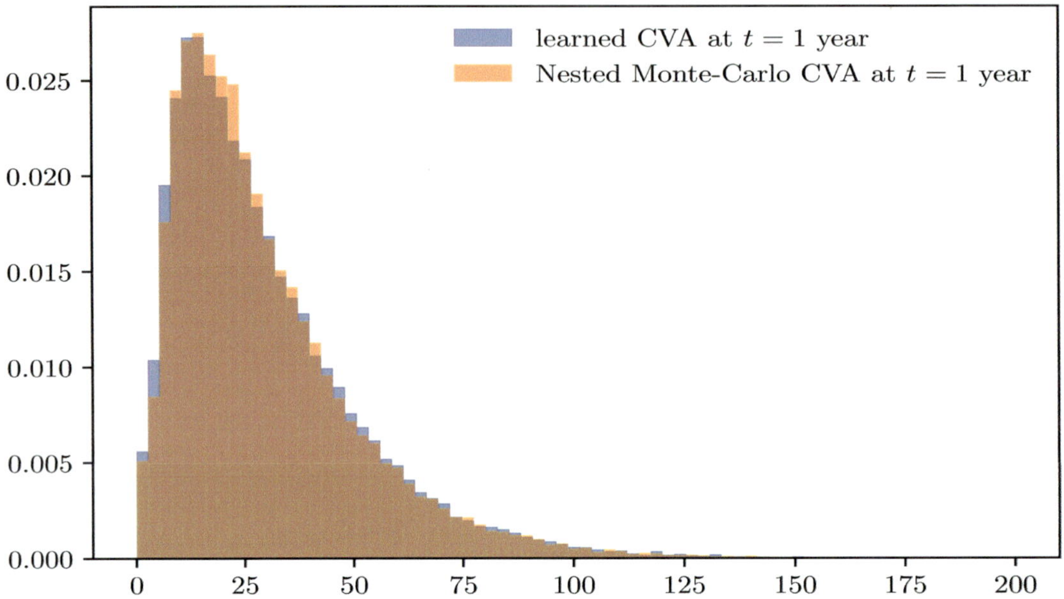

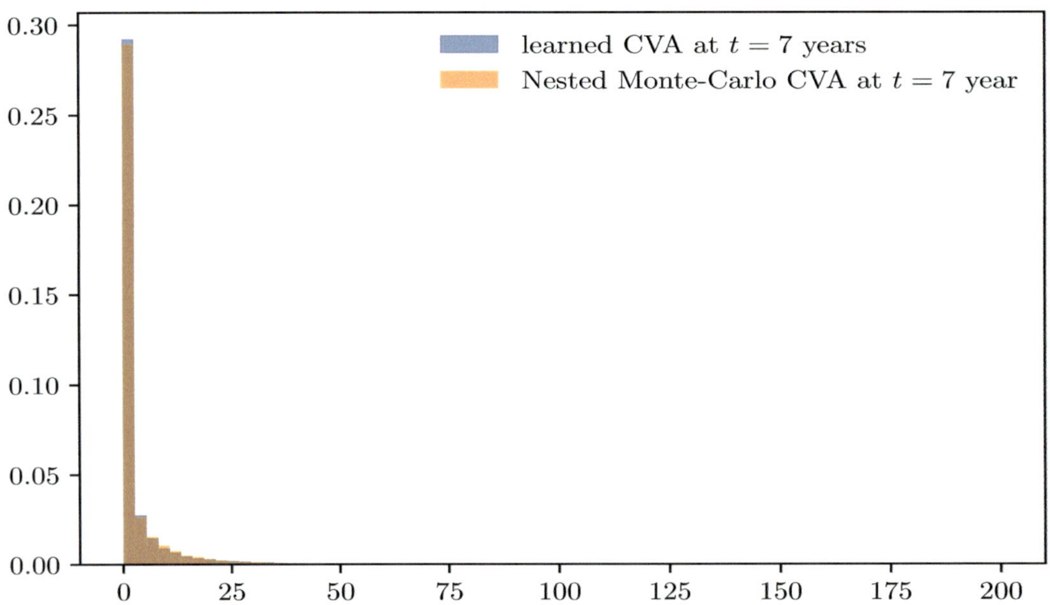

Figure 17: Random variables CVA_1 and CVA_7 observed after 1 year and 7 years (*bottom panel*), obtained by learning (blue histograms) versus nested Monte Carlo (orange histograms). All histograms are based on out-of-sample paths.

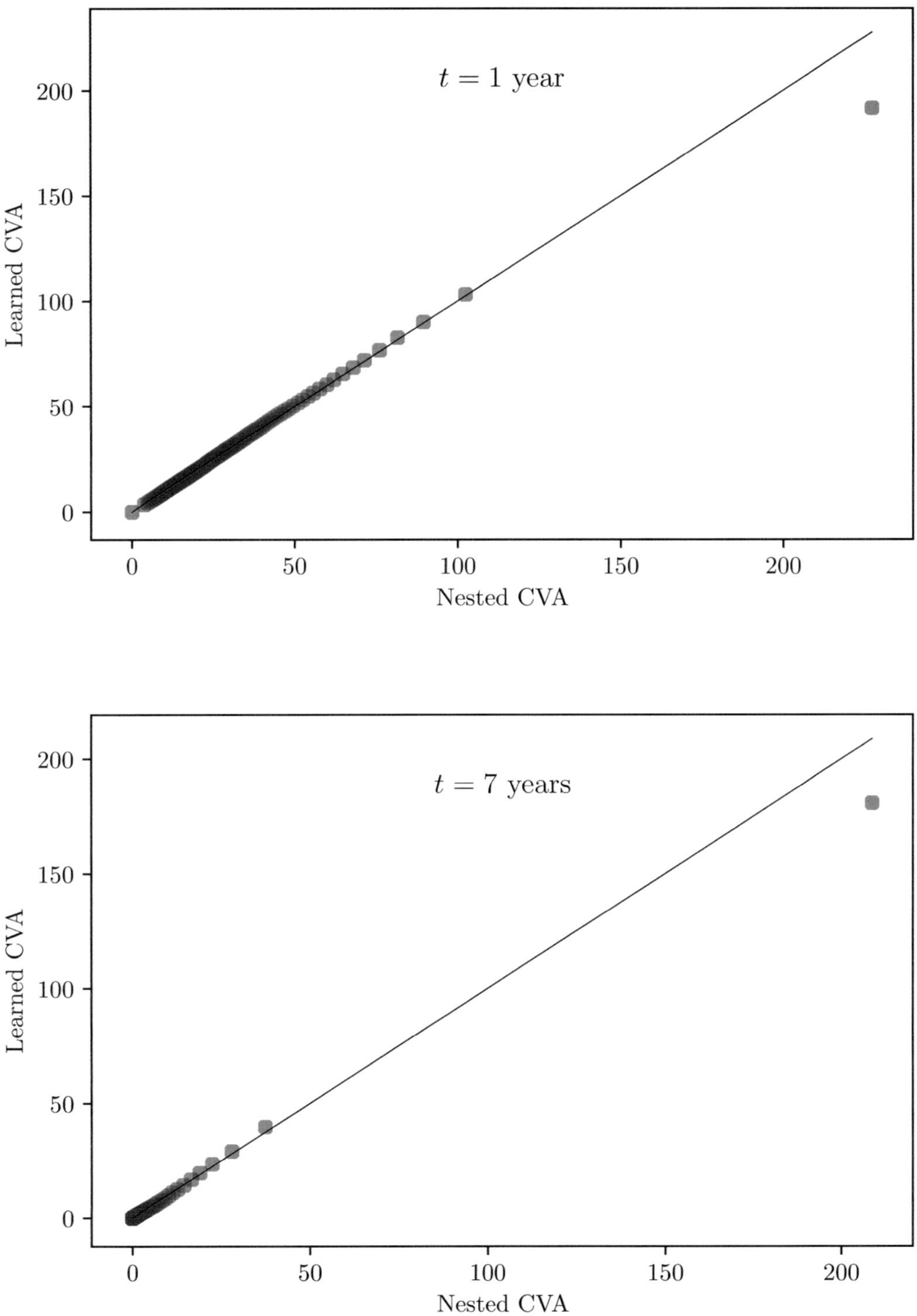

Figure 18: qq-plot of learned versus nested Monte Carlo CVA for the random variables CVA_1 (*top panel*) and CVA_7 (*bottom panel*). Paths are out-of-sample.

Table 6: Accuracy and computation times for the estimation of a CVA at a given pricing time-step i using the nested Monte Carlo procedure. The MSE here is the mean quadratic error between the nested Monte Carlo estimator and the labels $\widetilde{\xi}^{cva}_{i,n}$, and hence quantifies how well NMC is doing as a "regressor" of these labels.

# of inner paths	MSE vs. labels $\widetilde{\xi}^{cva}_{i,n}$	Computational time (seconds)
2	0.523	37.562
4	0.427	37.815
8	0.393	37.819
16	0.370	38.988
32	0.360	40.707
64	0.353	57.875
128	0.348	157.536
256	0.349	301.406
512	0.348	584.475
1024	0.348	1213.756

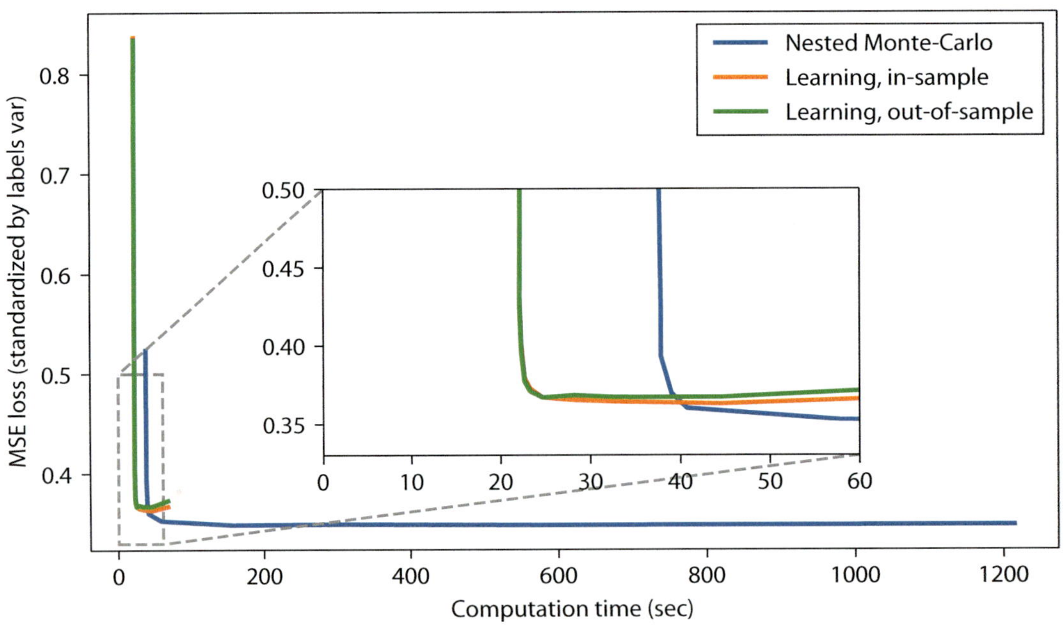

Figure 19: Speed versus accuracy for the CVA learned at a given pricing time i. We kept varying the number of inner paths for the nested Monte Carlo estimator and the number of epochs for the learning approach and recorded the computation time and the empirical quadratic loss.

Table 7: Accuracy and computation times for the calculation of a CVA at a given pricing time-step i using the learning approach. MSE against NMC CVA is the mean quadratic error between the CVA learned from default intensities and a CVA obtained using a nested Monte Carlo with 512 inner paths, while MSE against labels designates the mean quadratic error between the CVA learned from default intensities and the labels $\widetilde{\xi}_{i,n}^{cva}$ that were used during training and thus quantifies how well this learned CVA is doing as a regressor of these labels. Both errors are respectively normalized by the variances of the nested Monte Carlo estimator and of the labels. The paths used here are out-of-sample.

# of epochs	MSE vs. NMC CVA	MSE vs. labels $\widetilde{\xi}_{i,n}^{cva}$	Simulation time (sec)	Simulation time (sec)
1	0.977	0.979	21.992	0.880
2	0.729	0.729	21.992	0.434
4	0.423	0.425	21.992	0.524
8	0.399	0.401	21.992	0.719
16	0.371	0.369	21.992	1.088
32	0.369	0.365	21.992	1.800
64	0.370	0.363	21.992	3.243
128	0.371	0.363	21.992	6.227
256	0.370	0.361	21.992	10.883
512	0.370	0.362	21.992	20.096
1024	0.371	0.362	21.992	39.338

As with mainstream interpolation (as opposed to regression in our case), learning problems (see IV.§1.A.1), a good architecture is key to better learning and hence better approximation of our XVA metrics. As expected, increasing the model capacity reduces the in-sample error as shown in the top panel of Figure 20. Although fine-tuning in our case suggests a single layer yields the best out-of-sample performance for the CVA, a standard guess such as 3 layers can also be considered good enough, as shown in the bottom panel. Of course, such conclusions may depend on the complexity of the portfolio and the number of counterparties and risk factors.

D CVA Learning Results From Default Events

We now try to learn the CVA directly from the default events (as opposed to their intensities before) as per (10)-(11), an approach scalable to FVA and KVA computations later in Chapter VII.

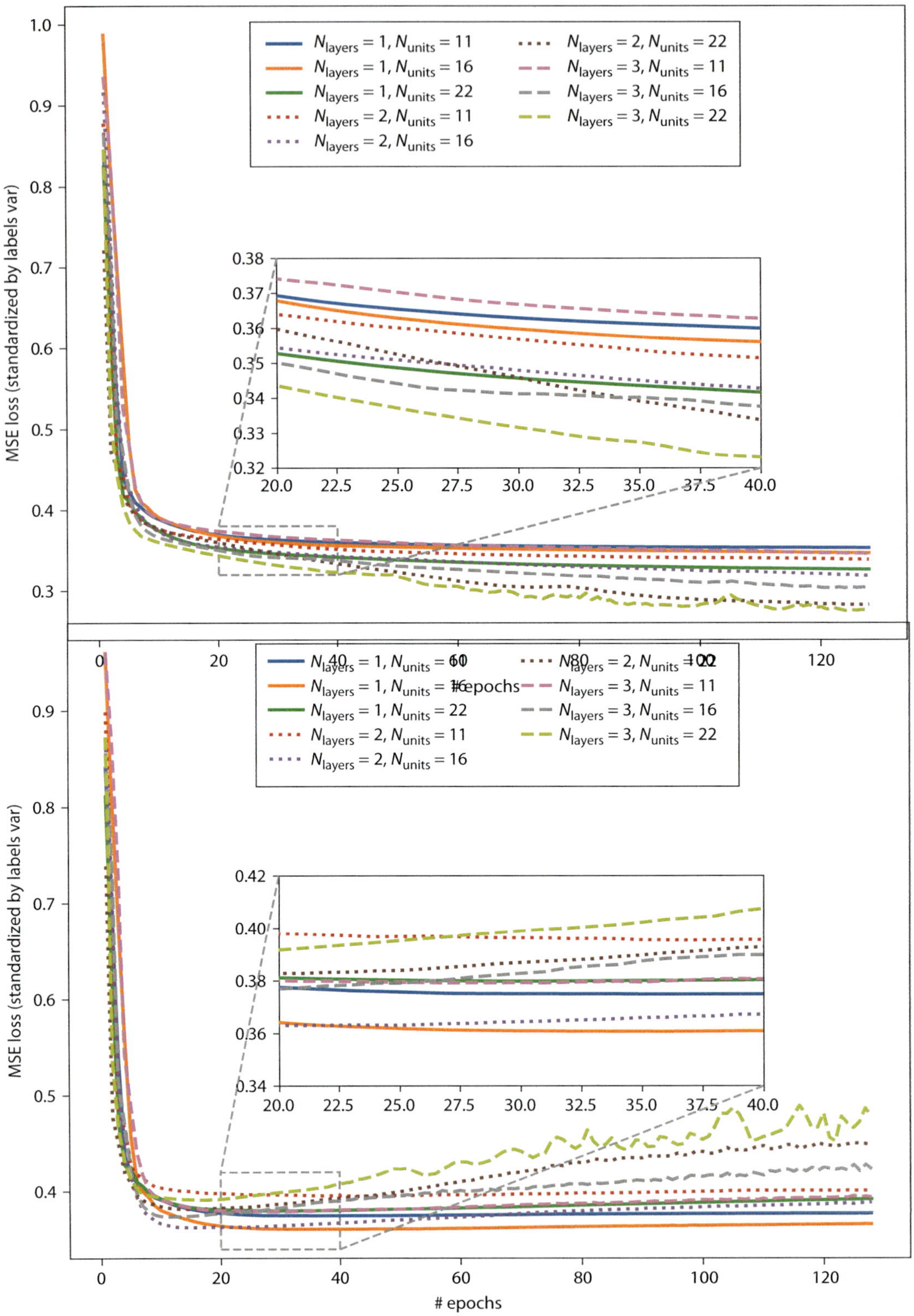

Figure 20: Empirical quadratic loss during CVA learning at time-step $t = 5$ years, standardized by the variance of the labels. *(Top)* Paths are in-sample. *(Bottom)* Paths are out-of-sample.

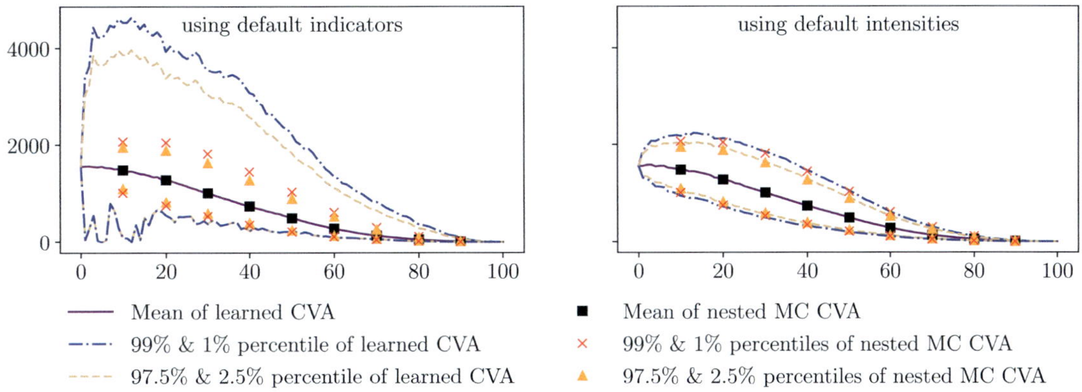

Figure 21: CVA learned using default indicators versus using default intensities (x axis pricing times, y axis CVA levels). Statistics computed using out-of-sample paths.

The comparison between the two panels of Figure 21 reveals a difficulty with the neural net learning approach of Algorithm 3 page 143 applied to the defaults-based formulation (10)-(11) implemented on the basis of i.i.d. simulated data. In this case, represented by the left panel in Figure 21, the network only learns a rather crude and noisy approximation of the CVA conditional to each training time: it is only on the mean that the learned CVA agrees with the nested Monte Carlo estimator; on the tails, it largely fails. As visible from the right panel, the CVA learned using the intensity-based formulation (12)-(13) of the same problem, instead, yields satisfactory results on a wide range of quantiles of the targeted distribution.

In order to improve the learning (11) of the defaults-based CVA (10), we apply to it the hierarchical simulation technique of §3. Let $(r^1, \chi^1, \gamma^1), \ldots, (r^M, \chi^M, \gamma^M)$, be i.i.d sample paths of the triple of vector-processes (r, χ, γ) in the model of IV.§4. Let $\{\epsilon^{k,l}, 1 \leq k \leq M, 1 \leq l \leq N\}$ be i.i.d samples of ϵ, the vector defined by the right-hand side in IV.(13) where b ranges over (bilateral) counterparties. Then we can define MN samples of the vector of the default indicator processes of the counterparties at every pricing time i based on IV.(13). Figure 22 illustrates the ensuing simulation scheme for the default indicator of a generic client of the bank, with sampled default times $\tau^{k,l}$.

We then learn the CVA process at different time steps for the whole portfolio at once based on (10)-(11), trying different combinations of the number of market paths M and of the hierarchical simulation factor N. Figures 23 and 24 show the relative RMSE of the trained neural network against the nested Monte Carlo benchmark[3], the simulation and training times on the GPU and the host RAM usage (see Subsection F below), as functions of the

[3]RMSE restricted to the realizations where the benchmark is non-zero.

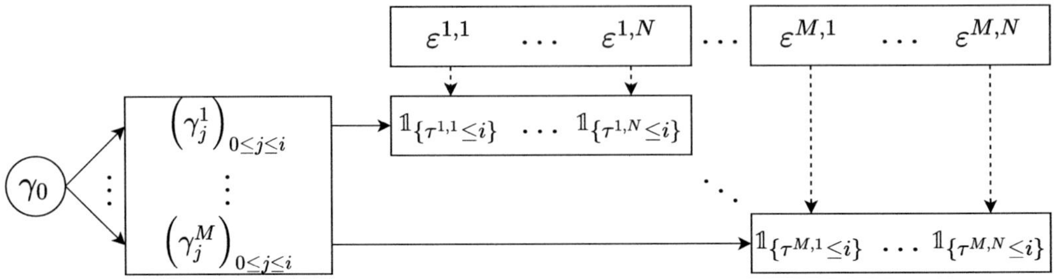

Figure 22: Default simulation scheme

number of diffusion paths M and of the hierarchical simulation factor N. For the execution times in Figure 24, the runs were done on a server with an Intel Xeon Gold 6248 CPU and 4 Nvidia Tesla V100 GPUs (out of which we used only one). For performance comparison reasons, we use for all (M, N) configurations the same number of epochs $E = 8$ and number of batches $|B| = 32$, which yields a total of 256 stochastic gradient descent steps during any training task. From Figures 23 and 24, we already see some configurations $(\frac{1}{2}M, N)$ being better than $(M, \frac{1}{2}N)$, as they achieve a similar accuracy with less memory footprint and computational time. For example, $(32768, 1024)$ is better than $(65536, 512)$, given that the former achieves a similar RMSE of 0.07 but is 30% faster to simulate and price, while also occupying 23% less CPU memory. In addition, to the credit of the twin Monte Carlo validation procedure, the comparison between Figures 23 and 25 shows that the error estimates provided by the latter are very much in line with the ones[4] provided by a much heavier nested Monte Carlo procedure (which would become unfeasible on more complex problems).

The dominance of the impact of the variance of the X_i on that of the $\xi_{i,n} := \mathrm{DL}_n - \mathrm{DL}_i$ has been demonstrated in Figure 15. Figure 26 shows the $\sqrt{\frac{Q_i^\theta}{R_i^\theta}}$ (cf. (7)) obtained in the base case $N = 1$. The values fluctuate quite significantly both in pricing time steps i and SGD iterations. For the purpose of applying Heuristic 1 page 147 (which can be applied for any N but the performance of which needs to be assessed carefully), we retain a rough average order of magnitude of the order of a few tens. To obtain from the $\sqrt{\frac{Q_i^\theta}{R_i^\theta}}$ the N_i^θ in (7), one needs to multiply them by $\sqrt{P}$ (e.g. if a market simulation is 100 times slower than an ensuing default simulation, then the factors displayed in Figure 26 must be multiplied by 10). Solving the equation (8) for P on the basis of the columns $M = 65536$ in Figures 23-24 yields $P \approx 497$. So the numbers in Figure 26 need to be multiplied by $\sqrt{497} \approx 22.3$ to get the

[4]in spite of the slight variation in the way these errors are computed (cf. the captions of Figures 23 and 25).

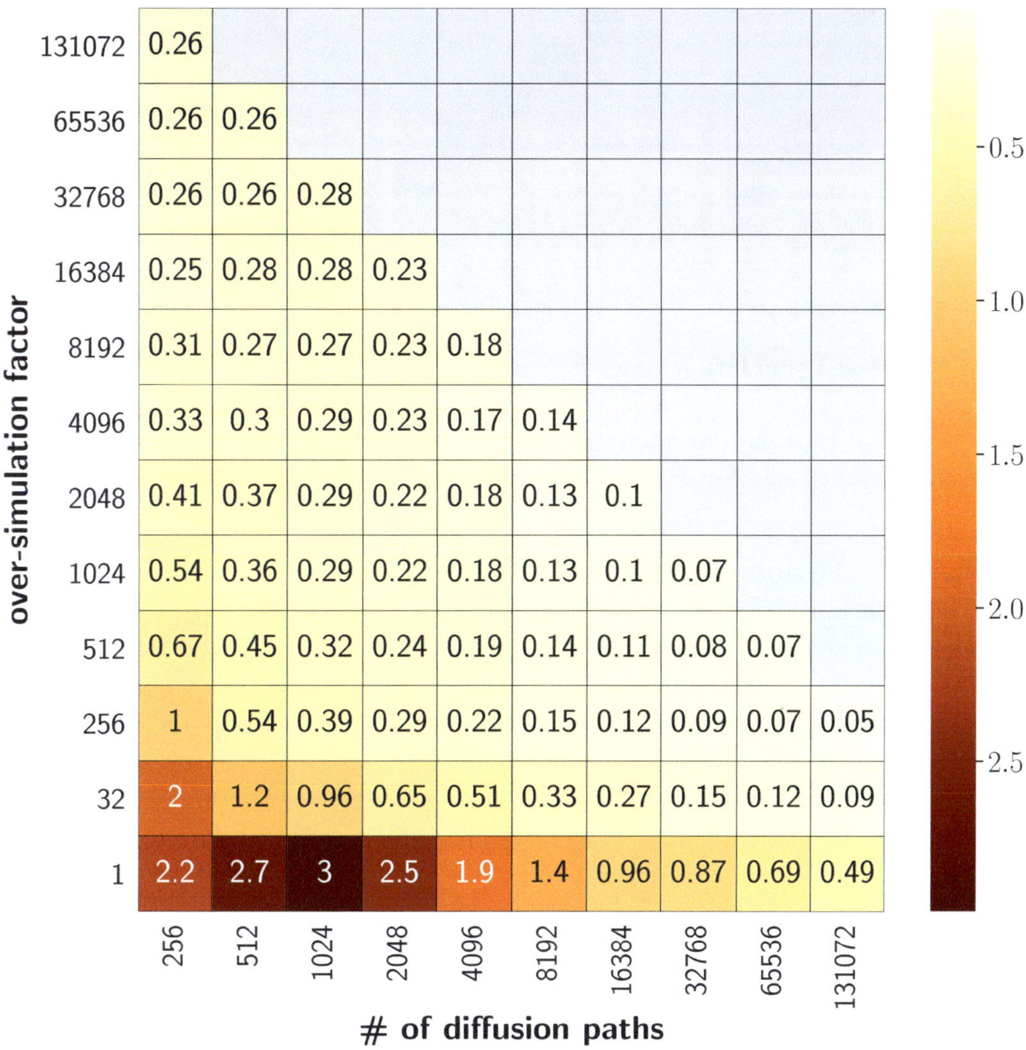

Figure 23: Relative RMSE of the CVA prediction against a nested Monte Carlo benchmark at the pricing time $i = 5$ years, for different combinations of the number of market paths M and of the hierarchical simulation factor N, when the nested Monte Carlo benchmark is non-zero. The error here is a Monte Carlo estimate of $\sqrt{\mathbb{E}\left[\left(\frac{\mathrm{CVA}_{\mathrm{pred}}-\mathrm{CVA}_{\mathrm{nested}}}{\mathrm{CVA}_{\mathrm{nested}}}\right)^2\right]}$ where $\mathrm{CVA}_{\mathrm{pred}}$ is the CVA estimate predicted by the considered neural network given a state of market and default factors, and $\mathrm{CVA}_{\mathrm{nested}}$ is a nested Monte Carlo estimator given the same state.

optimal N as per Heuristic 1 page 147. In view of this, we expect an optimal hierarchical simulation factor N of the order of a few hundreds.

Pathwise CVA estimators learned for $N = 1, 32, 64, 128, 512$ are shown in Figure 27, which are to be compared to the right plot in Figure 21 obtained when learning the CVA relying on the intensity-based formula (12).

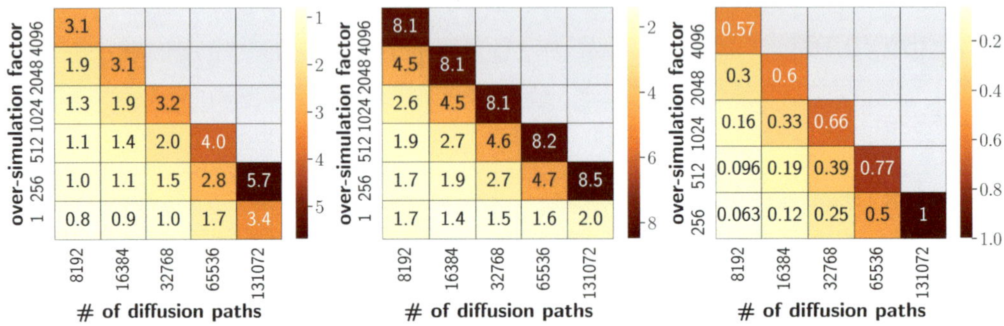

Figure 24: Model simulation times in seconds (*left*), CVA training times in minutes (*center*), and RAM usage as a % of its maximum usage over all the displayed experiments (*right*), for different combinations of the number of market paths M and of the hierarchical simulation factor N.

In line with the above expectations, one needs $N = 512$ in order to have a close enough match between the 1, 2.5, 97.5 and 99-th percentiles of the CVA learned from defaults and those of the nested Monte Carlo estimator (or of the intensity-based CVA learner represented by the right panel in Figure 21).

The above results show that hierarchical simulation is essential to a defaults-based CVA learner. The efficiency of Heuristic 1 is attested numerically on the double basis of our nested Monte Carlo benchmark (Figure 23) and of the twin Monte Carlo validation procedure of IV.§2.A (Figure 25). The fact that Heuristic 1 already has merit in such case (with not so stable $\sqrt{\frac{Q_i^\theta}{R_i^\theta}}$) can be put to the credit of the method.

E Conclusion

The bottom row of Figure 23 and the first plot ($N = 1$) of Figure 27 illustrate that a pathwise CVA cannot be learned based on the hybrid market and defaults formulation (10) without hierarchical simulation: For $M = 16384$ and 131072, the corresponding errors with respect to the benchmark nested Monte Carlo are 96% and 49%. However, increasing N from 1 (bottom row) to 256 brings these errors down to 11% and to 5%, while for $M = 1024$ and $N = 512$ the error is 1%. As visible from Figure 24, the simulation times are only marginally increased when increasing the hierarchical simulation factor N (while increasing the number of diffusion paths M increases the simulation time approximately proportionally). These results show that the hierarchical simulation technique is key to the success of a learning approach involving a combination of diffusive and default risk factors.

Even after writing an optimized GPU implementation for the nested Monte Carlo estimator, the latter takes at least 32 minutes on the same

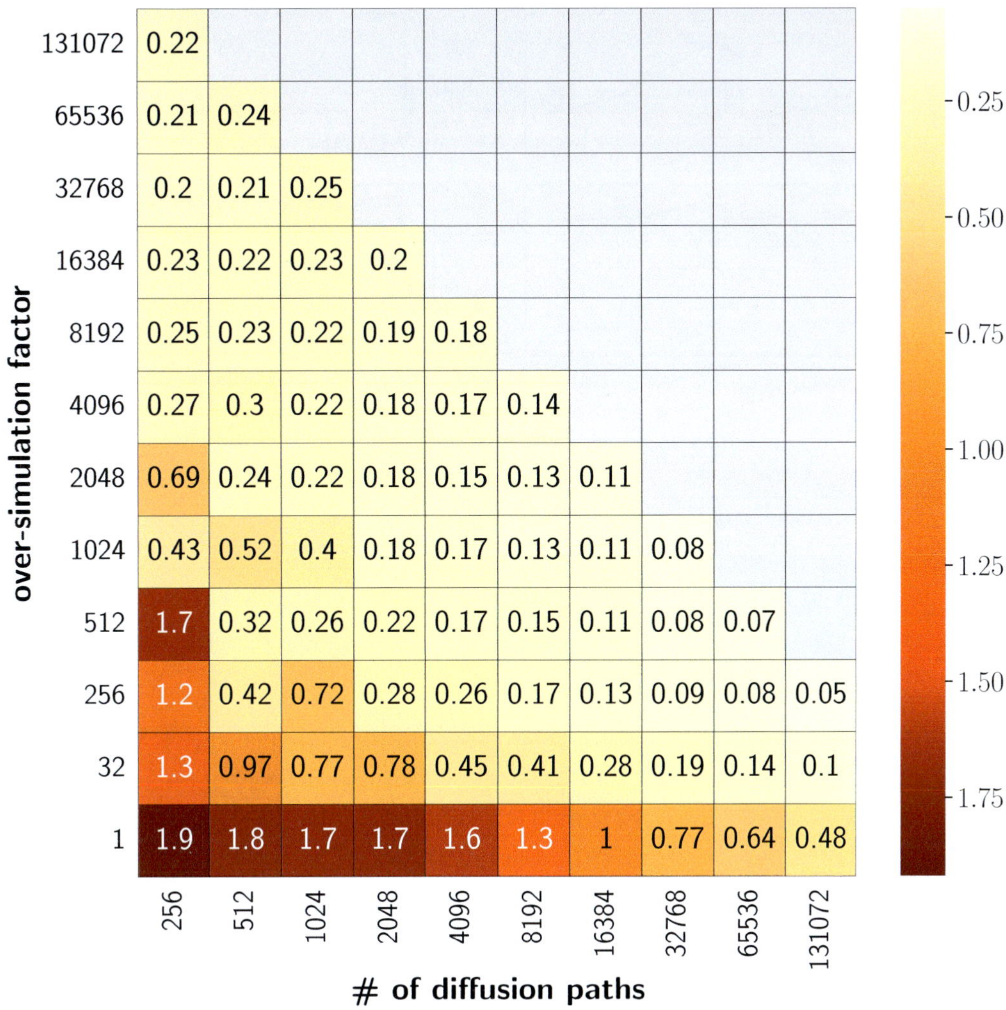

Figure 25: Relative RMSE of the CVA prediction against the ground-truth CVA at the pricing time $i = 5$ years, computed by the a posteriori twin Monte Carlo procedure of IV.§2.A for different combinations of the number of market paths M and of the hierarchical simulation factor N. The error here is a Monte Carlo estimate of $\sqrt{\frac{\mathbb{E}\left[\left(\mathrm{CVA}_{\mathrm{pred}}-\mathrm{CVA}_{\mathrm{exact}}\right)^2\right]}{\mathbb{E}\left[\mathrm{CVA}^2_{\mathrm{exact}}\right]}}$, where $\mathrm{CVA}_{\mathrm{pred}}$ is the CVA estimate predicted by the considered neural network given a state of market and default factors and $\mathrm{CVA}_{\mathrm{exact}}$ is the exact CVA (the value of which does not need to be computed, by virtue of IV.(5)) given the same state.

hardware as above to compute that estimator for $M = 16384$ and $\sqrt{M} = 128$ inner paths[5], compared to approximately 8 minutes in the case of the learning

[5] However, when doing the error computations and in all plots, we used 1024 inner paths to get benchmark CVAs that are sufficiently accurate *point-wise* and be able to get accurate

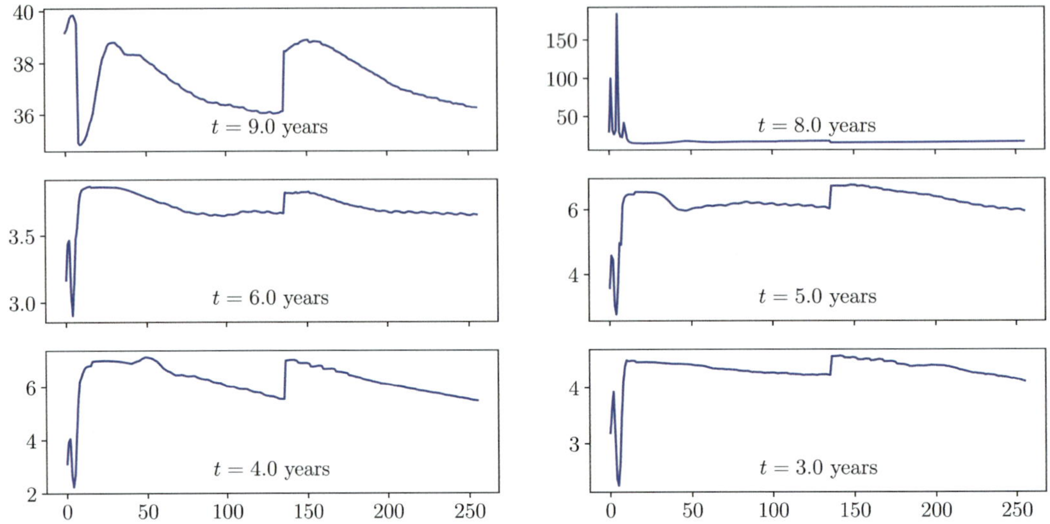

Figure 26: $\sqrt{\frac{Q_i^\theta}{R_i^\theta}}$ at different pricing time steps i (panels) and SGD iterations (x axes).

approach with a very high hierarchical simulation factor ($N = 2048$). Moreover, going to higher XVA layers such as the FVA and the KVA, a nested Monte Carlo approach would become $\sqrt{M}$ times slower per each new layer (Abbas-Turki et al., 2018, Section 3.3), whereas a regression approach would just become slower by a constant each time a new XVA layer is added. In addition, learned XVA metrics can be used in prediction at a very low cost (inference is very fast as it involves no automatic differentiation or stochastic gradient descent), whereas nested Monte Carlo numbers must be recomputed from scratch every time.

F Python/CUDA Optimized Implementation Using GPU (∗)

Contrary to most use-cases of machine learning, where the final product is the trained model and thus execution time is only critical during inference, in the case of learning from simulated data in XVA applications, the training process itself is part of the final product. Hence, particular care is needed when writing the training procedures.

We implemented Algorithm 3 on page 143 using Python programming with the CUDA API (application programming interface). Because the considered problem involves high variances (see §3) and thus requires a sufficiently large sample size, both training and inference are not easy to achieve in a reasonable execution time. First, we need to leverage the manycore parallel

tail estimates, and nested Monte Carlo simulation thus takes 8 times more computation time.

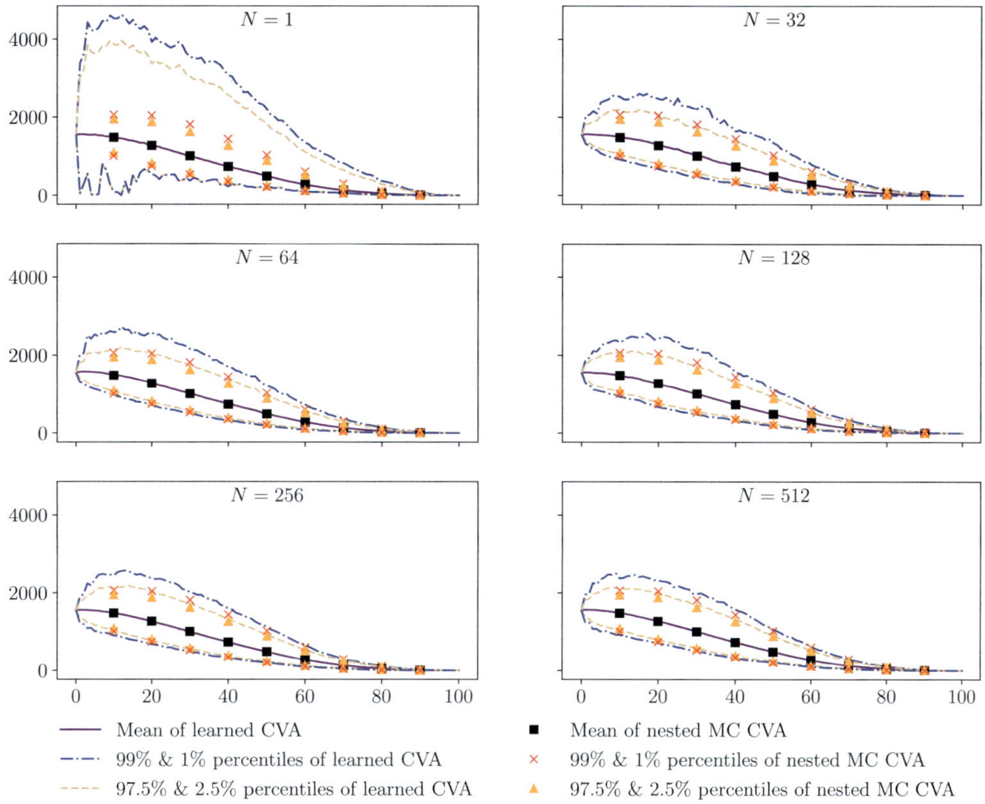

Figure 27: Learned and nested Monte Carlo CVA processes for various hierarchical simulation factors N (x-axis: pricing times, y-axis: CVA levels; $M = 16384$). Statistics computed using out-of-sample paths.

architecture of GPUs that involves streaming multiprocessors, which are used for the simulation, learning and inference phases. All phases are intertwined and performed in line. Hence, we need to carefully optimize each part of the algorithm.

On the simulation side, due to their intrinsically parallel nature, Monte Carlo simulations easily lend themselves to parallelization on GPUs. Nevertheless, various optimizations are needed to achieve a reasonable solution executed within a few seconds (cf. Figure 24 page 160). We chose to use Python and the CUDA kernels are compiled just-in-time using the module numba, which allows to dynamically generate CUDA kernels at run-time.

Regarding learning, we opted for PyTorch for its proximity to the CUDA programming model and its just-in-time compiler, allowing for static computation graphs and automatic fusion, whenever appropriate, of the kernels associated with the PyTorch operations used by the model.

We used most of the optimization techniques introduced in Abbas-Turki et al. (2018), except those related to classical regressions since these are replaced here by neural networks. We also introduced several additional optimizations, the most important one being to judiciously manage the CPU and GPU memories. A naive solution would involve the CPU/GPU virtual unified memory (NVIDIA Corporation, 2020) and let the compiler choose. However, this usually results in sub-optimal memory accesses. Our choice rather targets an efficient use of the GPU memory space, a reduction of CPU/GPU transfer and an optimized transfer when needed. These optimizations and implementation choices are developed in the accompanying Github repository[6].

Having in mind a portfolio of the order of one million trades spread over maturities ranging over 50 years and involving a few thousands clients, the computational CPU resources typically available in banks hardly allow computing (even overnight) a mark-to-market cube with more than 10^4 paths. Switching to GPU resources (as required anyway if training pathwise XVA metrics is envisioned) could allow computing a mark-to-market cube with 10^5 to 10^6 paths in about one hour of computations spread over a few GPUs.

In fact, while we performed our computations using only one GPU, we expect a bank to have access to more than just a single GPU. Monte Carlo simulations and stochastic gradient descent can easily be adapted to multi-GPU setups: see, for instance, Abbas-Turki, Vialle, Lapeyre and Mercier (2014) for a study of the parallelization of a Monte Carlo pricing procedure over multiple GPUs and nodes. As for training, the main parallelization issue is the ability of the optimization algorithm to scale over multiple GPUs or nodes: see in particular Recht, Re, Wright and Niu (2011) for an asynchronous SGD algorithm which does not require synchronization between the different workers involved. For a discussion on synchronous vs. asynchronous SGD in the context of distributed optimization or training, we refer the reader to Chen, Pan, Monga, Bengio and Jozefowicz (2016). Combining these approaches would allow for an implementation that can readily scale to multiple GPUs and nodes, reducing the computation times proportionally to the total number of GPU nodes that are available.

[6]https://github.com/BouazzaSE/NeuralXVA.

CHAPTER VI

CVA Sensitivities, Hedging and Risk

Default risk can hardly be hedged in practice (see I.§3.E.1 and II.§4), but CVA desks do hedge their market risk (see Burnett and Williams (2021) and Chapter XI). In this chapter we present a unified framework for computing CVA sensitivities, hedging the CVA, and assessing CVA risk, using probabilistic machine learning meant as refined regression tools on simulated data, validatable by low-cost companion Monte Carlo procedures. Various notions of sensitivities are introduced and benchmarked numerically. We identify the sensitivities representing the best practical trade-offs in downstream tasks including CVA hedging and risk assessment.

§1 INTRODUCTION

Probabilistic machine learning for CVA pricing was introduced in the previous chapter. Here we extend our approach to encompass CVA sensitivities and risk. The fact that probabilistic machine learning is performed on simulated data, which can be augmented at will, does not mean that there are no related data issues. As always with machine learning, the quality of the data is the first driver of the success of the approach. The variance of the training loss may be high and jeopardize the potential of a learning approach. This was first encountered in the CVA granular defaults pricing setup of the previous chapter, due to the scarcity of the default events compared with the diffusive scale of the risk factors in the model. Switching from prices to sensitivities in this chapter is another case of increased variance. But with probabilistic machine learning, we can also develop suitable variance reduction tools, namely oversimulation of defaults in Chapter V and common random numbers in this one. Another distinguishing feature of probabilistic machine learning, which is key for regulated banking applications, is the possibility to assess the quality of

DOI: 10.1201/9781003614944-10

the predictor of a conditional expectation by means of the low-cost companion Monte Carlo procedures of IV.§2.A.

A Outline and Generalities

§2 introduces different variants of bump sensitivities, benchmarked numerically in a CVA setup in §3. §4 shows how a conditional (e.g. future) CVA can be learned from simulated data (pricing model parameters and paths and financial derivative cash flows), both in a baseline calibrated setup and in an enriched setup, also accounting for recalibration model shifts. §5 and §6 develop a framework for internal modeling of CVA and/or counterparty default risks, entailing various notions of CVA sensitivities. §7 concludes as for which kind of sensitivity and numerical scheme provides the best practical trade-off for various downstream tasks, including CVA hedging and risk assessment.

A vector of partial derivatives with respect to x is denoted by ∂_x (or $\cdot'$ when x is clear from the context). All equations are written using the risk-free asset as a numéraire and are stated under the fininsurance probability measure advocated for XVA computations in the introduction to the Pricing part of the book or, more precisely, under the associated survival probability measure $\mathbb{Q}$ of a reference bank, with related expectation operator denoted below by $\mathbb{E}$. All cash flows are assumed to be square integrable. Some standing notation is listed in Table 8.

All neural network trainings are done using the PyTorch module and the Adam optimizer. Linear regressions are implemented using a truncated singular value decomposition (SVD) approach. Unless explicitly stated, we always include a ridge (i.e. Tikhonov) regularization term in the loss function

Table 8: Standing notation regarding CVA sensitivities.

ρ	model parameters, i.e. exogenous model parameters ϵ and initial conditions of model risk factors	Y	CVA diffusive risk factor process with initial condition y (if constant) or ι (if randomized)
ϱ	randomization of ρ	X	default indicator processes of clients
ω	stochastic drivers (e.g. Brownian paths)	Z	market price process, with initial condition z_0
ρ_0	calibrated value of ρ	$\xi(\rho;\omega)$	product payoff
$\bar{\rho}$	$2\rho_0 - \rho$	Π_0	$\mathbb{E}\xi(\rho)$
$\cdot^{\theta}$	neural net function with parameters θ	$\varsigma(\rho;\omega)$	$\xi(\rho;\omega) - \xi(\bar{\rho};\omega)$
ν	number of market instruments	$\Sigma_0(\rho)$	$\mathbb{E}\varsigma(\rho)$
m	number of Monte Carlo paths of the pricing model	Δ	CVA hedging ratio or linear coefficients of a CVA proxy
d	number of model parameters, i.e. dimension of ρ	Γ	diagonal quadratic coefficients of a CVA proxy

to stabilize training and regressions. Our computations are run on a server with an Intel(R) Xeon(R) Gold 5217 CPU and a Nvidia Tesla V100 GPU.

§2 FAST BUMP SENSITIVITIES

In this section we consider a time-0 option price $\Pi_0(\rho) = \mathbb{E}\xi(\rho)$, where the payoff $\xi(\rho) \equiv \xi(\rho;\omega)$ depends on constant model parameters ρ and (implicitly in the shorthand notation $\xi(\rho)$) on the randomness ω of the stochastic drivers of the model risk factors with respect to which the expectation is taken above. The model parameters ρ encompass the initial values of the risk factors of the pricing model, as well as all the exogenous (constant, in principle) model parameters, e.g. the value of the volatility in a Black-Scholes model. For each constant ρ, the price $\Pi_0(\rho)$ can be estimated by Monte Carlo. Our problem in this part is the estimation of the corresponding sensitivities $\partial_\rho \Pi_0(\rho_0)$, at a baseline (in practice, calibrated) value $\rho = \rho_0$ of the model parameters. Such sensitivities lie at the core of any related hedging scheme for the option. They are also key in many regulatory capital formulas.

Monte Carlo estimation of sensitivities in finance comes along three main streams (Crépey, 2013, Section 6.6): *(i)* differentiation of the density of the underlying process via integration by parts or more general Malliavin calculus techniques, assuming some regularity of this process; *(ii)* cash flows differentiation, assuming their differentiability, in chain rule with the stochastic flow of the underlying process; *(iii)* Monte Carlo finite differences, biased but generic, which are the Monte Carlo version of the industry standard bump sensitivities. But *(i)* suffers from intrinsic variance issues. In contemporary technology, *(ii)* appeals to adjoint algorithmic differentiation (AAD). A randomized version of this approach is provided by Sections 5.3–5.5 of Saadeddine (2022), targeted to model calibration, which requires sensitivities of vanilla options as a function of their model parameters. However, the embedded AAD layer can represent important implementation and memory costs on complex pricing problems at the portfolio level, such as CVA computations: see Capriotti, Jiang and Macrina (2017). An AAD greeking approach becomes even nearly unfeasible in the case of pricing problems embedding numerical optimization subroutines, e.g. the training of the conditional risk measures embedded in the refined CVA and higher-order XVA metrics of Chapter VII. Hence, we focus on *(iii)* hereafter.

A Common Random Numbers

Under the approach *(iii)*, first-order pointwise bump sensitivities are computed by relaunching the Monte Carlo pricing engine with common random numbers ω for values bumped by $\pm 1\%$ (typically and in relative terms) of each risk factor and/or model parameter of interest, then taking the accordingly normalized difference between the corresponding $\Pi_0(\rho)$ and $\Pi_0(\bar{\rho})$, where $\bar{\cdot}$ means symmetrization with respect to ρ_0, so

$$\frac{\rho+\bar{\rho}}{2}=\rho_0,\text{ i.e. }\bar{\rho}=2\rho_0-\rho. \tag{1}$$

This approach requires two Monte Carlo simulation runs per sensitivity, making it a robust but heavy procedure, which we try to accelerate by various means in what follows.

A predictor $\Pi_0^\theta(\rho)\approx\Pi_0(\rho)$ for the pricing function around $\rho=\rho_0$ within a suitable space of neural nets parameterized by θ readily leads to an AAD estimate $\partial_\rho\Pi_0(\rho_0)\approx\partial_\rho\Pi_0^\theta(\rho_0)$ for the corresponding sensitivities. However, even if ridge regularization may help in this regard, such an estimate, deemed naive AAD hereafter, may be bad as differentiation is not a continuous operator in the supremum norm (in other terms, functions may be arbitrarily close in sup norm, but their derivatives may be far from each other). Specifically, let $\mathcal{B}(\mathbb{R}^d)$ denote the space of the Borel measurable functions of ρ. The pricing function

$$\Pi_0(\rho)=\mathbb{E}\xi(\rho)=\mathbb{E}\Big(\xi(\varrho)\,\Big|\,\varrho=\rho\Big) \tag{2}$$

(for ϱ randomizing ρ around ρ_0) can be learned from simulated pairs $(\varrho,\xi(\varrho;\omega))$ based on the representation

$$\Pi_0(\cdot)=\underset{\Phi\in\mathcal{B}(\mathbb{R}^d)}{\operatorname{argmin}}\,\mathbb{E}\Big[\Big(\xi(\varrho)-\Phi(\varrho)\Big)^2\Big]. \tag{3}$$

To learn the function $\Pi_0(\cdot)$ around ρ_0, we can replace, in the optimization problem (3), $\mathcal{B}(\mathbb{R}^d)$ by a suitable space of neural nets and $\mathbb{E}$ by a simulated sample mean $\widehat{\mathbb{E}}$, with each "vertical" (time-0) draw of ϱ followed by an "horizontal" (across future times) draw of ω that is implicit in $\xi(\varrho)\equiv\xi(\varrho;\omega)$. The ensuing minimization problem for the weights θ of the neural net $\rho\mapsto\Pi_0^\theta(\rho)\approx\Pi_0(\rho)$ is then performed numerically by Adam mini-batch gradient descent. The corresponding sensitivities $\partial_\rho\Pi_0^\theta(\rho_0)$ are retrievable by AAD at negligible additional cost. However, as emphasized above, the ensuing **naive AAD sensitivities** $\partial_\rho\Pi_0^\theta$ may be a poor estimate of $\partial_\rho\Pi_0$.

In this learning setup, the corresponding instability reflects a variance issue. In order to cope with the increased variance due to the switch from prices to sensitivities, a trick is to introduce $\varsigma(\rho;\omega):=\xi(\rho;\omega)-\xi(\bar{\rho};\omega)$ (cf. (1)). We can then learn the sensitivity (in the sense here of finite differences) function $\rho\mapsto\Sigma_0(\rho):=\mathbb{E}\varsigma(\rho)$, which satisfies by linearity and chain rule (as $\bar{\rho}=2\rho_0-\rho$)

$$\Sigma_0'(\rho)=\Pi_0'(\rho)+\Pi_0'(\bar{\rho}),\text{ in particular }\partial_\rho\Sigma_0(\rho_0)=2\partial_\rho\Pi_0(\rho_0). \tag{4}$$

For learning the function $\Sigma_0(\cdot)$ locally around ρ_0, with ϱ randomizing ρ as above, we rely on the representation $\Sigma_0(\rho) = \mathbb{E}\Big(\varsigma(\varrho) \,\Big|\, \varrho = \rho\Big)$, i.e.

$$\Sigma_0(\cdot) = \underset{\Phi \in \mathcal{B}(\mathbb{R}^d)}{\operatorname{argmin}} \, \mathbb{E}\Big[\Big(\varsigma(\varrho) - \Phi(\varrho)\Big)^2\Big]. \tag{5}$$

Then we replace, in the optimization problem (5), $\mathcal{B}(\mathbb{R}^d)$ by a linear hypothesis space $\Sigma_0^\theta(\rho) = \theta^\top(\rho - \rho_0)$ (noting that $\Sigma_0(\rho_0) = 0$) and $\mathbb{E}$ by a simulated sample mean $\widehat{\mathbb{E}}$, with again each "vertical" draw of ϱ followed by one "horizontal" draw of ω that is implicit in $\varsigma(\varrho)$. This results in a linear least-squares problem for the weights θ, solved by SVD. The estimated weights $\theta/2$ are our **linear bump sensitivities** estimate for $\frac{1}{2}\partial_\rho \Sigma_0(\rho_0) = \partial_\rho \Pi_0(\rho_0)$. These sensitivity estimates are the slope coefficients of a multilinear regression, for which confidence intervals CI scaling in $1/\sqrt{|\text{sample size}|}$ are available (Matloff, 2017, Subsubsection 2.12.11). The use of each drawn set of model parameters ϱ twice, also via $\bar{\varrho}$ with a common ω in $\varsigma(\varrho;\omega) = \xi(\varrho;\omega) - \xi(\bar{\varrho};\omega)$, is a common random numbers variance reduction technique as in *(iii)* above. For well-chosen distributions of the randomization ϱ of ρ, this approach results in much more accurate sensitivities than the naive AAD approach. For simple parametric distributions of ϱ, the covariance matrix that appears in the regression for the first-order sensitivities is known and invertible in closed form, which reduces the linear regression (implemented without ridge regularization in this analytical case) to a standard Monte Carlo and a more robust CI. Nonlinear hypothesis spaces of neural networks trained the way described after (3) (just replacing $\xi(\varrho)$ and $\partial_\rho \Pi_0^\theta(\rho_0)$ there by $\varsigma(\varrho)$ and $\partial_\rho \Sigma_0^\theta(\rho_0)$ here) can also be used instead of the above linear model for $\Sigma_0(\rho)$. The **AAD bump sensitivities** are then obtained as the halved AAD sensitivities of the trained (no longer linear) network $\rho \mapsto \partial_\rho \Sigma_0^\theta$ at ρ_0 (but we lose the confidence interval CI in this case). Similar ideas can be applied to higher order sensitivities, using e.g. $\xi(\rho;\omega) - 2\xi(\rho_0;\omega) + \xi(\bar{\rho};\omega)$ instead of $\varsigma(\rho;\omega) = \xi(\rho;\omega) - \xi(\bar{\rho};\omega)$ to capture diagonal gammas. However, higher order means even more variance. Moreover, the Jacobian trick of Subsection C below to convert the model into market sensitivities only works for first-order sensitivities, hence our focus on the latter in this chapter.

An important ingredient in successful randomized (linear or AAD) bump sensitivities is the choice of appropriate distributions for ϱ. Since different parameters (components of ρ) may have very different magnitudes in values and price impact (see e.g. Figure 28 page 173 and Table 9 page 178), a linear regression may not be able to identify the individual effect of each parameter when all are bumped simultaneously. To address this issue, we divide the parameters into groups. The m simulated paths of the pricing model are then partitioned into blocks of paths such that only one group of parameters is

Algorithm 4: Fast bump sensitivities.

input : A (calibrated) baseline ρ_0 for the initial conditions of all risk factors and for the exogenous parameters of the pricing model, a number m of pricing model paths.

output: Estimated sensitivities $\theta/2$ (linear case) or $\frac{1}{2}\partial_\rho \Sigma_0^\theta(\rho_0)$ (more generally) $\approx \partial_\rho \Pi_0(\rho_0)$.

Draw m i.i.d. bumped model parameters ρ_j from some distribution randomizing ρ around ρ_0, e.g. $\mathcal{N}\Big(\rho_0, \mathrm{diag}\left(\sigma^2\rho_0 \odot \rho_0\right)\Big)$[a] with $\sigma = 1\%, 3\%, 5\%$[b] or, in the case of smart bump sensitivities, $\rho_j = \rho_0$ bumped by 1% on its $\left\lfloor \frac{jd}{m} \right\rfloor$-th component

Draw m i.i.d. stochastic drivers ω parameters $\{\omega_1, \dots, \omega_m\}$

for $(\rho, \omega) \in (\rho_1, \omega_1) .. (\rho_1, \omega_m)$ **do**

  Compute the payoffs $\xi(\rho; \omega)$ and $\xi(\bar{\rho}; \omega)$, where $\frac{\rho+\bar{\rho}}{2} = \rho_0$

  Compute $\varsigma(\rho; \omega) = \xi(\rho; \omega) - \xi(\bar{\rho}; \omega)$

end

if *AAD bump sensitivities* **then**

  Train a neural network $\Sigma_0^\theta(\rho)$ to regress $(\varsigma(\rho_1), \dots, \varsigma(\rho_m))$ against $(\rho_1, , \dots, \rho_m)$

  Retrieve $\partial_\rho \Sigma_0^\theta(\rho)$ by AAD and divide the obtained gradient by two for obtaining the AAD bump sensitivities.

else if *Linear bump sensitivities* **then**

  Regress linearly (without intercept) $(\varsigma(\rho_1), \dots, \varsigma(\rho_m))$ against $(\rho_1 - \rho_0, \dots, \rho_m - \rho_0)$ by SVD

  Divide the regression coefficients θ by two for obtaining the linear bump sensitivities

else if *Smart bump sensitivities* **then**

  Compute each sensitivity by dividing the average of ς over each block of size m/d by two times the corresponding bump size

end

---

[a] $\odot$ is the Hadamard (i.e. componentwise) product between vectors and diag(vec) is a diagonal matrix with diagonal "vec".

[b] The notation $\mathcal{N}\Big(\rho_0, \mathrm{diag}\left(\sigma^2\rho_0 \odot \rho_0\right)\Big)$ which is used for simplicity in this pseudo-code ignores the practically important block simulation trick mentioned in the end of this subsection A.

bumped in each block. In addition to this, one can use different distributions for each group or even for each parameter. Notably, we have observed numerically that slightly noiser distributions yield better sensitivity estimates for volatility-related parameters.

In addition to the above, we will also compute **benchmark bump sensitivities** as per *(iii)* in the above, obtained on the basis of d Monte Carlo repricings with m common random numbers ω and relative variations of $\pm 1\%$ of one model parameter in each Monte Carlo run, as well **smart bump sensitivities**, similar but only using m/d paths each, where d is the number of model parameters (dimension of ρ). Hence the time of computing all the smart bump sensitivities is of the same order of magnitude as the one of retrieving the linear bump sensitivities, which is also roughly twice the time of pricing $\Pi_0(\rho_0)$ by Monte Carlo with m paths. More precisely, each smart bump sensitivity uses m/d paths of two Monte Carlo simulation runs with m paths for each of the two. This is significantly more efficient than doing $2d$ Monte Carlo runs of size m/d each, especially in the GPU simulation environment of our CVA computations later below. Note that such smart bump sensitivities are also a special case of linear bump sensitivities with block simulation trick, for blocks of size m/d and deterministic bumps of relative size 1% on one and only one model parameter in each block, the linear regression degenerating in this case to a local sample mean over the m/d paths of each block: see Algorithm 4 page 170, which summarizes the above procedures for various fast bump sensitivities.

B Basket Black-Scholes Example

Let us consider a European call on the geometric average of q Black-Scholes assets, aiming for the corresponding deltas, vegas and (diagonal) gammas, which are known analytically in this lognormal setup. Regarding the above block simulation trick, the time-0 values of the assets are bumped in half of the paths, while the volatilities are bumped in the other half. We study two cases, with $q = 3$ and 10 and, respectively $m = 10^5$ and $m = 5 \times 10^5$ simulated Black-Scholes paths. In the $q = 3$ case, we implement all the fast bump approaches of Algorithm 4 and the naive AAD approach described after (3). The learning of the gammas is also reported in this low-dimensional experiment. In the $q = 10$ case, we skip the naive AAD approach as well as all the learned gamma results because of their poor performance. The hypothesis space used in all the AAD approaches is a vanilla multi-layer perceptron with two hidden layers and softplus activation functions.

The error bars in Figure 28 page 173 represent our 95% confidence intervals (CI) for the linear bump sensitivities (for $m = 10^5$ paths in both cases). Regarding the AAD bump approach, we train the neural network 100 times with different initializations to also get a 95% confidence interval, in a meaning weaker than CI though: as the data is not resampled in each run, the AAD confidence intervals ($\mathrm{CI}^\flat$) can only account for the randomness in training, not for the one of the simulated data. The plots of Figure 28(a) show the inaccuracy of the naive AAD sensitivities. The fast bump approaches consistently estimate deltas, but vegas and gammas appear to be more challenging. This is due to greater variance in the case of gammas, whereas in the case of vegas, the collusion between the noise of the random volatility coefficient and the one of the Brownian drivers ω makes the learning task more difficult. As should be, the CIs of the linear and smart bump sensitivities contain nearly all the exact sensitivities. This is also mostly the case for the $\mathrm{CI}^\flat$s of the AAD bump sensitivities, but in their case this comes without a theoretical guarantee. The benchmark bump sensitivities are exact with 2 significant digits (at least for deltas and vegas) in this Black-Scholes setup. They would be visually indistinguishable of the exact sensitivities if we added them to the graphs of Figure 28. The time of computing the smart bump sensitivities is of the same order of magnitude as the one of retrieving the linear bump sensitivities, with also similar accuracy as demonstrated in Figure 30(a) page 181, where both ratios between running times and errors of the linear and smart bump sensitivities with respect to the benchmark bump sensitivities stay close to each other. The complexities, describing how long it would take for each algorithm to achieve a relative error in the Black-Scholes case (or standard error in the CVA case) of 1%, are not significantly different for the linear, smart and benchmark bump sensitivities, for most of d; otherwise the smart ones beat by a small margin the linear ones, themselves a bit worse than the benchmark ones. The grey dashed lines in the right panels indicate the d^2 scaling of complexities expected for the benchmark bump sensitivities, which involve $2d$ simulation runs in dimension d.

In general, we observed that in high dimension the linear and smart bump sensitivities tend to be more stable and reliable than the AAD bump sensitivities. Hence, we forget AAD bump sensitivities hereafter.

C From Model to Market Sensitivities

The sensitivities $\partial_\rho \Pi_0(\rho_0)$ are sensitivities to model parameters. Practical hedging schemes require sensitivities to calibrated prices of hedging instruments. Hence, for hedging purposes, our sensitivities must be mapped to hedging ratios in market instruments. This can be done via the implicit function theorem, the way explained in Henrard (2011), Savine (2018) and Antonov, Issakov and McClelland (2018). In a nutshell, assume that, given market prices $z \in \mathbb{R}^\nu$ of suitable calibration (hedging) assets, the corresponding pricing model parameters $\rho(z)$ are obtained by a calibration procedure of the form

$$\rho(z) \in \operatorname*{argmin}_{\rho \in \mathbb{R}^d} \textit{cal-err}\,(z, \rho)\,, \tag{6}$$

where $\textit{cal-err}\,(z, \rho)$ quantifies the mean square error between the market prices z of the calibration instruments and their prices in the pricing model with parameters ρ. Note that, in practice, not all the pricing parameters are obtained via a minimization as in (6): some of them are bootstrapped or even directly

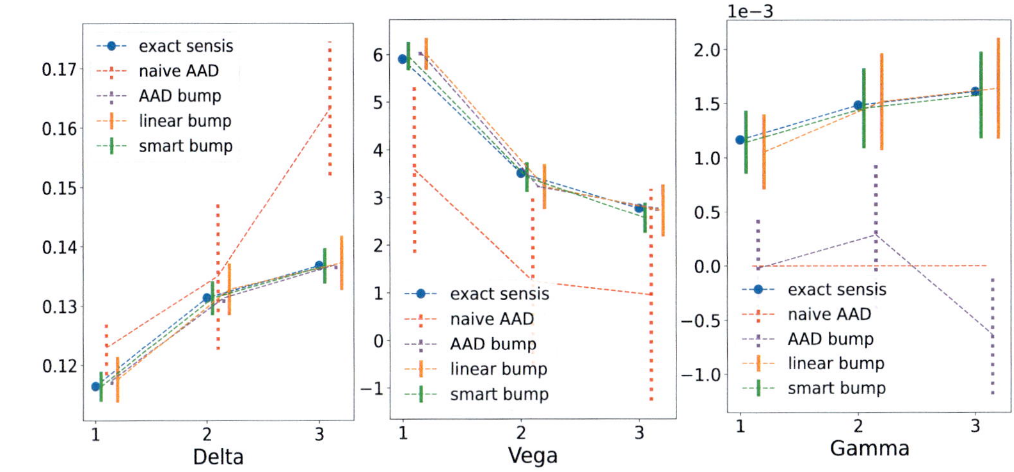

(a) $q = 3$ and $m = 10^5$ paths.

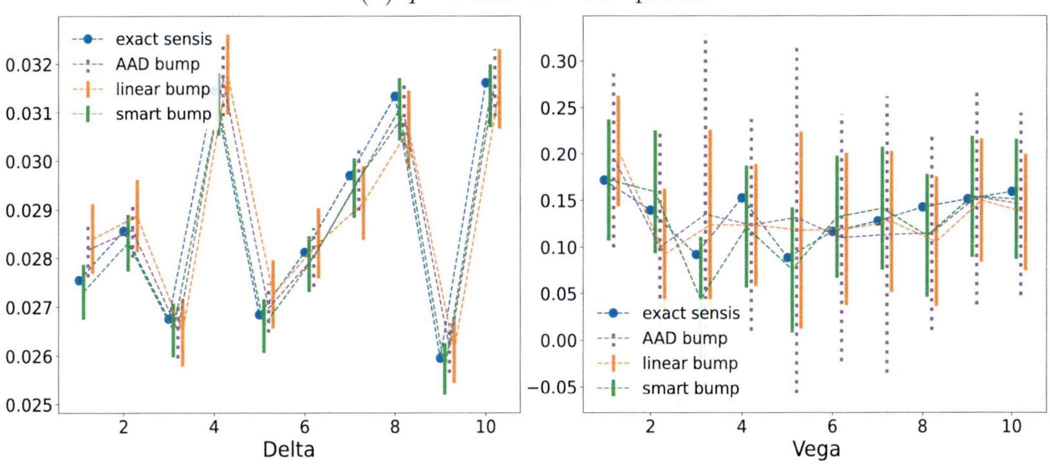

(b) $q = 10$ and $m = 5 \times 10^5$ paths.

Figure 28: Multi-asset geometric average call sensitivities on 3 (a) and 10 (b) independent Black-Scholes assets. The error bars represent confidence intervals CI or $\mathrm{CI}^\flat$, in dotted lines regarding the latter, which only reflect a training noise.

observed on the market, see Savine (2018) for a more detailed presentation and §3.A for an example.

Assume *cal-err* of class $\mathcal{C}^2$ in ρ. Let $\rho(z_0) = \rho_0$ be a solution of (6) associated with a particular $z = z_0$, hence $\partial_\rho \textit{cal-err}(z_0, \rho_0) = 0$. Denote by $\partial^2_{\rho,\rho}\textit{cal-err}$ the Hessian matrix of *cal-err* with respect to ρ, and assume that $\partial^2_{\rho,\rho}\textit{cal-err}(z_0, \rho_0)$ is invertible. Then, by the implicit function theorem applied to the function $\partial_\rho \textit{cal-err}$ of (z, ρ), there exists an open neighborhood $\mathcal{O} \in \mathbb{R}^\nu$ of z_0 on which (6) uniquely defines a function $\mathcal{O} \ni z \mapsto \rho(z)$, i.e.

$\partial_\rho cal\text{-}err(z, \rho(z)) = 0$ for all $z \in \mathcal{O}$. Moreover, if ρ is in $\mathcal{C}^1\left(\mathcal{O}; \mathbb{R}^d\right)$, then

$$\partial_z \rho(z) = -\left(\partial^2_{\rho,\rho} cal\text{-}err(z, \rho(z))\right)^{-1}\left(\partial^2_{\rho,z} cal\text{-}err(z, \rho(z))\right), \text{ for all } z \in \mathcal{O},$$

where we assume that the matrix of all the second derivatives of *cal-err* with respect to one component in ρ and one component in z, denoted by $\partial^2_{\rho,z} cal\text{-}err$, exists. From there, the chain rule

$$\begin{aligned}(\partial_z \Pi_0(z))^\top &= \left(\partial_\rho \Pi_0(\rho(z))\right)^\top \partial_z \rho(z) \\ &= -\left(\partial_\rho \Pi_0(\rho(z))\right)^\top \left(\partial^2_{\rho,\rho} cal\text{-}err(z, \rho(z))\right)^{-1}\left(\partial^2_{\rho,z} cal\text{-}err(z, \rho(z))\right), \text{ for } z \in \mathcal{O},\end{aligned} \tag{7}$$

allows deducing the market sensitivities $\partial_z \Pi_0$ from the model sensitivities $\partial_\rho \Pi_0$. When heavy Monte Carlo (such as CVA) pricing tasks are involved in the $\partial_\rho \Pi_0$ computations, the time of computing $\partial_z \Pi_0$ through (7) is dominated by the time of computing $\partial_\rho \Pi_0$.

Another way to compute market bump sensitivities is to bump each target calibration price and recompute the price Π_0 for pricing model parameters ρ recalibrated to each bumped calibrated data set. This direct approach does not need Jacobian transformations, and it is also amenable to second-order sensitivities. However, accounting for curves and surfaces of hedging assets, there may be much more market sensitivities than model sensitivities (as in our use case of §3.A). Moreover, direct market sensitivities require not only intensive repricings but also model recalibrations, as many as targeted market sensitivities. The direct approach is therefore typically much heavier than the one based on model bump sensitivities followed by Jacobian transformations. We therefore forget direct market bump sensitivities hereafter: by market bump sensitivities, we mean from now on (first-order) bump sensitivities with respect to model parameters, transformed to the corresponding first-order market sensitivities via (7).

§3 CREDIT VALUATION ADJUSTMENT AND ITS BUMP SENSITIVITIES

In the above Black-Scholes setup, bump sensitivities are useless because the exact greeking formulas are also faster. We now switch to CVA computations in the role of Π before, for which pricing and greeking can only be achieved by intensive Monte Carlo simulations implemented on a GPU. Since we are also interested in the risk of CVA fluctuations (if unhedged, or of fluctuations of a hedged CVA position more generally), we now consider the targeted price Π (CVA from now on) as a process. We denote by P^b, the counterparty-risk-free valuation of the portfolio of the bank with its (assumed all bilateral) counterparty b; τ_b, the counterparty b's default time, with intensity process

γ^b; X, with $X_0 \equiv 0$ (componentwise), the vector of the default indicator processes of the counterparties of the bank; Y, a diffusive vector process of model risk factors such that each P^b_t and γ^b_t is a measurable function of (t, Y_t), for $t \geq 0$ (in the case of credit derivatives with the counterparty b, P^b_t would also depend on X_t, which can be accommodated at no harm in our setup). The exogenous model parameters are denoted by ϵ. Let the baseline $\rho_0 = (y_0, \epsilon_0)$ denote a calibrated value of (y, ϵ), where y is used for referring to the initial condition of Y, whenever assumed constant. Let ι denote an initial condition for Y randomized around its baseline y_0, ε be likewise a randomization of ϵ around its baseline ϵ_0, and $\varrho_t = (Y_t, \varepsilon), t \geq 0$. Starting from the (random) initial condition $(0, \iota)$, the model (X, Y) is supposed to evolve according to some Markovian dynamics (e.g. the one of §3.A) parameterized by ε. This setup allows encompassing in a common formalism:

- the **baseline mode** of V.(12)-(13) where $\varrho_0 \equiv \rho_0$;
- the **risk mode** where $Y_0 \equiv y_0$;
- the **sensis mode**, or general ϱ_0 case, used with an exogenous distribution of ι in Subsection A below and with ι distributed as the diffusive risk factors of the CVA simulation engine in the risk mode at a specified risk horizon in §6.

Again, the CVA engine in the baseline mode $\varrho_0 \equiv \rho_0$ is nothing but the CVA learner of V.(12)-(13). The risk and sensis mode also incorporating a randomization ε of the exogenous model parameters ϵ, and of Y_0 in the sensis mode, are novelties of the present chapter. The distribution of ε can be set exogenously or (in the risk mode in particular) assessed statistically on a historical basis, as done in the SIMM context of Albanese, Caenazzo and Syrkin (2017).

We restrict ourselves to an uncollateralized CVA for notational simplicity. Given n pricing time steps of length h such that $n\mathrm{h} = T$, the final maturity of the derivative portfolio of the bank, let, at each $t = i\mathrm{h}$,

$$\mathrm{DL}_t = \sum_b \sum_{j=0}^{i-1} (P^b_{j\mathrm{h}})^+ \mathbb{1}_{\{j\mathrm{h} < \tau_b \leq (j+1)\mathrm{h}\}}. \tag{8}$$

We recall the following default-based and intensity-based formulations of the (time-discretized) CVA of a bank with bilateral counterparties b, at the pricing time $t = i\mathrm{h}$ (cf. Eqns. V.(10)-(12)):

$$\mathrm{CVA}_t(x,\rho) = \mathbb{E}\Big[\underbrace{\sum_b \sum_{j=i}^{n-1} (P^b_{j\mathrm{h}})^+ \mathbb{1}_{\{j\mathrm{h}<\tau_b\le(j+1)\mathrm{h}\}}}_{\mathrm{DL}_T-\mathrm{DL}_t} \Big| X_{i\mathrm{h}} = x, \varrho_{i\mathrm{h}} = \rho\Big]$$
$$= \mathbb{E}\Big[\underbrace{\sum_b \sum_{j=i}^{n-1} (P^b_{j\mathrm{h}})^+ (e^{-\int_{i\mathrm{h}}^{j\mathrm{h}}\gamma^b(s)ds} - e^{-\int_{i\mathrm{h}}^{(j+1)\mathrm{h}}\gamma^b(s)ds})\mathbb{1}_{\{\tau_b>i\mathrm{h}\}}}_{=:\xi^{cva}_{t,T}} \Big| X_{i\mathrm{h}} = x, \varrho_{i\mathrm{h}} = \rho\Big], \quad (9)$$

where each coordinate of x is 0 or 1. From a numerical viewpoint, the second line of (9) entails less variance than the first one (see Figure 15 page 145). Hence we rely for our CVA computations on this second line (after time-discretization of the integrals under the exponentials, cf. VI.(13)). At the initial time 0, as $X_0 \equiv 0$, we can restrict attention to the origin $x = 0$ and skip the argument x as well as the conditioning by $X_{i\mathrm{h}} = x$ in (9). In the sensis mode, one is in the setup (2) for $\xi = \xi^{cva}_{0,T}$, which implicitly depends on $\varrho_0 = (\iota, \varepsilon)$, i.e. $\mathrm{CVA}_0(\varrho_0) = \mathbb{E}\big(\xi(\varrho_0)\,\big|\,\varrho_0\big)$. The baseline CVA sensitivities can thus be computed by Algorithm 4 page 170, with CVA, (y,ϵ) and (y_0,ϵ_0) here in the role of Π, ρ and ρ_0 there. In the baseline mode where $\varrho_0 \equiv \rho_0$, $\mathrm{CVA}_0(\varrho_0)$ is constant, equal to the corresponding

$$\mathrm{CVA}_0(\rho_0) = \mathbb{E}\,\mathrm{DL}_{n\mathrm{h}} = \mathbb{E}\xi^{cva}_{0,T}, \quad (10)$$

which is computed by Monte Carlo based on the second line of (9) for $t = 0$ there, as a sample mean of $\xi^{cva}_{0,T}$, along with the corresponding 95% confidence interval.

All our CVA computations are done on a GPU (whereas the previous Black-Scholes calculations were done on CPU, except for neural net training on a GPU).

A CVA Lab

Our numerics below are performed in the market and credit model of IV.§4. We have 10 economies. For each of them we have a short-term interest rate driven by a Vasicek diffusion and, except for the reference economy, a Black-Scholes exchange rate with respect to the currency of the reference economy. The reference bank has 8 counterparties with corresponding default intensity processes driven by CIR diffusions. We thus have 8 default indicator processes X of the counterparties and $10+9+8 = 27$ diffusive risk factors Y. This results in a Markovian model (X,Y) of dimension 35, entailing $d = 90$ parameters corresponding to the 27 initial conditions of the Y processes plus their 63 exogenous parameters (see Table 9 page 178). This model is only for illustrative purposes: the methodology of this chapter can be applied to any Markovian model (X, Y) of client defaults and diffusive (or jump-diffusive if wished) risk factors.

A "reasonably stressed" but arbitrary baseline ρ_0 (see after (12)) plays the role of calibrated model parameters in our numerics. In the above pricing model, we consider the CVA on a portfolio of interest rate swaps with characteristics generated randomly much like in V.§4.C-D. The portfolio consists of 500 interest rate swaps with random characteristics (maturity $\leq T = 10$ years, notional, currency and counterparty) and strikes such that the swaps are worth 0 in the baseline model (i.e. for $\varrho_t \equiv \rho_0$) at time 0. The swaps have analytic counterparty-risk-free valuation in our pricing model (see IV.§4.B). Their price processes are converted into the reference currency and aggregated into the corresponding clients P^b processes.

We simulate using an Euler scheme $m = 2^{17} \approx 1.3 \times 10^5$ paths of the pricing model (X, Y), with $n = 100$ pricing time steps of length $\mathrm{h} = 0.1$ and 25 Euler simulation sub-steps per pricing time step (referred to as daily basis). A Monte Carlo computation of (10) in the baseline mode then yields $\mathrm{CVA}_0(\rho_0) \in 5,027 \pm 18$ with 95% probability (computed in about 30s). A randomization of Y_0 and ϵ in the sensis mode is used for deriving CVA linear bump sensitivities as per §2.A. Table 9 presents benchmark versus fast (linear and smart) bump sensitivities with respect to the $d = 90$ model parameters. The CIs of the linear and smart bump sensitivities consistently cover the benchmark bump sensitivities, which, with regard to its much smaller confidence intervals, serve as a reference for these sensitivities (even if biased with respect to the exact $\partial_\rho \mathrm{CVA}_0$). Regarding the subset simulation trick exposed after Algorithm 4 page 170, we divided the model parameters into 10 groups separated by horizontal lines in Table 9.

For converting the model sensitivities into market sensitivities by the Jacobian method of §2.C, we first need to specify the market instruments. In this case, each of the 10 zero-yield curves has the 14 pillars $0.01, 0.1, 0.2, 0.5, 1, \ldots, 10$ years, each of the 9 FX forward curves has 4 pillars $0.01, 0.1, 0.2, 0.5$ year, and each of the 8 CDS curves (with monthly payments and loss-given-default parameter set to 60% for each counterparty) has the 10 pillars $1, \ldots, 10$ years, resulting in a total of $\nu = 256$ market instruments and first-order market sensitivities. As we do not introduce any options as hedging assets, we freeze the volatility model parameters (in orange in Table 9) and only consider the calibration error as a function of the initial conditions and drift parameters of the model risk factors. The reason why our FX curves are restricted to 4 points is that we thus only have one FX-related model (spot exchange rate) parameter for each foreign currency. Figure 29 page 180 displays some of the interest-rate, credit and FX market sensitivities deduced from the model sensitivities of Table 9 by Jacobian transformation, way explained in

Table 9: CVA sensitivities and 95% confidence intervals (CI) with respect to the $d = 90$ parameters of the model, estimated by the benchmark, linear and smart bump approaches. The notation for the parameters follows Eqns. IV.(10)-(11). The horizontal lines separate groups of parameters bumped simultaneously under the linear bump approach, the way explained after Algorithm 4 page 170. The model volatility parameters left aside from the calibration process (see the above comments to Figure 29) are in orange.

param.	benchmark bump	linear bump	smart bump	param.	benchmark bump	linear bump	smart bump
$r_0^{\langle 0 \rangle}$	-12,354± 41	-14,426± 1,426	-12,310± 384	$b^{\langle 8 \rangle}$	-37,295± 487	-43,620± 10,327	-40,903± 4,519
$r_0^{\langle 1 \rangle}$	-4,761± 57	-4,349± 1,331	-4,597± 514	$b^{\langle 9 \rangle}$	94,235± 760	89,363± 11,404	93,950± 6,802
$r_0^{\langle 2 \rangle}$	10,715± 92	12,060± 1,438	11,010± 859	$\sigma^{r,\langle 0 \rangle}$	23,850± 209	25,610± 5,852	24,283± 1,979
$r_0^{\langle 3 \rangle}$	1,433± 37	1,331± 1,315	1,521± 353	$\sigma^{r,\langle 1 \rangle}$	23,563± 311	22,360± 4,287	21,755± 2,460
$r_0^{\langle 4 \rangle}$	14,712± 62	14,648± 1,350	15,054± 573	$\sigma^{r,\langle 2 \rangle}$	33,945± 392	32,240± 4,193	36,699± 4,635
$r_0^{\langle 5 \rangle}$	24,539± 146	26,762± 1,697	25,057± 1,424	$\sigma^{r,\langle 3 \rangle}$	14,402± 191	16,502± 4,445	13,434± 1,671
$r_0^{\langle 6 \rangle}$	15,100± 96	15,450± 1,416	14,744± 901	$\sigma^{r,\langle 4 \rangle}$	20,347± 292	18,847± 3,576	20,866± 2,828
$r_0^{\langle 7 \rangle}$	29,368± 161	30,239± 1,612	29,637± 1,398	$\sigma^{r,\langle 5 \rangle}$	36,305± 500	38,439± 5,455	34,401± 4,262
$r_0^{\langle 8 \rangle}$	5,930± 66	5,410± 1,372	6,264± 689	$\sigma^{r,\langle 6 \rangle}$	26,597± 400	23,572± 4,182	26,593± 3,329
$r_0^{\langle 9 \rangle}$	5,132± 57	7,117± 1,347	4,879± 484	$\sigma^{r,\langle 7 \rangle}$	31,233± 644	29,380± 5,086	32,828± 7,071
$\chi_0^{\langle 1 \rangle}$	151± 3	188± 80	164± 26	$\sigma^{r,\langle 8 \rangle}$	28,051± 391	29,169± 4,154	25,456± 2,634
$\chi_0^{\langle 2 \rangle}$	733± 7	709± 86	757± 78	$\sigma^{r,\langle 9 \rangle}$	24,085± 322	22,284± 3,722	24,944± 3,326
$\chi_0^{\langle 3 \rangle}$	123± 2	134± 77	134± 21	$\sigma^{\chi,\langle 1 \rangle}$	292± 10	194± 211	364± 103
$\chi_0^{\langle 4 \rangle}$	816± 6	790± 81	808± 56	$\sigma^{\chi,\langle 2 \rangle}$	406± 21	317± 198	378± 181
$\chi_0^{\langle 5 \rangle}$	829± 8	932± 85	941± 83	$\sigma^{\chi,\langle 3 \rangle}$	224± 8	229± 189	185± 49
$\chi_0^{\langle 6 \rangle}$	835± 9	861± 96	852± 89	$\sigma^{\chi,\langle 4 \rangle}$	300± 18	-31± 251	247± 139
$\chi_0^{\langle 7 \rangle}$	1,030± 11	1,036± 92	979± 92	$\sigma^{\chi,\langle 5 \rangle}$	460± 23	426± 205	526± 233
$\chi_0^{\langle 8 \rangle}$	243± 4	213± 77	243± 32	$\sigma^{\chi,\langle 6 \rangle}$	543± 29	440± 200	487± 220
$\chi_0^{\langle 9 \rangle}$	583± 6	543± 85	559± 50	$\sigma^{\chi,\langle 7 \rangle}$	458± 36	394± 229	211± 191
γ_0^1	2,201± 15	2,371± 266	2,155± 131	$\sigma^{\chi,\langle 8 \rangle}$	402± 13	405± 210	307± 66
γ_0^2	1,528± 12	1,554± 261	1,498± 109	$\sigma^{\chi,\langle 9 \rangle}$	344± 20	401± 214	459± 200
γ_0^3	3,097± 24	2,843± 267	3,133± 228	α_0^1	86± 1	86± 19	82± 11
γ_0^4	1,250± 10	1,447± 255	1,280± 92	α_0^2	69± 1	66± 19	68± 9
γ_0^5	1,473± 12	1,466± 263	1,384± 103	α_0^3	143± 2	140± 21	152± 19
γ_0^6	2,982± 15	2,964± 276	2,937± 136	α_0^4	38± 1	45± 14	41± 5
γ_0^7	6,068± 32	6,122± 321	6,001± 306	α_0^5	45± 1	58± 18	48± 6
γ_0^8	5,887± 27	5,976± 310	5,976± 258	α_0^6	154± 1	158± 18	165± 11
$a^{\langle 0 \rangle}$	-1,125± 5	-1,142± 91	-1,116± 47	α_0^7	336± 3	336± 23	315± 28
$a^{\langle 1 \rangle}$	-823± 10	-831± 103	-838± 83	α_0^8	285± 2	283± 19	270± 25
$a^{\langle 2 \rangle}$	133± 9	162± 102	120± 92	δ^1	6,386± 53	5,818± 1,185	6,032± 437
$a^{\langle 3 \rangle}$	-240± 4	-249± 83	-239± 38	δ^2	6,737± 53	6,831± 1,236	6,556± 440
$a^{\langle 4 \rangle}$	570± 7	506± 91	519± 72	δ^3	8,693± 91	9,617± 1,212	8,476± 738
$a^{\langle 5 \rangle}$	1,093± 11	1,055± 123	1,097± 98	δ^4	6,096± 42	6,372± 1,183	5,928± 358
$a^{\langle 6 \rangle}$	660± 9	681± 95	728± 81	δ^5	5,888± 36	5,708± 1,276	5,846± 305
$a^{\langle 7 \rangle}$	1,377± 13	1,462± 118	1,421± 117	δ^6	14,539± 67	14,824± 1,185	15,038± 686
$a^{\langle 8 \rangle}$	-482± 11	-430± 112	-509± 91	δ^7	23,261± 128	21,714± 1,336	23,014± 1,082
$a^{\langle 9 \rangle}$	-68± 7	-56± 106	-69± 73	δ^8	31,441± 144	31,949± 1,484	31,938± 1,558
$b^{\langle 0 \rangle}$	-166,788± 437	-169,247± 10,994	-166,618± 4,384	ν^1	-38± 8	-78± 132	16± 84
$b^{\langle 1 \rangle}$	-31,802± 406	-38,577± 9,149	-31,678± 4,753	ν^2	-47± 8	-53± 132	-147± 59
$b^{\langle 2 \rangle}$	78,709± 823	81,995± 10,636	86,000± 8,440	ν^3	-57± 15	-87± 138	6± 132
$b^{\langle 3 \rangle}$	-6,206± 341	-2,364± 10,461	-3,459± 3,561	ν^4	-26± 6	-10± 126	-37± 62
$b^{\langle 4 \rangle}$	140,127± 683	150,195± 10,716	138,407± 6,189	ν^5	-35± 6	45± 128	-44± 49
$b^{\langle 5 \rangle}$	114,437± 914	109,486± 10,779	121,754± 9,471	ν^6	-66± 13	-60± 151	-100± 118
$b^{\langle 6 \rangle}$	127,783± 1,108	120,891± 9,923	123,214± 9,430	ν^7	-151± 23	-244± 146	-150± 209
$b^{\langle 7 \rangle}$	191,031± 1,373	188,073± 12,629	196,130± 14,048	ν^8	-161± 24	-230± 154	-129± 221

§2.C. We note the consistency between the orders and magnitudes, but also term structure profiles (except at the ends of the second credit curve), of the market sensitivities deduced from the benchmark vs. fast (linear and smart) bump sensitivities, with again a slight advantage of smart over linear bump sensitivities.

The times taken to generate the model sensitivities of Table 9 and the corresponding market sensitivities are recorded in Table 10. The linear and smart bump calculations use the same m Brownian driving paths ω of the pricing model as the one used for the Monte Carlo computation of $\mathrm{CVA}_0(\rho_0)$. The speedups of the linear and smart bump sensitivities are almost identical, of about 90 with respect to the benchmark bump sensitivities. In Figure 30(b) page 181, the increasing curves in the left panel highlight the almost linear growth of the speedup of the linear and smart bump sensitivities with respect to the benchmark ones when the number d of pricing model parameters increases, but with also increasing errors displayed in the middle panel. The smart bump sensitivities have smaller confidence interval than the linear bump sensitivities for all tested d. Combining with the timing result, we conclude that the smart bump sensitivities outperform the linear bump sensitivities in this CVA use case. This is confirmed by the right panel of Figure 30(b), where for all tested d the complexity of linear bump sensitivities is higher than the one of smart bump sensitivities, itself very close (as expected) to the one of the benchmark bump sensitivities.

Table 10: Computation times (and speedups "×" in parentheses) for CVA bump sensitivities are shown in Table 9. The time of the benchmark bump sensitivities contains diagonal gamma estimation (needed for the "bench. bump sensis w/ Γ" in Table 13 page 190), which amounts to an extra 5 seconds needed for computing the baseline $\mathrm{CVA}_0(\rho_0)$ (the $\mathrm{CVA}_0(\varrho_0)$ and $\mathrm{CVA}_0(\bar{\varrho}_0)$ being already needed for the deltas). Here and in Figure 30(b) page 181 likewise, the times exclude the infrastructure initialization times taken once for all computations (allocating GPU memory, compiling CUDA kernel, etc.), which amount to 10 to 30 seconds depending on d.

	Market sensitivities				
	Parameter sensitivities			Jacobian	Total
	Simul.	Lin. regr.	Total		
bench.	12m48s	N/A	12m48s(1×)	30s	13m18s(1×)
linear	8.6s	0.1s	8.7s(88.5×)		38.7s(20.6×)
smart	8.5s	N/A	8.5s(90.0×)		38.5s(20.7×)

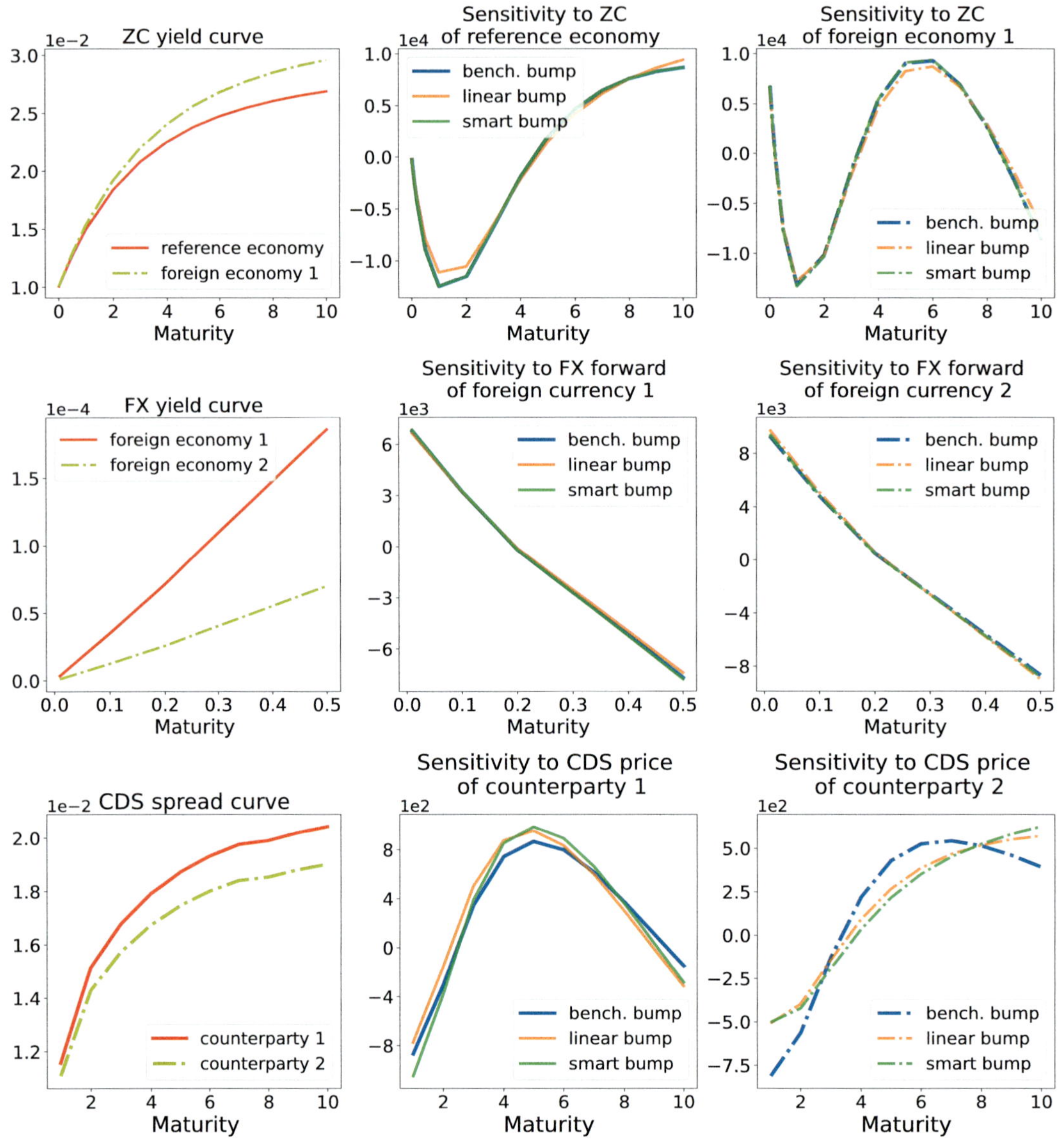

Figure 29: CVA market sensitivities to prices of selected zero-coupons (ZCs) (*top*), FX forward contracts (*middle*), and CDS contracts (*bottom*), estimated by Jacobian transformations from benchmark and fast (linear and smart) bump sensitivities to the pricing model parameters. ZC and FX yield curves are constructed from ZC and FX forward prices through the transformation $\cdot \mapsto -\frac{\log(\cdot)}{\tau}$, where τ denotes the time-to-maturity.

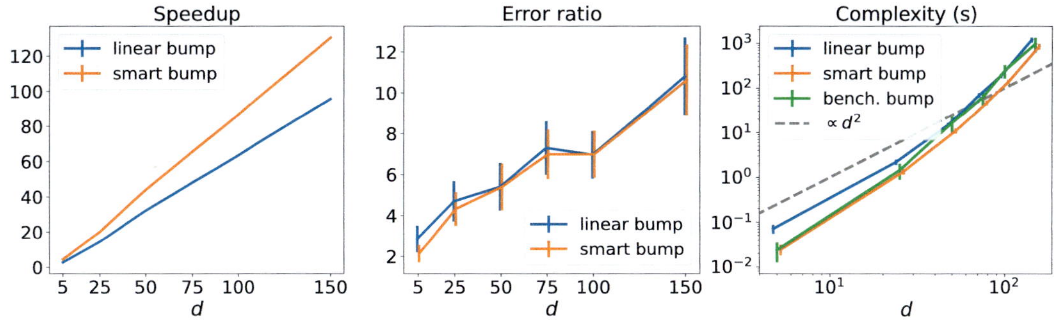

(a) Performance in Black-Scholes computations

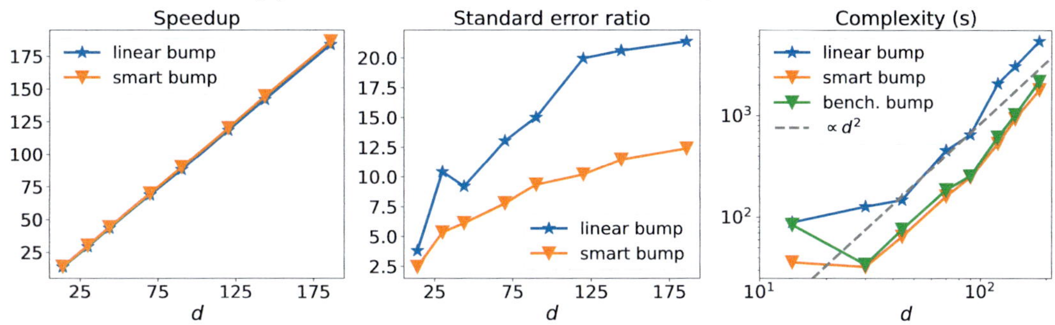

(b) Performance in CVA computations

Figure 30: Comparison of the performance of linear and smart bump sensitivities as a function of the number of model parameters d fixing $m = 2^{16}$ in (a) and $m = 2^{17}$ in (b), using benchmark bump sensitivities as references. The speedups displayed in the left panels are obtained by dividing the execution times for benchmark sensitivities by the execution times for linear or smart bump sensitivities. For each value of d, the errors displayed in the middle panel are: in the Black-Scholes case, the mean of the d ratios between the relative error of the linear (resp. smart) bump sensitivity and the relative error of the benchmark sensitivity; in the CVA case, the median of the d ratios between the Monte Carlo standard errors of the linear (resp. smart) bump sensitivity and of the benchmark sensitivity. The complexities in the right panels are defined as the projected times of reaching a relative error (for Black-Scholes) or relative standard error (for CVA) of 1% given the $1/\sqrt{|\text{sample size}|}$ scaling of the error and standard error, with $|\text{sample size}| = m$ for linear and benchmark bumps and m/d for smart bumps; that is, projected time (i.e. complexity) = exec. time $\times\, 10^4 \left(\text{relative (standard) error}\right)^2$. The $\sqrt{|\text{sample size}|}$ above is justified as explained after (5) for linear bumps and by the central limit theorem for benchmark and smart bumps. The performance metrics of the Black-Scholes computations rely on 64 Monte Carlo simulation runs per d, with associated confidence intervals computed according to the principle of uncertainty propagation in (Lee and Forthofer, 2005, Chapter 4 page 22). The performances of CVA sensitivities are produced from a single Monte Carlo simulations run per d (as for a given run, we only have access to CVA sensitivities errors in the sense of standard errors, which cannot be aggregated across different runs).

§4 LEARNING THE FUTURE CVA

Equivalently to the second line in (9), we have for $t > 0$:

$$\mathrm{CVA}_t(\cdot) = \underset{\Phi \in \mathcal{B}(\mathbb{R}^p \times \mathbb{R}^d)}{\operatorname{argmin}} \mathbb{E}\Big[\Big(\xi^{cva}_{t,T} - \Phi(X_t, \varrho_t)\Big)^2\Big], \tag{11}$$

where $\mathcal{B}(\mathbb{R}^p \times \mathbb{R}^d)$, in which p is the dimension of x (number of default indicators in the model), is the set of the Borel measurable functions of (x, ρ). We denote by $\mathrm{CVA}^\theta_t(X_t, \varrho_t)$ the conditional CVA at time $t = ih > 0$ learned by a neural network with parameters θ on the basis of simulated pairs (X_t, ϱ_t) and cash flows $\xi^{cva}_{t,T}$. The conditional CVA pricing function $\mathrm{CVA}^\theta_t(x, \rho)$ is obtained by replacing $\mathbb{E}$ by a simulated sample mean $\widehat{\mathbb{E}}$ and $\mathcal{B}(\mathbb{R}^p \times \mathbb{R}^d)$ by a neural net (or linear as a special case) search space in the optimization problem (11). The latter is then addressed numerically by Adam mini-batch gradient descent on the basis of simulated pairs (X_t, ϱ_t) as features and $\xi^{cva}_{t,T}$ as labels: see Algorithm 5, in the baseline and risk modes.

A Twin Monte Carlo Validation

In the context of this chapter, it is important to emphasize that the twin errors of the a posteriori twin simulation procedure of Algorithm 1 page 129 measure the performance of the predictor of a conditional expectation, but not of the associated sensitivities, as a low error of the predictor induces no guarantee on the error for the derivatives.

Table 11 shows the twin scores of neural network and linear regression versus nested Monte Carlo predictors of $\mathrm{CVA}_t(X_t, \varrho_t)$ in the baseline and risk modes. In both modes, all three methods have comparable performance for small t (0.01 and 0.1 years). The linear learning model for $\mathrm{CVA}^\theta_t(x, \rho)$ does not require sophisticated training like the neural network but only linear algebra, it converges equally well for $t = 0.01$ and 0.1, but falls short at one year, where the nonlinearity of $\mathrm{CVA}_t(x, \rho)$ becomes significant: for $t = 1$ and 3.5 years, the neural network outperforms the linear regression significantly. As illustrated by the left panel of Figure 14 page 142, this nonlinearity and the necessity of a neural network would become more stringent with option portfolios, possibly for lower t already. In the baseline mode, the learned $\mathrm{CVA}_t(X_t, \varrho_t)$ can occasionally be more accurate than its nested Monte Carlo counterpart, but the latter is also much slower: for a given t, the nested Monte Carlo CVA (with $m = 2^{17}$ outer paths and 1024 inner paths throughout the chapter) takes approximately 40 minutes, while simulating the data and training a neural net CVA with 2 hidden layers and softplus activation (resp. regressing a linear CVA) takes roughly 30 (resp. 25) seconds. In the risk mode, the neural network is less accurate: the input dimension of the neural network becomes 35 for (x, y) plus 63 for ϵ in the risk mode versus 35 simply in the baseline mode,

---

Algorithm 5: Learning $\mathrm{CVA}_t(X_t, \varrho_t)$ at some future time $t > 0$ in the baseline or risk mode. The distribution of ε in the risk mode can be set exogenously, as displayed here for simplicity, or assessed statistically on a historical basis.

---

input : Calibrated pricing model parameters $\rho_0 = (y_0, \epsilon_0)$ (with client default indicators X_0 all equal to 0), a CVA pricing time $t = i\mathrm{h} > 0$, a number of paths m with n pricing time steps (and daily Euler simulation step) of the pricing model, a number of training epochs E (e.g. 1000) of the (neural network or linear) learning model for $\mathrm{CVA}_t(X_t, \varrho_t)$, a partition B of $1\,..\,m$ into mini-batches. Adam optimizer is set by default.

output: one trained set θ of the parameters of the predictor $\mathrm{CVA}_t^\theta(x, y) \approx \mathrm{CVA}_t(x, y)$ in the baseline mode or $\mathrm{CVA}_t^\theta(x, y, \epsilon) \approx \mathrm{CVA}_t(x, y, \epsilon)$ in the risk mode.

if *baseline mode* **then**
  Set $\epsilon_j = \epsilon_0$, for $j = 1\,..\,m$
else if *risk mode* **then**
  Draw m i.i.d. bumped exogenous model parameters ϵ_j from the distribution $\mathcal{N}\Big(\epsilon_0, \mathrm{diag}\big(\sigma^2\epsilon_0 \odot \epsilon_0\big)\Big)$
end
For each $j \in 1\,..\,m$ with associated $\rho_j = (y_0, \epsilon_j)$ and driving Brownian path ω_j, simulate one pricing model path $(X, Y)(\rho_j; \omega_j)$ starting from $(0, y_0)$ with exogenous model parameters ϵ_j and compute the corresponding $\xi_{t,T}^{cva}(\rho_j; \omega_j)$
Initialize a neural network $\mathrm{CVA}_t^\theta(x, y)$ in the baseline mode or $\mathrm{CVA}_t^\theta(x, y, \epsilon)$ in the risk mode
Define the loss function $\mathcal{L}(\theta, batch)$ = sample mean over trajectories in $batch \in B$ of the (cf. (11)) $\Big(\xi_{t,T}^{cva}(\rho_j; \omega_j) - \Phi_j^\theta\Big)^2$, with $\Phi_j^\theta = \mathrm{CVA}_t^\theta\Big(X_t(\rho_j; \omega_j), Y_t(\rho_j; \omega_j\Big)$ in the baseline mode or $\mathrm{CVA}_t^\theta\Big(X_t(\rho_j; \omega_j), Y_t(\rho_j; \omega_j), \epsilon_j\Big)$ in the risk mode
for *epoch* $= 1, \ldots, E$ **do**
  for *batch* $\in B$ **do**
    $\theta \leftarrow \mathrm{AdamStep}(\mathcal{L}(\theta, batch))$
  end
end

---

hence training $\mathrm{CVA}_t^\theta(x, \rho)$ becomes harder (or would require more data) and the resulting predictor becomes less accurate. For $t = 3.5$ the neural network prediction is bad, but not as bad as the linear prediction. A better neural

Table 11: The mean and (95% upper bound) for the relative twin error normalized by $\mathrm{CVA}_0(\rho_0)$ (cf. Algorithm 1 page 129) of CVA_t nested Monte Carlo, neural network, and linear regression predictors, for $t = 0.01, 0.1, 1$, and $3.5 \approx T/3$ years (where the CVA of a portfolio of swaps is deemed the most volatile), in the baseline and risk modes.

t (in yr)	Baseline mode			Risk mode		
	Nested MC	NN	Linear	Nested MC	NN	Linear
0.01	N/A(4.3%)	N/A(4.3%)	N/A(4.5%)	N/A(4.3%)	2.1%(5.0%)	3.0%(5.5%)
0.1	6.6%(8.0%)	5.7%(7.3%)	6.8%(8.2%)	6.6%(8.0%)	9.8%(10.8%)	9.8%(10.8%)
1	9.2%(10.3%)	9.7%(10.8%)	10.4%(11.4%)	9.4%(10.5%)	22.1%(22.6%)	22.3%(22.9%)
3.5	8.8%(10.1%)	12.0%(13.5%)	15.7%(17.3%)	9.7%(11.5%)	26.0%(27.0%)	27.0%(28.2%)

network predictor might be achievable by fine-tuning the SGD or enriching the simulated dataset and/or the architecture of the network. Since we only need the risk mode for $t \leq 1$ in the downstream tasks of §5-§6, we do not venture more in these directions.

§5 RUN-OFF CVA RISK

An economic (or "internal") view gained from simulating the movements of model or/and market risk factors and obtaining risk measures of CVA fluctuations is an important dimension of the CVA capital regulatory requirements of a bank, in the context of its supervisory review and evaluation process (SREP). Quoting https://www.bankingsupervision.europa.eu/legalframework/publiccons/html/icaap_ilaap_faq.en.html (last accessed June 6, 2024), "the risks the institution has identified and quantified will play an enhanced role in, for example, the determination of additional own funds requirements on a risk-by-risk basis."

The focus of this section is on

$$\delta\mathrm{CVA}_t^\theta + \mathrm{DL}_t, \text{ where } \delta\mathrm{CVA}_t^\theta = \mathrm{CVA}_t^\theta(X_t, \varrho_t) - \mathrm{CVA}_0(\rho_0) \tag{12}$$

(cf. (8)-(9)), assessed in the risk mode, where $\varrho_t = (Y_t(y_0, \varepsilon), \varepsilon)$ and ε follows $\mathcal{N}\Big(\epsilon_0, \mathrm{diag}\left(t(1\%)^2\epsilon_0 \odot \epsilon_0\right)\Big)$. The random variable $(\delta\mathrm{CVA}_t^\theta + \mathrm{DL}_t)$ reflects a dynamic but also run-off view on CVA and counterparty default risk altogether, as opposed to the stationary run-on CVA risk view of §6. We assess CVA risk, counterparty default risk, and both risks combined, on the basis of value-at-risk (VaR) and expected shortfall (ES) of $\delta\mathrm{CVA}_t^\theta$ and DL_t in the risk mode, reported for $t = 0.01, 0.1, 1$ yr and various quantile levels $\alpha = 95, 97.5$ or 99% in Table 12. The middle-point of 97.5% for the ES fits a nowadays reference 99.9% value-at-risk reference level, via the conventional mapping between a 97.5% ES and a 99% VaR in a Gaussian setup, also considering that our baseline parametrization reflects market conditions at a 90% level of stress (cf.

the high CDS spreads visible in the bottom left panel of Figure 29 page 180). The results of Table 12 emphasize that CVA and counterparty default risks assessed on a run-off basis are primarily driven by client defaults, especially at higher quantile levels. As visible in the right plots, for $t = 0.01$ and 0.1, there are few client defaults, and the right tail of the distribution of $\delta\mathrm{CVA}_t^\theta + \mathrm{DL}_t$ is dominated by the term $\delta\mathrm{CVA}_t^\theta$. For $t = 1$, instead, the DL_t term takes the lead, significantly shifting the right tail of the distribution upward.

A Run-off CVA Hedging

Let (in vector form) CF represent the cumulative cash flows process of the market instruments of §3.A, with price process $Z_t = \mathbb{E}(\mathrm{CF}_T - \mathrm{CF}_t | X_t, \varrho_t), t \in 0\,..\,nh = T$, and let

$$\delta Z_t = Z_t(X_t, \varrho_t) - z_0 \tag{13}$$

denote the difference between the market prices $Z_t(X_t, \varrho_t)$ and their time-0 baseline price z_0. By loss L_t^θ, we mean the following hedged loss (and profit) of the CVA desk over the risk horizon t:

$$L_t^\theta = \delta\mathrm{CVA}_t^\theta + \mathrm{DL}_t - (\delta Z_t + \mathrm{CF}_t)^\top \Delta - c, \tag{14}$$

where the hedging ratio $\Delta \in \mathbb{R}^\nu$ is treated as a free parameter, while c is deduced from Δ through the constraint that $\mathbb{E}L_t^\theta = 0$ (or $\widehat{\mathbb{E}}L_t^\theta = 0$ in our numerics). The constant c, which is equal to 0 (modulo the numerical noise) in the baseline mode where $\mathrm{CVA} + \mathrm{DL}$ and $Z + \mathrm{CF}$ are both martingales, can be interpreted in terms of a hedging valuation adjustment (HVA) in the spirit of Chapter XI, i.e. a provision for model risk. Chapter XI develops how, such a provision having been set apart in a first stage, the loss L_t^θ, thus centered via the "HVA trend" c (i.e. $c = -\mathbb{E}\delta\mathrm{HVA}_t = \mathrm{HVA}_0 - \mathbb{E}\mathrm{HVA}_t$), deserves an economic capital, which we quantify below as an expected shortfall of L_t^θ.

Bump sensitivities $\Delta = \Delta^{bump}$ can be used in (14) but they are expected to be inappropriate for dealing with client defaults. As an alternative approach yielding hedging ratios, HVA trend and economic capital at the same time, one can use

$$\Delta^{ec} = \underset{\Delta \in \mathbb{R}^\nu}{\operatorname{argmin}}\, \mathbb{ES}\left(L_t^\theta\right). \tag{15}$$

Following Rockafellar and Uryasev (2000), (15) can be reformulated as the following convex optimization problem:

$$(\Delta^{ec}, q^{ec}) = \underset{\Delta \in \mathbb{R}^\nu, q \in \mathbb{R}}{\operatorname{argmin}}\; q + (1-\alpha)^{-1}\mathbb{E}\left[\left(L_t^\theta - q\right)^+\right], \tag{16}$$

where q^{ec} is then the value-at-risk (quantile of level α) of the corresponding L_t^θ. The constraint $\mathbb{E}L_t^\theta = 0$, motivated financially after (14), is also necessary numerically to stabilize the training of these EC sensitivities. Note that one

Table 12: $\delta\mathrm{CVA}_t^\theta$, DL_t and their sum ($\mathrm{CVA}_0(\rho_0) = 5,027$). (*Left*) Risk measures. (*Right*) qq plots against $\delta\mathrm{CVA}_t^\theta$ used as a benchmark (corresponding to the diagonal). The red circles correspond to the VaRs 95%, 97.5% and 99%.

$t = 0.01$	$\delta\mathrm{CVA}_t^\theta$	DL_t	$\delta\mathrm{CVA}_t^\theta + \mathrm{DL}_t$
Expectation	-9	0.28	-8
VaR 95%	268	0	268
VaR 97.5%	323	0	323
VaR 99%	388	0	389
ES 95%	344	0.28	345
ES 97.5%	395	0.28	397
ES 99%	460	0.28	464

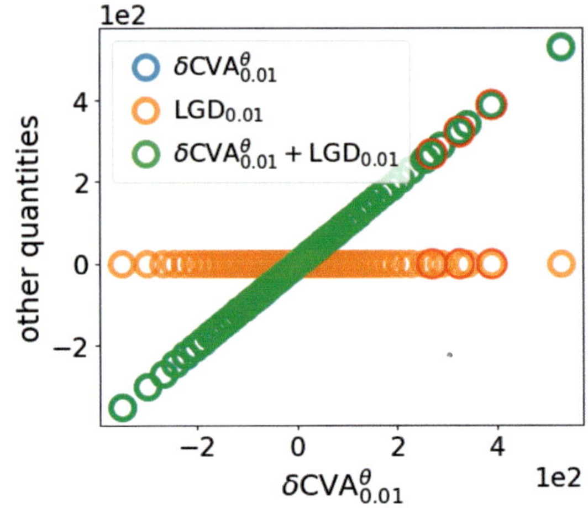

$t = 0.1$	$\delta\mathrm{CVA}_t^\theta$	DL_t	$\delta\mathrm{CVA}_t^\theta + \mathrm{DL}_t$
Expectation	56	7	63
VaR 95%	939	0	953
VaR 97.5%	1,124	0	1,145
VaR 99%	1,347	0	1,389
ES 95%	1,192	7	1,243
ES 97.5%	1,361	7	1,445
ES 99%	1,571	7	1,740

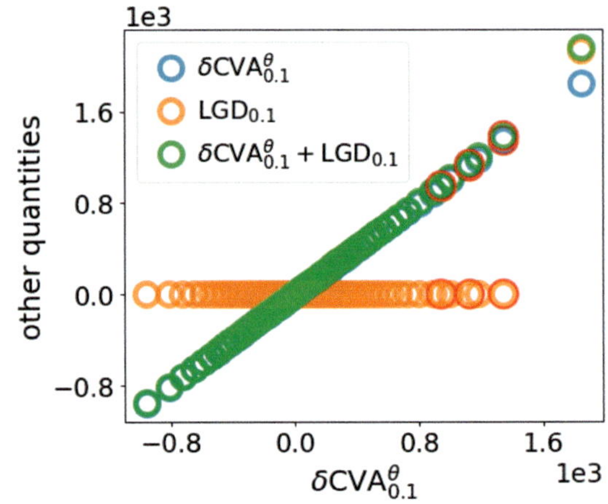

$t = 1$	$\delta\mathrm{CVA}_t^\theta$	DL_t	$\delta\mathrm{CVA}_t^\theta + \mathrm{DL}_t$
Expectation	75	502	578
VaR 95%	2,942	3,621	4,597
VaR 97.5%	3,746	7,309	7,190
VaR 99%	4,796	11,997	11,757
ES 95%	4,122	8,846	8,980
ES 97.5%	4,953	12,383	12,297
ES 99%	6,102	17,004	17,048

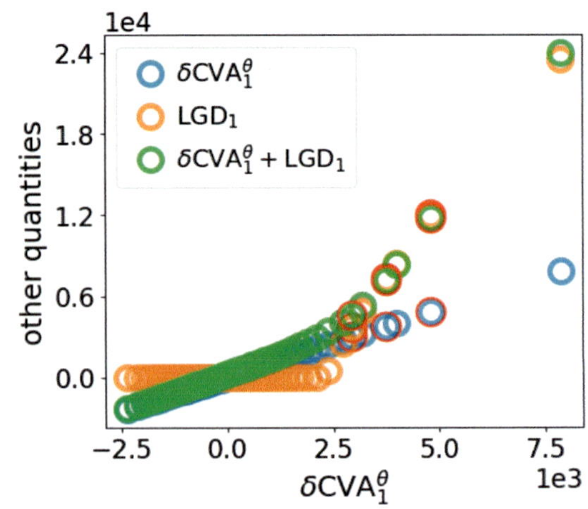

could also easily account for transaction costs in this setup, which is comparable to a deep hedging approach over one time step for computing the hedging policy Δ^{ec} and the HVA trend c compressing the ensuing economic capital and its cost (see Buehler (2019), Buehler, Gonon, Teichmann and Wood (2019)).

As a possible (simpler) variation on the **economic capital (EC) run-off sensitivities** (15)-(16), we also consider the following **PnL explain (PLE) run-off sensitivities**:

$$\Delta^{ple} = \underset{\Delta \in \mathbb{R}^\nu}{\operatorname{argmin}} \, \mathbb{E}[(L_t^\theta)^2]. \tag{17}$$

Once $\mathrm{CVA}_t^\theta(X_t, \varrho_t)$ is learned from the cash flows $\xi_{t,T}^{cva}(\varrho_t)$, the way described in Algorithm 5, run in the risk mode, the EC sensitivities (15) are computed by stochastic gradient descent based on an empirical version of (16); the PLE sensitivities are computed by performing a linear regression corresponding to an empirical version of (17): cf. lines 9–13 of Algorithm 6 page 192 (which corresponds to the CVA run-on setup of §6).

By unexplained PnL UPL (resp. economic capital EC), we mean the standard error (resp. expected shortfall) of L_t^θ. As performance metrics, we consider a backtesting, out-of-sample UPL (resp. EC with $\alpha = 95\%$) for $\Delta = 0$, divided by UPL (resp. EC with $\alpha = 95\%$) for each considered set of sensitivities: the higher the corresponding "compression ratios", the better the corresponding sensitivities. For each simulation run below (here and in §6), we use $m = 2^{17}$ paths to estimate the PLE and EC sensitivities and we generate other $m = 2^{17}$ paths for our backtest.

The left panels in Figure 31 compare the hedging performance of benchmark bump versus EC or PLE sensitivities for $t = 0.01, 0.1$, and 1 yr. The unhedged case corresponds to the red horizontal dash-dot lines. Consistent with their definitions, the PLE sensitivities always (even though we are out-of-sample) yield the highest unexplained PnL compression ratios, while the EC sensitivities, except for $t = 0.01$, yield the highest EC compression ratio. As expected, bump sensitivities provide very poor hedging performance (we omitted fast bump sensitivities, which provide results similar to the benchmark ones). At the risk horizon $t = 0.01$, client defaults are rare (only occurring on 0.017% of the scenarios) and a bump sensitivities hedge reduces risk, but much less so than the EC or PLE sensitivities hedges. For $t = 0.1$ and 1 yr, hedging by bump sensitivities is even counterproductive, worsening both unexplained PnL and economic capital compared to the unhedged case; EC and PLE sensitivities hedges achieve significant unexplained PnL and economic capital compression, but this comes along with very high HVA trends c.

The expected conclusion of this part is that for properly hedging CVA in the run-off mode, one should first replicate the impact of the defaults with

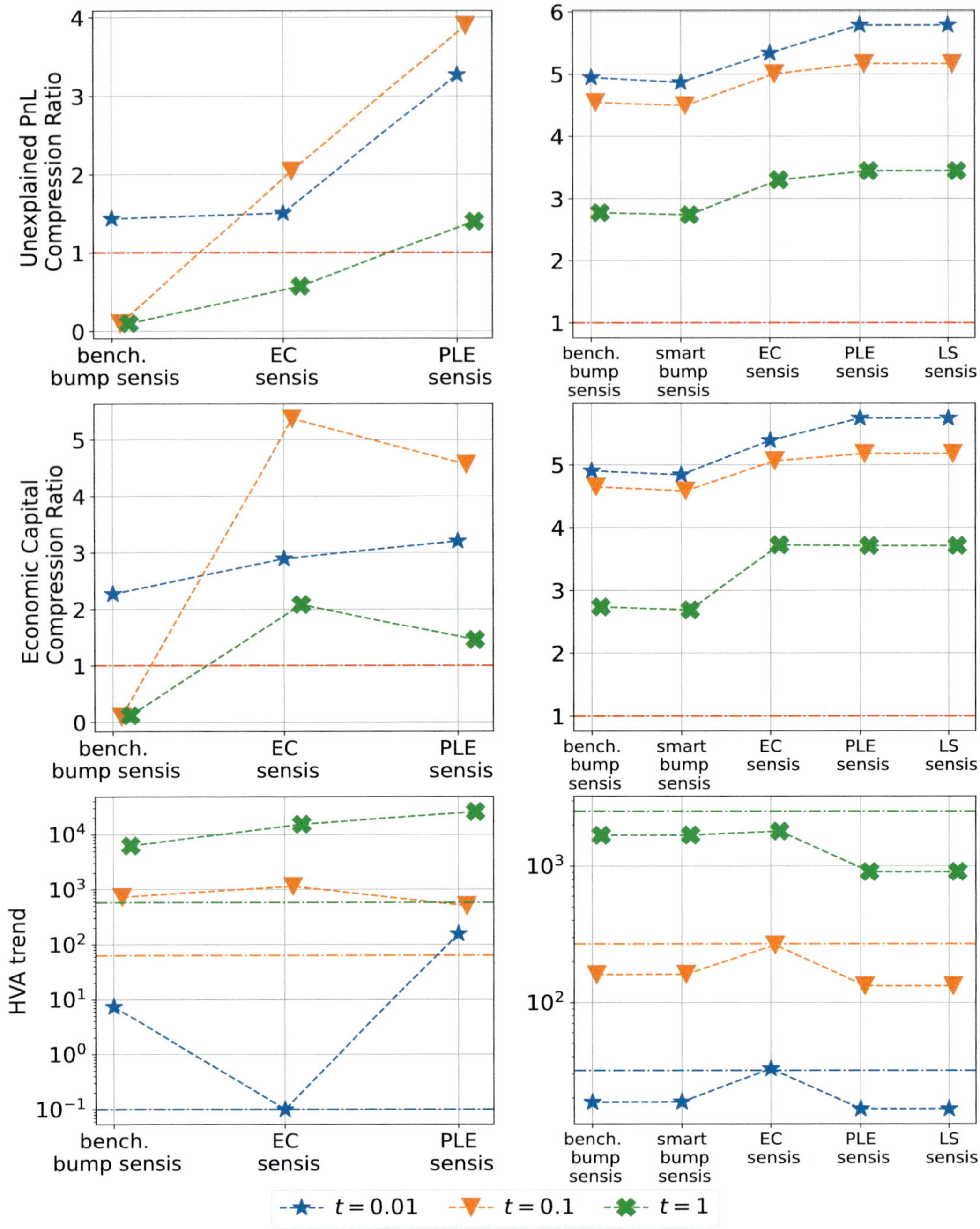

Figure 31: Compression ratios of UPL (*top*), EC (*middle*), and HVA trends c (*bottom*), for various $\delta\mathrm{CVA}_t^\theta + \mathrm{DL}_t$ hedging approaches in §5.A (*left*), and $\delta\mathrm{CVA}_{(t)}^\theta$ hedging approaches in §6.A (*right*). Horizontal dash-dot lines correspond to the unhedged case with $\Delta = 0$ in (14) and (19).

appropriate CDS positions (or decide to warehouse default risk, especially if CDSs are not liquidly available), rather than trying to hedge "on average", which only makes sense for CVA assessed in the run-on mode. The latter corresponds to the right panels in Figure 31, to be commented upon in §6.A.

§6 RUN-ON CVA RISK

With $\cdot_{(t)}$ in $\varepsilon_{(t)}$ referring to the dependence of the variance σ^2 of $\varepsilon - \epsilon_0$ (in the notation of Algorithm 5 page 183) in t in what follows, namely $\mathbb{V}\mathrm{ar}(\varepsilon_{(t)}) = t \times 0.01\%$, denoting by $Y_t(y_0, \varepsilon_{(t)})$ the Y process at time t starting from y_0 at 0 and for model parameters set to $\varepsilon_{(t)}$, and by $\varrho_{(t)} = \big(Y_t(y_0, \varepsilon_{(t)}), \varepsilon_{(t)}\big)$, let (cf. (13))

$$\begin{aligned}&\delta\varrho_{(t)} = (Y_t(y_0, \varepsilon_{(t)}) - y_0, \varepsilon_{(t)} - \epsilon_0) = \varrho_{(t)} - \rho_0,\ \delta Z_{(t)} = Z_0(\varrho_{(t)}) - z_0,\\&\delta\mathrm{CVA}_{(t)} = \mathrm{CVA}_0(\varrho_{(t)}) - \mathrm{CVA}_0(\rho_0).\end{aligned} \tag{18}$$

The fact that we consider the time-0 $\mathrm{CVA}_0(\varrho_{(t)})$ (see after (9)) and likewise $Z_0(\varrho_{(t)})$ here, as opposed to $\mathrm{CVA}_t(X_t, \varrho_t)$ and $Z_t(X_t, \varrho_t)$ in §5, is in line with an assessment of risk on a run-on portfolio and customers basis and with a siloing of CVA vs. counterparty default risks, which have both become standard in regulation and market practice. Various predictors of $\delta\mathrm{CVA}_{(t)}$ can be learned directly from the simulated model parameters $\varrho_{(t)}$ and cash flows $\xi^{cva}_{0,T}(\varrho_{(t)}) - \xi^{cva}_{0,T}(\rho_0)$ (as opposed to and better than learning $\delta\mathrm{CVA}_{(t)}$ via $\mathrm{CVA}_0(\varrho_{(t)})$, which would involve more variance): nested Monte Carlo estimator, neural net regressor $\delta\mathrm{CVA}^{\theta}_{(t)}$, and linear(-diagonal quadratic) regressors against $\delta\varrho_{(t)}$ or $\delta Z_{(t)}$ referred to as LS (for "least squares") below. The neural network used for training $\delta\mathrm{CVA}^{\theta}_{(t)}$ based on simulated data $(\varrho_{(t)} - \rho_0, \xi^{cva}_{0,T}(\varrho_{(t)}; \omega) - \xi^{cva}_{0,T}(\rho_0; \omega))$ as per line 8 of Algorithm 6 page 192 has one hidden layer with two hundred hidden units and softplus activation functions.

Table 13 displays twin upper bounds (cf. Algorithm 1 page 129) and risk measures of $\delta\mathrm{CVA}_{(t)}$ computed with these different approximations, as well as with linear(- diagonal quadratic) Taylor expansions in $\delta\varrho_{(t)}$ or $\delta Z_{(t)}$ with coefficients estimated as benchmark or smart bump sensitivities.

In terms of the twin upper bounds, the nested CVA has the best accuracy, but (for a given risk horizon t) it takes about 2 hours, versus about one minute of simulation time for generating the labels, plus 30 seconds for training by neural networks and 2-3 seconds for LS regression. The neural network excels at large t, where the non-linearity becomes significant, while being outperformed by the linear methods at small t, where $\delta\mathrm{CVA}_{(t)}$ is approximately linear. With diagonal gamma (Γ) elements taken into account, the performance of the LS regressor improves at large t. The linear quadratic Taylor expansions show relatively good twin upper bounds for small t, but worsen for large t.

Table 13: Risk measures of $\delta\mathrm{CVA}_{(t)}$ computed by Monte Carlo using $\delta\mathrm{CVA}_{(t)}$ simulated by various predictors. The three lowest (i.e. best) twin errors (without normalization, cf. Algorithm 1, page 129) and the three highest (i.e. most conservative) risk estimates on each row are emphasized in bold.

t	risk measure	nonparametric		linear(-quadratic) in $\delta\varrho_{(t)}$			linear(-quadratic) in $\delta Z_{(t)}$		
		nested $\delta\mathrm{CVA}_{(t)}$	$\delta\mathrm{CVA}^{\theta}_{(t)}$	bench. bump sensis w/ Γ	smart bump sensis	LS sensis w/ Γ	bench. bump sensis	smart bump sensis	LS sensis w/ Γ
0.01	twin-ub	**11**	29	**13**	**15**	18	23	23	18
	VaR 95%	**364**	**315**	312	306	302	**315**	312	308
	VaR 97.5%	**431**	**369**	366	358	354	**370**	367	362
	VaR 99%	**510**	**431**	429	418	415	**435**	431	424
	ES 95%	**454**	**387**	384	375	372	**389**	385	380
	ES 97.5%	**514**	**434**	432	421	417	**437**	433	427
	ES 99%	**585**	**491**	489	476	473	**495**	490	484
0.1	twin-ub	**59**	103	110	111	109	**101**	104	**94**
	VaR 95%	**1,223**	**1,189**	**1,154**	1,101	1,148	1,147	1,138	1,151
	VaR 97.5%	**1,431**	**1,383**	**1,340**	1,266	1,327	1,330	1,318	1,332
	VaR 99%	**1,686**	**1,618**	**1,562**	1,463	1,539	1,554	1,542	1,556
	ES 95%	**1,504**	**1,449**	**1,403**	1,321	1,386	1,396	1,383	1,396
	ES 97.5%	**1,693**	**1,622**	**1,570**	1,467	1,544	1,559	1,545	1,561
	ES 99%	**1,923**	**1,830**	**1,769**	1,638	1,731	1,757	1,740	1,757
1	twin-ub	**325**	**693**	1,244	1,307	1,113	932	943	**743**
	VaR 95%	**7,097**	**6,992**	6,365	5,300	6,440	5,867	5,831	**6,676**
	VaR 97.5%	**8,433**	**8,199**	7,451	5,897	7,432	6,737	6,686	**7,838**
	VaR 99%	**10,333**	**9,887**	8,991	6,646	8,812	7,914	7,846	**9,493**
	ES 95%	**9,163**	**8,805**	8,077	6,142	7,956	7,162	7,097	**8,492**
	ES 97.5%	**10,654**	**10,090**	9,309	6,715	9,032	8,075	7,988	**9,792**
	ES 99%	**12,781**	**11,869**	11,161	7,456	10,598	9,339	9,216	**11,677**

Regarding now the risk measures, the nested $\delta\mathrm{CVA}_{(t)}$ and the neural network $\delta\mathrm{CVA}^{\theta}_{(t)}$ provide more conservative VaR and ES estimates than any linear(-quadratic) proxy in all cases. Surprisingly, even though $\mathrm{CVA}_0(\varrho_{(t)})$ in $\delta\mathrm{CVA}_{(t)}$ is a function of $\varrho_{(t)}$, for $t = 1$ the linear(-quadratic) proxies in $\delta Z_{(t)}$ outperform those in $\varrho_{(t)}$, in terms of both twin error and of consistency of the ensuing risk measures with those provided by the nonparametric (neural net and nested) references. Also note that, when compared with the nonparametric approaches again, the smart bump sensitivities proxy in $\delta Z_{(t)}$ yields results almost as good as the much slower benchmark bump sensitivities proxy in $\delta Z_{(t)}$ (see Tables 10 and 14 pages 179 and 192).

A Run-on CVA Hedging

Let

$$L^{\theta}_{(t)} = \delta\mathrm{CVA}^{\theta}_{(t)} - (\delta Z_{(t)})^{\top}\Delta - c \tag{19}$$

(cf. (13) and (18)). As in §5, the "HVA trend" c (here $c = -\mathbb{E}\delta\mathrm{HVA}_{(t)} = \mathrm{HVA}_0 - \mathbb{E}\mathrm{HVA}_{(t)}$, where $\mathrm{HVA}_{(t)}$ is the HVA analog of $\mathrm{CVA}_{(t)}$), is deduced from Δ through the constraint that $\mathbb{E}L^{\theta}_{(t)} = 0$ (or $\widehat{\mathbb{E}}L^{\theta}_{(t)} = 0$ in the numerics). By **EC and PLE run-on sensitivities**, we mean

$$\Delta^{ec} = \underset{\Delta \in \mathbb{R}^{\nu}}{\operatorname{argmin}}\, \mathbb{ES}\left(L^{\theta}_{(t)}\right) \text{ and } \Delta^{ple} = \underset{\Delta \in \mathbb{R}^{\nu}}{\operatorname{argmin}}\, \mathbb{E}[(L^{\theta}_{(t)})^2], \tag{20}$$

where $\mathbb{ES}$ means 95% expected shortfall as in §5.A. Once $\delta\mathrm{CVA}^{\theta}_{(t)}$ learned from simulated $\varrho_{(t)}$ and $\xi^{cva}_{0,T}(\varrho_{(t)}) - \xi^{cva}_{0,T}(\rho_0)$, the way mentioned after (18), these sensitivities are computed much like their run-off counterparts of §5. Even simpler (but still optimized) **LS (run-on) sensitivities** are obtained without prior learning of $\delta\mathrm{CVA}_{(t)}$, just by regressing linearly $\xi^{cva}_{0,T}(\varrho_{(t)}) - \xi^{cva}_{0,T}(\rho_0)$ against $\delta Z_{(t)}$ the way explained after (18) (purely linear LS regression here as opposed to linear diagonal quadratic LS regression in Table 13, due to our hedging focus of this part). These LS sensitivities are thus obtained much like the linear bump sensitivities of Algorithm 4 page 170, except for the scaling of the bumps that are used in the corresponding $\varrho_{(t)}$, and the fact that these LS sensitivities are computed directly in the market variables, without Jacobian transformation. The derivation of the LS, EC, and PLE run-on sensitivities is summarized in Algorithm 6. Their computation times are reported in Table 14. The right panels in Figure 31 page 188 show the run-on CVA hedging performance of different candidate sensitivities. All the risk compression ratios decrease with the risk horizon t. All sensitivities reduce both the unexplained PnL and economic capital by at least 2.5 times for $t = 0.01$ and 4.5 times for $t = 0.1$ and 1. Since client defaults are skipped in the run-on mode, the efficiency of bump sensitivities hedges is understandable. For each risk horizon and performance metric, the optimized sensitivities always have better results than (benchmark or smart) bump sensitivities. Among those, the PLE and LS sensitivities hedges display the highest risk compression ratios and the lowest HVA trend c. Unlike what we observed in the run-off mode, most sensitivities (except for EC sensitivities when $t = 0.01$ or 0.1) also compress the HVA trend c compared to the unhedged case.

Algorithm 6: LS, EC and PLE run-on sensitivities.

input : A set of calibrated initial values and model parameters $\rho_0 = (y_0, \epsilon_0)$ (client default indicators X_0 all set to 0), a time horizon t, a number of exposure paths m with n pricing time steps.

output: Estimated sensitivities Δ.

1 Simulate m cash flows $\xi^{cva}_{0,T}(\rho_0; \omega_j)$ as per line 6 of Algorithm 5 in the baseline mode

2 Simulate m realizations ϵ_j of $\varepsilon_{(t)}$ as per line 2 of Algorithm 5 with $\sigma = 1\%\sqrt{t}$ and $y_j = Y^j_t(y_0, \epsilon_j; \omega_j)$ as per line 6 of Algorithm 5, for Brownian drivers ω_j drawn independently from the ω_j, and set $\rho_j = (y_j, \epsilon_j)$

3 Simulate m realizations $\xi^{cva}_{0,T}(\rho_j; \omega_j)$ as per line 6 of Algorithm 5 in the risk mode (with common random numbers, meaning here the same ω_j as in line 1)

4 For each ρ_j, $j = 0..m$, compute the corresponding market hedging instrument prices scenario z_j

5 **if** LS **then**

6   Regress linearly the $\xi^{cva}_{0,T}(\rho_j; \omega_j) - \xi^{cva}_{0,T}(\rho_0; \omega_j)$ against the $z_j - z_0$ by SVD. The obtained coefficients are the LS sensitivities.

7 **else**

8   Train a neural network $\delta\text{CVA}^{\theta}_{(t)}(y, \epsilon)$ to regress the $\xi^{cva}_{0,T}(\rho_j; \omega_j) - \xi^{cva}_{0,T}(\rho_0; \omega_j)$ against the $\rho_j - \rho_0$.

9   **if** EC **then**

10   Solve the left-hand side in (20) by Adam stochastic gradient descent.

11   **else if** PLE **then**

12   Solve the right-hand side in (20) by SVD linear regression.

13   **end**

14 **end**

Table 14: Computation times for learning $\delta\text{CVA}^{\theta}_{(t)}$ and the related sensitivities. The speedups measure the ratios of the total time taken by the benchmark bump sensitivities approach to the total time taken by each of the sensitivities.

[Smart] bump sensitivities		Regression sensitivities					Speedup		
Model sensis	Jacobian transform	MtM simulation	LS	$\delta\text{CVA}^{\theta}_{(t)}$ learning	PnL regression EC	PnL regression PLE	LS	EC	PLE
12min48s [8.5s]	30s	27s	1s	6s	31s	1s	28.5 [1.4]	12.5 [0.6]	23.5 [1.1]

§7 CONCLUSION

Table 15 synthesizes our findings regarding CVA (or more general, regarding columns 1 to 3) sensitivities, as far as their approximation quality to corresponding partial derivatives (for bump sensitivities) and their hedging abilities (regarding also the optimized sensitivities) are concerned. The winner that emerges as the best trade-off for each downstream task in **blue** **green** **red** in the first row is identified by the same color in the list of sensitivities. **bench. bump** plays the role of market standard. Less common sensitivities are emphasized in yellow : smart bump sensitivities essentially mean standard bump sensitivities with less paths, but with the important implementation caveat mentioned at the end of §2.A; PLE sensitivities were already introduced in the different context of SIMM computations in Albanese, Caenazzo and Syrkin (2017); EC sensitivities were introduced in Rockafellar and Uryasev (2000) and are also considered in Buehler (2019).

Table 15: Conclusions regarding CVA sensitivities and hedging. By local accuracy of a bump sensitivity, we mean the accuracy of the approximation it provides for the corresponding partial derivative. *Italics* means tested but not reported in tables or figures in this chapter.

	sensitivities	speed	stability	**local accuracy**	**CVA run-off hedge**	**CVA run-on hedge**
	bench. bump	very slow	very stable	benchmark	bad	good
fast bump sensis	naive AAD bump	fast	fragile	*bad*	*bad*	*bad*
	AAD bump	fast	fragile	*average*	*bad*	*bad*
	linear bump	fast	average	good	*bad*	*good*
	smart bump	fast	stable	good	*bad*	good
optimized sensis	EC sensis	fast	average	not applicable	good	very good
	PLE sensis	very fast	stable	not applicable	good	excellent
	LS w/o Γ	very fast	stable	not applicable	not applicable	excellent

Regarding the assessment of CVA risk, in the run-on CVA case (see Table 16 using the same color code as Table 15), we found out that neural net regression of conditional CVA results in likely more reliable (judging by the twin scores of the associated CVA learners) and also faster value-at-risk and expected shortfall estimates than CVA Taylor expansions based on bump sensitivities (such as the ones that inspire certain regulatory CVA capital charge formulas). But an LS proxy, linear diagonal quadratic in market bumps, provides an even quicker (as it is regressed without training) and almost equally reliable view on CVA risk as the neural net CVA. In the **run-off CVA** case (not represented in the table), the **neural net learner of CVA$_t$** in the risk mode (or nested CVA learner alike but in much longer time) allows one to get a consistent and dynamic view on CVA and counterparty default risk altogether.

Table 16: Conclusions regarding run-on CVA risk.

	δCVA$_{(t)}$ learners	speed	stability	**twin accuracy**	δCVA$_{(t)}$ VaR and ES
nonparametric	nested MC	very slow	stable	very good	very conservative
	neural net	fast	average	good	conservative
linear(-quadratic) in market bumps	**bench. bump**	very slow	very stable	good/average for small/large t	aggressive
	LS w/ Γ in $\delta Z_{(t)}$	very fast	stable	good	conservative

CHAPTER VII

Regressing Pathwise FVA, Economic Capital and KVA

Set on a forward/backward SDE formulation, the cost-of-capital XVA equations can be solved efficiently using GPU computing combined with neural regression methods in a whole bank balance sheet context. This chapter is a numerical case study on large, multi-counterparty portfolios of interest rate swaps, in an anticipated BSDEs setup also accounting for the fungibility of capital at risk with variation margin. The results emphasize the workability and added value of the corresponding pathwise XVA computations.

§1 THE CONTINUOUS-TIME XVA ANTICIPATED BSDES

In preparation of the numerical implementation of the case study of this chapter, we streamline the XVA reduced equations of Theorem III.1 in the special case of purely bilateral transactions and their dynamic hedge (without central clearing). On the one hand, we rid the equations of negligible terms. On the other hand, we refine the funding policy, accounting for the fungibility of capital at risk and variation margin as introduced in the static setup in I.§3.D.

We suppose that the derivative portfolio of the bank is partitioned into bilateral netting sets of contracts which are jointly collateralized and liquidated upon bank or counterparties default. Given such a netting set b of the bank portfolio, we denote by:

- $\mathcal{P}^b$ and P^b, the corresponding contractually promised cash flows and (counterparty-risk-free) value processes;
- τ_b and R_b, the default times, survival indicators, and recovery rates of the corresponding counterparties, whereas τ and R are the analogous data regarding the bank itself, with bank credit spread process $\lambda = (1 - R)\gamma$ taken as a proxy of its unsecured funding spread process, while $\widetilde{\lambda} \leq \lambda$

 DOI: 10.1201/9781003614944-11

denotes a blended funding spread for the IM posted by the bank (see III.§2.C);

- $\tau_b^\delta = \tau_b + \delta$ and $\tau^\delta = \tau + \delta$, where δ is a positive margin period of risk, in the sense that the liquidation of the netting set b happens at time $\tau_b^\delta \wedge \tau^\delta$;

- VM^b, the variation margin (re-hypothebable collateral) exchanged between the bank and the counterparty of the netting set b, counted positively when received by the bank;

- PIM^b and RIM^b, the related initial margin (segregated collateral) posted and received by the bank;

- RC and CR, the reserve capital and capital at risk of the bank.

The contractually promised cash flows are supposed to be hedged out by the bank, but one conservatively assumes no XVA hedge. In this practical chapter, we distinguish between a (strict) FVA, in the strict sense of the cost of raising variation margin, and an MVA for the cost of raising initial margin (see Remark 1 page 23). The (other than K)VA equations are then

$$\mathrm{RC} = \mathrm{CA} = \mathrm{CVA} + \mathrm{FVA} + \mathrm{MVA}, \tag{1}$$

the so-called "contra-assets valuation" sourced from the clients of the bank and deposited in the reserve capital account of the bank at portfolio inception (see I.§2). The related reduction of filtration issues has been detailed in the BSDE setup of Chapters II-III. Below, as in I.§4, we restrict ourselves to a pedagogical presentation of the valuation equations under the bank survival measure $\mathbb{Q}$ associated with a reference fininsurance probability measure $\mathbb{Q}^*$ (see the introduction to the Pricing part of this book). In particular, seen under $\mathbb{Q}$, the MtM process reduces to

$$\sum_b (\mathbb{1}_{[\![0,\tau_b[\![} P^b + \mathbb{1}_{[\![\tau_b,\tau_b^\delta[\![} P^b_{\tau_b-}) \tag{2}$$

(cf. III.(16) with τ formally set to $+\infty$ there).

For simplicity, we neglect the corrective term $\sum_b \mathcal{X}^b$ in the CVA formula III.(68) (here restricted to its bilateral netting set components), see Remark III.9 page 113). But we do account for the fungibility of capital at risk and variation margin, which translates into the terms $(-\mathrm{CA} - \mathrm{CR})$ in the funding costs below (cf. I.§3.D). By considerations similar to the ones having led to Theorem III.1 and modulo the above-mentioned simplification, we obtain : for

$t < \tau$,

$$\mathrm{CVA}_t = \mathbb{E}_t \sum_{t<\tau_b^\delta} (1 - R_b)\Big((P^b + \mathcal{P}^b)_{\tau_b^\delta} - (\mathcal{P}^b + \mathrm{VM}^b + \mathrm{RIM}^b)_{\tau_b -} \Big)^+$$

$$\mathrm{FVA}_t = \mathbb{E}_t \int_t^T \lambda_s \Big(\sum_b (\mathbb{1}_{[\![0,\tau_b[\![} P^b + \mathbb{1}_{[\![\tau_b,\tau_b^\delta[\![} P^b_{\tau_b -}) - \sum_b \mathrm{VM}^b - \mathrm{CA} - \mathrm{CR}\Big)_s^+ ds \quad (3)$$

$$\mathrm{MVA}_t = \mathbb{E}_t \int_t^T \widetilde{\lambda}_s \sum_b \mathbb{1}_{\{s<\tau_b\}} \mathrm{PIM}_s^b ds.$$

The corresponding trading loss process $\mathcal{L}$ of the bank satisfies

$$\begin{aligned}
&\mathcal{L}_0 = 0 \text{ and, for } t < \tau,\\
&d\mathcal{L}_t = \sum_b (1 - R_b)\Big((P^b + \mathcal{P}^b)_{\tau_b^\delta} - (\mathcal{P}^b + \mathrm{VM}^b + \mathrm{RIM}^b)_{\tau_b -} \Big)^+ \boldsymbol{\delta}_{\tau_b^\delta}(dt)\\
&+ \lambda_t \Big(\sum_b (\mathbb{1}_{[\![0,\tau_b[\![} P^b + \mathbb{1}_{[\![\tau_b,\tau_b^\delta[\![} P^b_{\tau_b -}) - \sum_b \mathrm{VM}^b - \mathrm{CA} - \mathrm{CR}\Big)_t^+ dt \qquad (4)\\
&+ \widetilde{\lambda}_t \sum_b \mathbb{1}_{\{t<\tau_b\}} \mathrm{PIM}_t^b dt\\
&+ d\mathrm{CA}_t.
\end{aligned}$$

By Lemma I.3, $\mathcal{L}$ is a $\mathbb{Q}$ martingale. Capital at risk (CR), economic capital (EC) and the KVA of the bank follow as per Definitions I.5 and I.6, i.e., given a positive target hurdle rate h,

$$\begin{aligned}
&\mathrm{CR} = \max(\mathrm{EC}, \mathrm{KVA}) \text{ with, for } t \in [0, T],\\
&\mathrm{EC}_t = \mathbb{ES}_t(\mathcal{L}_{t+1} - \mathcal{L}_t) \text{ and } \mathrm{KVA}_t = \mathbb{E}_t \int_t^T h e^{-h(s-t)} \max(\mathrm{EC}_s, \mathrm{KVA}_s) ds,
\end{aligned} \qquad (5)$$

where $\mathbb{ES}_t$ denotes the $(\mathfrak{G}_t, \mathbb{Q})$ conditional expected shortfall at the $\alpha = 97.5\%$ confidence level.

A The XVA Anticipated BSDEs Are Well-Posed

The CVA in (3) is essentially the same as the bilateral CVA terms in Theorem III.1 (modulo the neglected $\sum_b \mathcal{X}^b$). But as capital at risk is fungible with variation margin (cf. I.§3.D), i.e. in consideration of the CR term as per (5) in (3)–(4), we here deal with an (FVA, KVA) *system* of equations. As EC depends on $\mathcal{L}$ that itself involves the FVA process via CA as per (1), we even deal with a forward-backward system for the forward loss process $\mathcal{L}$ and the backward pair of processes (FVA, KVA). The BSDE-trained reader will recognize a system of *anticipated BSDEs* (ABSDEs), in the line of Pardoux and Peng (1990), for the FVA and the KVA. The coupling between (FVA, KVA) and $\mathcal{L}$ can be disentangled by the following Picard iteration:

- Let CVA and MVA be as in (3), $\mathcal{L}^{(0)} = \mathrm{KVA}^{(0)} = 0$, and, for $t < \tau$,

$$\mathrm{FVA}_t^{(0)} = \tag{6}$$

$$\mathbb{E}_t \int_t^T \lambda_s \Big(\sum_b (\mathbb{1}_{[\![0,\tau_b[\![} P^b + \mathbb{1}_{[\![\tau_b,\tau_b^\delta[\![} P^b_{\tau_b-}) - \sum_b \mathrm{VM}^b - \mathrm{CA}^{(0)} \Big)_s^+ ds, \tag{7}$$

where $\mathrm{CA}^{(0)} = \mathrm{CVA} + \mathrm{FVA}^{(0)} + \mathrm{MVA}$;

- For $k \geq 1$, writing explicitly $\mathrm{EC} = \mathrm{EC}(L)$ to emphasize the dependence of EC on a loss process L, so $\mathrm{EC}(L)_t = \mathbb{ES}_t(L_{t+1} - L_t)$, let $\mathcal{L}_0^{(k)} = 0$ and, for $t < \tau$,

$$\begin{aligned}
d\mathcal{L}_t^{(k)} &= \sum_b (1 - R_b)\Big((P^b + \mathcal{P}^b)_{\tau_b^\delta} - (\mathcal{P}^b + \mathrm{VM}^b + \mathrm{RIM}^b)_{\tau_b-} \Big)^+ \boldsymbol{\delta}_{\tau_b^\delta}(dt) \\
&\quad + \lambda_t \Big(\sum_b (\mathbb{1}_{[\![0,\tau_b[\![} P^b + \mathbb{1}_{[\![\tau_b,\tau_b^\delta[\![} P^b_{\tau_b-}) \\
&\qquad - \sum_b \mathrm{VM}^b - \mathrm{CA}^{(k-1)} - \max\Big(\mathrm{EC}(\mathcal{L}^{(k-1)}), \mathrm{KVA}^{(k-1)}\Big)\Big)_t^+ dt \\
&\quad + \widetilde{\lambda}_t \sum_b \mathbb{1}_{\{t<\tau_b\}} \mathrm{PIM}_t^b dt + d\mathrm{CA}_t^{(k-1)}, \\
\mathrm{KVA}_t^{(k)} &= h\mathbb{E}_t \int_t^T e^{-h(s-t)} \max\Big(\mathrm{EC}_s(\mathcal{L}^{(k)}), \mathrm{KVA}_s^{(k)}\Big) ds, \\
\mathrm{CA}_t^{(k)} &= \mathrm{CVA}_t + \mathrm{FVA}_t^{(k)} + \mathrm{MVA}_t \text{ where } \mathrm{FVA}_t^{(k)} = \\
&\mathbb{E}_t \int_t^T \lambda_s \Big(\sum_b (\mathbb{1}_{[\![0,\tau_b[\![} P^b + \mathbb{1}_{[\![\tau_b,\tau_b^\delta[\![} P^b_{\tau_b-}) \\
&\qquad - \sum_b \mathrm{VM}^b - \mathrm{CA}^{(k)} - \max\Big(\mathrm{EC}(\mathcal{L}^{(k)}), \mathrm{KVA}^{(k)}\Big)\Big)_s^+ ds.
\end{aligned} \tag{8}$$

Referring the reader to Crépey, Sabbagh and Song (2020) for the embedded ABSDE technicalities:

Theorem 4.1 in Crépey, Sabbagh and Song (2020). *Assuming square integrable data, the FVA and KVA anticipated BSDEs* (3)–(5) *that account for the fungibility of capital at risk with variation margin are well-posed within a square integrable solution. Moreover, the above Picard iteration converges to their unique square integrable solution.*

B Collateralization Schemes

We denote by $\Delta_t^b = \mathcal{P}_t^b - \mathcal{P}_{(t-\delta)-}^b$ the cumulative contractual cash flows with the counterparty b accumulated over a past period of length δ. In our case study, we consider both "no CSA"[1] netting sets b, with $\mathrm{VM}^b = \mathrm{RIM}^b = \mathrm{PIM}^b = 0$, and "CSA" netting sets b, with $\mathrm{VM}_t^b = P_t^b$ and, for $t \le \tau_b$,

$$\mathrm{RIM}_t^b = \mathbb{V}\mathrm{a}\mathbb{R}_t\Big((P_{t^\delta}^b + \Delta_{t^\delta}^b) - P_t^b\Big), \quad \mathrm{PIM}_t^b = \mathbb{V}\mathrm{a}\mathbb{R}_t\Big(-(P_{t^\delta}^b + \Delta_{t^\delta}^b) + P_t^b\Big) \quad (9)$$

(assumed nonnegative), for some PIM^b and RIM^b quantile levels a_{pim} and a_{rim} (and $t^\delta = t + \delta$).

The following result can be derived by similar computations as the ones in Albanese, Armenti and Crépey (2020, Section A).

Proposition 1. *In a common shock default model of the counterparties and the bank itself (see after IV.(13)), with pre-default intensity processes γ^b of the counterparties and γ of the bank,* CVA, *then* $\mathrm{CVA} = \mathrm{CVA}^{no\ csa} + \mathrm{CVA}^{csa}$, *where, for $t < \tau$,*

$$\begin{aligned}\mathrm{CVA}_t^{no\ csa} &= \sum_{b\ no\ csa} \mathbb{1}_{t<\tau_b}(1-R_b)\mathbb{E}_t \int_t^T (P_{s^\delta}^b + \Delta_{s^\delta}^b)^+ \gamma_s^b e^{-\int_t^s \gamma_u^b du} ds \\ &+ \sum_{b\ no\ csa} \mathbb{1}_{\tau_b < t < \tau_b^\delta}(1-R_b)\mathbb{E}_t(P_{\tau_b^\delta}^b + \Delta_{\tau_b^\delta}^b)^+, \end{aligned} \quad (10)$$

$$\begin{aligned}\mathrm{CVA}_t^{csa} &= \sum_{b\ csa} \mathbb{1}_{t<\tau_b}(1-R_b)(1-a_{rim}) \times \\ &\mathbb{E}_t \int_t^T (\mathbb{ES}_s - \mathbb{V}\mathrm{a}\mathbb{R}_s)\Big((P_{s^\delta}^b + \Delta_{s^\delta}^b) - P_s^b\Big)\gamma_s^b e^{-\int_t^s \gamma_u^b du} ds \quad (11) \\ &+ \sum_{b\ csa} \mathbb{1}_{\tau_b < t < \tau_b^\delta}(1-R_b)\mathbb{E}_t\Big((P_{\tau_b^\delta}^b + \Delta_{\tau_b^\delta}^b) - (P_{\tau_b}^b + \mathrm{RIM}_{\tau_b}^b)\Big)^+, \end{aligned}$$

where ($\mathbb{ES}_s - \mathbb{V}\mathrm{a}\mathbb{R}_s$) *in* (11) *is computed at the a_{rim} confidence level. Moreover,* $\mathrm{MVA} = \mathrm{MVA}^{csa}$, *where, for $t < \tau$,*

$$\mathrm{MVA}_t^{csa} = \sum_{b\ csa} \mathbb{1}_{\{t<\tau_b\}}\mathbb{E}_t \int_t^T \widetilde{\lambda}_s \mathrm{PIM}_s^b e^{-\int_t^s \gamma_u^b du} ds. \ \square \quad (12)$$

§2 NEURAL NETWORK REGRESSION XVA FRAMEWORK

To implement (6)–(12) numerically, we simulate the risk factors and we calculate the mark-to-market cube $(P^b)_b$ using GPU computing, going forward in time, and the XVA cube C/F/M/K VA using neural net (NN) regression and quantile regression methods, going backward (at each Picard iteration). This involves no nested Monte Carlo (other than for benchmarking and validation purposes), each successive layer (from right to left) beyond the base

[1]CSA stands for credit support annex, the document that specifies the collateralization scheme in derivatives transactions.

MtM layer in the XVA dependence tree of Figure 10 page 121 is "learned" instead, see Algorithm 7, which complexity is linear in the number of XVA layers, pricing time steps, and Picard iterations (the latter being typically fixed to a small value such as 3 or 4). At each pricing time (of each Picard iteration), going backward, we perform the required NN regressions and/or quantile regressions by solving optimization (least-squares or pinball training) problems as per IV.§2-§3. This "multiple local trainings" strategy is in line with Huré, Pham and Warin (2020) and contrasts with E, Han and Jentzen (2017), Gnoatto, Picarelli and Reisinger (2023), which "global single training" approach would face GPU memory issues if implemented at the scale of a realistic banking derivatives portfolio (unless costly memory transfers between CPU RAM, GPU RAM, and even disks are considered).

Algorithm 7 yields a fully (time and space) discrete scheme for simulating the Picard iteration (8) until numerical convergence to the XVA processes. As in Algorithm 3 page 143, one can do transfer learning as per V.§2.C, starting the learning at T and then at every pricing time step, going backward, reusing the previous solution as an initialization of the training algorithm (within each Picard iteration). This accelerates the convergence of the learning procedure. As local minima associated with successive time steps will be close to each other, it also improves the smoothness over time of the learned functions. See also Abbas-Turki, Crépey, Li and Saadeddine (2025) for an alternative, explicit scheme not requiring Picard iterations. We emphasize that, in any case, for obtaining stable economic capital and KVA results, it is crucial to use the hierarchical simulation scheme of Chapter V for generating the default times of the counterparties. For validation purposes, fault of a feasible nested Monte Carlo benchmark beyond the CVA cases of Chapters V-VI, we advocate the a posteriori twin Monte Carlo procedures of IV.§2.A and IV.§3.A.

§3 SWAP PORTFOLIO CASE STUDY

We consider an interest rate swap portfolio case study in the market and credit model of IV.§4, with counterparties in different economies, first involving 10 one-factor Hull White interest-rates, 9 Black-Scholes exchange rates, and 11 Cox-Ingersoll-Ross default intensity processes. The default times of the counterparties and the bank itself are jointly modeled by a "common shock" or dynamic Marshall-Olkin copula model as per Crépey, Bielecki and Brigo (2014, Chapter 8–10). This whole setup results in about 40 risk factors used as deep learning features (including the counterparty default indicators). By default we use Monte Carlo simulation with 50K paths of 16 coarse (pricing) and 32 fine (risk factors) time steps per year.

Algorithm 7: XVAs neural regression algorithm.

- Simulate forward m realizations (Euler paths) of the market risk factor processes and of the counterparty survival indicator processes (i.e. default times) on a refined time grid;
- For each pricing time $t = t_i$ of a pricing time grid, with coarser time step denoted by h, and for each counterparty b:
  - Learn the corresponding $\mathbb{V}\mathrm{a}\mathbb{R}_t$ and $\mathbb{ES}_t$ terms visible in (9) or (under the time-discretized outer integral in) (11);
  - Learn the corresponding $\mathbb{E}_t$ terms visible in (10) through (12);
  - Compute the ensuing pathwise CVA and MVA as per (10)–(12);
- For $\mathrm{FVA}^{(0)}$, consider the following time discretization of (6) with time step h:

$$\begin{aligned}\mathrm{FVA}^{(0)}_t \approx \mathbb{E}_t[\mathrm{FVA}^{(0)}_{t+\mathrm{h}}] + \mathrm{h}\lambda_t\Big(\sum_b(\mathbb{1}_{\{0\leq t<\tau_b\}}P^b_t + \mathbb{1}_{\{\tau_b\leq t<\tau^\delta_b\}}P^b_{\tau_b-})\\ - \sum_b \mathrm{VM}^b_t - \mathrm{CVA}_t - \mathrm{MVA}_t - \mathrm{FVA}^{(0)}_t\Big)^+\end{aligned} \tag{13}$$

  and, for each $t = t_i$, learn the corresponding $\mathbb{E}_t$ in (13), then solve the semi-linear equation for $\mathrm{FVA}^{(0)}_t$;
- For each Picard iteration k (until numerical convergence), simulate forward $\mathcal{L}^{(k)}$ as per the first line in (8) (which only uses known or already learned quantities), and:
  - For economic capital $\mathrm{EC}^{(k)}$, for each $t = t_i$, learn $\mathbb{ES}_t\big(\mathcal{L}^{(k)}_{t+1} - \mathcal{L}^{(k)}_t\big)$;
  - $\mathrm{KVA}^{(k)}$ and $\mathrm{FVA}^{(k)}$ then require a backward recursion solved by neural net regression much like the one for $\mathrm{FVA}^{(0)}$ above.

A Validation Results

In the above model we first consider a bank portfolio of 10K randomly generated swap trades, with

- trade currency and counterparty both uniform on $[1, 2, 3\ldots, 10]$,
- notional uniform on $[10K, 20K, \ldots, 100K]$,
- collateralization (cf. §1.B): either "no CSA counterparty" without initial

margin nor variation margin, or "CSA counterparty" with $\mathrm{VM}^b = P^b$ and posted initial margin (PIM^b) pledged at 99% gap risk value-at-risk, received initial margin (RIM^b) covering 75% gap risk and leaving excess as residual gap CVA (and we just take $\widetilde{\lambda} = \lambda$, the unsecured funding spread $(1 - R)\gamma$ of the bank),

- for economic capital, 97.5% expected a shortfall of the 1-year ahead trading losses of the bank shareholders.

Figure 32 shows the learned $\mathrm{FVA}^{(0)}$ profile as per (13). The orange FVA curve represents the mean FVA originating cash flows, which, in principle as on the picture, matches the blue mean FVA itself learned from these cash flows. The 5th and 95th percentiles of FVA estimates are a bit less smooth in time than the mean profiles, as expected.

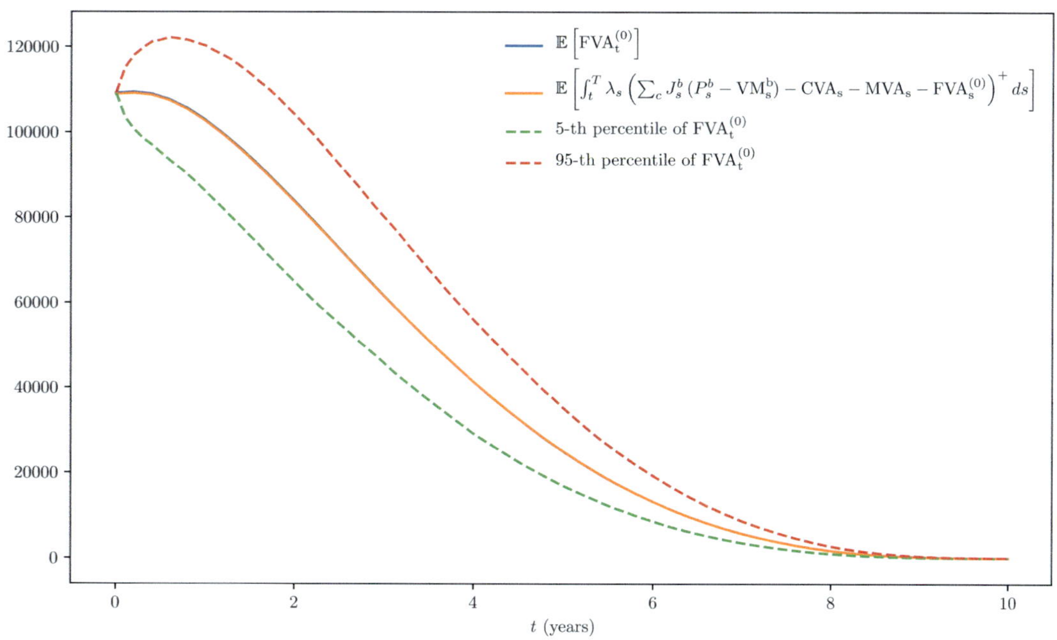

Figure 32: Learned $\mathrm{FVA}^{(0)}$.

Figure 33 *(top)* is a sanity check that the profiles of the successive iterates $\mathcal{L}^{(k)}$ of the shareholder trading loss process $\mathcal{L}$ in Algorithm 7 converge rapidly with k. Figure 33 *(bottom)* shows the loss process $\mathcal{L}^{(3)}$, displayed as its mean and mean $\pm$ 2 stdev profiles. Consistent with its martingale property, the loss process $\mathcal{L}^{(3)}$ appears numerically centered around zero. The latter holds, at

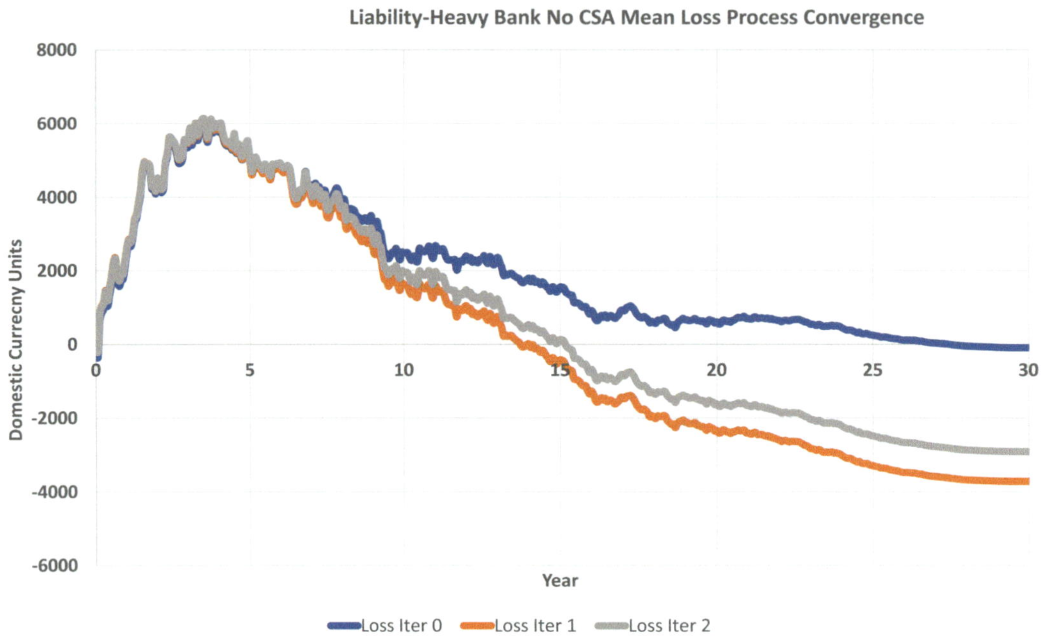

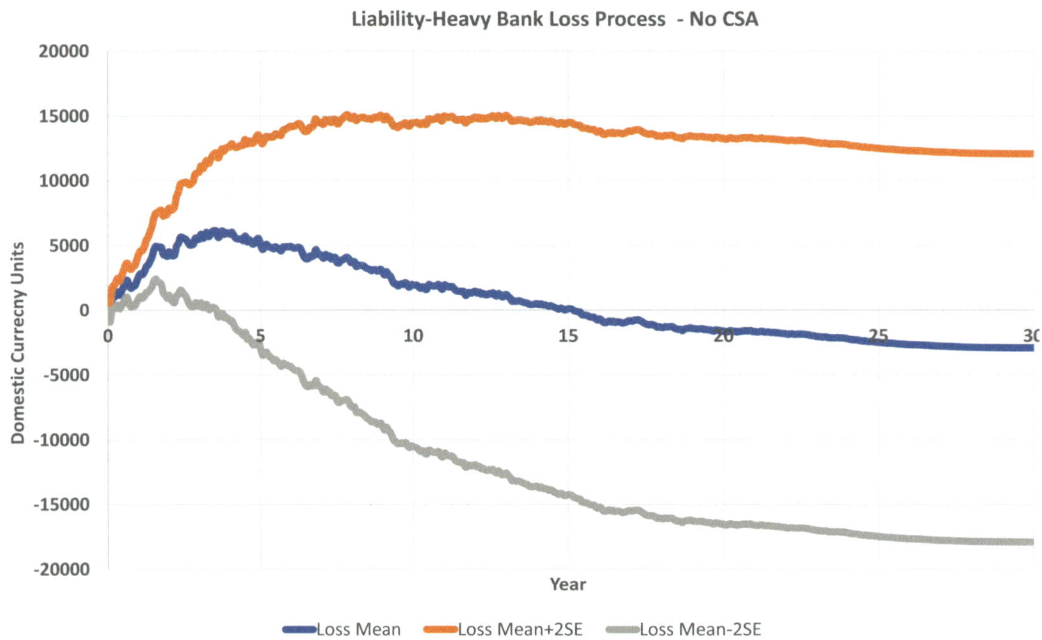

Figure 33: *(Top)* Profiles of the processes $\mathcal{L}^{(k)}$, for $k = 1, 2, 3$. *(Bottom)* Mean $\pm$ 2 stdev profiles of the process $\mathcal{L}^{(3)}$.

least, beyond $t \sim 5$ years. For earlier times, the regression errors, accumulated backward across pricing times since the final maturity of the portfolio, induce a non-negligible bias (the corresponding confidence intervals no longer contain 0). This is the reason why we use a coarser pricing time step than the simulation time step in Algorithm 7.

Twin Monte Carlo validation for FVA and KVA computations can be done the way explained in IV.§2.A.1. This is detailed and illustrated numerically in the context of Abbas-Turki et al. (2025, Section 4 and Subsection 6.2).

B Financial Case Study

For the financial case study that follows, we consider

- swap rates uniformly distributed on $[0.005, 0.05]$ (hence swaps already in-the-money or out-of-the-money at time 0),
- number of six-monthly coupon resets uniform on $[5 \dots 60]$ (final maturity of the portfolio $T = 30$ years),
- portfolio direction: either "asset heavy" bank mostly in receivables in the future, or "liability-heavy" bank mostly in the payables in the future[2] (respectively corresponding, with our data, to a bank 75% likely to pay fixed in the swaps, or 75% likely to receive fixed).

The figures that follow only display profiles, i.e. term structures, that is, expectations as a function of time of the corresponding processes. But all these processes are computed pathwise, based on the neural regression and quantile regression methodology of IV.§2-§3 allowing for all XVA inter-dependencies. XVA profiles (or pathwise XVAs if wished) are much more informative than the spot XVA values (or time 0 confidence intervals) returned by most XVA systems.

Assuming 10 counterparties, Figure 34 shows the GPU generated profiles of the MtM process (2) in the case of the asset-heavy portfolio and of the liability-heavy portfolio.

Figure 35 shows the porftolio-wide XVA profiles in the cases of *(top to bottom panels)* asset-heavy no CSA, asset-heavy CSA, liability-heavy no CSA, and liability-heavy CSA portfolios. Obviously, asset–heavy or no CSA means more CVA. The correponding curves also emphasize the transfer from counterparty credit into liquidity funding risk prompted by extensive collateralisation. Yet FVA/MVA risk is ignored in current derivatives capital regulation.

Figure 36 shows that *(top)* capital at risk as funding (cf. §1 and I.§2.D) has a material impact on the already (reduced FVA capital as funding) *(bottom)* treating KVA as a risk margin (cf. I.(22)) gives a huge discounting impact.

Figure 37 shows that *(top)* learning detects material initial margin convexity in the asset-heavy CSA portfolio, and *(bottom)* learning detects material economic capital convexity in the asset-heavy no CSA portfolio.

The above findings demonstrate the relevance of pathwise capital and margin calculations for accurate FVA, MVA, and KVA calculations.

[2]cf. the bottom, dashed boxes in Figure 3 page 28.

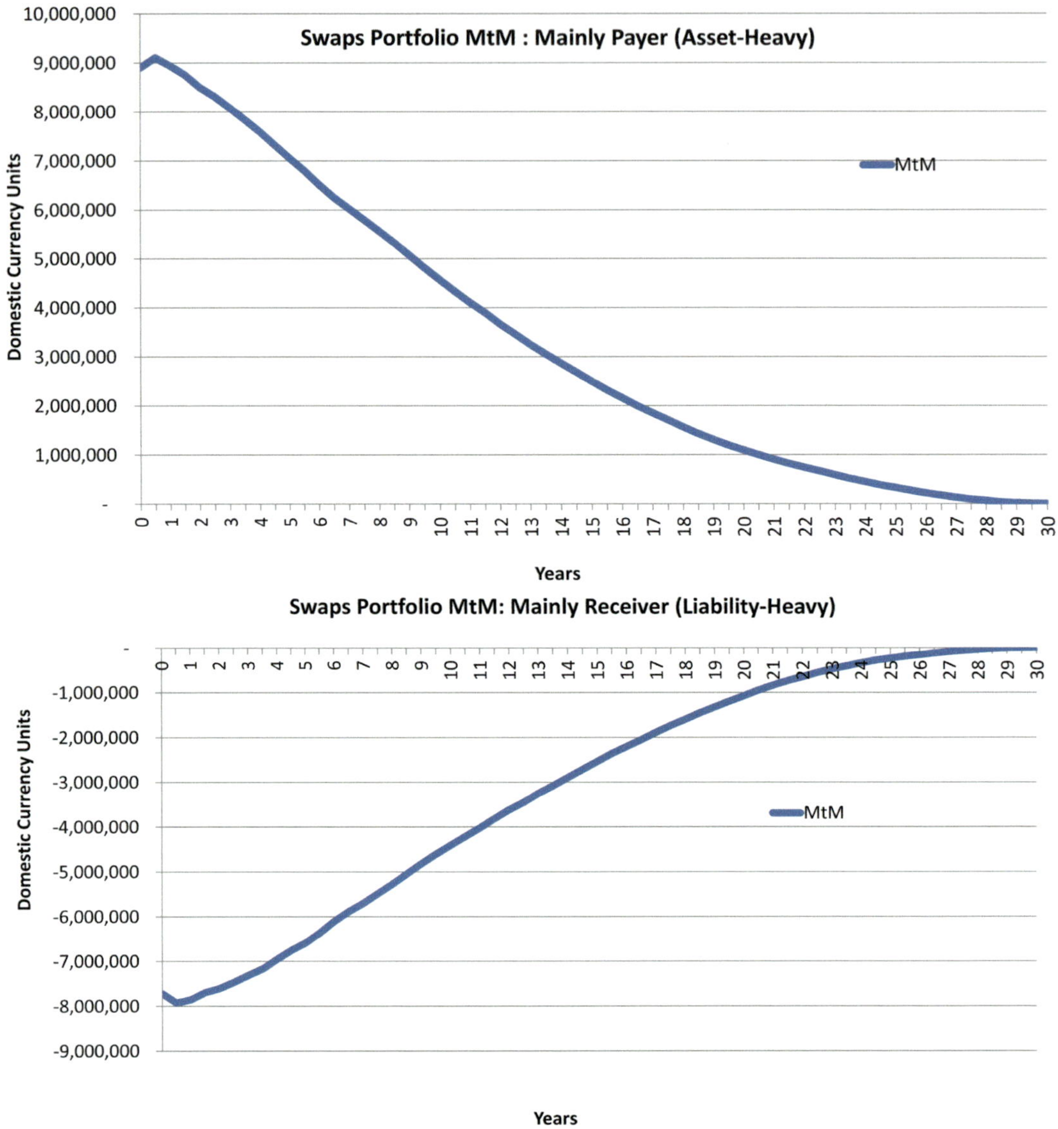

Figure 34: MtM profiles. *(Top)* Asset-heavy portfolio. *(Bottom)* Liability-heavy portfolio.

C Trade Incremental XVA Profiles

Next, we consider, on top of the previous portfolios, an incremental trade given as a par 30 year (receive fix or pay fix) swap with 100K notional. Figure 38 shows the trade incremental XVA profiles produced by our XVA learning approach in four cases. The FTP related to the new trade can be positive or negative, depending on the data, as can only be inferred by a refined XVA computation. Note that, for obtaining such smooth incremental profiles, it has been key to use common random numbers, as much as possible

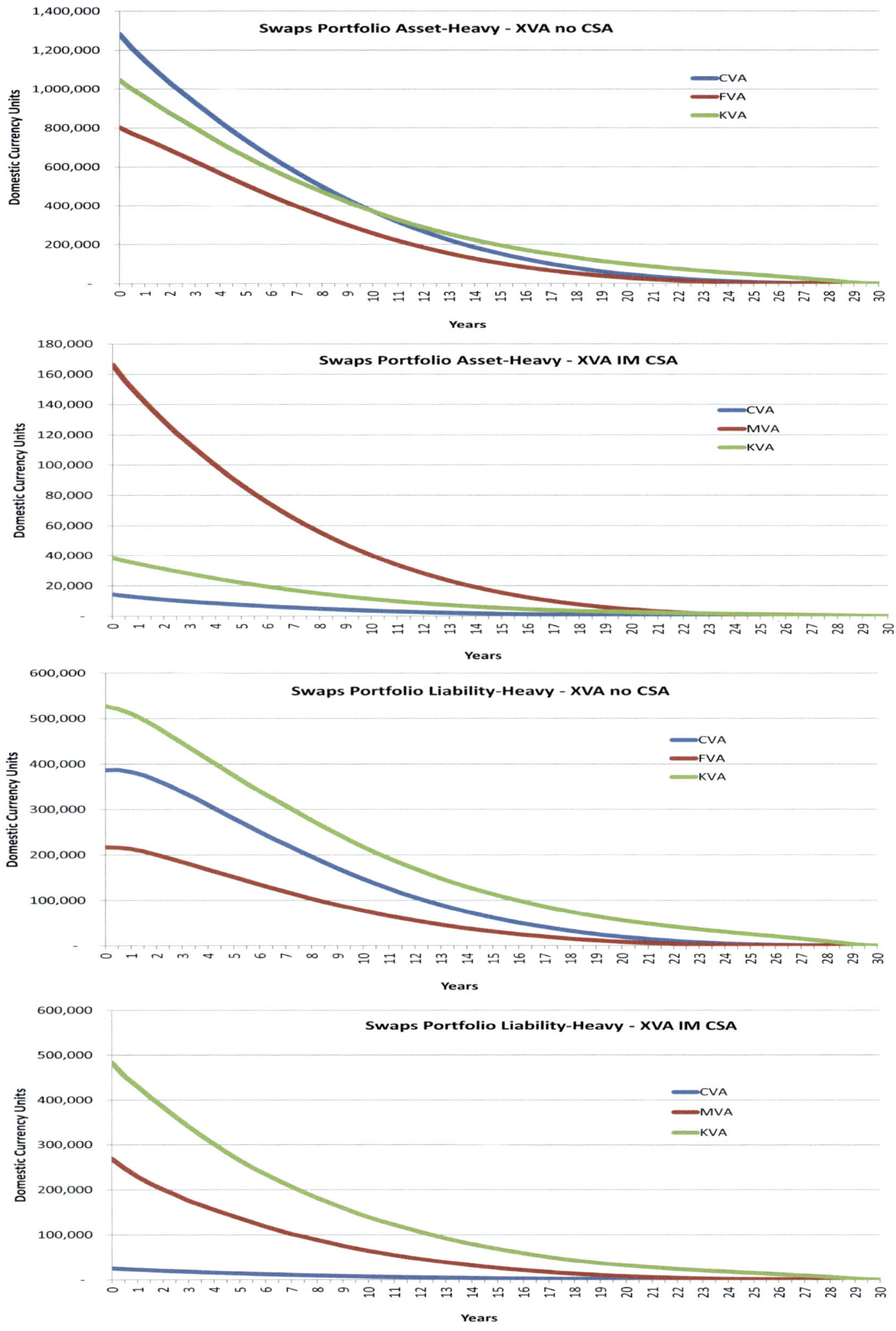

Figure 35: Porftolio-wide XVA profiles.

XVA Analysis

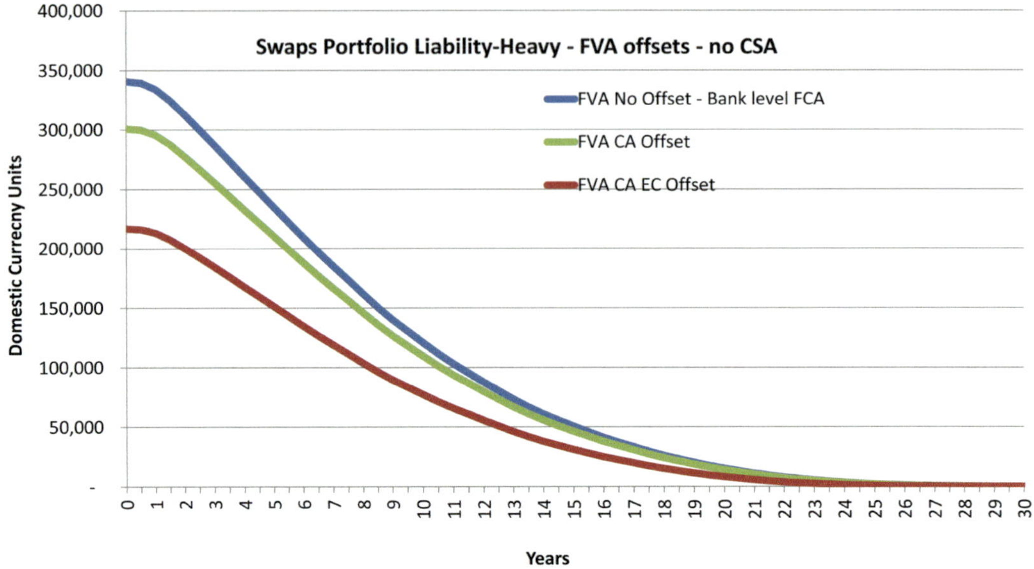

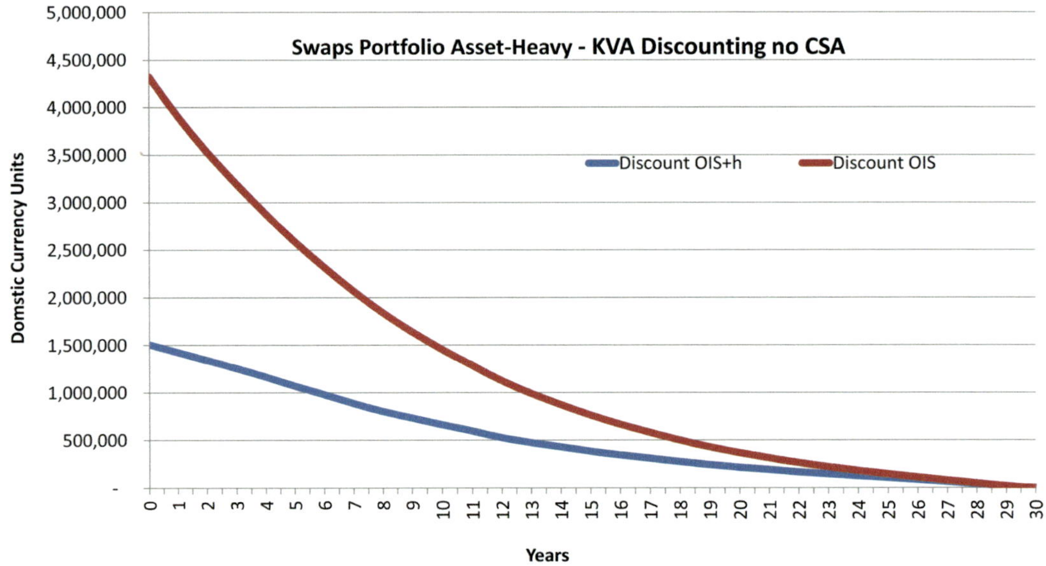

Figure 36: Portfolio-wide XVA profiles cont'd. *(Top)* FVA ignoring the off-setting impact of reserve capital and capital at risk, cf. I.§2.D (blue), $\mathrm{FVA}\equiv\mathrm{FVA}^{(0)}$ as per (6) accounting for the off-setting impact of reserve capital but ignoring the one of capital at risk (green), refined FVA as per (3) accounting for both impacts (red). *(Bottom)* KVA ignoring the off-setting impact of the risk margin, i.e., ignoring the discount factor $e^{-h(s-t)}$ in (5) (red), refined KVA as per (5) (blue).

(cf. VI.§2.A), between the original portfolio XVA computations and the ones regarding the portfolio expanded with the new trade.

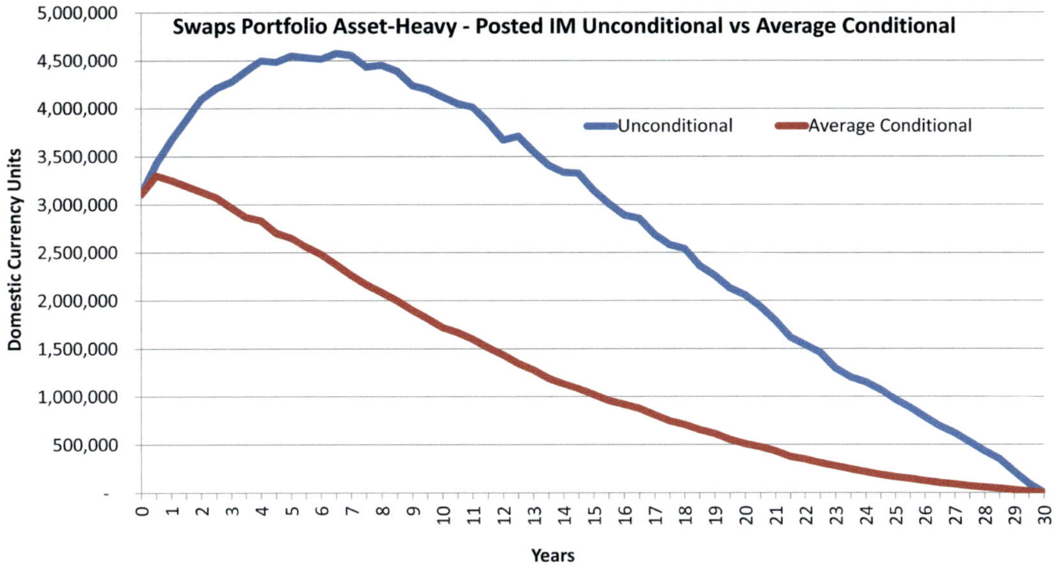

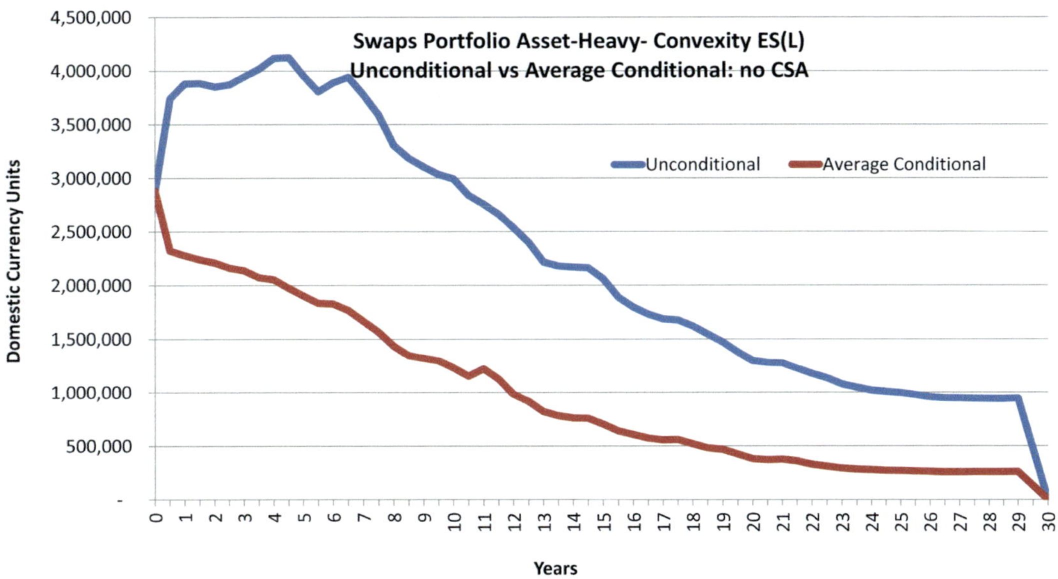

Figure 37: Portfolio-wide XVA profiles cont'd. *(Top)* In the case of the asset-heavy portfolio under CSA, unconditional PIM profile, i.e. with $\mathbb{V}\mathrm{a}\mathbb{R}_t$ replaced by $\mathbb{V}\mathrm{a}\mathbb{R}$ in (9) (blue), vs. pathwise PIM profile, i.e. mean of the pathwise PIM process as per (9) (red). *(Bottom)* In the asset-heavy portfolio no CSA case, unconditional economic capital profile, i.e. EC profile ignoring the words "time-t conditional" in Definition I.5 (blue), vs. pathwise economic capital profile, i.e. mean of the pathwise EC process in (5) (red).

D Trade and Hedge Incremental XVA Profiles

Our model assumes the market risk of trades to be fully hedged (see after Remark I.2). In Subsection C, the new swap was implicitly meant to be hedged,

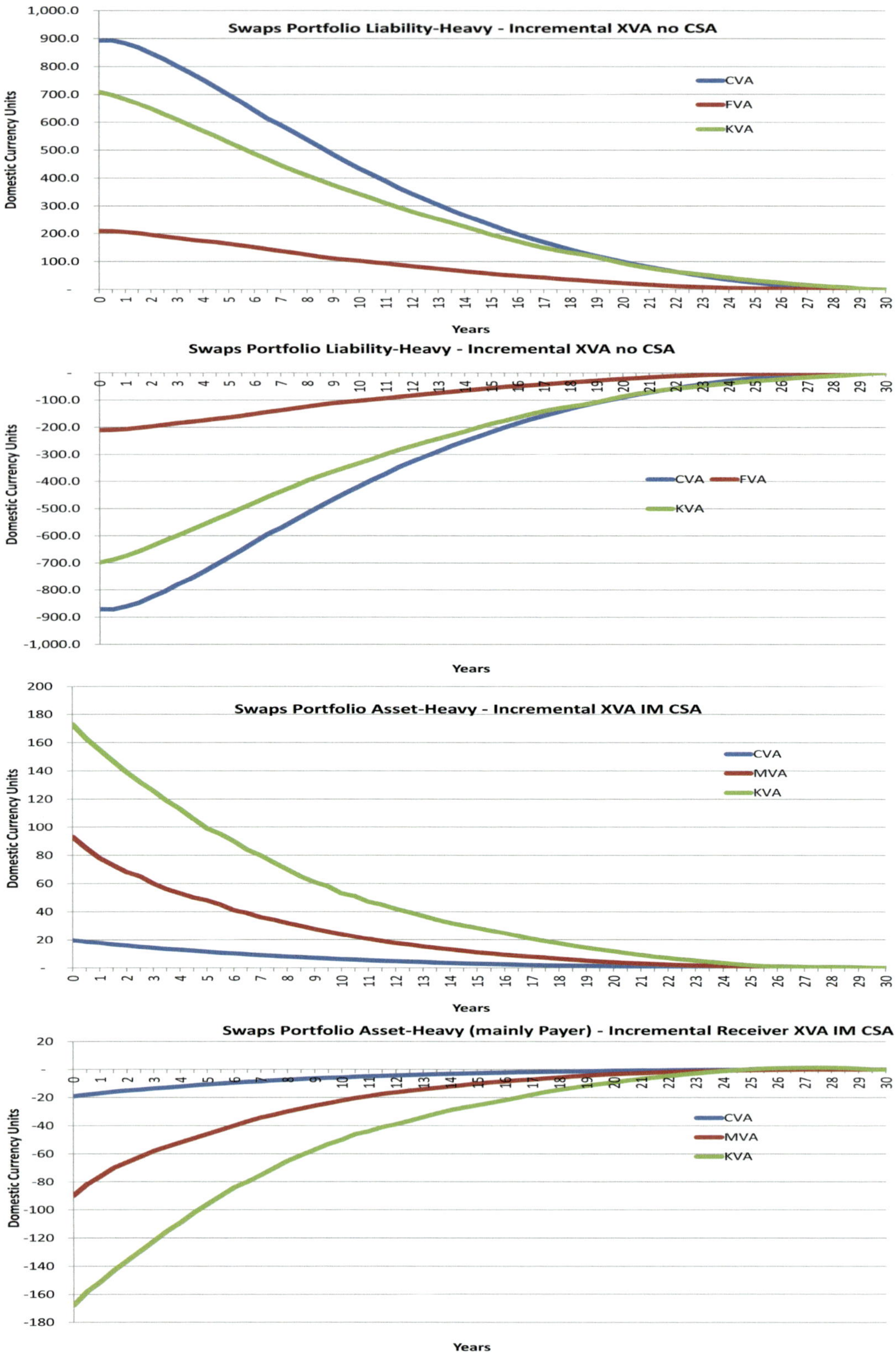

Figure 38: Trade incremental XVA profiles.

in terms of market risk, by the trading desks, through an accordingly modified dynamic hedging loss process $\mathcal{H}$ (cf. II.(12)). Here we consider an alternative situation where the market risk of the new swap is back-to-back hedged via a financial, hedge counterparty (as in the one-period example of I.§3). Specifically, we deal with

- 10 counterparties: 8 no CSA clients and 2 bilateral VM/IM CSA hedge counterparties,
- portfolios of 5K randomly generated swap trades as before, plus 5K corresponding hedge trades,
- an incremental trade given as a par 30-year swap with 100K notional, along with the corresponding hedge trade.

In particular, $\mathrm{MtM}_0 = 0$ (cf. (2)), in both portfolios excluding or including the new swap. In case a client or hedge counterparty defaults, the corresponding market hedge is assumed to be unwound through the trading desks via an accordingly modified hedging loss process $\mathcal{H}$.

The eight no CSA counterparties are primarily asset or liability-heavy. One bilateral CSA hedge counterparty is asset-heavy, and one liability-heavy. Figure 39 provides the trade incremental XVA profiles of the bilateral hedge alternatives in combination with those for the initial counterparty trade. The main XVA impact of the hedge is then a corresponding incremental MVA term, which can contribute to making the global FTP related to the trade+hedge package more or less positive or negative, depending on the data (cf. the four panels in Figure 39), as can only be inferred by a refined XVA computation.

In the above, we do not include the XVA costs/benefits of the bilateral hedge counterparty itself (cf. Remark I.3). If the latter is also a bank, it would attribute its XVA costs to its own end-clients.

Many hedge trades now face central instead of bilateral counterparties. This occurs at additional XVA costs for the client of the initial swap that can be computed based on the XVA equations of Theorem III.1. For risk management (as opposed to front-office pricing) applications, a one-period alteration of the setup may be enough and is of course much lighter to handle numerically, as developed in Chapters VIII-IX.

E Scalability

The neural regression XVA implementation of this chapter uses CNTK, the Microsoft Cognitive Toolkit. CNTK is written in core C++/CUDA (with wrappers for Python, C#, and Java). This is convenient for XVA applications, which are usually developed in C++: CNTK automatic differentiation

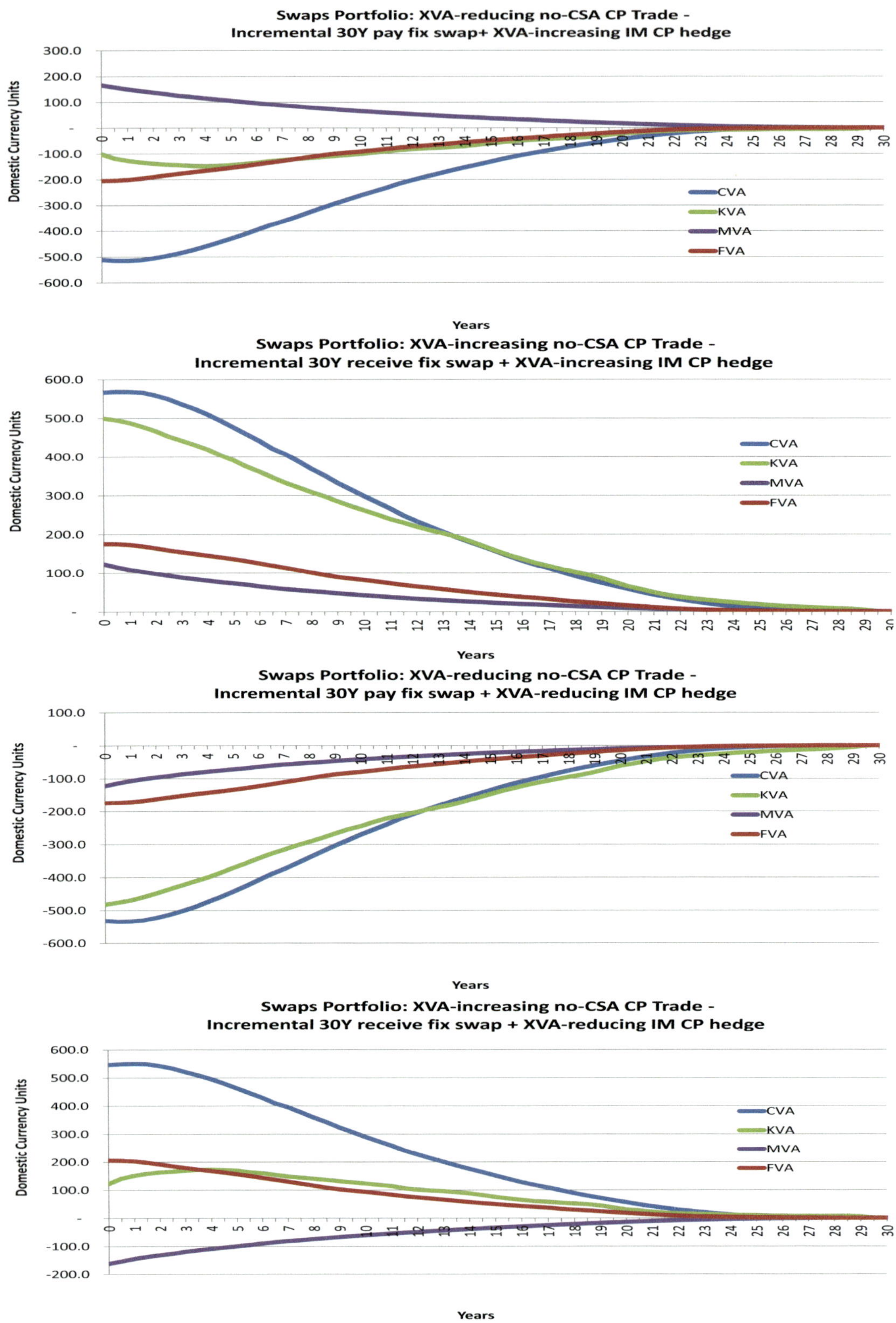

Figure 39: Trade and hedge incremental XVA profiles.

in C++/CUDA enables C++ in-process training. This allows embedding the learning task within XVA processing.

Table 17 sets out computation times, including additional results obtained by doubling the numbers of counterparties and risk factors (to 20 counterparties and 80 risk factors). All these results were computed on a Lenovo P52 laptop with NVidia Quadro P3200 GPU @ 5.5 Teraflops peak FP32 performance, and 14 streaming multiprocessors.

Table 17: XVA neural regression computation timings (seconds).

	10 CP 40 risk factors		20 CP 80 risk factors	
	No CSA	IM CSA	No CSA	IM CSA
Initial risk factor & trade pricing simulation Cuda	352	352	426	426
Counterparty and bank level learning calculations	4,529	13,466	19,154	59,342
Total initial batch	4,881	13,818	19,580	59,768
Re-simulate 1 counterparty trade pricing Cuda	40	40	51	51
Counterparty and bank level learning calculations	2,785	2,736	7,694	6,628
Total incremental trade	2,825	2,776	7,745	6,679

The computations for 20 counterparties took more than twice as long as those for 10 counterparties. However, the neural regressions achieved around 80 to 90% Cuda occupancy for 10 counterparties, and at times fell to half that level for 20 counterparties. Scaling to realistically high dimensions should be achievable, but acceptable trade-off incremental pricing performance in production would require server-grade GPU hardware, performance tuning for high GPU utilisation, and, possibly, caching computations.

RISK

Following the subprime crisis, whose roots date back to the American real estate bubble in the 2000s leading to the bankruptcy of Lehman Brothers in September 2008, a crisis of liquidity and solvency of financial institutions, but also of countries in the Eurozone, affected the entire financial sector until the end of 2010. Repercussions followed during 2011 and 2012 via the European debt crisis with Greece's difficulties in paying its debts, the nationalization of several banks in Ireland and rescue plans with significant liquidity funding via facilitated loans from the European Union for Portugal, Spain and Cyprus (€77 billion, €41 billion and €7.3 billion respectively, with liquidity funds from EU and IMF funds, see Gourinchas, Martin and Messer (2023, Table 1)). As a response, international standardization bodies, such as the Basel Committee on Banking Supervision and the Committee on Payments and Market Infrastructures established, in 2008, instructions to banking establishments in order to guarantee better robustness of the financial system. These instructions aim to limit the risks of contagion of defaults of players, mainly banks, in the financial network, both regarding the lack of liquidity of financial establishments and the impact on capital or equity levels required by these establishments. The United States adopted the Dodd-Frank Wall Street Reform and Consumer Protection Act (The United States Congress, 2010), while the European Union adopted regulations of the European Market Infrastructure Regulation (European Parliament, 2012). This was followed by the updates of the Basel texts that guarantee international agreements governing the solvency ratios of financial institutions. An important component of this regulatory toolbox consists of conducting annual stress exercises (also called stress tests) by the authorities of banking institutions such as the European Banking Authority and the United States Federal Reserve. If the objective of the regulation is to be able to limit systemic and contagion risks from non-favorable economic and financial events, these exercises must be able to capture and identify the signals of weaknesses due to these risks. This implies in particular, for the banking institution which conducts the exercise, to include the various interactions between the financial and economic actors constituting the network of the financial system.

As detailed in III.§2.A, the counterparty of the bank in a netting set can be a corporate client, another bank, or a group of financial institutions pooled in the form of a central clearinghouse counterparty (CCP). Regarding the concentration of collaterals, although these allow a certain guarantee of financial operations continuity for the CCPs and their members following the default of one of the clearing members, they pressure the smallest actors in the network, not members of these CCPs, with significant liquidity needs. Hence counterparty risk has transformed, with the growing importance of CCPs, into a liquidity risk, which consists for a financial actor of the risk of not being able to respect one or more of its financial commitments, including providing collateral, additional calls either due to a lack of means to ensure them or due to reduced and expensive accessibility for some of these collaterals.

The regulatory instructions have taken into account the recent transformations of the financial network and its risks by recommending their adequate representation and quantification for stress test exercises. These transformations also constitute a key point of attention for the executive committee of financial institutions. It is thus critical to develop tools that can capture these financial network mechanisms and its risks, with the support of numerical quantification, having targeted approaches and implementation that can be scalable to a larger and more complete financial network, yet with limited computing power and time.

This last part of the book proposes XVA metrics for that matter. This is first done in the numerically tractable one-period setups of Chapters VIII-IX: Chapter VIII models the various mechanisms and risks of the financial markets for both clearing and bilateral OTC activities, along with the different related actors and resulting costs, allowing various risk management exercises; Chapter IX combines the counterparty credit risk approach of Chapter VIII with game-theoretical equilibrium considerations to dig into the intertwined liquidity issue. Chapter X provides a bottom-up and dynamical setup, slightly simplified with respect to the full-flesh XVA pricing setup of Chapters II-III, specialized for quantitative reverse stress testing analysis.

CHAPTER VIII

Derivatives' Risks as Costs in a One-Period Setup

The XVA metrics have been introduced and traditionally used by investment banks for accounting, pricing, and collateral/capital optimization purposes. We demonstrate in this chapter that they can be fruitfully used for risk management, suggesting a sounder approach to regulatory requirements. We present a one-period cost-of-capital XVA setup extending the one in I.§3 to encompass bilateral and centrally cleared trading in a unified framework. We illustrate possible uses of this framework for running stress test exercises on financial networks with one and two clearinghouses from a clearing member's perspective or for optimizing the porting of the portfolio of a defaulted clearing member using Monte Carlo techniques in elliptical models.

§1 INTRODUCTION

In the wake of the 2007–09 global financial crisis, clearing through central counterparties (CCPs) has become mandatory for standardized derivatives, other ones remaining under bilateral setup with higher capital requirements.

One role of the CCPs is to provide to their clearing members fully collateralized hedges of their market risk with their clients. But this comes at XVA costs to the clearing members, which should pass it to their corporate clients in the form of XVA add-ons. In III.§5, we delivered the corresponding XVA analysis in continuous time. For the sake of numerical tractability, this is mimicked here in a stylized one-period setup, which may be sufficient for applications including risk assessment in the context of stress test exercises[1] or optimizing the porting of the portfolio of defaulted clearing members.

The first type of application is motivated by the default in 2020 of Ronin Capital, a broker/dealer firm that had clearing exposures on both CCP

[1]as required by (European Parliament, 2013, Article 302).

DOI: 10.1201/9781003614944-13

services Fixed Income Clearing Corporation (FICC) GSD[2] segment (123 members) and CME Futures (56 members of which 24 common with FICC GSD). If all members are assumed to be only exposed to these CCPs and their cleared clients, we can illustrate these relationships by the network depicted in Figure 40. Any common member of those two CCPs needs to ensure conservative risk assessment that can be achieved in the proposed framework by accounting for common memberships on the two CCPs. If such common memberships are ignored, they can lead to lower loss estimates giving wrong risk view on potential losses.

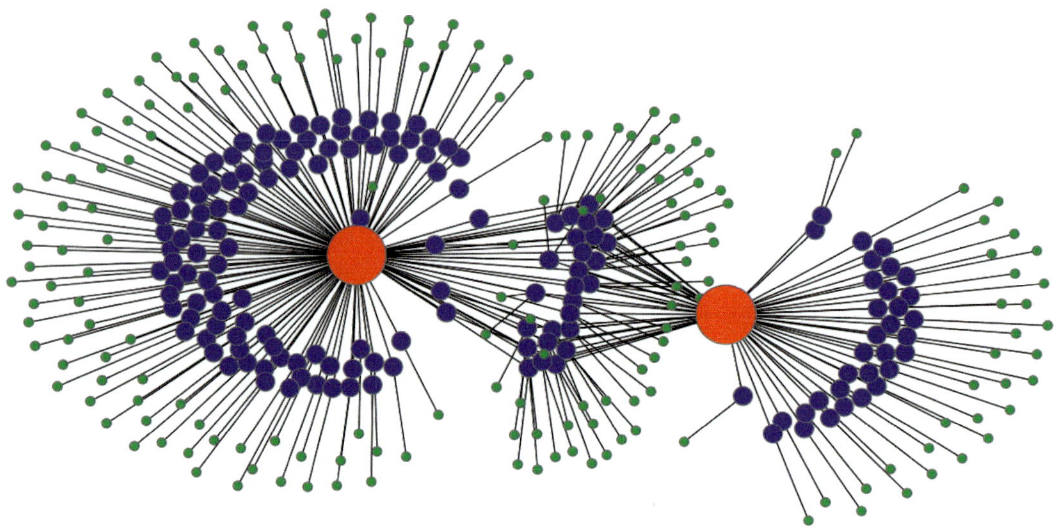

Figure 40: Network consisting of two CCPs (in red), 123 members for CCP1 seen on the left-hand side, and 56 members for CCP2 on the right-hand side, with 24 common members displayed as the group of members in the middle of the two CCPs (155 members in total, in blue), and with 179 cleared clients (in green).

The second type of application is an illustration of the results of defaulted portfolio porting as it has been the case for the trader Einer Aas on NASDAQ OMX[3] that has defaulted on 2018 with a loss spill-over effect on surviving members.

This chapter is outlined as follows. §2 sets the stage. §3 develops the corresponding XVA analysis. §4 sets up an elliptical market and credit model amenable to efficient XVA computations. §5 introduces the case studies. §6 provides numerical results of stress test exercises. §7 shows how to use our XVA framework for optimizing the porting of defaulted members portfolios. §8 concludes.

[2]Government Securities Division.

[3]Optionsmäklarna/Helsinki Stock Exchange.

§2 GENERAL SETUP

We consider a finite set of market participants, also susceptible to serve as clearing members of CCPs. Derivative transactions can then be concluded either *(i)* between two individual participants, or *(ii)* between a set of participants (two or more, in practice from a few units to a few hundred), pooled in the form of a CCP, and a clearing member of this CCP.

The trades of a clearing member bank with a CCP are partitioned between proprietary trades, which are in effect hedges of the bilateral trading exposure of the bank, and back-to-back hedges of so-called cleared client trades, through which non-member clients gain access to the clearing services of a CCP: see the respective cash flows $\mathcal{P}$ and $\overline{\mathcal{P}}$ in Figure 41. As illustrated on the figure, the contractual cash flows from cleared and bilateral clients to a reference clearing member, dubbed the bank hereafter, are promised in successive turns from the bank to the CCP, from the CCP to other clearing members, and from the latter to their own clients. As a consequence, the CCP is flat in terms of market risk, as is also each of the clearing members. To mitigate the liquidity issues posed by clearing member default resolution procedures (explored in Chapter IX), CCPs are typically siloed into different services, each devoted to a specific class of derivatives. We first consider a setup with a single CCP service, the extension to several CCPs and/or CCP services being done in §3.C.

A Defaults Settlement Rules

Regarding the liquidation cash flows, we go by Principle I.2 page 34, the zero recovery specification of Principle III.1 page 86, further specialized to central clearing by Assumption 2 page 224 below.

Remark 1. In this and the next chapter, like in I.§3, we assume zero recoveries. For our purposes in this part of the book, random recoveries could be used for emphasizing some extra dependencies via correlations with other random modeling features. We refrain from doing so for avoiding to blur the main features.

Principle I.2, or its nonzero recovery extension Principle III.1, apply not only to bilateral trades as in I.§3, but also to the netting set of transactions between a clearing member and a CCP (cf. the specialization of Principle III.1 to central clearing by Assumption III.2 in the continuous-time setup). Like in continuous time in Chapter III, in our stylized setup, a CCP is nothing but the collection of its clearing members. Our CCP has no resources of its own (in particular, it cannot post any default fund contribution, or

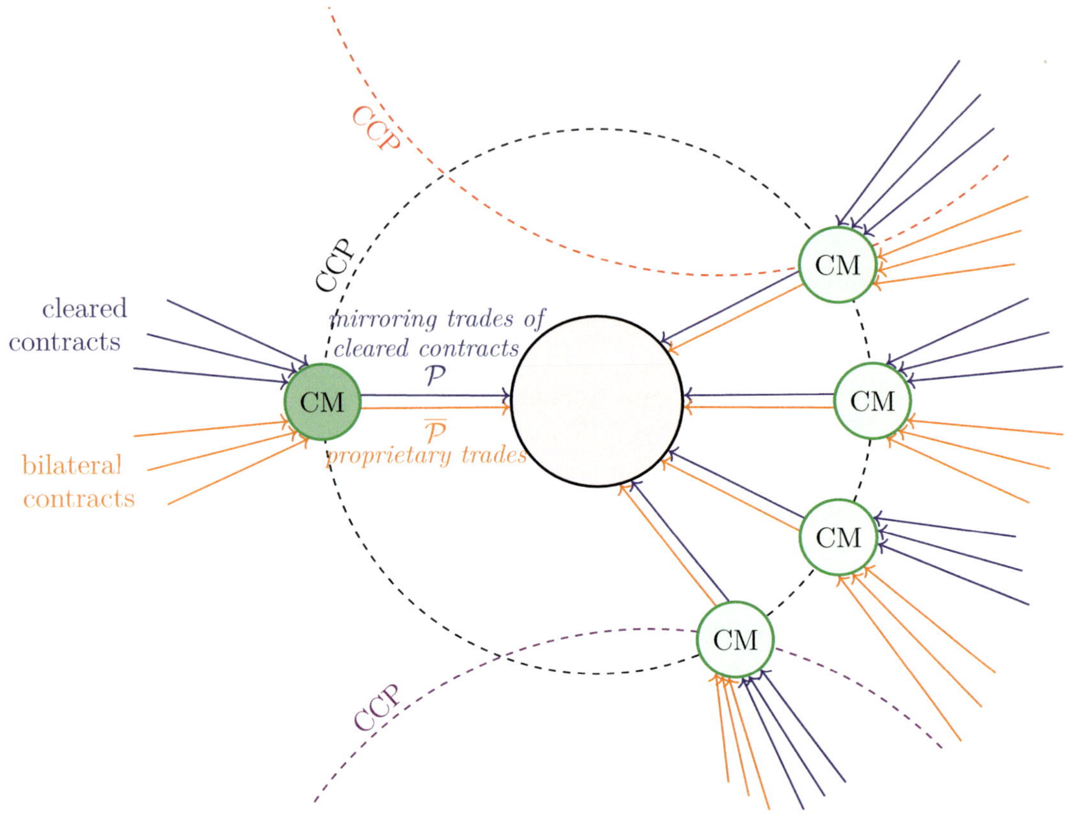

Figure 41: Promised cash flows between market participants. The reference clearing member bank is on the left.

"skin-in-the-game"[4]). As long as it is non-default, i.e. as long as at least one of its clearing members is non-default, our CCP can only handle the losses triggered by the defaults of some of its clearing members by redirecting these losses on the surviving ones. This participation of the surviving members in the losses triggered by the defaults of the other members corresponds in our framework to the usage by the CCP of the default fund, both funded and unfunded. As will be detailed in equations below (see also III.§4), the funded default fund contributions (DFC) are used for covering losses triggered by the defaults of clearing members over their margins. The unfunded default fund contributions correspond to additional refills that can be required by the CCP,

[4]such an additional protection layer, though quite common in practice, is of marginal magnitude compared to the other protection layers. By omitting skin-in-the-game component, the obtained results are conservative in terms of risk management and the various formulations are simplified, see Remark III.7 page 98.

often up to some cap in principle, without bounds in our model, in case the funded default fund contributions of the surviving members are not enough.

B XVA Framework

As established in the Pricing part of the book, the main XVA metrics that deserve to be accounted for in entry prices are the CVA, the FVA/MVA and the KVA (but not the DVA and the FDA contra-liabilities), where

i. the CVA is the expected cost for the bank of the default risk of its clients,

ii. the FVA/MVA is the expected cost for the bank of its own default risk or, more precisely, of the implications of this risk in terms of rehypothecable/segregated collateral funding spreads for the bank;

iii. the KVA is the cost for the bank of having to remunerate its shareholders at some hurdle rate for their capital at risk, capital which is required by the regulator as a provision against the residual risk left uncovered by i. and ii. (as default risk cannot be hedged by the bank),

In more details, assume that at time 0 all the banking participants, including the reference clearing member bank (cf. Figure 41), with no prior endowments, enter transactions with their clients and hedge their positions, both bilaterally between them and through the CCP. As seen above, the CCP and each bank are flat in terms of market risk. However, as market participants are assumed to be defaultable with zero recovery, in order to account for counterparty credit risk and its funding and capital consequences, the reference bank (and each clearing member bank alike) requires from its (nonbanking) corporate clients a pricing rebate (considering conventionally the bank as the "buyer") with respect to the counterparty-risk-free valuation of the deals. The corporate clients of the bank are assumed to absorb the ensuing prices via their corporate business, which is their primary motivation for these deals.

A fininsurance probability measure $\mathbb{Q}^*$, relevant for grounding both stress test exercises and risk management analysis, such as economic capital calculation, with corresponding expectation operator, denoted by $\mathbb{E}^*$, is used for the linear valuation of cash flows, using the risk-free asset as numéraire.

Following the general XVA guidelines of II.§1, the XVA pricing rebate required by the reference clearing member bank from its corporate clients, dubbed funds transfer price (FTP), comes in two parts: first, the expected counterparty default losses and funding expenditures of the bank, an amount that flows into the bank liabilities, and which we refer to as contra-asset valuation ($\mathrm{CA} = \mathrm{CVA} + \mathrm{FVA} + \mathrm{MVA}$ as we will see); second, a cost of capital risk premium (KVA), which is used by the management of the bank as retained earnings for remunerating the shareholders of the bank for their capital at risk within the bank. All in one, the bank buys the deals from its clients at the (aggregated) price (MtM − FTP), where MtM is their counterparty-risk-free value and

$$\mathrm{FTP} = \underbrace{\mathrm{CA}}_{\text{Expected costs}} + \underbrace{\mathrm{KVA}}_{\text{Risk premium}} . \tag{1}$$

The KVA risk premium is not a liability (see Figure 3 page 28), hence it is also loss-absorbing and therefore part of capital at risk. Let EC denote an economic capital of the bank corresponding to the minimum level of capital at risk that the bank should hold from a regulatory (i.e. solvency) perspective. If KVA < EC, then the bank shareholders need to provide the missing amount (EC − KVA) of capital at risk, so that the actual level of capital at risk of the bank is

$$\max(\mathrm{EC}, \mathrm{KVA}),$$

while *shareholder* capital at risk reduces to

$$\max(\mathrm{EC}, \mathrm{KVA}) - \mathrm{KVA} = (\mathrm{EC} - \mathrm{KVA})^+. \tag{2}$$

§3 THEORETICAL XVA ANALYSIS

In this section we detail each term in the equations above, in the case of a bank involved into an arbitrary combination of bilateral and centrally cleared portfolios, in a tractable one-period setup with period length T. In the one-period XVA model of I.§3, there were no CCPs, and the bank was supposed to have found a bilateral hedge counterparty assumed to be default-free, so that the hedging side of the position of the bank induced no additional contra-asset. In practice, banks find hedge counterparties on exchanges and their hedging portfolios are centrally cleared via CCPs. The hedging counterparties are default-prone, so that the hedging side of the bank's portfolio also triggers XVA costs (as already in III.§4-§5, cf. also VII.§3.D). In this chapter we analyze these costs in a numerically tractable extension of the introductory one-period model of I.§3 also encompassing centrally cleared trading (corresponding to a one-period alteration of Theorem III.1).

A Cash Flows

Given disjoint sets of indices $I \ni 0$, C, and B for the clearing members (including the reference bank labeled by 0) and for the respective cleared and bilateral netting sets of the bank with its individual counterparties, we denote by:

- J_0, shortened as J, and J_i, $i \in I \setminus \{0\}$, the survival indicator random variables of the bank and of the other clearing members at time 1; $\gamma = \mathbb{Q}^*(J = 0)$, the default probability of the bank;

- $\mathcal{J} = \max_i J_i$, the survival indicator random variable of the CCP (i.e. of at least one clearing member),
- $\mathcal{P}_i$, $P_i = \mathbb{E}^*\mathcal{P}_i$, and PIM_i, $i \in I$, the contractual cash flows, variation margin, and initial margin from the clearing member i to the CCP corresponding to the cleared clients account of the member i;
- $\overline{\mathcal{P}}_i$, $\overline{P}_i = \mathbb{E}^*\overline{\mathcal{P}}_i$, and $\overline{\mathrm{PIM}}_i$, $i \in I$, the contractual cash flows, variation margin, and initial margin from the clearing member i to the CCP corresponding to the proprietary (also dubbed house) account of the clearing member i;
- DFC_i, $i \in I$, the funded default fund contribution posted by the clearing member i to the CCP;
- J_b, $b \in B$, the survival indicator random variable of the counterparty of the bilateral netting set b of the reference bank; $\mathcal{P}_b$, VM_b, and RIM_b, the associated contractual cash flows, variation margin, and initial margin from the corresponding counterparty to the bank; and PIM_b, the initial margin from the bank to this counterparty;
- J_c, $c \in C$, the survival indicator random variable of the client of the cleared trading netting set c of the bank, and $\mathcal{P}_c$, $P_c = \mathbb{E}^*\mathcal{P}_c$, and RIM_c, the associated contractual cash flows, variation margin, and initial margin from the corresponding client to the bank (we reassert that a bank does not post any initial margin on its cleared client netting sets);
- Λ, the loss of the CCP, i.e. the loss triggered by the defaults of its clearing members beyond their posted collateral[5], which is borne by the surviving members (if any, see §2.A);
- $\mu = J\mu$, the proportion of these losses allocated to the reference clearing member bank (based on remaining survivors).

Moreover, in case $i = 0$ (so regarding the reference clearing member bank), we typically skip the index i (as in $J_0 = J$).

The above specifications $P_i = \mathbb{E}^*\mathcal{P}_i$, $\overline{P}_i = \mathbb{E}^*\overline{\mathcal{P}}_i$, and $P_c = \mathbb{E}^*\mathcal{P}_c$ reflect the fact that the CCP is fully collateralized in its trading with the clearing members, as are also the latter in their trading with cleared clients. In addition:

Assumption 1. $\sum_i(\mathcal{P}_i + \overline{\mathcal{P}}_i) = 0$ (the CCP is flat in terms of market risk), $\sum_c \mathcal{P}_c = \mathcal{P}_0$ (by definition of cleared trades and of their mirroring trades), and $\sum_b \mathcal{P}_b = \overline{\mathcal{P}}_0$ (the reference bank is flat in terms of market risk).

[5]variation margin, initial margin, and (funded) default fund contributions.

Assumption 1 yields the clearing conditions regarding the contractually promised cash flows, which applies to each banking participant (written there for the reference bank) and to the CCP.

Also, in line with Principle I.2 that monitors the default liquidation cash flows in the present zero-recovery setup:

Assumption 2. On the CCP survival event $\{\mathcal{J}=1\}$, the CCP receives from each clearing member i

$$\begin{aligned} &J_i(\mathcal{P}_i+\overline{\mathcal{P}}_i)+(1-J_i)\Big(\mathcal{P}_i\wedge(P_i+\mathrm{PIM}_i)+\overline{\mathcal{P}}_i\wedge(\overline{P}_i+\overline{\mathrm{PIM}}_i)\\ &\quad+\big((\mathcal{P}_i-(P_i+\mathrm{PIM}_i))^+ +(\overline{\mathcal{P}}_i-(\overline{P}_i+\overline{\mathrm{PIM}}_i))^+\big)\wedge\mathrm{DFC}_i\Big). \end{aligned} \tag{3}$$

On the bank survival event $\{J=1\}$ $(\subset\{\mathcal{J}=1\})$, the bank receives on each cleared netting set c and bilateral netting set b

$$J_c\mathcal{P}_c+(1-J_c)\big(\mathcal{P}_c\wedge(P_c+\mathrm{RIM}_c)\big) \text{ and } J_b\mathcal{P}_b+(1-J_b)\big(\mathcal{P}_b\wedge(\mathrm{VM}_b+\mathrm{RIM}_b)\big), \tag{4}$$

whereas it pays to the CCP

$$\sum_c \mathcal{P}_c+\sum_b \mathcal{P}_b=\sum_c\big(J_c\mathcal{P}_c+(1-J_c)\ \mathcal{P}_c\big)+\sum_b\big(J_b\mathcal{P}_b+(1-J_b)\mathcal{P}_b\big). \tag{5}$$

We need one more condition, regarding the funding side of the problem:

Assumption 3. At time 0 the amounts CA and KVA sourced from the corporate clients of the bank are deposited on the reserve capital and capital at risk accounts of the bank. The bank can use the amounts CA and max(EC, KVA) on its reserve capital and capital at risk accounts for its variation margin borrowing purposes. Funds needed beyond CA + max(EC, KVA) for variation margin posting purposes are borrowed by the bank at its credit spread γ above the risk-free rate. The bank must borrow entirely the amounts to post as initial margin and funded default fund contributions, but this can be achieved at some blended funding spread $\widetilde{\gamma}\le\gamma$.

The rationale for funding variation margin but not initial margin from CA + max(EC, KVA) is set out in Remark I.12. The motivation for the assumption $\widetilde{\gamma}\le\gamma$ is provided in III.§2.C.

Lemma 1. *The borrowing needs of the bank for rehypothecable and segregated collateral amount to, respectively,*

$$\begin{aligned} &\Big(\sum_b(P_b-\mathrm{VM}_b)-\mathrm{CA}-\max(\mathrm{EC},\mathrm{KVA})\Big)^+ \textit{ and}\\ &\mathrm{PIM}+\overline{\mathrm{PIM}}+\mathrm{DFC}+\sum_b\mathrm{PIM}_b. \end{aligned} \tag{6}$$

***Proof*.** On the bilateral trades of the bank and their hedges, the Treasury of the bank receives $\sum_b \mathrm{VM}_b$ of variation margin from its counterparties and has to post an aggregated amount $\sum_b P_b$ of variation margin. Assumption 3 then leads to (6). □

Lemma 2. *On the bank survival event $\{J = 1\}$, the counterparty default losses $\mathcal{C}$ and the funding expenses $\mathcal{F}$ of the bank are given by*

$$\mathcal{C} = \sum_b (1-J_b)(\mathcal{P}_b - \mathrm{VM}_b - \mathrm{RIM}_b)^+ + \sum_c (1-J_c)(\mathcal{P}_c - P_c - \mathrm{RIM}_c)^+ + \mu\Lambda, \quad (7)$$

where

$$\Lambda = \sum_i (1-J_i)\Big((\mathcal{P}_i - P_i - \mathrm{PIM}_i)^+ + (\overline{\mathcal{P}}_i - \overline{P}_i - \overline{\mathrm{PIM}}_i)^+ - \mathrm{DFC}_i\Big)^+, \quad (8)$$

and

$$\begin{aligned}\mathcal{F} = \tilde{\gamma}\Big(\mathrm{PIM} + \overline{\mathrm{PIM}} + \mathrm{DFC}\Big) + \tilde{\gamma}\sum_b \mathrm{PIM}_b + \\ \gamma\Big(\sum_b (P_b - \mathrm{VM}_b) - \mathrm{CA} - \max(\mathrm{EC}, \mathrm{KVA})\Big)^+.\end{aligned} \quad (9)$$

***Proof*.** In view of Lemma 1 and Assumption 3, the (risky) funding expenses of the bank correspond to the formula (9) for $\mathcal{F}$.

Regarding $\mathcal{C}$, on the CCP survival event $\{\mathcal{J} = 1\}$, the CCP receives, by Assumption 2,

$$\begin{aligned}\sum_i \Big(J_i(\mathcal{P}_i + \overline{\mathcal{P}}_i) + (1-J_i)\Big(\mathcal{P}_i \wedge (P_i + \mathrm{PIM}_i) + \overline{\mathcal{P}}_i \wedge (\overline{P}_i + \overline{\mathrm{PIM}}_i) \\ + \big((\mathcal{P}_i - (P_i + \mathrm{PIM}_i))^+ + (\overline{\mathcal{P}}_i - (\overline{P}_i + \overline{\mathrm{PIM}}_i))^+\big) \wedge \mathrm{DFC}_i\Big)\Big).\end{aligned} \quad (10)$$

By the CCP clearing condition in Assumption 1,

$$0 = \sum_i (\mathcal{P}_i + \overline{\mathcal{P}}_i) = \sum_i \Big(J_i(\mathcal{P}_i + \overline{\mathcal{P}}_i) + (1-J_i)(\mathcal{P}_i + \overline{\mathcal{P}}_i)\Big).$$

Hence (10) is equal to

$$-\sum_i (1-J_i)\Big((\mathcal{P}_i - P_i - \mathrm{PIM}_i)^+ + (\overline{\mathcal{P}}_i - \overline{P}_i - \overline{\mathrm{PIM}}_i)^+ - \mathrm{DFC}_i\Big)^+ = -\Lambda, \quad (11)$$

by definition (8) of Λ.

On the bank survival event $\{J = 1\}$ ($\subset \{\mathcal{J} = 1\}$), by the respective Assumptions 2 and 1, the bank receives from its clients and counterparties

$$\begin{aligned}\sum_c \Big(J_c\mathcal{P}_c + (1-J_c)\big(\mathcal{P}_c \wedge (P_c + \mathrm{RIM}_c)\big)\Big) + \\ \sum_b \Big(J_b\mathcal{P}_b + (1-J_b)\big(\mathcal{P}_b \wedge (\mathrm{VM}_b + \mathrm{RIM}_b)\big)\Big),\end{aligned} \quad (12)$$

respectively pays to the CCP

$$\sum_c \mathcal{P}_c + \sum_b \mathcal{P}_b = \sum_c \Big(J_c\mathcal{P}_c + (1-J_c)\mathcal{P}_c\Big) + \sum_b \Big(J_b\mathcal{P}_b + (1-J_b)\mathcal{P}_b\Big). \quad (13)$$

Subtracting (12) from (13), we obtain

$$\sum_c (1-J_c)(\mathcal{P}_c - P_c - \mathrm{RIM}_c)^+ + \sum_b (1-J_b)(\mathcal{P}_b - \mathrm{VM}_b - \mathrm{RIM}_b)^+.$$

On top of this comes the participation $\mu\Lambda$ of the bank in the CCP default losses, which yields the formula (7) for $\mathcal{C}$. □

Remark 2. In the present one-period, single-CCP setup, the random variables Λ and $\mu\Lambda$ in Lemma 2 correspond to the unfunded default fund of the CCP and to the corresponding contribution of the reference bank (cf. Remark III.(8) and Eqn. III.(36) in continuous time).

B Valuation

Let $\mathbb{E}$ denote the expectation with respect to the bank survival measure $\mathbb{Q}$ associated with $\mathbb{Q}^*$, i.e., for any random variable $\mathcal{Y}$,

$$\mathbb{E}\mathcal{Y} = (1-\gamma)^{-1}\mathbb{E}^*[J\mathcal{Y}]. \quad (14)$$

(expectation of $\mathcal{Y}$ conditional on the survival of the bank, as seen in Lemma I.1).

Under a one-period cost-of-capital XVA approach (see I.§3), the bank charges its future losses to its corporate clients at a CA level making $J\mathcal{L} = J(\mathcal{C} + \mathcal{F} - \mathrm{CA})$ centered under $\mathbb{Q}^*$, where $\mathcal{L}$ is the bank trading loss, so that $J\mathcal{L}$ is the trading loss of the shareholders of the bank. In addition, given a target hurdle rate h assumed in $[0, 1]$ (and typically of the order of 10%), the management of the bank ensures to the bank shareholders receive dividends at the height of h times their capital at risk $(\mathrm{EC} - \mathrm{KVA})^+$ (cf. (2)), where we model EC as $\mathbb{ES}(J\mathcal{L})$, the expected shortfall of $J\mathcal{L}$ computed under the bank survival measure $\mathbb{Q}$ at a quantile level[6] α (e.g. $\alpha = 99\%$ and $\alpha = 99.75\%$ in our experiments).

The definitions of the XVA metrics corresponding to the above specifications are given in Table 18. Hence in view of (7) and (9), we have

$$\mathrm{CA} = \mathbb{E}^*\Big[J\big(\mathcal{C} + \mathcal{F}\big) + (1-J)\mathrm{CA}\Big], \quad (15)$$

[6]under normal distribution assumptions, such ES at the quantile level 99.75% allows reaching similar loss level as with a VaR (quantile) risk metric at the level 99.9%. In practice, regulatory and economic capital may indeed aim at capturing extreme losses that can occur once every 1000 years, cf. Paragraph 5.1 from the Basel Committee on Banking Supervision (2005).

Table 18: Definitions of the XVA metrics in the single-CCP static setup (with $\mathcal{C}$, $\mathcal{F}$ and Λ given by Lemma 2).

XVA	Expression	Full name and description
KVA	$\mathbb{E}^*\Big[Jh(\mathrm{EC}-\mathrm{KVA})^+ + (1-J)\mathrm{KVA}\Big]$, where $\mathrm{EC}=\mathbb{ES}\big(J(\mathcal{C}+\mathcal{F}-\mathrm{CA})\big)$	capital valuation adjustment
CA	CVA + MVA + FVA	contra-asset valuation
CVA	BCVA + CCVA	credit valuation adjustment
BCVA	$\mathbb{E}^*\Big[J\sum_b(1-J_b)(\mathcal{P}_b-\mathrm{VM}_b-\mathrm{RIM}_b)^+ \quad +(1-J)\mathrm{BCVA}\Big]$	credit valuation adjustment for bilateral exposures
CCVA	$\mathbb{E}^*\Big[J\sum_c(1-J_c)(\mathcal{P}_c-P_c-\mathrm{RIM}_c)^+ + \mu\Lambda + (1-J)\mathrm{CCVA}\Big]$	credit valuation adjustment for clearing activity exposures
MVA	BMVA + CMVA	margin valuation adjustment
BMVA	$\mathbb{E}^*\Big[J\tilde{\gamma}\sum_b \mathrm{PIM}_b + (1-J)\mathrm{BMVA}\Big]$	margin valuation adjustment for bilateral exposures
CMVA	$\mathbb{E}^*\Big[J\tilde{\gamma}\big(\mathrm{PIM}+\overline{\mathrm{PIM}}+\mathrm{DFC}\big)+(1-J)\mathrm{CMVA}\Big]$	margin valuation adjustment for clearing activity exposures
FVA	$\mathbb{E}^*\Big[J\gamma\Big(\sum_b(P_b-\mathrm{VM}_b)-\mathrm{CA} - \max(\mathrm{EC},\mathrm{KVA})\Big)^+ + (1-J)\mathrm{FVA}\Big]$	funding valuation adjustment

i.e. $\mathbb{E}^*\Big[J\big(\mathcal{C}+\mathcal{F}-\mathrm{CA}\big)\Big]=0$, as desired, while the above-mentioned hurdle rate condition is represented in Table 18 by the KVA equation

$$\mathrm{KVA}=\mathbb{E}^*\Big[Jh(\mathrm{EC}-\mathrm{KVA})^+ + (1-J)\mathrm{KVA}\Big]. \tag{16}$$

As explained after Definition I.2 page 36, the terminal cash flows of the form $(1-J)\times\cdots$ in Table 18 are in line with the shareholder centric perspective relevant for the determination of entry prices. Because of these implicit terminal conditions, the definitions of Table 18 are in fact a fixed-point system of equations. The split of the CA and KVA equations (15)-(16) into the collection of equations in Table 18 is motivated by both interpretation and numerical considerations. From an interpretation viewpoint, it is useful to provide the more granular view on the costs of the bank provided by the split of the global CA amount between, on the one hand, bilateral and centrally cleared trading default risk components BCVA and CCVA, and, on the other hand, bilateral and centrally cleared trading funding risk components, BMVA and CMVA for segregated initial margin, whereas the FVA cost of funding the variation margin is holistic in nature (can only be apprehended at the level of the bank balance-sheet as a whole), via the feedback impact of CA + max(EC, KVA) into the FVA. From a numerical viewpoint, the collection of smaller problems in Table 18 may be easier to address than the global equations (15)-(16). Each of the smaller problems can also be handled by a dedicated desk of the bank, namely, the CVA desk for the BCVA and CCVA, the Treasury of the bank for the BMVA, CMVA and the FVA, and the management of the bank for the KVA.

Passing in the above equations into the bank survival measure $\mathbb{Q}$ based on Lemma I.1 page 33 shows that the corresponding fixed point problem is in fact well-posed and yields explicit formulas for all the quantities at hand.

Theorem 1. *The explicit XVA formulas of Table 19 hold, and the bank trading loss $\mathcal{L}$ satisfies*

$$J\mathcal{L} = J(\mathcal{C} - \mathrm{CVA}) = J\Big(\sum_c (1 - J_c)(\mathcal{P}_c - P_c - \mathrm{RIM}_c)^+ \\ + \mu\Lambda - \mathrm{CCVA} + \sum_b (1 - J_b)(\mathcal{P}_b - \mathrm{VM}_b - \mathrm{RIM}_b)^+ - \mathrm{BCVA}\Big). \tag{17}$$

In particular, all the XVA (and also EC) numbers are nonnegative, in line with the general principles of II.§1 (see also II.§5.A).

Proof. By the results recalled in Prerequisites in the general introduction to the book, EC is nonnegative as an expected shortfall under $\mathbb{Q}$ of the random variable $J(\mathcal{C} + \mathcal{F} - \mathrm{CA})$, which is centered under $\mathbb{Q}^*$ and therefore under $\mathbb{Q}$, by (14). The first four formulas in Table 19 directly follow from the definitions of Table 18 and Lemma I.1, which also implies that $\mathrm{KVA} = \mathbb{E}\big[h(\mathrm{EC} - \mathrm{KVA})^+\big] = h(\mathrm{EC} - \mathrm{KVA})^+$. As h is nonnegative, this KVA semilinear equation is equivalent to

$$(\mathrm{KVA} > \mathrm{EC} \text{ and } \mathrm{KVA} = 0) \text{ or } (\mathrm{KVA} \le \mathrm{EC} \text{ and } \mathrm{KVA} = \frac{h}{1+h}\mathrm{EC}),$$

where ($\mathrm{KVA} > \mathrm{EC}$ and $\mathrm{KVA} = 0$) would contradict the nonnegativity of EC, whereas, for $h \in [0, 1]$ as assumed and $\mathrm{EC} \ge 0$, $\mathrm{KVA} = \frac{h}{1+h}\mathrm{EC}$ implies $\mathrm{KVA} \le \mathrm{EC}$, i.e. $\max(\mathrm{EC}, \mathrm{KVA}) = \mathrm{EC}$. This and Lemma I.1 yield

$$\mathrm{FVA} = \mathbb{E}\Big[\gamma\Big(\sum_b (P_b - \mathrm{VM}_b) - \mathrm{CA} - \mathrm{EC}\Big)^+\Big] = \gamma\Big(\sum_b (P_b - \mathrm{VM}_b) - \mathrm{CA} - \mathrm{EC}\Big)^+.$$

As $\mathrm{CA} = \mathrm{CCVA} + \mathrm{CMVA} + \mathrm{BCVA} + \mathrm{BMVA} + \mathrm{FVA}$, this is an FVA semilinear equation, which, as γ is nonnegative, is equivalent to the FVA formula

$$\mathrm{FVA} = \frac{\gamma}{1+\gamma}\Big(\sum_b (P_b - \mathrm{VM}_b) - (\mathrm{CCVA} + \mathrm{CMVA} + \mathrm{BCVA} + \mathrm{BMVA}) - \mathrm{EC}\Big)^+.$$

Last, we have $\mathrm{EC} = \mathbb{ES}(J(\mathcal{C} + \mathcal{F} - \mathrm{CA}))$, where the identity $\mathcal{C} + \mathcal{F} - \mathrm{CA} = \mathcal{C} - \mathrm{CVA}$ and the formula for $J(\mathcal{C} - \mathrm{CVA})$ in (17) are obtained by substituting the already derived XVA formulas in (7) and (9). □

Table 19: XVA explicit formulas in the single-CCP static setup (with $\mathcal{C}$, $\mathcal{F}$ and Λ given by Lemma 2).

XVA	Explicit formula
CCVA	$\mathbb{E}\left[\sum_c (1-J_c)(\mathcal{P}_c - P_c - \mathrm{RIM}_c)^+ + \mu\Lambda\right]$
CMVA	$\widetilde{\gamma}\big(\mathrm{IM} + \overline{\mathrm{IM}} + \mathrm{DFC}\big)$
BCVA	$\mathbb{E}\left[\sum_b (1-J_b)(\mathcal{P}_b - \mathrm{VM}_b - \mathrm{RIM}_b)^+\right]$
BMVA	$\widetilde{\gamma}\sum_b \mathrm{PIM}_b$
EC	$\mathbb{ES}\big(J(\mathcal{C} - \mathrm{CVA})\big)$
FVA	$\frac{\gamma}{1+\gamma}\left(\sum_b (P_b - \mathrm{VM}_b) - (\mathrm{CCVA} + \mathrm{CMVA} + \mathrm{BCVA} + \mathrm{BMVA}) - \mathrm{EC}\right)^+$
KVA	$\frac{h}{1+h}\mathrm{EC}$

C Extension to Several CCPs or CCP Services

In the realistic case where the reference bank is a clearing member of several services of one or several CCPs, we index all the CCP-related quantities in the above by an additional index ccp in a finite set disjoint from $I \cup C \cup B$. Then, with CA = CCVA + CMVA + BCVA + BMVA + FVA as before:

Theorem 2. *On the bank survival event $\{J = 1\}$, the counterparty default loss $\mathcal{C}$ across several counterparties and several CCPs is given by*

$$\mathcal{C} = \sum_{ccp,c}(1-J_c)(\mathcal{P}_c^{ccp} - P_c^{ccp} - \mathrm{RIM}_c^{ccp})^+ + \sum_{ccp}\mu^{ccp}\Lambda^{ccp} + \sum_b (1-J_b)(\mathcal{P}_b - \mathrm{VM}_b - \mathrm{RIM}_b)^+, \tag{18}$$

where

$$\Lambda^{ccp} = \sum_i (1-J_i)\Big((\mathcal{P}_i^{ccp} - P_i^{ccp} - \mathrm{PIM}_i^{ccp})^+ + (\overline{\mathcal{P}}_i^{ccp} - \overline{P}_i^{ccp} - \overline{\mathrm{PIM}}_i^{ccp})^+ - \mathrm{DFC}_i^{ccp}\Big)^+; \tag{19}$$

the funding expenses $\mathcal{F}$ across several CCPs and several counterparties are given by

$$\begin{aligned}\mathcal{F} = \widetilde{\gamma}\sum_{ccp}\Big(\mathrm{PIM}^{ccp} + \overline{\mathrm{PIM}}^{ccp} + \mathrm{DFC}^{ccp}\Big) + \widetilde{\gamma}\sum_b \mathrm{PIM}_b \\ + \gamma\Big(\sum_b (P_b - \mathrm{VM}_b) - \mathrm{CA} - \max(\mathrm{EC}, \mathrm{KVA})\Big)^+.\end{aligned} \tag{20}$$

The XVA definitions and explicit formulas are the same as in Tables 18 and

19, but for $\mathcal{C}$ and $\mathcal{F}$ generalized as above, and for CCVA and CMVA as detailed in Tables 20 and 21. Moreover, the bank's trading loss $\mathcal{L}$ satisfies

$$J\mathcal{L} = J(\mathcal{C} + \mathcal{F} - \mathrm{CA}) = J(\mathcal{C} - \mathrm{CVA}) =$$
$$= J\Big(\sum_{ccp,c}(1-J_c)(\mathcal{P}_c^{ccp} - P_c^{ccp} - \mathrm{RIM}_c^{ccp})^+ + \sum_{ccp}\mu^{ccp}\Lambda^{ccp} - \mathrm{CCVA}$$
$$+ \sum_b (1-J_b)(\mathcal{P}_b - \mathrm{VM}_b - \mathrm{RIM}_b)^+ - \mathrm{BCVA}\Big).$$

Table 20: CCVA and CMVA definitions with several CCPs (for $\mathcal{C}$ and $\mathcal{F}$ is now given by Theorem 2, as also Λ^{ccp}).

XVA	Expression	Full name and description
CCVA	$\mathbb{E}^*\Big[J\sum_{ccp,c}(1-J_c)(\mathcal{P}_c^{ccp} - P_c^{ccp} - \mathrm{RIM}_c^{ccp})^+ + \sum_{ccp}\mu^{ccp}\Lambda^{ccp} + (1-J)\mathrm{CCVA}\Big]$	credit valuation adjustment for clearing activity exposures
CMVA	$\mathbb{E}^*\Big[J\sum_{ccp}\tilde{\gamma}\big(\mathrm{PIM}^{ccp} + \overline{\mathrm{PIM}}^{ccp} + \mathrm{DFC}^{ccp}\big) + (1-J)\mathrm{CMVA}\Big]$	margin valuation adjustment for clearing activity exposures

Table 21: CCVA and CMVA explicit formulas with several CCPs (for $\mathcal{C}$ and $\mathcal{F}$ is now given by Theorem 2, as also Λ^{ccp}).

XVA	Explicit formula
CCVA	$\mathbb{E}\Big[\sum_{ccp,c}(1-J_c)(\mathcal{P}_c^{ccp} - P_c^{ccp} - \mathrm{RIM}_c^{ccp})^+ + \sum_{ccp}\mu^{ccp}\Lambda^{ccp}\Big]$
CMVA	$\sum_{ccp}\tilde{\gamma}\big(\mathrm{PIM}^{ccp} + \overline{\mathrm{PIM}}^{ccp} + \mathrm{DFC}^{ccp}\big)$

Proof. In the case of several CCP services, the second line in (6) must be turned into $\sum_{ccp}\big(\mathrm{PIM}^{ccp} + \overline{\mathrm{PIM}}^{ccp} + \mathrm{DFC}^{ccp}\big) + \sum_b \mathrm{PIM}_b$; the terms in the first lines of (7) and (9) must be summed over the various CCP services in which the bank is involved as a clearing member. The rest of the analysis proceeds as before. □

§4 MARKET AND CREDIT MODEL

We introduce a market and credit model, written under $\mathbb{Q}^*$, with parameters that can capture dependence between portfolio changes, joint defaults, and possible adverse exacerbated changes of the portfolio due to their owner's default known as wrong-way risk (see Crépey, Bielecki and Brigo (2014), Crépey and Song (2017a, 2016)).

Let $\mathcal{T}$ denote the Student-t cdf with 3 degrees of freedom and Δ_l be positive liquidation period accounting for the time taken by the CCP to novate or liquidate defaulted portfolios (cf. §7 and Chapter IX). For any $j \in I \cup B \cup C$, we denote F_j the marginal cdf of a financial participant j's default time τ_j, $\Delta\mathcal{P}_j := \mathcal{P}_j - P_j$, nom_j a (signed) nominal of the portfolio of j, σ_j its annualized relative volatility, and we define

$$\begin{cases} \tau_j = F_j^{-1}\left(\mathcal{T}(X_j)\right), \\ \\ \dfrac{\Delta\mathcal{P}_j}{\mathrm{nom}_j \sigma_j \sqrt{\Delta_l}} = Y_j, \end{cases} \tag{21}$$

where

$$\begin{cases} X_j = \sqrt{\mathcal{K}_j}\,(\,\sqrt{\rho^{cr}}\vartheta - \sqrt{\rho_j^{wwr}}\chi_j + \sqrt{1-\rho^{cr}-\rho_j^{wwr}}\vartheta_j\,) \\ \\ Y_j = \sqrt{\rho^{mkt}}\varepsilon + \sqrt{\mathcal{K}_j}\sqrt{\rho_j^{wwr}}\chi_j + \sqrt{1-\rho^{mkt}-\rho_j^{wwr}}\varepsilon_j. \end{cases} \tag{22}$$

Here ρ^{cr}, ρ^{mkt} and the ρ_j^{wwr} are positive credit/credit, market/market and credit/ market correlation coefficients, while ϑ and the ϑ_j, χ_j are i.i.d. random variables following Gaussian distributions; ε and the ε_j are i.i.d. random variables following Student-t distributions with degree of freedom 3; the $3/\mathcal{K}_j$ are i.i.d. random variables following independent chi-squared distributions of degree of freedom 3, so that the $\chi_j\sqrt{\frac{3}{3/\mathcal{K}_j}} = \sqrt{\mathcal{K}_j}\chi_j$ follow Student-t distributions of degree of freedom 3 (Johnson, Kotz and Balakrishnan, 1995, Eqn. (28.1), page 362). From a financial interpretation viewpoint:

- ϑ represents the common systemic factor for default times across members,
- ε represents the common systemic factor for portfolio variations across members,
- $\sqrt{\mathcal{K}_j}\chi_j$ is the common factor co-driving portfolio variations and default time of market participant j,
- ϑ_j is the idiosyncratic factor for market participant j's default time,
- ε_j is the idiosyncratic factor for market participant j's portfolio variations.

Remark 3. In practice, margin computations rely on historical estimates based on several market-stressed periods. Our approach, instead, aims at reflecting extreme market shocks with heavy-tailed Student-t distributions of degree of freedom 3, and volatility level within a reasonable range of [20%, 40%]. Our static formulation depicts stationary increments of the defaulted portfolios' value over the liquidation period.

In view of the above, the setup is well defined if and only if

$$\rho_j^{wwr} < \min\left(1 - \rho^{cr}, 1 - \rho^{mkt}\right); \tag{23}$$

otherwise, the model for both default time and portfolio variation factors is undefined due to their idiosyncratic coefficient term $\sqrt{1 - \rho^{cr} - \rho_j^{wwr}}$ and $\sqrt{1 - \rho^{mkt} - \rho_j^{wwr}}$. We also discard the limit cases where $\rho_j^{wwr} = 1 - \rho^{cr}$ or $\rho_j^{wwr} = 1 - \rho^{mkt}$ as they lead to a zero contribution of the idiosyncratic factors, which would be unrealistic. The "minus" sign in front of the common credit-market factor ($-\sqrt{\rho_j^{wwr}}$) in the first line of (22) ensures that the corresponding common factor accelerates defaults, whilst increasing the market exposure due to the $\sqrt{\rho_j^{wwr}}$ factor in the second line.

Remark 4. Andersen and Dickinson (2019) introduce a dynamic model locally elliptical in the sense of elliptical on each next time step given the information at the beginning of the time step. Under simplifying assumptions including their equation (32) and assumption 2, they obtain (in our notation) analytical CMVA and approximate CCVA formulas. In their case, defaults are triggered by $\Delta\mathcal{P}_j$ (in our notation) falling below a Merton-like threshold. In our static setup with extra latent variables for defaults, we do not have such analytical formulas. However, Monte Carlo simulation is quite efficient and required anyway for stress test exercises that aim at identifying scenarios leading to extreme losses with adequate description such as the identification of defaulted members and their corresponding losses.

Hereafter, we describe two possible applications of our XVA framework which will be illustrated by numerical case studies in the above model. To these ends, two networks will be defined to serve the numerical illustrations, one rather educational on the use of the XVA metrics and the other one reflecting the more realistic situation depicted by Figure 40 page 218.

In the numerical applications that follow, all members play in turn the role of the reference bank in the theoretical XVA framework of §2-§3. The CVA and KVA computations require a Monte Carlo routine run under $\mathbb{Q}^\star$ in combination with a rejection technique in order to yield simulations under the survival measures associated with different clearing members. For obtaining confidence intervals regarding the expected shortfalls that are embedded in the KVA computations, the simulations are split into several batches, from which the mean of the (partial) EC estimates yields the final EC estimate, while their standard deviation is used to define a confidence interval.

§5 CASE STUDIES SETUP

In the examples that follow, market participants are identified by a number and can then be included in one of several of the considered CCPs. We restrict ourselves to cleared client trades, so that the nonvanishing XVA metrics reduce to the CCVA, the CMVA, and the KVA.

A Single CCP Setup and Initial XVA Costs

We consider a single CCP service with 20 members labeled by $i \in 0..n = 19$, only trading for cleared clients (i.e. without bilateral or centrally cleared proprietary trading). Each member faces one client. The ensuing financial network is depicted by Figure 42.

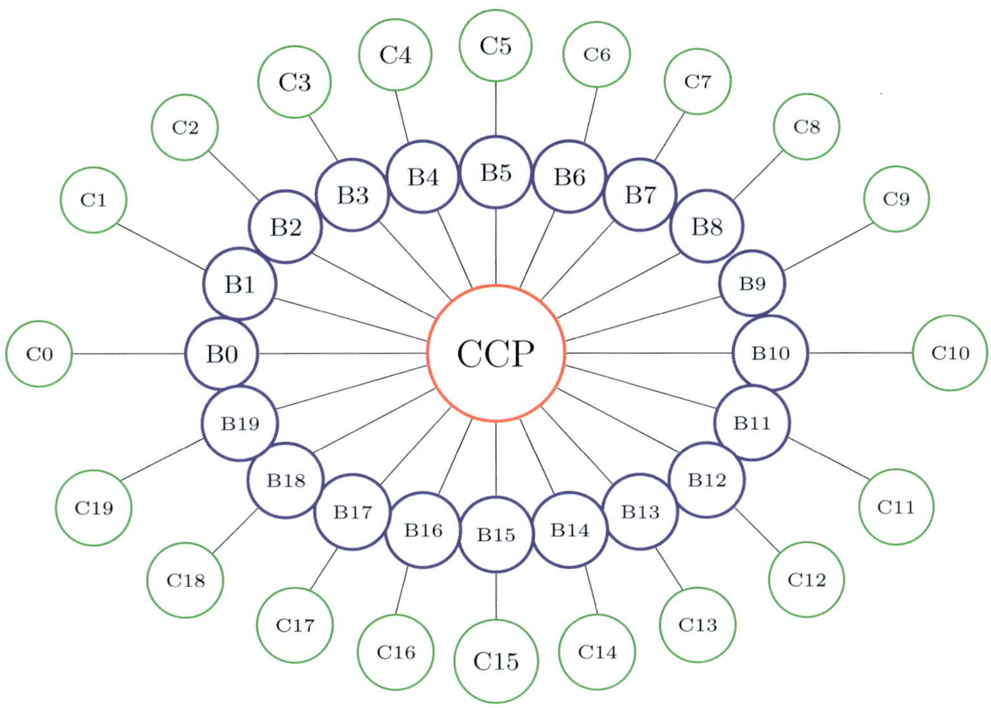

Figure 42: Financial network composed of 1 CCP, its 20 member banks (labeled by B) and one client per member (labeled by C).

All clients are assumed to be risk-free. For any member i, its posted IM to the CCP is calculated based on the idea of a VM call not fulfilled over a time period $\Delta_s < \Delta_l$ at a confidence level $\alpha \in (1/2, 1)$, using a VaR metric applied to the non-coverage of VM call also taken with cdf $\mathcal{T}$, assessed under the reference member survival measure as

$$\text{PIM}_i = \mathbb{V}\text{a}\mathbb{R}\Big(\text{nom}_i \sigma_i \sqrt{\Delta_s} \mathcal{T}\Big) = |\text{nom}_i|\, \sigma_i \sqrt{\Delta_s} \mathcal{T}^{-1}(\alpha). \tag{24}$$

The (funded) default fund is calculated at the CCP level as

$$\mathrm{Cover2} = \mathrm{SLOIM}_{(0)} + \mathrm{SLOIM}_{(1)}, \tag{25}$$

for the two largest stressed losses over IM (SLOIM_i) among members, identified with subscripts (0) and (1), where SLOIM is calculated as the value-at-risk $\mathbb{V}\mathrm{a}\mathbb{R}'$ at a confidence level $\alpha' > \alpha$ of the loss over IM, i.e.

$$\mathrm{SLOIM}_i = \mathbb{V}\mathrm{a}\mathbb{R}'\Big(\mathrm{nom}_i\sigma_i\sqrt{\Delta_s}\mathcal{T} - \mathrm{PIM}_i\Big) = |\mathrm{nom}_i|\,\sigma_i\sqrt{\Delta_s}\Big(S_\nu{}^{-1}(\alpha') - \mathcal{T}^{-1}(\alpha)\Big). \tag{26}$$

The total amount (25) is then allocated between the clearing members to define their (funded) default fund contributions as $\mathrm{DFC}_i = \dfrac{\mathrm{SLOIM}_i}{\sum_j \mathrm{SLOIM}_j}\mathrm{Cover2}$.

The nom_j of other clearing members are not observable by a given one. However, following Murphy and Nahai-Williamson (2014) and Lipton (2018), $|\mathrm{nom}|_{(i)}$ denoting the i-th largest absolute nominal amount for $i \in 0\,..\,n = 19$, a parameterization of the form

$$|\mathrm{nom}|_{(i)} = \beta e^{-\beta'(i+1)}\,,\ \beta, \beta' > 0 \tag{27}$$

can be fit to the total default fund held by the CCP[7] and the sum of its five largest default fund contributions[8], made public each quarter for most of the CCPs. The parameters β and β' inferred from the default fund data are used to depict a similar pattern on the absolute nominal sizes (as if the funded default fund amounts were proportional to the portfolio sizes). The participants and portfolios parameter inputs are detailed in Table 22, where (from now on) id is the identifier of the CM, DP stands for the one-year probability of default of the member expressed in percentage points, size represents the overall portfolio size of the member detained within the CCP, and vol is the annual volatility used for the portfolio variations.

The portfolios listed in Table 22 relate to the members towards the CCP (which are mirroring the ones between the members and their clients). The signs of the nominals are distributed so that $\sum_j \mathrm{nom}_j = 0$, consistent with the CCP clearing condition (first identity in Assumption 1, here without proprietary trades).

[7]item referenced as 4.3.15 in Bank of International Settlements and OICV-IOSCO (2015), *Value of pre-funded default resources (excluding initial and retained variation margin) held for each clearing service in total, post-haircut,* in the quantitative disclosure documents.

[8]item referenced as 18.4.2 in Bank of International Settlements and OICV-IOSCO (2015):*For each segregated default fund with 25 or more members; Percentage of participants contributions to the default fund contributed by the largest five clearing members in aggregate.*; or item referenced 18.4.1 for CCP services with less than 25 members

Table 22: CPP member characteristics and portfolio parameters, ordered by decreasing member size.

CM id	0	1	2	3	4	5	6	7	8	9
DP (bps)	50	60	70	80	90	200	190	180	170	160
size	-242	184	139	105	-80	-61	-46	35	26	-20
vol (%)	20	21	22	23	24	25	26	27	28	29

CM id	10	11	12	13	14	15	16	17	18	19
DP (bps)	150	140	130	120	110	100	90	80	70	60
size	-15	-11	-9	-6	5	-4	-3	2	2	-1
vol (%)	30	31	32	33	34	35	36	37	38	39

The parameters of the XVA costs calculations are summarized in the Table 23. Note that most OTC derivatives have a maturity of less than 5 years (Bank for International Settlements, 2022, Graphs A.2–4), hence the chosen period length of $T = 5$ years covers the bulk (if not the final maturity) of most realistic CCP portfolios. The confidence level of 97% used for SLOIM in DFC calibration allows for a ratio of funded default fund over the initial margin of about 10% in our calculations, a ratio (of this level or less) often observed in practice.

Table 23: XVAs calculation configuration.

One-period length T	5 years
Liquidation period at default Δ_l	5 days
Portfolio variations correlation ρ^{mkt}	30%
Credit factors correlation ρ^{cr}	20%
Correlation between credit factors and portfolio variations ρ^{wwr}	20%
IM covering period (MPoR) Δ_s	2 days
IM quantile level	95%
Funding blending ratio $\widetilde{\gamma}/\gamma$	25%
SLOIM calculation for DFC Cover-2	$\mathbb{V}\mathrm{a}\mathbb{R}$ 97%
DFC allocation rule	$\propto$ SLOIM
Unfunded default fund Λ allocation rule ($\ni \mu$)	$\propto \mathrm{DFC}_i$
Quantile level used for clearing members EC calculation	99.75%
Hurdle rate h used for KVA computations	10.0%
Number of Monte Carlo simulation (for CCVA and KVA computations)	10M
Number of batches (for KVA computations)	100

For each member, the CCVA, CMVA and KVA costs are calculated and reported in Table 24. For KVA, two calculations have been performed, one based on $\mathbb{ES}$ at the 99th percentile level and another one based on the 99.75th

Table 24: Initial XVA costs: estimates, [value-at-risk underlying the KVA estimate] and (95% confidence level errors).

CM id	CMVA	CCVA	KVA (99%)	KVA (99.75%)
0	0.0687	0.1413 (0.3%)	0.2631 [0.1327] (0.7%)	0.8512 [0.5129] (1.0%)
1	0.0656	0.1154 (0.3%)	0.3088 [0.1373] (0.7%)	0.7818 [0.453] (1.1%)
2	0.0604	0.0932 (0.3%)	0.2505 [0.1103] (0.7%)	0.6606 [0.3855] (1.1%)
3	0.0544	0.0750 (0.4%)	0.2023 [0.0886] (0.8%)	0.546 [0.3172] (1.3%)
4	0.0485	0.0604 (0.4%)	0.1562 [0.0646] (0.8%)	0.445 [0.2572] (1.2%)
5	0.0834	0.0440 (0.4%)	0.1303 [0.0540] (0.8%)	0.3578 [0.2108] (1.1%)
6	0.0623	0.0358 (0.4%)	0.1044 [0.0447] (0.8%)	0.2843 [0.1663] (1.1%)
7	0.0467	0.0291 (0.4%)	0.0839 [0.0371] (0.7%)	0.228 [0.1321] (1.4%)
8	0.0341	0.0230 (0.4%)	0.0645 [0.0286] (0.7%)	0.1768 [0.1019] (1.4%)
9	0.0256	0.0187 (0.4%)	0.0514 [0.0227] (0.7%)	0.1425 [0.0811] (1.4%)
10	0.0187	0.0147 (0.4%)	0.0398 [0.0177] (0.7%)	0.1124 [0.0629] (1.8%)
11	0.0132	0.0113 (0.4%)	0.0301 [0.0134] (0.7%)	0.086 [0.0476] (1.5%)
12	0.0104	0.0097 (0.4%)	0.0254 [0.0113] (0.7%)	0.0731 [0.0401] (1.6%)
13	0.0066	0.0068 (0.6%)	0.0174 [0.0077] (0.7%)	0.0536 [0.0276] (2.8%)
14	0.0052	0.0059 (0.6%)	0.0149 [0.0066] (0.7%)	0.0472 [0.0237] (2.8%)
15	0.0039	0.0050 (0.8%)	0.0122 [0.0054] (0.8%)	0.042 [0.0195] (3.4%)
16	0.0027	0.0039 (0.9%)	0.0094 [0.0042] (0.7%)	0.0341 [0.0151] (3.7%)
17	0.0017	0.0028 (1.1%)	0.0064 [0.0029] (0.7%)	0.0251 [0.0103] (4.3%)
18	0.0015	0.0029 (1.4%)	0.0066 [0.0029] (0.7%)	0.0275 [0.0106] (4.9%)
19	0.0007	0.0016 (2.5%)	0.0033 [0.0015] (0.7%)	0.0194 [0.0055] (8.6%)

percentile level. The amount in the square bracket is the corresponding quantile level from which the average is calculated, and numbers in parenthesis represent the 95% confidence interval in relative difference from calculated metric for both CCVA and KVA. All the XVA numbers decrease with the member size. To assess the average behavior w.r.t. ρ^{cr}, ρ^{mkt} and ρ^{wwr} of the CCVA and KVA, we vary these correlations between 5% and 95% with 5% step and display in Figures 43 and 44 the corresponding metrics, aggregated over all clearing members successively, considered as the reference bank. For such tests, the default correlation ρ^{cr} and ρ^{mkt} are both set to 4% when they are not changed between 5% and 95%. This is to allow for runs with $\rho_i^{wwr} = 95\%$ satisfying the condition (23).

The KVA depicts an increase w.r.t. ρ^{cr} but also w.r.t. ρ^{wwr} and very limited change w.r.t. ρ^{mkt}. The correlation ρ^{wwr} has more impact than ρ^{cr} and ρ^{mkt} (right panels in Figures 43 and 44). As seen in the left panels of Figures 43 and 44, there are very marginal changes for the aggregated CCVA w.r.t. ρ^{cr} and ρ^{mkt}, but a significant positive impact of ρ^{wwr}. This is understandable for the sensitivity to ρ^{cr} and ρ^{mkt} as, apart for modulations of the measure with respect to which each individual CCVA is assessed, the CCVA aggregated over clearing members is essentially an expectation of the CCP loss Λ (cf. the first line of Table 19). The individual CCVAs (as per the first line of Table 19) of each clearing member, however, may depend on ρ^{cr} and ρ^{mkt} (on top of ρ^{wwr}) in a strong and nontrivial manner, via the allocation coefficient μ.

Remark 5. The monotonicity features that are apparent numerically in the Figures 43 and 44 are established mathematically in Bastide and Crépey (2025). These results support the use of such setups for computing credit provisions and economic capital or for conducting stress test exercises and risk management analysis.

B Two CCPs Network Setup

We now consider the case of Figure 40 page 218, where there are two CCPs with some common members and stress test is considered from the perspective of one of these common members. The motivation for this case is to provide a realistic example mimicking, in a simplified way, the trading firm Ronin Capital, which had memberships in both FICC GSD segments, hereafter denominated by CCP1, and the CME Futures segment, hereafter denominated by CCP2, in March 2020. It is well known that a VaR-type risk measure is not sub-additive, in particular for credit portfolios as illustrated in Example 5.4 in Acerbi and Tasche (2002) and Example 2.25 in McNeil et al. (2015) for a portfolio of defaultable bonds, so that for a common member adding VaR estimates of trading losses on two CCPs separately can lead to underestimated levels with respect to the actual VaR of the global exposition of the member. As such, stress test exercises accounting for common memberships could reveal a larger value-at-risk compared to the exercise where stress tests are conducted separately on each CCP.

To perform the analysis, the following setup is considered:

- all members have only clearing client positions[9], with 123 members on CCP1 and 56 members on CCP2, out of which 24 are common to both CCPs,
- all clients are assumed default free,
- both CCPs use the configuration as per Table 23,
- the sizes of the positions are assumed to be exponentially distributed in the sense that from the most exposed member to the least one, the absolute value of positions decreases exponentially with the form in (27) as depicted by Figures 45 and 46 respectively,

[9] Ronin Capital had in fact only a house account and was thus not clearing any client position.

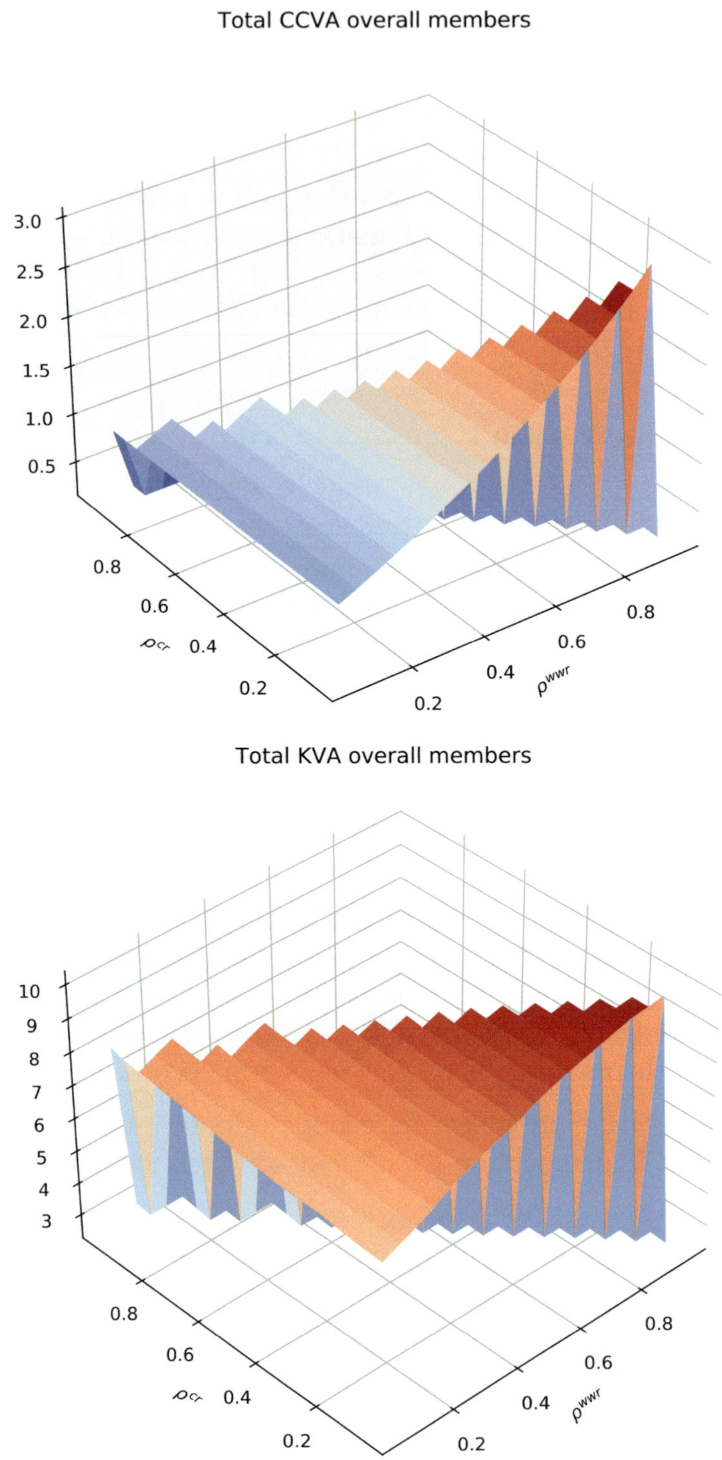

Figure 43: CCVA and KVA w.r.t. credit factors correlation and credit and portfolio variation factors correlation.

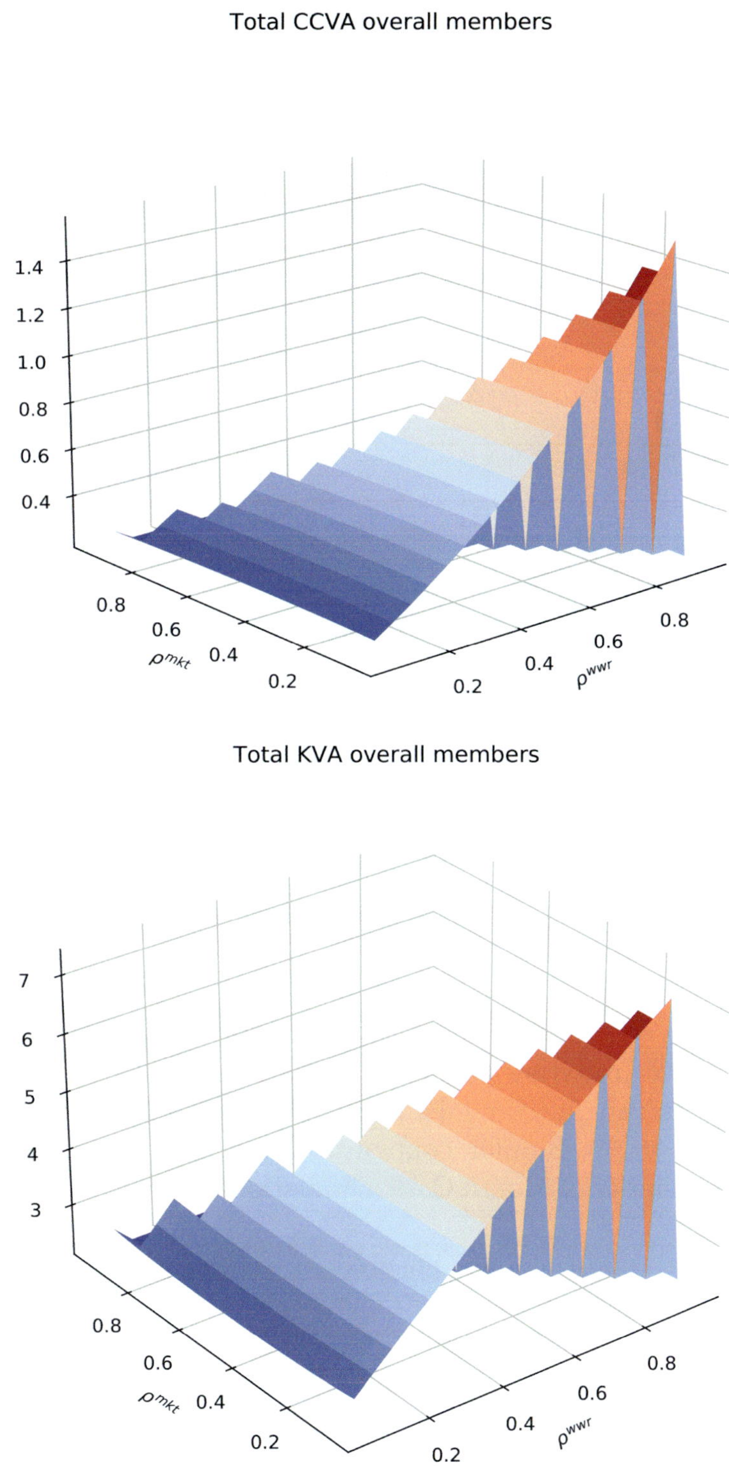

Figure 44: CCVA and KVA w.r.t. market factors correlation and credit, and portfolio variation factors correlation.

- the proportion of the default fund detained by the 5 biggest members is 25% for CCP1 and 61% for CCP2[10],
- the size of the default fund of CCP1 is assumed to be twice that one of the default fund of CCP2.

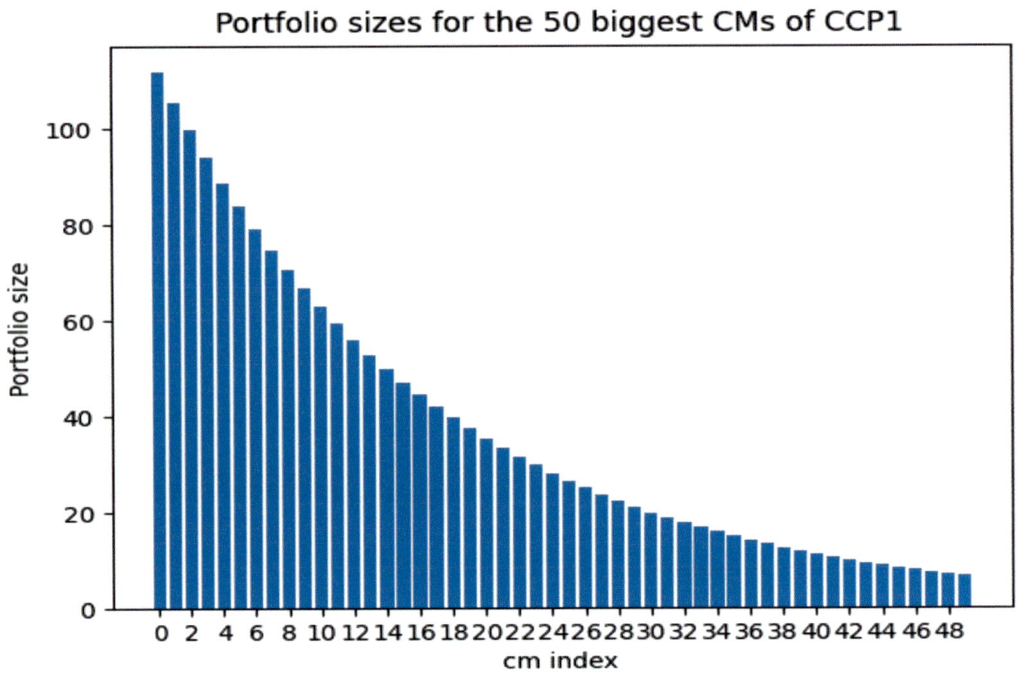

Figure 45: Decreasing absolute nom_i per member for CCP1

All data used are either public sources or have been anonymized, with default intensities ranging from 10 bps to 400 bps and portfolio volatilities ranging from 20 to 30. Similar configuration as given in Table 23 is used, apart from the number of Monte Carlo simulations reduced to 2M for memory capacity reasons.

The clearing conditions are ensured by setting the sum of the portfolio sizes nom_i to zero on each CCP. The situation of member 3, exposed to both CCPs, as the defaulting member, corresponds roughly to the situation of Ronin Capital in 2018. In particular, an annual probability of default of 0.1% corresponds roughly to a BBB rating, which was assigned to Ronin Capital in 2018 for its issuances see https://www.spglobal.com/marketintelligence/en/news-insights/blog/banking-essentials-newsletter-july-edition-2.

[10]taken from the quantitative disclosure of both CCPs as of the third quarter of 2020.

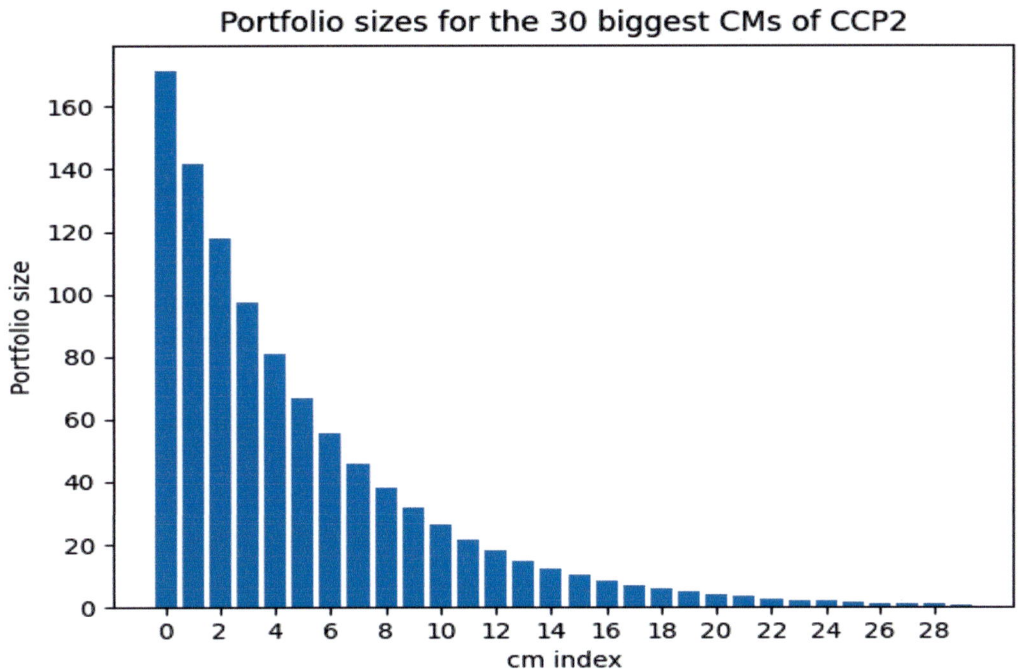

Figure 46: Decreasing the absolute nom_i per member for CCP2

§6 STRESS TEST EXERCISES

As outlined in the capital requirements regulation detailed in the European Parliament (2013) article 290, financial institutions must conduct regular stress test (ST) exercises of their credit and counterparty exposures. Paragraph 8 of this article also stipulates the reverse stress test (RST, see dedicated definition page 12 in Board of Governors of the Federal Reserve System (2012) and articles 97, 98 page 37 in European Central Bank (2018) for official regulatory definitions) requirement to

> [...] identify extreme, but plausible, scenarios that could result in significant adverse outcomes.

This is complemented by article 302 on the exposure of financial institutions may have towards CCPs:

> Institutions shall assess, through appropriate scenario analysis and stress testing, whether the level of own funds held against exposures to a CCP, including potential future credit exposures, exposures from default fund contributions and, where the institution is acting as a clearing member, exposures resulting from contractual

arrangements as laid down in Article 304, adequately relate to the inherent risks of those exposures.

In practice, stress test exercises aim at assessing the capacity of financial institutions to absorb financial and economic shocks. In regular exercises, such as the ones conducted by the European Banking Authority, the shocks are usually considered under so-called *central* and *baseline* macro-economic scenarios corresponding to a median quantile and *adverse* scenario usually taken as a 90th percentile reflecting severe yet plausible scenario that can occur once every 10 years[11]. Additionally, extreme scenarios can be considered for measuring the capital adequacy[12] for absorbing extremely severe losses around a confidence level at 99.9%. From a clearing member perspective, this requires to have the capacity of scanning certain points of its trading loss distribution. In our framework, this boils down to identifying particular levels of the distribution of the bank shareholders' trading loss $J(\mathcal{C} - \text{CCVA} - \text{BCVA})$ of the reference clearing member bank, where the different terms are detailed in Theorem 2.

The other type of stress test exercises, referenced as reverse stress test[13] (Bellini et al., 2021), consists in identifying the probability of reaching a given loss level as well as describing the scenario configuration, such as projected defaults and loss magnitude leading to such loss levels. The distribution must span a sufficiently large spectrum of losses, including the ones targeted by the exercise, but it also has to be sufficiently rich numerically to allow identifying combinations of events leading to such losses.

Confidence intervals of corresponding extreme scenario probabilities should complement the analysis to ensure the reliability of the used model and numerical methods.

Regulators have the ability to challenge financial institutions on these elements and demand for improvements. This may entail reassessment of the Pillar 2 guidance, additional capital requirement set in the annual Supervisory Review and Evaluation Process reported by Banks, cf. European Central Bank (2021) for a brief definition and use and the Basel Committee on Banking Supervision (2019) for more extensive details as well as Board of Governors of the Federal Reserve System (2020) for similar requirements.

[11]such confidence levels are suggested, by the Federal Reserve, outlining page 10 in Board of Governors of the Federal Reserve System (2013) the various recession periods of the United States listed in their Table 1 page 14; the 2021 instructions in European Bank Authority (2021) also indicate page 72 that stressed market risk factors are based on shocks specified in European Systemic Risk Board (2013), citing Dordu et al. (2017, p. 29), with the US recessions periods as stressful economic episodes.

[12]cf. Paragraph 5.1 page 11 from Basel Committee on Banking Supervision (2005).

[13]see also the dedicated definition on page 12 in the Board of Governors of the Federal Reserve System (2012) and articles 97, 98 page 37 in European Central Bank (2018) for official regulatory definitions.

A Scenarios Identification for Reverse Stress Test

We now briefly explain how to identify and exploit the scenarios leading to contribute the most to economic capital (cf. X.§3.A in a multi-period setup). We denote by M the number of Monte Carlo scenarios for which $J = 1$, i.e. survival of the reference bank. Its trading loss $\mathcal{C} - \mathrm{CVA}$ for a simulation m is given by $\mathcal{C}^m - \mathrm{CVA}$, where $m \in 1 \cdots M$ enumerates the simulated scenarios for which the reference member bank ends up in survival state.

To get an estimate of the economic capital based on expected shortfall, relying on Acerbi and Tasche (2002, Definition 2.6 and Proposition 4.1), we calculate, for a high confidence level $\alpha \in (\frac{1}{2}, 1)$ and $[x]$ denoting the integer part of any real x,

$$\widehat{\mathbb{ES}}\,(\mathcal{C} - \mathrm{CVA}) := \frac{1}{M - [\alpha M]} \sum_{m=[\alpha M]+1}^{M} (\mathcal{C}^{(m)} - \mathrm{CVA}), \tag{28}$$

where the $\mathcal{C}^{(m)} - \mathrm{CVA}$'s are the simulated trading losses of the reference bank ranked in increasing order.

To obtain the contribution of any simulated scenario m (with $\mathcal{C}^m \geq \mathcal{C}^{([\alpha M])}$) to the economic capital estimated by (28), we compute

$$\begin{aligned}&\widehat{\mathbb{ES}}^{-m}(\mathcal{C} - \mathrm{CVA})\\ &:= \frac{1}{M - 1 - [\alpha(M-1)]}\Big(\big(M - [\alpha M]\big)\widehat{\mathbb{ES}}\,(\mathcal{C} - \mathrm{CVA}) - (\mathcal{C}^m - \mathrm{CVA})\Big).\end{aligned} \tag{29}$$

The contribution $\delta_m\widehat{\mathbb{ES}}\,(\mathcal{C} - \mathrm{CVA})$ of scenario m to $\widehat{\mathbb{ES}}\,(\mathcal{C} - \mathrm{CVA})$ is then given by

$$\delta_m\widehat{\mathbb{ES}}\,(\mathcal{C} - \mathrm{CVA}) = \widehat{\mathbb{ES}}\,(\mathcal{C} - \mathrm{CVA}) - \widehat{\mathbb{ES}}^{-m}(\mathcal{C} - \mathrm{CVA})\,. \tag{30}$$

B Numerical Results

To illustrate the various flavors of stress test exercises that can be conducted by a CCP member, we report numerical results for the two network examples introduced in §5.

B.1 Single CCP Results

We start with a reverse stress test exercise on the example covered by Table 22 page 235. For this first illustration, a specific extreme loss is targeted and the corresponding probability of loss reaching at least such a target level is estimated.

Table 25: Stress test extreme quantile, 1.5× ST extreme quantile and reverse stress test probability to breach 1.5 times the 99.9th quantile loss level, for each member, based on 10M simulations (in parentheses: corresponding 95% confidence intervals).

CM id	99.9%	1.5 × 99.9%	RST scenario probability
0	4.7322 (-0.9%, 1.0%)	7.0984	0.0395% (3.2%)
1	5.9679 (-0.9%, 0.9%)	8.9518	0.0382% (3.4%)
2	4.8541 (-0.9%, 1.0%)	7.2812	0.0386% (3.2%)
3	3.9294 (-0.9%, 0.9%)	5.8942	0.0380% (3.7%)
4	3.0931 (-0.9%, 1.1%)	4.6396	0.0388% (3.6%)
5	2.5726 (-0.9%, 1.0%)	3.8590	0.0385% (3.7%)
6	2.0423 (-1.1%, 1.0%)	3.0635	0.0382% (3.6%)
7	1.6224 (-0.9%, 1.0%)	2.4336	0.0380% (3.4%)
8	1.2488 (-0.9%, 1.0%)	1.8732	0.0379% (3.5%)
9	0.9938 (-0.9%, 0.9%)	1.4906	0.0379% (3.6%)
10	0.7682 (-1.0%, 0.9%)	1.1523	0.0379% (3.6%)
11	0.5811 (-1.0%, 0.9%)	0.8716	0.0384% (3.6%)
12	0.4900 (-0.9%, 0.9%)	0.7351	0.0381% (3.6%)
13	0.3353 (-0.9%, 1.0%)	0.5029	0.0379% (3.8%)
14	0.2862 (-0.9%, 0.9%)	0.4294	0.0383% (3.7%)
15	0.2356 (-0.9%, 1.0%)	0.3534	0.0382% (3.8%)
16	0.1809 (-0.8%, 1.0%)	0.2714	0.0382% (3.7%)
17	0.1235 (-0.8%, 1.0%)	0.1852	0.0381% (3.7%)
18	0.1264 (-0.9%, 1.0%)	0.1896	0.0379% (3.6%)
19	0.0645 (-0.9%, 1.0%)	0.0967	0.0382% (3.7%)

In Table 25, we report, for the example summarized in Table 22, the 99.9th percentile trading loss levels, referenced as extreme quantile, with corresponding (asymmetric) confidence intervals based on the approach proposed in Meeker et al. (2017, Subsection G.2). This is done for every clearing member (CM) successively playing the role of the reference bank in the setup of §2-§3. We also compute the probabilities of reaching a loss equal to 1.5 times the obtained extreme quantile level, referenced as RST scenario, with corresponding confidence levels. The calculation of these confidence intervals for the probability of being above a quantile relies on the same numerical approach based on batches used for KVA calculations. This batch approach leads to reasonably tight confidence intervals for the RST scenario probabilities.

Our description of the scenarios leading to such losses includes the identified defaulted members, the generated losses and the allocated loss coefficient

Table 26: Economic capital 20 worst scenarios details for CM1 in decreasing order of total loss. μ indicates the loss coefficient allocated to CM1. n is the number of defaults within the scenario.

Rank	Loss	n	μ	Defaulters	Losses triggered by defaulters
1	10.04	1	0.21	cm0	518
2	8.92	2	0.21	cm0, 19	459.21, 0.18
3	8.67	5	0.23	cm0, 8, 10, 15, 18	411.41, 0, 1.83, 0, 0
4	8.30	3	0.19	cm6, 8, 12	489.31, 0.17, 0.79
5	7.97	1	0.21	cm0	411.19
6	7.81	1	0.21	cm0	403.05
7	7.48	4	0.20	cm3, 7, 12, 17	409.76, 0, 0, 0
8	7.35	2	0.24	cm0, 4	339.10, 0
9	6.96	2	0.23	cm0, 7	340.43, 0
10	6.94	3	0.29	cm0, 2, 4	258.70, 0, 1.01
11	6.91	3	0.24	cm0, 6, 8	318.76, 0, 0
12	6.84	1	0.21	cm0	352.86
13	6.60	2	0.22	cm0, 11	334.17, 0.10
14	6.59	2	0.21	cm2, 5	344.84, 0
15	6.47	1	0.21	cm0	333.87
16	6.27	1	0.21	cm0	323.79
17	5.97	1	0.21	cm0	308.19
18	5.75	2	0.22	cm0, 8	284.80, 0
19	5.70	5	0.28	cm0, 5, 6, 7, 8	222.31, 0, 0.63, 0, 0
20	5.50	3	0.23	cm0, 7, 9	258.51, 1.12, 0.29

of the reference clearing member (CM1 in this example, the second biggest member). Its theoretical number of scenarios above the RST loss level is the number of MC simulations of 10M multiplied by CM1's survival probability over 5 years and by CM1's RST loss level probability estimated in Table 25 as 0.0382%, so 3709 scenarios. This is, of course, far too many to report. Table 26 provides the description of the 20 worst ones, contributing the most to the EC estimation for CM1. A focus on these 20 worst scenarios already illustrates the type of information that can be exploited for such exercises. Most of these scenarios are driven by significant losses stemming from CM0's default, reflecting the highly concentrated position of CM0. We observe that several scenarios illustrate the cases where more than one clearing member default, such as the 2nd to 4th scenarios, for which not only CM0 generates most of the loss, but other defaulting members generate significant losses (yet of less magnitude compared to CM0).

From the CM1 viewpoint (i.e. with CM1 in the role of the reference clearing member), 17 scenarios entail significant losses over the collateral posted by the defaulted CM0 (positive first entries in the last column of Table 26). CM0 bears a very large concentrated position compared to other members. Even if CM0 has more IM and default fund requirements than others, this is still not enough: this example highlights that the employed default fund allocation rules in this example dilute the default fund collateral requirements for concentrated positions. It also illustrates that scenarios with multiple defaults do not necessarily lead to extreme losses, due to the fact that members with medium or small positions have large default fund contributions stemming from others' concentrated positions.

B.2 Two CCPs Results

We then consider the example illustrated by Figure 40 page 218, where projected loss levels for specific confidence levels are indicated for the members with common memberships on the two CCPs.

In Table 27, we report, for the example illustrated by Figure 40 with 2 CCPs, the trading loss levels (value-at-risks) at confidence levels 90% and 99.9%, for the 24 common members on the two CCPs. The corresponding numbers in the case where the two CCPs would be considered separately is reported in columns g and i. For members with very low size on one of the two CCPs compared to the other, considering the common memberships or not does not affect the loss estimates, as expected (a CCP with a very low size compared to another has marginal impact). For other members, however, at 90% confidence levels, the value-at-risks are significantly higher (compare columns g and h in Table 27) when the common membership are considered compared to the stand-alone value-at-risks calculation conducted on each CCP and summed, especially for the first ten members. On the contrary, at the confidence level 99.9%, the sum of the stand-alone value-at-risks is well above the value-at-risk when common memberships are taken into consideration (columns i and j in Table 27). These two situations illustrate that a regulator and the board (top management) of the bank could equally and rightfully criticise a simplistic standalone approach, too aggressive in some cases (making it unacceptable by the regulator) and over-conservative in others (making it unacceptable by the board).

§7 OPTIMIZING THE PORTING OF DEFAULTED CLIENT PORTFOLIOS

In case a clearing member defaults, the CCP tentatively novates part of the CCP portfolio of the defaulted member through auctions among the surviving clearing members (Default Risk Management Working Group, 2016; Bank for International Settlements, 2019a), and it liquidates the residual on

Table 27: Quantile loss levels (confidence errors) for 90% and 99.9% confidence levels across members for the example with 2 CCPs and 155 members including 24 common members. Legend for column headers: a. Member Id, b. DP (%), c. Size on CCP1, d. Volatility on CCP1, e. Size on CCP2, f. Volatility on CCP2, g. 90th Perc. stand-alone, h. 90th Perc., i. 99.9th Perc. stand-alone, j. 99.9th Perc.

a	b	c	d	e	f	g	h	i	j
3	0.1	19.90	21	-97.48	23	0.0761 (-1.5%, 1.7%)	0.0788 (-1.2%, 1.2%)	3.4710 (-1.8%, 2.0%)	3.1277 (-2.0%, 1.8%)
4	0.1	80.79	24	-18.79	22	0.0679 (-1.6%, 1.5%)	0.0724 (-1.1%, 1.0%)	2.9993 (-2.1%, 2.1%)	2.6544 (-1.9%, 2.2%)
9	3.1	-31.58	29	17.74	23	0.0477 (-1.4%, 1.5%)	0.0619 (-0.8%, 0.8%)	1.8253 (-2.4%, 2.2%)	1.4621 (-2.0%, 2.2%)
12	0.1	17.97	21	-16.75	24	0.0340 (-1.0%, 1.0%)	0.0418 (-0.7%, 0.7%)	0.8706 (-1.9%, 2.1%)	0.5990 (-1.8%, 2.1%)
13	0.1	-14.90	22	15.81	25	0.0328 (-1.0%, 1.1%)	0.0398 (-0.6%, 0.6%)	0.7960 (-1.9%, 2.2%)	0.5434 (-2.3%, 2.0%)
14	0.2	12.34	23	-14.93	26	0.0313 (-1.0%, 1.0%)	0.0379 (-0.6%, 0.6%)	0.7302 (-1.8%, 2.2%)	0.4916 (-1.9%, 2.1%)
15	0.1	-10.23	24	14.09	27	0.0300 (-0.9%, 1.0%)	0.0358 (-0.6%, 0.6%)	0.6683 (-1.9%, 2.2%)	0.4469 (-1.6%, 1.7%)
17	0.3	-7.03	26	-13.30	28	0.0283 (-0.9%, 0.9%)	0.0330 (-0.6%, 0.6%)	0.5767 (-1.8%, 2.0%)	0.3891 (-1.7%, 1.7%)
19	0.2	-4.83	28	12.56	29	0.0267 (-0.9%, 0.9%)	0.0300 (-0.7%, 0.6%)	0.5018 (-1.8%, 2.3%)	0.3495 (-1.5%, 2.0%)
22	3.9	2.75	20	-11.86	30	0.0289 (-0.9%, 0.8%)	0.0303 (-0.7%, 0.7%)	0.4437 (-1.8%, 2.2%)	0.3593 (-2.1%, 1.7%)
26	0.1	1.30	24	11.20	20	0.0152 (-0.8%, 0.8%)	0.0156 (-0.6%, 0.7%)	0.2345 (-1.8%, 2.3%)	0.1930 (-1.9%, 2.4%)
27	0.1	1.07	25	-10.57	21	0.0150 (-0.8%, 0.8%)	0.0152 (-0.7%, 0.7%)	0.2265 (-1.8%, 2.3%)	0.1903 (-1.7%, 2.3%)
28	1.5	0.89	26	9.98	22	0.0157 (-0.7%, 0.8%)	0.0159 (-0.7%, 0.7%)	0.2328 (-1.8%, 2.0%)	0.1997 (-1.8%, 2.1%)
31	0.1	-0.51	29	-9.42	23	0.0143 (-0.7%, 0.8%)	0.0144 (-0.7%, 0.7%)	0.2049 (-1.7%, 2.4%)	0.1845 (-1.9%, 2.4%)
34	0.1	0.29	21	8.89	24	0.0139 (-0.7%, 0.8%)	0.0139 (-0.7%, 0.7%)	0.1902 (-1.8%, 2.4%)	0.1819 (-2.0%, 2.3%)
35	0.1	-0.24	22	-8.40	25	0.0137 (-0.7%, 0.7%)	0.0137 (-0.7%, 0.7%)	0.1859 (-1.7%, 2.4%)	0.1787 (-1.9%, 2.3%)
36	0.1	0.20	23	7.93	26	0.0134 (-0.7%, 0.7%)	0.0134 (-0.7%, 0.7%)	0.1823 (-1.9%, 2.3%)	0.1758 (-1.9%, 2.3%)
39	0.1	-0.11	26	-7.48	27	0.0131 (-0.7%, 0.7%)	0.0131 (-0.7%, 0.7%)	0.1760 (-1.8%, 2.3%)	0.1719 (-1.8%, 2.3%)
40	0.5	0.09	27	7.07	28	0.0131 (-0.7%, 0.8%)	0.0131 (-0.7%, 0.7%)	0.1747 (-1.7%, 2.1%)	0.1713 (-1.8%, 2.0%)
44	0.1	0.04	20	-6.67	29	0.0125 (-0.7%, 0.7%)	0.0125 (-0.7%, 0.7%)	0.1656 (-1.7%, 2.4%)	0.1645 (-1.7%, 2.4%)
49	0.1	-0.02	25	6.30	30	0.0122 (-0.7%, 0.7%)	0.0122 (-0.7%, 0.7%)	0.1614 (-1.7%, 2.4%)	0.1607 (-1.7%, 2.4%)
50	0.1	0.01	26	-5.95	20	0.0077 (-0.7%, 0.7%)	0.0077 (-0.7%, 0.7%)	0.1016 (-1.7%, 2.4%)	0.1012 (-1.7%, 2.4%)
51	0.1	-0.01	27	5.61	21	0.0076 (-0.7%, 0.7%)	0.0076 (-0.7%, 0.7%)	0.1009 (-1.9%, 2.3%)	0.1005 (-1.9%, 2.2%)
55	0.1	-0.01	20	-5.30	22	0.0075 (-0.7%, 0.7%)	0.0075 (-0.7%, 0.7%)	0.0996 (-1.8%, 2.4%)	0.0993 (-1.8%, 2.4%)

the market. A natural baseline is that the CCP novates (auctions among surviving members) client trades and their mirroring client account positions, collectively dubbed client positions for brevity hereafter, whereas the house account positions are liquidated.

The liquidation side of the procedure cannot be handled in our modeling setup, which does not embed the fundamentals of price formation (our counterparty-risk-free prices P_i are assumed to be exogenously given): see Chapter IX for that matter. On the other hand, an XVA-based procedure can be used for rendering what would be the output of an idealized, efficient auction, assuming a large number of clearing members (Oleschak et al., 2019, Section 3.3). Namely, supposing that the reference clearing member, labeled by 0 in §2-§3, defaults at time 0, i.e. just after that all portfolios have been settled, for each surviving member CM$*$ successively envisioned as a potential taker of the defaulted (client) positions of CM0, one computes, for each surviving member (CM$*$ included), the incremental (Δ) XVAs of porting the defaulted positions to CM$*$ (note that all members are impacted by the recalibration of their DFC by the CCP, whereas only the member taker of the portfolio sees in addition its IM adjusted). The corresponding incremental XVA numbers are then summed over metrics and survivors, resulting in the funds transfer price (FTP$*$) of porting defaulted client positions to CM$*$. The effective taker is then the surviving member for which the ensuing FTP$*$ is the smallest (or, indifferently in case of multiple minima, one of the minimizing FTP$*$ members). See Albanese, Chataigner and Crépey (2020, Section 5.2) for more details on such "XVA Pareto optimally driven" novation procedures.

In what follows, based on the example of Table 22 page 235 (which only involves client positions), we analyze from this perspective a first scenario of a single default on the CCP.

Taking the first case with a single default, we first assume the scenario whereby CM0 defaults at time 0. Table 28 summarizes the total ΔXVA$*$ aggregated over survivors, across members $*$ from 1 to 19, in increasing order of the FTP$*$ indicated in the last column. Based on the results of Table 28, CM1 appears to be the potential taker leading to the least overall FTP costs across all surviving members. This is understandable as this member's portfolio size (184 in Table 22) nets the most the defaulted member's portfolio size (-242), with volatility and credit default probability similar to (in particular, not significantly higher than) the ones of the defaulted member.

By offset we refer to risk reduction when taking over some additional position. The effect of correlation is such that an opposite sign in portfolio size does not imply an equal offset of the risk of the aggregated positions. For instance, even with opposite sizes and same volatilities but for $\rho^{mkt} \in (0, 1/2)$, the member ends up with more risk. As CM1 concentrates more risks due in particular to non-perfect offset between its prior positions and the defaulting

Table 28: Total ΔXVA∗ aggregated over survivors corresponding to the different surviving CM∗, i.e. for ∗ other than 0, assuming an instant default of CM0 at time 0. In parenthesis, the contributions to ΔXVA∗ of CM∗ itself.

Surv. member ∗	Total ΔCMVA∗	Total ΔCCVA∗	Total ΔKVA∗	Total FTP∗
1	0.0768 (0.0295)	-0.0428 (-0.0045)	-0.4165 (-0.1065)	-0.3826 (-0.0815)
19	0.1298 (0.0818)	-0.0278 (0.0371)	-0.4284 (0.2790)	-0.3264 (0.3979)
2	0.0921 (0.0428)	-0.0392 (0.0018)	-0.3530 (-0.0391)	-0.3001 (0.0055)
18	0.1417 (0.0939)	-0.0230 (0.0363)	-0.3898 (0.2737)	-0.2710 (0.4038)
3	0.1054 (0.0576)	-0.0323 (0.0080)	-0.2953 (0.0281)	-0.2221 (0.0937)
17	0.1549 (0.1070)	-0.0178 (0.0363)	-0.3284 (0.2759)	-0.1913 (0.4192)
4	0.1525 (0.1022)	-0.0513 (0.0158)	-0.2429 (0.1050)	-0.1416 (0.2230)
16	0.1688 (0.1208)	-0.0145 (0.0359)	-0.2720 (0.2760)	-0.1177 (0.4327)
15	0.1814 (0.1334)	-0.0109 (0.0354)	-0.2262 (0.2739)	-0.0557 (0.4426)
14	0.1903 (0.1426)	-0.0048 (0.0345)	-0.1907 (0.2665)	-0.0052 (0.4437)
13	0.2061 (0.1582)	-0.0030 (0.0344)	-0.1204 (0.2704)	0.0827 (0.4630)
12	0.2171 (0.1692)	-0.0021 (0.0332)	-0.0905 (0.2635)	0.1245 (0.4659)
11	0.2285 (0.1807)	0.0013 (0.0325)	-0.0344 (0.2622)	0.1954 (0.4753)
10	0.2385 (0.1908)	0.0010 (0.0309)	-0.0151 (0.2520)	0.2244 (0.4738)
9	0.2478 (0.2003)	0.0013 (0.0292)	0.0375 (0.2421)	0.2865 (0.4715)
8	0.2340 (0.1881)	0.0149 (0.0262)	0.0515 (0.2057)	0.3004 (0.4200)
7	0.2327 (0.1876)	0.0182 (0.0233)	0.1009 (0.1825)	0.3518 (0.3934)
6	0.2687 (0.2225)	-0.0074 (0.0213)	0.1390 (0.1876)	0.4004 (0.4314)
5	0.2728 (0.2274)	-0.0146 (0.0171)	0.1746 (0.1491)	0.4327 (0.3936)

Table 29: Standard deviation across surviving members ∗ of the ΔXVA∗ for the example with 1 CCP and 20 members, assuming an instant default of CM0 at time 0.

ΔCMVA	ΔCCVA	ΔKVA
0.0593	0.0189	0.1964

one, there is an increase of its IM reflected through an increase of CMVA. But the new risk of CM1 is less than the sum of the former risks of CM0 and CM1, hence the CCVA aggregated across surviving members is reduced. This only happens when CM1 takes over the defaulting portfolio; other potential takers leading to an overall increase of the CCVA. As for the KVA, there is a reduction effect for CM1 when CM1 is the taker (see the term in parentheses in Table 28), and an overall decrease in the total KVA (aggregated over all surviving members), which is also the case for most members. Having CM1 as the taker allows to obtain the most significant decrease in ΔKVA.

As expected, among the three XVA components, KVA is the main determinant of the optimal taker: see Table 29.

Once the CCP has re-allocated all defaulted client positions, the resulting financial network formerly depicted in Figure 42 page 233 becomes the network with 19 members shown in Figure 47. The thick lines represent the new portfolio exposures for CM1 and the pale dashed lines show the defaulted CM0 positions.

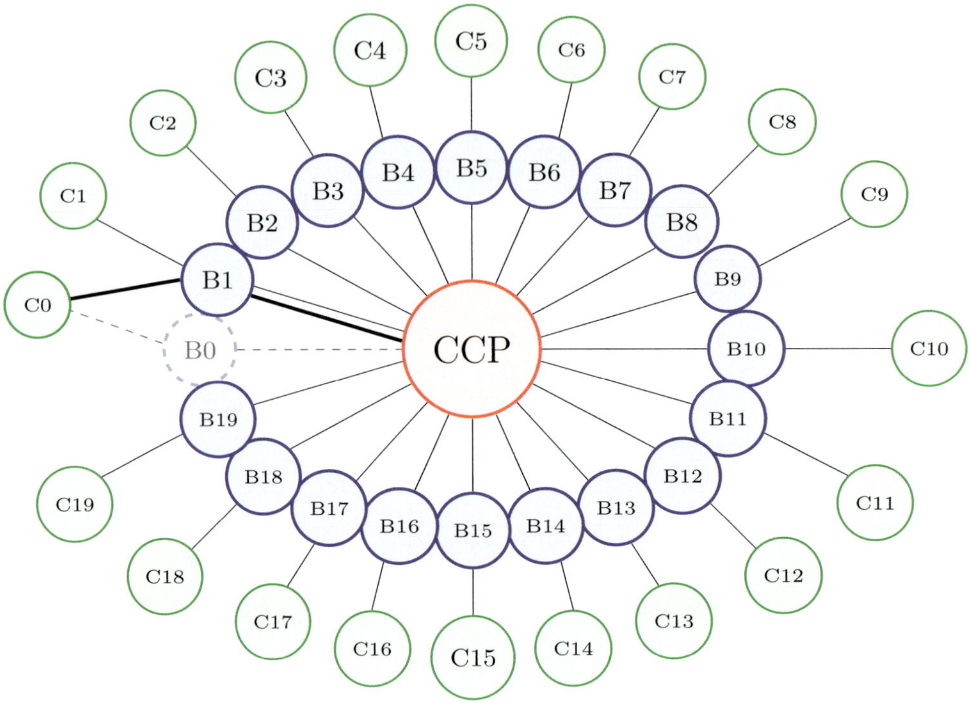

Figure 47: The 1-CCP, a former 20-member financial network with 19 members post CM0 default. Defaulted CM0, labeled "B0" in the presented network, is represented as pale dashed node with pale dashed links to reflect former exposures to its client and toward the CCP. The optimal porting of the CM0 portfolio with CM1, labeled "B1", is outlined with bold links to reflect the new exposures for CM1.

§8 CONCLUSION

In this chapter we have proposed a fully integrated risk management framework that can be used for stress test analysis, including reverse stress test in line with regulatory requirements, or for optimizing the porting of defaulted portfolios, in a setup encompassing all the trades (bilateral as centrally cleared and their hedges) of a reference bank. The framework includes dependence between financial participants portfolios, joint defaults, and a configurable wrong-way risk feature. This is done in a numerically tractable static setup, although already quite demanding on large financial networks . The dynamic extension considered in III.§5 is only workable at a much higher computational burden, using the simulation and learning techniques of the Numerical

Methods part of the book (see, however, Chapter X for a slightly simplified dynamic setup is amenable to simply nested Monte Carlo computations for the whole XVA suite).

A possible improvement would be to incorporate regulatory constraints such as minimum regulatory capital requirements and liquidity leverage ratios. More fundamentally, in this chapter, we tackle the derivatives risk problem from a pure counterparty credit risk viewpoint: if members, clients and counterparties are all default free, then in view of Theorem 2 all considered XVAs are zero, so that our setup becomes trivial. Another dimension to the problem is liquidity (Amini, Filipovic and Minca, 2020; Faruqui, Huang and Takats, 2018). Depending on the considered applications, credit or liquidity is the main force at hand. The following chapter is an attempt to integrate liquidity to our setup.

CHAPTER IX

Resolving a Clearing Member's Default: A Radner Equilibrium Approach

For vanilla derivatives, which form the core of investment banks' hedging portfolios, central clearing via central counterparties (CCPs) has become the prevailing standard. A fundamental role of a CCP is to ensure an efficient and effective resolution process in the event of a clearing member's default. Upon such a default, the CCP is tasked with hedging and subsequently auctioning or liquidating the defaulted positions. While the counterparty credit risk associated with auctioning has been examined in VIII.§7 from a valuation adjustments (XVA) perspective, the present chapter also considers the costs associated with hedging or liquidation. This is achieved by contrasting pre- and post-default market equilibria through a Radner equilibrium framework applied to portfolio allocation and price discovery in different default resolution scenarios. We establish the unique existence of Radner equilibria and provide both analytical and numerical solutions within elliptically distributed market settings. These insights equip CCPs with a rational basis for determining whether to hedge, auction, or liquidate defaulted portfolios in specific markets. Clearing members can leverage these findings to conduct what-if analyses, addressing inquiries from senior management and regulatory bodies.

§1 INTRODUCTION

Exchanges serve as platforms where financial actors can find counterparties for their transactions. In the context of derivatives, the exchange is supported by a central counterparty (CCP). As explained in Subsection III.§2.A and Chapter VIII, a CCP is a financial institution that acts as an intermediary in derivatives and securities markets, standing between two trading parties to reduce counterparty credit risk. When two parties engage in a transaction, the CCP

DOI: 10.1201/9781003614944-14

becomes the buyer to the seller and the seller to the buyer, effectively guaranteeing the trade's completion even if one party defaults. This mechanism ensures that the failure of one market participant does not directly impact others. The CCP manages this risk by netting positions across multiple trades, requiring members to post collateral (margin) and maintaining a default fund composed of contributions from its clearing members. The primary challenge for a CCP is to effectively handle the default of a clearing member by liquidating the member's portfolio within a few days of the default. In the event of a default, the CCP follows a specific procedure: first, it hedges and auctions large portions of the portfolio to the remaining clearing members; second, it sells the remaining part of the portfolio on the exchange. Any shortfall arising during this fire sale procedure is covered sequentially by the collateral of the defaulted member, the private fund of the CCP (skin in the game), and finally the default fund (Gregory, 2014; Cont, 2017; Biais, Heider and Hoerova, 2016; Oleschak et al., 2019).

From this simplified description, it is evident that the impact of fire selling the portfolio in a short amount of time through auction and liquidation should be reflected in the amount of collateral required from participants and in their contributions to the default fund. Several studies have addressed the auctioning aspect (Oleschak et al., 2019; Ferrara and Li, 2020), as well as the liquidation process (Avellaneda and Cont, 2013; Vicente, Cerezetti, Faria, Iwashita and Pereira, 2015). While the auctioning impact is typically derived endogenously from a game-theoretical perspective (see also VIII.§7 for an XVA take on that matter), the liquidation impact has mainly been examined from an exogenous financial engineering perspective, considering liquidation scheduling.

In this chapter, using a one-period Radner equilibrium approach, we measure the equilibrium price movement between pre- and post-default resulting from the sudden fire sale of a defaulted member's portfolio. We provide existence and uniqueness for the specific objective measures of the participants in the CCP in the general case. We further derive explicit solutions for such equilibrium in an elliptical market case, which allows us to compare quantitatively the impact of liquidation on the market. This perspective is however complicated by the following factors: first, a single CCP can clear on different exchanges, while multiple CCPs can operate on the same exchange; second, although a CCP is theoretically market-neutral by not holding any positions, in practice, it may engage in proprietary trading during fire sale situations by hedging the portfolio of the defaulted member. We compare different default resolution scenarios in our study, whereby the CCP liquidates the portfolio on the defaulter's exchange or on an external one, or/and engages in proprietary trading by hedging the position. To measure the impact of these different liquidation procedures, we assess the resulting price impact from the equilibrium in each scenario.

As surveyed in Magill and Shafer (1991) or Radner (1982), the use of competitive Walrasian equilibrium in economics to derive asset prices and optimal allocation has a long history. To derive equilibrium prices specific to CCPs, we specify to our one-period setup the methods developed in a dynamic setting by Cheridito, Horst, Kupper and Pirvu (2015). Although the mathematical theory is well established, the study of equilibrium disruption resulting from a default is an original approach. We further provide precise quantification of this price impact by deriving explicit solutions to the equilibrium in a realistic context, using monetary risk measures that are established regulatory instruments to determine the required amount of collateral.

A Outline and Standing Notation

§2 provides our Radner equilibrium market model. §3 introduces the related comparative statics approach for the analysis of the market costs of hedging or liquidating a defaulted clearing member portfolio, either on the exchange of the CCP of the defaulter, or on an external exchange. §4 computes the ensuing costs in the case of expected shortfall risk measures. §5 analyzes the additional XVA impact of counterparty credit risk, which is assessed numerically in Section §6. Section §7 concludes.

Given vectors $x \in \mathbb{R}^m$ and $y \in \mathbb{R}^n$ (understood as column matrices), $x^\top$ is the transpose of x and (x, y) is the vector of $\mathbb{R}^{m+n}$ formed by stacking x above y. We denote by $\mathcal{N}_n(\mu, \Gamma)$, the n-variate Gaussian distribution with mean μ and covariance matrix Γ; by $\mathcal{E}_n(\mu, \Gamma, \psi)$, the n-variate elliptical distribution with mean μ, covariance matrix Γ, and characteristic generator function ψ, and by $\mathcal{T}_n(\mu, \Gamma, \nu)$, the n-variate Student t-distribution of degree of freedom ν with mean μ and covariance matrix Γ. Throughout this chapter, $(\Omega, \mathcal{A}, \mathbb{Q})$ denotes a fixed probability space, with expectation, variance and covariance operators $\mathbb{E}$, $\mathbb{V}$ar and $\mathbb{C}$ov; $\mathbb{L}^0$ and $\mathbb{L}^1$ respectively denote the space of all the measurable and integrable random variables (identified in the $\mathbb{Q}$ almost sure sense), $\mathbb{X}$ is a linear subspace of $\mathbb{L}^1$ containing the constants. **Bold letters** refer to the solution of a Radner equilibrium.

B Convex Analysis Reminder

For a function $f\colon \mathbb{R}^m \to \mathbb{R}$, its directional derivative at x in the direction of y is defined as

$$D_y f(x) = \lim_{\epsilon \searrow 0} \frac{f(x + \epsilon y) - f(x)}{\epsilon}; \tag{1}$$

for f convex, a point $y \in \mathbb{R}^m$ is said to be a subgradient of f at x, denoted $y \in \partial f(x)$, if

$$f(z) \geq f(x) + y^\top (z - x), \quad z \in \mathbb{R}^m; \tag{2}$$

the convex conjugate f^* of f is defined as

$$f^*(y) = \sup \left\{ y^\top x - f(x);\, x \in \mathbb{R}^m \right\}, \quad y \in \mathbb{R}^m. \tag{3}$$

Let $f_e \colon \mathbb{R}^m \to \mathbb{R}$ be convex functions, for e in a finite set E. The inf-convolution f of the f_e is defined as

$$f(x) = \Big\{ \sum_{e \in E} f_e(x_e);\, \sum_{e \in E} x_e = x \Big\}, \quad x \in \mathbb{R}^m. \tag{4}$$

A function $\rho : \mathbb{X} \to \mathbb{R}$ is called a risk measure if it is
• monotonous: $\rho(\mathcal{X}) \leq \rho(\mathcal{Y})$ for all $\mathcal{X}$ and $\mathcal{Y}$ in $\mathcal{X}$ such that $\mathcal{X} \leq \mathcal{Y}$;
• convex: $\rho(\lambda\mathcal{X} + (1-\lambda)\mathcal{Y}) \leq \lambda\rho(\mathcal{X}) + (1-\lambda)\rho(\mathcal{Y})$ for all $\mathcal{X}$ and $\mathcal{Y}$ in $\mathcal{X}$, and $0 \leq \lambda \leq 1$;
• translation equivariant: $\rho(\mathcal{X} - m) = \rho(\mathcal{X}) - m$ for all $\mathcal{X}$ in $\mathcal{X}$ and m in $\mathbb{R}$.
Such a risk measure is called normalized if $\rho(0) = 0$ and law invariant if $\rho(\mathcal{X}) = \rho(\mathcal{Y})$ whenever X and $\mathcal{Y}$ have the same law. A coherent risk measure is a normalized risk measure satisfying the positive homogeneity condition $\rho(\lambda\mathcal{X}) = \lambda\rho(\mathcal{X})$, $\lambda \in \mathbb{R}_+$. Expected shortfall is an example of a coherent and law-invariant risk measure, whereas value-at-risk is not a risk measure in the above sense, fault of convexity (see Prerequisites in the preliminaries of the book).

A risk measure ρ is *differentiable* at $\mathcal{X} \in \mathbb{X}$ if there exists a random variable $\mathcal{W} \in \mathbb{X}^*$ (the dual space of $\mathbb{X}$) such that (cf. (1))

$$\lim_{\epsilon \searrow 0} \frac{\rho(\mathcal{X} + \epsilon\mathcal{Y}) - \rho(\mathcal{X})}{\epsilon} = \mathbb{E}[\mathcal{Y}\mathcal{W}],\ \mathcal{Y} \in \mathbb{X}; \tag{5}$$

in this case, we write $\mathcal{W} = \nabla\rho(\mathcal{X})$.

§2 RADNER EQUILIBRIUM MARKET MODEL

We consider a one-period model of a financial exchange, with trading participants indexed by a finite set E. A unitary position in each of the assets traded on the exchange pays a vector of random payoffs $\mathcal{P} \in \mathbb{R}^m$ (with $m \geq 1$) at the terminal time T. We assume that each participant e in E is endowed with a real-valued random receivable $\mathcal{R}_e$, i.e. $\mathcal{R}_e \geq 0$ means a cash flow promised to the participant e. Unhedged market risk can generate significant regulatory capital requirements, rendering the contractual commitments non-viable for the trading participant (Bank for International Settlements, 2019b, Section MAR23, pages 64 and 93). Accordingly, for each $e \in E$, the participant e

hedges $\mathcal{R}_e$ by entering a portfolio $\mathbf{q}_e \in \mathbb{R}^m$ of traded assets. The corresponding market loss of member e is

$$-\mathcal{R}_e + \mathbf{q}_e^\top(\mathbf{p} - \mathcal{P}), \tag{6}$$

where $\mathbf{p} \in \mathbb{R}^m$ is the vector of prices of the traded assets at the initial time: see Table 30.

For the monetary valuation of the risk of the participant e, we consider a normalized law invariant risk measure ρ_e. We want to determine the portfolios $\mathbf{q}_e$ and the prices $\mathbf{p}$ endogenously as a Radner equilibrium driven by the offer and demand of all the participants to the exchange:

Definition 1. A matrix of positions $(\mathbf{q}_e)_{e\in E}$ and a price vector $\mathbf{p} \in \mathbb{R}^m$ form a Radner equilibrium on E if
- (optimality condition relative to each market participant $e \in E$)

$$\rho_e(-\mathcal{R}_e + \mathbf{q}_e^\top(\mathbf{p} - \mathcal{P})) \le \rho_e(-\mathcal{R}_e + q_e^\top(\mathbf{p} - \mathcal{P})),\ q_e \in \mathbb{R}^m, \tag{7}$$

- (zero clearing condition)

$$\sum_{e\in E} \mathbf{q}_e = 0. \tag{8}$$

Table 30: Main notation relative to an exchange E.

$\mathbf{p}$	equilibrium prices of the traded (hedging) assets
$\mathbf{q}_e$	equilibrium positions of participant e in the traded (hedging) assets
$\mathcal{P}$	a vector of random payoff at terminal time T
$\mathcal{R}_e$	receivable to be hedged by the participant $e \in E$
ρ_e	risk measure of the participant e

Remark 1. Since $\mathcal{R}_e$ is assumed to be exogenously given, the price of the corresponding receivable to the participant e is not part of the equilibrium. We say nothing on this price (assumed exogenously given and in fact implicitly part of $\mathcal{R}_e$ itself in our setup), nor on the way it could be impacted (in our setup it is simply not) by the instant default of a participant to the exchange.

A Generic Results

For each trading participant e in E, we consider the convex function $r_e\colon \mathbb{R}^m \to \mathbb{R}$ defined as

$$r_e(q_e) = \rho_e(-\mathcal{R}_e - q_e^\top \mathcal{P}),\ q_e \in \mathbb{R}^m. \tag{9}$$

By translation equivariance of ρ_e, the member e optimality condition (8) can be rewritten as

$$r_e(q_e) \ge r_e(\mathbf{q}_e) + (-\mathbf{p})^\top(q_e - \mathbf{q}_e),\ q_e \in \mathbb{R}^m,$$

i.e., by (2),

$$-\mathbf{p} \in \partial r_e(\mathbf{q}_e). \tag{10}$$

By Rockafellar (1970, Theorem 23.5), this is in turn equivalent to

$$r_e(\mathbf{q}_e) = -\mathbf{q}_e^\top \mathbf{p} - r_e^*(-\mathbf{p}), \tag{11}$$

where r_e^* is the convex conjugate (3) of r_e. Note that

$$\begin{gathered} -q_e^\top p - r_e^*(-p) \le r_e(q_e),\ q_e, p \in \mathbb{R}^m, \\ \text{with equality if and only if } -p \in \partial r_e(q_e), \end{gathered} \tag{12}$$

by Rockafellar (1970, Theorem 23.5).

Lemma 1 and Theorems 1-2 below are variants, for a single-period model but with unbounded $(\mathcal{R}_e, \mathcal{P})$ (as we want to endorse elliptical factor models later below), of Cheridito, Horst, Kupper and Pirvu (2015, Theorems 1 and 2). Radner equilibria admit the following dual characterization in terms of the inf-convolution (4) r of the r_e.

Lemma 1. *A matrix of positions $(\mathbf{q}_e)_{e\in E}$ and a price vector $\mathbf{p} \in \mathbb{R}^m$ form a Radner equilibrium on E if and only if*
(i) $-\mathbf{p} \in \partial r(0)$,
(ii) $r(0) = \sum_{e\in E} r_e(\mathbf{q}_e)$, *and*
(iii) $\sum_{e\in E} \mathbf{q}_e = 0$.

Proof. Let $((\mathbf{q}_e)_{e\in E}, \mathbf{p})$ be a Radner equilibrium as per Definition 1. The zero clearing condition (8) yields (iii). By Rockafellar (1970, Theorem 16.4, page 145), the convex conjugate of the inf-convolution of proper convex functions is the sum of the corresponding conjugates, i.e.

$$r^*(-\mathbf{p}) = \sum_{e\in E} r_e^*(-\mathbf{p}).$$

Summing the expression (11) across all r_e and using (iii) gives

$$\sum_{e\in E} r_e(\mathbf{q}_e) = 0 - r^*(-\mathbf{p}) \le r(0),$$

where the inequality holds by definition (3) of the convex conjugate of r. By definition of $r(0)$, the above inequality becomes equality, i.e.

$$r(0) = -r^*(-\mathbf{p}) = \sum_{e\in E} r_e(\mathbf{q}_e).$$

Hence (ii) holds and so does also (i), in view of the equivalence between (10) and (11), here applied to r (instead of r_e there).

Conversely, suppose that $((\mathbf{q}_e)_{e\in E}, \mathbf{p})$ satisfies (i)–(iii). (iii) is the zero clearing condition (8), whereas (i) implies via (11) applied to r that

$$r(0) = -r^*(-\mathbf{p}) = \sum_{e\in E} \left(-\mathbf{q}_e^\top \mathbf{p} - r_e^*(-\mathbf{p})\right). \tag{13}$$

By (12) and (13), $-\mathbf{p} \notin \partial r_e(\mathbf{q}_e)$ for some $e \in E$ would imply that $r(0) < \sum_{e\in E} r_e(\mathbf{q}_e)$, contradicting (ii).

Hence (10), which is equivalent to the member e optimality condition (8), holds for each $e \in E$. □

Since the subgradient of a real-valued convex function is non-empty, Lemma 1 implies that a Radner equilibrium exists if and only if the inf-convolution r of the r_e is attained at 0. It also implies that, whenever a Radner equilibrium exists, the optimal price is unique if and only if r is differentiable at 0.

Theorem 1. *If ρ_e is sensitive to large losses, i.e.*[1] $\lim_{\lambda\to\infty} \rho_e(\lambda\mathcal{X}) = \infty$ *for all $\mathcal{X} \in \mathbb{X}$ such that $\mathbb{Q}[\mathcal{X} > 0] > 0$, e in E, then there exists a Radner equilibrium on E.*

Proof. Let Q be the set of vectors of positions $(q_e)_{e\in E}$ satisfying the zero clearing condition, i.e.

$$Q = \left\{q \in \mathbb{R}^{m|E|} \colon q^\top b^k = 0, k = 1, \dots, m\right\},$$

where b^k is a vector in $\mathbb{R}^{m|E|}$ such that, for all $a = 0, \dots, |E| - 1$, the $k + am$ entries of b^k equal 1 and all the other entries of b^k are 0 . Note that Q is a non-empty closed convex polyhedral subset of $\mathbb{R}^{m|E|}$. Let $\mu = \mathbb{E}[\mathcal{P}]$. By Dalang, Morton and Willinger (1990, Theorem 3.3), $q_e^\top(\mu - \mathcal{P}) = 0$ almost surely holds or $\mathbb{Q}[q_e^\top(\mu - \mathcal{P}) > 0] > 0$ holds, for any q_e in $\mathbb{R}^m$. The closed proper convex function

$$\mathbb{R}^{m|E|} \ni q = (q_1, \dots, q_{|E|}) \overset{\beta}{\mapsto} \sum_{e\in E} \rho_e(-\mathcal{R}_e + q_e^\top(\mu - \mathcal{P})) \in \mathbb{R} \tag{14}$$

is such that

$$\inf_{q\in Q} \beta(q) = \inf_{\sum_{e\in E} q_e = 0} \sum_{e\in E} r_e(q_e).$$

In view of the comment preceding the statement of the theorem, it suffices to show that β attains its minimum on Q. Let $S = \{q \in \mathbb{R}^{m|E|} \colon \beta(q) \le \beta(0)\}$, with recession cone $0^+S = \{y \in \mathbb{R}^{m|E|} \colon b + \lambda y \in S, \forall\lambda \ge 0, \forall b \in S\}$. Let $0^+\beta = \{y \in \mathbb{R}^{m|E|} \colon \beta 0^+(y) \le 0\}$ denote the recession cone of β, where $\beta 0^+$ is its recession function, i.e. the map defined on $\mathbb{R}^{m|E|}$ as $\beta 0^+(y) = \inf\left\{m \in \mathbb{R} \colon (y, m) \in 0^+\text{epi } \beta\right\}$, where epi β =

[1] see Cheridito et al. (2015, Section 2.3).

$\{(q,n) \in \mathbb{R}^{m|E|} \times \mathbb{R}\colon n \geq g(q)\}$. By Rockafellar (1970, Theorem 8.7, page 70), $0^+S = 0^+\beta$. Since S is a closed convex set containing the origin, Rockafellar (1970, Corollary 8.3.2, page 64) yields

$$0^+S = \{y \in \mathbb{R}^{m|E|}\colon \lambda y \in S \quad \forall \lambda > 0\}. \tag{15}$$

Let $y = (y_1, \ldots, y_{|E|}) \in \mathbb{R}^{m|E|} \setminus \{0\}$ (where each y_e is in $\mathbb{R}^m$). (i) If $\mathbb{Q}[y_e^\top(\mu - \mathcal{P}) > 0] > 0$ holds for some $e \in E$, then, by the sensitivity to large losses condition on ρ_e, $\beta(\lambda y)$ goes to infinity as λ goes to infinity, which implies that $y \notin 0^+S$. (ii) If, instead, $y_e^\top(\mu - \mathcal{P}) = 0$ holds for all $e \in E$, then $\pm y \in 0^+S$ hold by definitions (14) of β, S and (15) of 0^+S. In particular, for any $y \in \mathbb{R}^{m|E|} \setminus \{0\}$ such that $y \in 0^+\beta$, we have $-y \in 0^+\beta$. Hence, by Rockafellar (1970, Corollary 8.6.1, page 69), every direction of recession is a direction in which β is constant. Following Rockafellar (1970, Theorem 27.3, page 267), in either case (i) or (ii), β attains its minimum over Q. □

Regarding the uniqueness of an optimal solution:

Theorem 2. *Let $((\mathbf{q}_e)_{e\in E}, \mathbf{p})$ be a Radner equilibrium on E.*
(i) *If r_e is differentiable at $\mathbf{q}_e$ (cf. (5)) for some e in E, then $\mathbf{p}$ is unique.*
(ii) *For any $e \in E$, if r_e is differentiable and strictly convex on $\mathbb{R}^m$ then $\mathbf{q}_e$ is unique.*

Proof. Let $((\mathbf{q}_e)_{e\in E}, \mathbf{p})$ be optimal. If r_e is differentiable at $\mathbf{q}_e$, then Lemma 1(i) together with Rockafellar (1970, Theorems 23.2, p. 216 and 25.2, p. 244) yield

$$D_x r_e(\mathbf{q}_e) = x^\top(-\mathbf{p}) \leq D_x r(0), \tag{16}$$

where $D_x r_e(\mathbf{q}_e)$ is the directional derivative (1) of r_e at $\mathbf{q}_e$ along x. Take $\hat{q}_e = \mathbf{q}_e + \epsilon x$ and $\hat{q}_a = \mathbf{q}_a$ for all $a \neq e$. By definition (4) of the inf-convolution,

$$r(\epsilon x) \leq \sum_{a\in E} r_a(\hat{q}_a) = \sum_{a\neq e} r_a(\mathbf{q}_a) + r_e(\mathbf{q}_e + \epsilon x).$$

This together with Lemma 1(ii) yields

$$D_x r(0) = \lim_{\epsilon\searrow 0} \frac{r(\epsilon x) - r(0)}{\epsilon} \leq \lim_{\epsilon\searrow 0} \frac{\sum_{a\in E} r_a(\hat{q}_a) - r(0)}{\epsilon} = D_x r_e(\mathbf{q}_e) \leq D_x r(0),$$

where the second inequality is due to (16). Hence $x \mapsto D_x r(0)$ is linear. Thus, by Rockafellar (1970, Theorem 25.2, page 244), r is differentiable at 0 , i.e. $\partial r(0)$ is a singleton, which, in view of Lemma (1) (ii) implies (i). As for (ii), if r_e is a strictly convex and differentiable on $\mathbb{R}^m$, then it is closed and proper. Following Rockafellar (1970, Corollary 26.3.1, page 254), $\partial r_e^*(-\mathbf{p})$ is a singleton and $\mathbf{q}_e = \nabla r_e^*(-\mathbf{p})$ is unique, by (11). □

Remark 2. Let $((\mathbf{q}_e)_{e\in E}, \mathbf{p})$ be optimal. Following Cheridito et al. (2015, Proposition 2), if a risk measure ρ_e is differentiable at $-\mathcal{R}_e - \mathbf{q}_e^\top\mathcal{P}$ for some e in E, then r_e is differentiable at $\mathbf{q}_e$, hence the optimal price $\mathbf{p}$ is unique.

B Results Specific to Expected Shortfall Risk Measures

With regulatory standards in view, from now on, ρ_e is an expected shortfall risk measure $\rho_e = \mathbb{ES}_{\alpha_e}$ at some confidence level $0 \leq \alpha_e < 1$ (see Prerequisites in the general introduction to the book). In elliptical markets, coherent risk measures (e.g. expected shortfall, see after (4)) lead to analytical expressions for equilibria. A summary of the notations that will be used hereafter is provided in Table 31, where further computations yield

$$\Gamma_e = \begin{bmatrix} \mathbb{V}\mathrm{ar}(\mathcal{R}_e) & \mathrm{cov}_e \\ \mathrm{cov}_e & \Gamma. \end{bmatrix}. \tag{17}$$

Table 31: Summary of the main notations in Proposition 1.

μ	the vector $\mathbb{E}[\mathcal{P}]$
Γ	the matrix $\mathbb{C}\mathrm{ov}(\mathcal{P})$
μ_e	the vector $\mathbb{E}[(\mathcal{R}_e, \mathcal{P})]$
cov_e	the vector $\big(\mathbb{C}\mathrm{ov}(\mathcal{R}_e, \mathcal{P}_1), \ldots, \mathbb{C}\mathrm{ov}(\mathcal{R}_e, \mathcal{P}_m)\big)$
cov	$\sum_{e \in E} \mathrm{cov}_e$
Γ_e	the covariance matrix $\mathbb{C}\mathrm{ov}((\mathcal{R}_e, \mathcal{P}))$, i.e. of the vector $(\mathcal{R}_e, \mathcal{P})$
α_e	confidence level for an expected shortfall risk measure of the participant e
$\mathbb{ES}_{\alpha_e}$	an expected shortfall risk measure with confidence level α_e

Proposition 1. *If $(\mathcal{R}_e, \mathcal{P}) \sim \mathcal{N}_{m+1}(\mu_e, \Gamma_e)$ and $\rho_e = \mathbb{ES}_{\alpha_e}$ for some $0 \leq \alpha_e < 1$, $e \in E$, then there exists a Radner equilibrium with a unique equilibrium price. If Γ_e is further positive definite, $e \in E$, then the Radner equilibrium is unique.*

Proof. We divide the proof in three steps.

Existence In view of (17),

$$-\mathcal{R}_e - \mathbf{q}_e^\top \mathcal{P} \sim \mathcal{N}_1\Big(- \mathbb{E}[\mathcal{R}_e] - \mathbf{q}_e^\top \mu, \mathbb{V}\mathrm{ar}(\mathcal{R}_e) + 2\mathbf{q}_e^\top \mathrm{cov}_e + \mathbf{q}_e^\top \Gamma \mathbf{q}_e \Big). \tag{18}$$

For a univariate normally distributed r.v. $\mathcal{X}$, McNeil, Frey and Embrechts (2015, Example 2.14, page 70) yields $\mathbb{ES}_\alpha(\mathcal{X}) = \mathbb{E}[\mathcal{X}] + \sqrt{\mathbb{V}\mathrm{ar}(\mathcal{X})}\, \mathbb{ES}_\alpha(\varepsilon)$ with $\varepsilon \sim \mathcal{N}_1(0,1)$, $\mathbb{ES}_\alpha(\varepsilon) = \frac{\phi(\mathcal{N}^{-1}(\alpha))}{1-\alpha}$. This and (18) yield

$$r_e(q_e) = -\mathbb{E}[\mathcal{R}_e] - q_e^\top \mu + \mathbb{ES}_{\alpha_e}(\varepsilon)\sqrt{\mathbb{V}\mathrm{ar}(\mathcal{R}_e) + 2q_e^\top \mathrm{cov}_e + q_e^\top \Gamma q_e}, \quad q_e \in \mathbb{R}^m. \tag{19}$$

By Rockafellar (1970, Corollary 8.5.2), the recession function of r_e is given by

$$(r_e 0^+)(y) = \lim_{\lambda \searrow 0} \lambda r_e(y/\lambda) = -y^\top \mu + \sqrt{y^\top \Gamma y}\, \mathbb{ES}_{\alpha_e}(\varepsilon).$$

Let $q_1, \ldots, q_{|E|}$ be vectors in $\mathbb{R}^m$ such that

$$\sum_{e \in E} (r_e 0^+)(q_e) \leq 0 \quad \text{and} \quad \sum_{e \in E} (r_e 0^+)(-q_e) > 0,$$

i.e.

$$-\mu^\top \left(\sum_{e \in E} q_e \right) < \sum_{e \in E} \sqrt{q_e^\top \Gamma q_e}\, \mathbb{ES}_{\alpha_e}(\varepsilon) \leq \mu^\top \left(\sum_{e \in E} q_e \right).$$

Thus $\sum_{e \in E} q_e \neq 0$. By Rockafellar (1970, Corollary 9.2.1, page 76), the inf-convolution of real-valued convex functions is a real-valued convex function. Hence the inf-convolution r is attained on $\mathbb{R}^m$ and, by Lemma 1, there exists a Radner equilibrium $((\mathbf{q}_e)_{e \in E}, \mathbf{p})$.

Unique price $-\mathcal{R}_e - \mathbf{q}_e^\top \mathcal{P}$ is a continuous random variable, $e \in E$. By Kalkbrener (2005, Theorem 4.3 and Section 5.2), an expected shortfall is differentiable at continuous random variables. Therefore, by Remark 2, the optimal price $\mathbf{p}$ is unique.

Unique portfolio If Γ_e is positive definite, then,

$$(1, q_e)^\top \Gamma_e (1, q_e) = \mathbb{V}\mathrm{ar}(\mathcal{R}_e) + 2 q_e^\top \mathrm{cov}_e + q_e^\top \Gamma q_e > 0, \quad q_e \in \mathbb{R}^m.$$

This and (19) imply that r_e is differentiable such that

$$\boldsymbol{\nabla} r_e(q_e) = -\mu + \frac{\mathbb{ES}_{\alpha_e}(\varepsilon)}{\sqrt{(1, q_e)^\top \Gamma_e (1, q_e)}} (\mathrm{cov}_e + \Gamma q_e), \quad q_e \in \mathbb{R}^m.$$

Following Rockafellar and Wets (1998, Theorem 2.14, page 47), the strict convexity of r_e is equivalent to

$$r_e(y) > r_e(q_e) + \boldsymbol{\nabla} r_e(q_e)^\top (y - q_e), \quad q_e \neq y.$$

A simple computation reduces this first order condition to

$$\sqrt{\left[(1, q_e)^\top \Gamma_e (1, q_e)\right] \left[(1, y)^\top \Gamma_e (1, y)\right]} > (1, q_e)^\top \Gamma_e (1, y), \quad q_e \neq y. \tag{20}$$

If $y \neq q_e$, then $(1, y)$ is not colinear to $(1, q_e)$. Hence, by Johnson and Wichern (2007, Eqn. (2.49), page 79) applied with $\mathbf{b} = (1, q_e)$, $\mathbf{d} = \Gamma_e(1, y)$ and $\mathbf{B} = \Gamma_e$ (hence $(1, q_e) = \mathbf{b} \neq c\mathbf{B}^{-1}\mathbf{d} = c(1, y)$ for any constant c),

$$\left[(1, q_e)^\top \Gamma_e (1, q_e)\right] \left[(1, y)^\top \Gamma_e (1, y)\right] > \left[(1, q_e)^\top \Gamma_e (1, y)\right]^2 \quad \text{holds for any} \quad q_e \neq y.$$

This in turn yields (20). Hence by Theorem 2 there exists a unique equilibrium. □

Remark 3. If $\mathbb{V}\mathrm{ar}(\mathcal{R}_e) > 0$, $\mathrm{cov}_e = 0$ and Γ is invertible, then $z^\top \Gamma_e z = z_1^2 \mathbb{V}\mathrm{ar}(\mathcal{R}_e) + 2 z_1 \hat{z}^\top \mathrm{cov}_e + \hat{z}^\top \Gamma \hat{z} > 0$ holds for any $z = (z_1, \ldots, z_{m+1}) \in \mathbb{R}^{m+1} \setminus \{0\}$, where $\hat{z} = (z_2, \ldots, z_{m+1})$. Hence Γ_e is positive definite as assumed in the last part of Proposition 1.

Instead, Γ_e is not positive definite when $\mathbb{V}\mathrm{ar}(\mathcal{R}_e) = \mathrm{cov}_e^\top \Gamma^{-1} \mathrm{cov}_e$, because $z^\top \Gamma_e z = 0$ for $z = (-1, \Gamma^{-1}\mathrm{cov}_e)$. This is for instance the case when $\mathcal{R}_e$ is in the span of $\mathcal{P}$, i.e. $\mathcal{R}_e = a_e^\top \mathcal{P} + b_e$ for some constants $a_e \in \mathbb{R}^m$ and $b_e \in \mathbb{R}$, whence $\mathrm{cov}_e = \Gamma a_e$ and $\mathbb{V}\mathrm{ar}(\mathcal{R}_e) = a_e^\top \Gamma a_e = \mathrm{cov}_e^\top \Gamma^{-1} \mathrm{cov}_e$.

Remark 4. An n-variate random vector $\mathcal{X}$ has an elliptical distribution written as $\mathcal{X} \sim \mathcal{E}_n(\mu, \Gamma, \psi)$ if its characteristic function is expressed as

$$\mathbb{E}\Big[\exp(\mathrm{i}\, z^\top \mathcal{X})\Big] = \exp(\mathrm{i}\ z^\top \mu)\psi\left(\frac{1}{2} z^\top \Gamma z\right), \quad z \in \mathbb{R}^n,$$

for $\mu = \mathbb{E}[\mathcal{X}], \Gamma = \mathbb{C}\mathrm{ov}(\mathcal{X})$ and a function $\psi\colon [0,\infty) \to \mathbb{R}$ such that $-\psi'(0) = -1$. As is well known (Landsman and Valdez, 2003; McNeil, Frey and Embrechts, 2015), if $\mathcal{X} \sim \mathcal{E}_n(\mu, \Gamma, \psi)$ (or, more specifically[2], $\mathcal{T}_n(\mu, \Gamma, \nu)$), then $y^\top \mathcal{X} \stackrel{d}{=} y^\top \mu + \sqrt{y^\top \Gamma\, y}\ \varepsilon$, where $\varepsilon \sim \mathcal{E}_1(0, 1, \psi)$ (specifically, $\mathcal{T}_1(0, 1, \nu)$). Hence, for any coherent risk measure ρ,

$$\rho(y^\top \mathcal{X}) = y^\top \mu + \rho(\varepsilon)\sqrt{y^\top \Gamma y}. \tag{21}$$

Assuming $(\mathcal{R}_e, \mathcal{P}) \sim \mathcal{E}_{m+1}(\mu_e, \Gamma_e, \psi)$ (e.g. $\mathcal{T}_{m+1}(\mu_e, \Gamma_e, \nu)$) with Γ_e positive definite, $e \in E$, the above implies that

$$r_e(q_e) = -\mathbb{E}[\mathcal{R}_e] - q_e^\top \mu + \rho_e(\varepsilon)\sqrt{\mathbb{V}\mathrm{ar}(\mathcal{R}_e) + 2 q_e^\top \mathrm{cov}_e + q_e^\top \Gamma q_e}. \tag{22}$$

Proposition 1 thus holds more generally for any law invariant and coherent risk measure ρ_e differentiable on the linear space spanned by the components of $(\mathcal{R}_e, \mathcal{P})$, e in E.

§3 THE COMPARATIVE STATICS APPROACH FOR DEFAULT RESOLUTION ANALYSIS

A Methodology

Let an index d represent a clearing member of a CCP defaulting instantaneously (and in isolation[3]) at time 0. We want to analyze and compare different close-out procedures, of the "hedging or not and liquidation or auctioning"

[2] a Student t-distribution is elliptical (Gaunt, 2021).

[3] considering several instant defaulters would mainly mean replacing $\mathbf{q}_d$ by $\sum_d \mathbf{q}_d$ hereafter.

types (one of the contributions of this chapter being to propose mathematical specifications of these notions), for the CCP portfolio of member d. Our methodology is based on a comparative statics analysis, whereby, for each envisioned default resolution strategy, we compute the sum of the associated risk incrementals of market participants, dubbed funds transfer price (FTP). Following an idea of Pareto optimality, we then elect the strategy that minimises the funds transfer price.

Until §4, we describe the market cost (MC) component of the FTP, deferring to §5 the credit cost component calculation. Along with auctions that will also be considered in §5, a default resolution strategy conducted by a CCP consists in two types of operations: liquidating or hedging. The CCP can execute these operations either on its own exchange or an external exchange. Liquidation involves the CCP selling the defaulted positions on an exchange, whereas hedging requires the CCP to retain those positions and act as a participant on an exchange to eliminate the market risk stemming from maintaining those positions. Performing these operations individually represents only the extreme ends of the default resolution strategies spectrum. In practice, a CCP would implement a combination of these operations on both its own and external exchanges: see Bank of England (2018) and Bank For International Settlements (2020).

Let D be the exchange of the CCP itself, and E label any exchange, for instance D but without restriction to it. For each (pre-default) exchange E, we denote by E' its advent in the wake of the instant default of d, depending on the settlement procedure implemented by the CCP. If the CCP of an exchange D faces the default of a clearing member d, then this CCP can envision different default resolution procedures, impacting possibly different exchanges E' (starting with D$'$), for the CCP portfolio $\mathbf{q}_d$ of the defaulter (in a pre-default equilibrium on D). For each considered default resolution strategy, each of the impacted exchanges E' (or their corresponding CCPs) would compute its corresponding market cost MC_E as per (29) and communicate it to the CCP of D. The ensuing $\mathrm{MC} = \sum_E \mathrm{MC}_E$ of the strategy is the price that the markets would charge to the CCP of D, should the latter choose this strategy for resolving $\mathbf{q}_d$. The CCP of D would then choose the most efficient strategy, i.e. the one minimizing MC (or a broader FTP also encompassing a credit cost the way detailed in §5).

We assume that the different exchanges E trade the same assets with terminal payoff $\mathcal{P}$, possibly at different initial prices $\mathbf{p}$ (interpreted in this setup as "time $0-$", pre-default prices), reflecting different market equilibria. Hedging procedures involve the CCP as a new trading participant, represented for this purpose by a new index ccp (not involved in any exchange E). We use similar notation for d and ccp as for participants e of E in §2 (see Tables 30 and 31 pages 256 and 260). Although other choices could be used without methodological change in what follows, and due to non-suitable calibration data in this regard, we assume that any data other than $\mathbf{p}$ and $\mathbf{q}_e$ in Tables 30-31 are not affected by the instant default of d. The only exception is given in (28) below for $\mathcal{R}_{ccp}$, which represents the post-default receivable (further specified by Eqn. (35) below) of the hedging CCP due to the portfolio of the defaulted member d taken over by the hedging CCP. As a CCP should not bear any position except for the ones inherited from defaulted market participants during the close-out period of their portfolios, the pre-default receivable of the CCP is zero.

The pair $((\mathbf{q}'_e)_{e\in E'}, \mathbf{p}')$ relative to any post-default exchange E' involved in the settlement of the defaulted portfolio is derived using a Radner equilibrium in E'. Note that all the receivables and equilibrium portfolios and prices implicitly depend on the corresponding exchange. Regarding prices and also the quantities q_d^E, ℓ_d^E and h_d^E below, we make this dependence explicit hereafter, denoting by $\mathbf{p}^E$ a pre-default ("time 0−") equilibrium price on E and by $\mathbf{p}'^E$ a post-default ("time 0") equilibrium price on E'.

B Market Cost

We define $\mathbf{q}_e = 0, e \in E' \setminus E$ and $\Delta\mathbf{q}_e = \mathbf{q}'_e - \mathbf{q}_e, e \in E' \cup E$, hence

$$\sum_{e\in E'\cap E} \mathbf{q}_e + \sum_{e\in E'} \Delta\mathbf{q}_e = \sum_{e\in E'} \mathbf{q}_e + \sum_{E'} \Delta\mathbf{q}_e = \sum_{e\in E'} \mathbf{q}'_e. \tag{23}$$

If the CCP chooses to liquidate a portion ℓ_d^E of $\mathbf{q}_d$ and hedge $h_d^E = q_d^E - \ell_d^E$ on each exchange E, with $\sum_E q_d^E = \mathbf{q}_d$, then the incremental positions of each participant e to the post-default exchange E' can be split as

$$\Delta\mathbf{q}_e = \Delta\mathbf{q}_e^\ell + \Delta\mathbf{q}_e^h, \tag{24}$$

where $\Delta\mathbf{q}_e^\ell$ and $\Delta\mathbf{q}_e^h$ are the increments implied by the liquidation and hedging legs of the strategy (see e.g. Subsubsection C.4 below)–with always in particular

$$\Delta\mathbf{q}_{ccp}^\ell = 0, \tag{25}$$

as a CCP does not take part as a participant to a liquidation. Since the amount demanded should be equal to the amount supplied on both legs of the strategy, we have

$$\sum_{e\in E'} \Delta\mathbf{q}_e^\ell = \ell_d^E \quad \text{and} \quad \sum_{e\in E'} \Delta\mathbf{q}_e^h = 0, \text{ hence } \sum_{e\in E'} \Delta\mathbf{q}_e = \ell_d^E. \tag{26}$$

Using (23) and $\Delta\mathbf{q}_e^\ell + \Delta\mathbf{q}_e^h = \Delta\mathbf{q}_e$, the post-default market loss of any trading participant $e \in E'$ is

$$\begin{aligned}
&\underbrace{-\mathcal{R}_e + \mathbf{q}_e^\top(\mathbf{p}^E - \mathcal{P})}_{\text{pre-default market loss}} + (\Delta\mathbf{q}_e^h)^\top(\mathbf{p}'^E - \mathcal{P})\\
&\qquad + (\Delta\mathbf{q}_e^\ell)^\top(\mathbf{p}^E - \mathcal{P}) \pm (\Delta\mathbf{q}_e^\ell)^\top(\mathbf{p}'^E - \mathcal{P})\\
&= -\mathcal{R}_e + \mathbf{q}_e^\top(\mathbf{p}^E - \mathcal{P}) + \Delta\mathbf{q}_e^\top(\mathbf{p}'^E - \mathcal{P}) + \left(\Delta\mathbf{q}_e^\ell\right)^\top(\mathbf{p}^E - \mathbf{p}'^E) \pm \mathbf{q}_e^\top(\mathbf{p}'^E - \mathcal{P})\\
&= -\mathcal{R}_e + (\mathbf{q}_e')^\top(\mathbf{p}'^E - \mathcal{P}) + \underbrace{(\mathbf{q}_e + \Delta\mathbf{q}_e^\ell)^\top(\mathbf{p}^E - \mathbf{p}'^E)}_{\mathrm{LC}_e}.
\end{aligned} \tag{27}$$

As reflected in the left-hand side of (27), the contracts $\Delta\mathbf{q}_e^\ell$ involved in the liquidation leg of the strategy are old ("time 0−") contracts with the pre-default prices $\mathbf{p}^E$, while the new ("time 0") contracts $\Delta\mathbf{q}_e^h$ involved in the hedging leg of the strategy are post-default contracts with the new prices $\mathbf{p}'^E$.

Hence, by translation equivariance of ρ_e, the post-default risk of participant e is

$$\rho_e\Big(-\mathcal{R}_e + (\mathbf{q}_e')^\top(\mathbf{p}'^E - \mathcal{P})\Big) + \mathrm{LC}_e.$$

The risk incremental of participant e is therefore

$$\mathrm{LC}_e + \underbrace{\rho_e\Big(-\mathcal{R}_e + (\mathbf{q}_e')^\top(\mathbf{p}'^E - \mathcal{P})\Big) - \rho_e\Big(-\mathbf{1}_{e\neq ccp}\mathcal{R}_e + \mathbf{q}_e^\top(\mathbf{p}^E - \mathcal{P})\Big)}_{=:\Delta\rho_e},\ e \in E'. \tag{28}$$

Accordingly, we assess the market cost (MC) of a default resolution strategy by

$$\mathrm{MC} = \sum_E \mathrm{MC}_E, \text{ where } \mathrm{MC}_E = \mathrm{LC}_E + \sum_{e\in E'} \Delta\boldsymbol{\rho}_e \text{ with } \mathrm{LC}_E = \sum_{e\in E'} \mathrm{LC}_e. \tag{29}$$

B.1 Liquidity cost

One can interpret LC_E as a liquidity cost, generated by price moves, which is part of the market cost. In fact, as $\mathbf{q}_e = 0, e \in E' \setminus E$, we have by definition of LC_e in (27) and LC_E in (29):

$$\begin{aligned}
\mathrm{LC}_E &= \sum_{e\in E'\cap E} \mathbf{q}_e^\top(\mathbf{p}^E - \mathbf{p}'^E) + \sum_{e\in E'} (\Delta\mathbf{q}_e^\ell)^\top(\mathbf{p}^E - \mathbf{p}'^E)\\
&= \sum_{e\in E'\cap E} \mathbf{q}_e^\top(\mathbf{p}^E - \mathbf{p}'^E) + \ell_d^E(\mathbf{p}^E - \mathbf{p}'^E),
\end{aligned} \tag{30}$$

which can be interpreted as margin payments (like in futures markets) by market participants at time 0 in response to the default settlement procedure of d, i.e. the price they have to pay for the transition from the pre-default to the post-default exchanges. As reflected in the first line of (30), it is only the old ("time 0−") contracts, with the pre-default prices $\mathbf{p}^E$, that deserve margin payments, while the new ("time 0") contracts $\Delta\mathbf{q}_e^h$ involved in the hedging leg of the strategy are post-default contracts with the new prices $\mathbf{p}'^E$. However, the following reformulation of LC_E in terms of the $\mathbf{q}_e + \Delta\mathbf{q}_e = \mathbf{q}'_e$ is possible:

Lemma 2. *On each exchange E,*

$$\mathrm{LC}_E = \sum_{e\in E'} (\mathbf{q}'_e)^\top(\mathbf{p}^E - \mathbf{p}'^E). \tag{31}$$

Proof. By (26), $\sum_{e\in E'} \Delta\mathbf{q}_e^h = 0$. Hence the definition of LC_E via (29)-(27) yields

$$\mathrm{LC}_E = \sum_{e\in E'} \left(\mathbf{q}_e + \Delta\mathbf{q}_e^\ell + \Delta\mathbf{q}_e^h\right)^\top (\mathbf{p}^E - \mathbf{p}'^E),$$

where $\Delta\mathbf{q}_e^\ell + \Delta\mathbf{q}_e^h = \Delta\mathbf{q}_e$. □

C Examples of Default Resolution Strategies

The pre-default equilibria $((\mathbf{q}_e)_{e\in E}, \mathbf{p}^E)$ involved in (29) are obtained by direct application of the results of §2. We now detail the corresponding post-default Radner equilibria $((\mathbf{q}'_e)_{e\in E'}, \mathbf{p}'^E)$ in different cases (without post-default new invited participants other than the CCP itself in the hedging cases, though; extra new invited participants will be considered in §4). The member optimality condition for the post-default market participant $e \in E'$ is always stated as

$$\rho_e\Big(-\mathcal{R}_e + (\mathbf{q}'_e)^\top(\mathbf{p}'^E - \mathcal{P})\Big) \le \rho_e\Big(-\mathcal{R}_e + q_e^\top(\mathbf{p}'^E - \mathcal{P})\Big),\ q_e \in \mathbb{R}^m. \tag{32}$$

The clearing condition, instead, depends on the considered default resolution strategy. We first examine strategies in which the CCP exclusively operates on its own exchange.

C.1 The CCP Fully Liquidates on Its Own Exchange

As a first default resolution alternative, the CCP may want to liquidate the defaulter's position $\mathbf{q}_d$ on its own exchange D. Then $\mathrm{MC}_E = 0, E \neq \mathrm{D}$ and $\sum_{e\in \mathrm{D}'=\mathrm{D}\setminus\{d\}} \Delta\mathbf{q}_e = \mathbf{q}_d$. As $\sum_{e\in \mathrm{D}\setminus\{d\}} \mathbf{q}_e + \mathbf{q}_d = 0$, we obtain a post-default equilibrium clearing condition

$$\sum_{e\in \mathrm{D}'=\mathrm{D}\setminus\{d\}} \mathbf{q}'_e = 0 \tag{33}$$

and

$$\mathrm{LC} = \mathrm{LC}_\mathrm{D} = 0,\ \mathrm{MC} = \mathrm{MC}_\mathrm{D} = \sum_{e\in \mathrm{D}'=\mathrm{D}\setminus\{d\}} \Delta\boldsymbol{\rho}_e. \tag{34}$$

C.2 The CCP Fully Hedges on Its Own Exchange

If $\mathbf{q}_d$ is not instantaneously liquidated upon the default of member d at time 0, then the CCP ccp of d endorses at time 0 the receivable

$$\mathcal{R}_{ccp} = \mathbf{q}_d^\top(\mathcal{P} - \mathbf{p}^{\mathrm{D}}), \tag{35}$$

which it can hedge by holding on its own exchange D a portfolio $\Delta\mathbf{q}_{ccp}$ minimizing some risk measure ρ_{ccp}. The corresponding member optimality condition (32) for the CCP ccp, playing the role of a new post-default trading participant, is

$$\begin{aligned}\rho_{ccp}\Big(\mathbf{q}_d^\top(\mathbf{p}^{\mathrm{D}} - \mathcal{P}) + (\mathbf{q}'_{ccp})^\top(\mathbf{p}'^{\mathrm{D}} - \mathcal{P})\Big) \leq \\ \rho_{ccp}\Big(\mathbf{q}_d^\top(\mathbf{p}^{\mathrm{D}} - \mathcal{P}) + q_{ccp}^\top(\mathbf{p}'^{\mathrm{D}} - \mathcal{P})\Big),\ q_{ccp} \in \mathbb{R}^d.\end{aligned} \tag{36}$$

In this case, $\mathrm{MC}_E = 0, E \neq \mathrm{D}$ and $\sum_{e\in\mathrm{D}'=(\mathrm{D}\setminus\{d\})\cup\{ccp\}}\Delta\mathbf{q}_e = 0$ (as, in this hedging case, on the post-default market D′, the amount demanded must be equal to the amount supplied). Since $\sum_{e\in\mathrm{D}\setminus\{d\}}\mathbf{q}_e = -\mathbf{q}_d$ and $\Delta\mathbf{q}'_{ccp} = \mathbf{q}'_{ccp}$, we obtain a post-default equilibrium clearing condition

$$\sum_{e\in\mathrm{D}'=(\mathrm{D}\setminus\{d\})\cup\{ccp\}}\mathbf{q}'_e = -\mathbf{q}_d \tag{37}$$

and

$$\mathrm{MC} = \mathrm{MC}_{\mathrm{D}} = \underbrace{-\mathbf{q}_d^\top(\mathbf{p}^{\mathrm{D}} - \mathbf{p}'^{\mathrm{D}})}_{\mathrm{LC}_{\mathrm{D}}} + \sum_{e\in\mathrm{D}'=(\mathrm{D}\setminus\{d\})\cup\{ccp\}}\Delta\boldsymbol{\rho}_e.$$

Remark 5. By change of variable $\mathbf{z}'_e = \Delta\mathbf{q}_e, e \in (\mathrm{D}\setminus\{d\})\cup\{ccp\}$, $\mathcal{R}'_e = \mathcal{R}_e + \mathbf{q}_e^\top\mathcal{P}$, $e \in \mathrm{D}\setminus\{d\}$) and $\mathcal{R}'_{ccp} = \mathcal{R}_{ccp}$, the clearing condition (37) relative to the post-default equilibrium $((\mathbf{q}'_e)_{e\in\mathrm{D}'}, \mathbf{p}'^{\mathrm{D}})$ can be converted to a zero clearing condition as per Definition 1 on D′.

C.3 The CCP Fully Replicates on Its Own Exchange

By replication on its own exchange, we refer to a default resolution strategy whereby the CCP ccp replicates the portfolio $\mathbf{q}_d$ (if not liquidated) that the CCP inherits from d by mirroring position $\mathbf{q}'_{ccp} = -\mathbf{q}_d$ on D. In this case $\mathrm{MC}_E = 0, E \neq \mathrm{D}$, and we have $\mathrm{D}' = (\mathrm{D}\setminus\{d\})\cup\{ccp\}, \Delta\mathbf{q}_{ccp} = -\mathbf{q}_d$ (in this replication case, the post-default position of the CCP is $-\mathbf{q}_d$ on D′), $\Delta\boldsymbol{\rho}_{ccp} = \rho_{ccp}\Big(\mathbf{q}_d^\top(\mathbf{p}^{\mathrm{D}} - \mathcal{P}) - \mathbf{q}_d^\top(\mathbf{p}'^{\mathrm{D}} - \mathcal{P})\Big) = \mathbf{q}_d^\top\Big(\mathbf{p}^{\mathrm{D}} - \mathbf{p}'^{\mathrm{D}}\Big)$. On the post-default exchange D′ where the hedge is implemented, the amount demanded must be equal to the amount supplied, i.e. $\sum_{e\in\mathrm{D}'=(\mathrm{D}\setminus\{d\})\cup\{ccp\}}\Delta\mathbf{q}_e = 0$, whence the post-default clearing condition

$$\sum_{e\in\mathrm{D}'=(\mathrm{D}\setminus\{d\})\cup\{ccp\}}\mathbf{q}'_e = \underbrace{\sum_{e\in\mathrm{D}\setminus\{d\}}\mathbf{q}'_e}_{0} + \underbrace{\mathbf{q}'_{ccp}}_{-\mathbf{q}_d} = -\mathbf{q}_d. \tag{38}$$

Therefore

$$\mathrm{MC} = \mathrm{MC}_{\mathrm{D}} = \underbrace{-\mathbf{q}_d^\top(\mathbf{p}^{\mathrm{D}} - \mathbf{p}'^{\mathrm{D}})}_{\mathrm{LC}=\mathrm{LC}_{\mathrm{D}}} + \sum_{e\in\mathrm{D}\setminus\{d\}} \Delta\boldsymbol{\rho}_e + \underbrace{\Delta\boldsymbol{\rho}_{ccp}}_{\mathbf{q}_d^\top(\mathbf{p}^{\mathrm{D}}-\mathbf{p}'^{\mathrm{D}})} = \sum_{e\in\mathrm{D}\setminus\{d\}} \Delta\boldsymbol{\rho}_e.$$

The market cost is the same as in the liquidation case of Subsubsection C.1 (note that the embedded post-default Radner equilibria are the same), but its split between LC and $\sum_{e\in\mathrm{D}'} \Delta\boldsymbol{\rho}_e$ is different, see Table 32.

Remark 6. As in Remark 5, by change of variable $\mathbf{z}'_e = \Delta\mathbf{q}_e, e \in (\mathrm{D}\setminus\{d\}) \cup \{ccp\}$, $\mathcal{R}'_e = \mathcal{R}_e + \mathbf{q}_e^\top\mathcal{P}$, $e \in \mathrm{D}\setminus\{d\}$), and $\mathcal{R}'_{ccp} = \mathcal{R}_{ccp}$, the clearing condition (38) relative to the post-default equilibrium $((\mathbf{q}'_e)_{e\in\mathrm{D}'}, \mathbf{p}'^{\mathrm{D}})$ can be converted to a zero clearing condition as per Definition 1 on D'.

C.4 The CCP Partially Liquidates and Hedges on Its Own Exchange

The CCP can also liquidate a portion ℓ_d^{D} of the defaulted position $\mathbf{q}_d$ and hedge the remaining $h_d^{\mathrm{D}} = \mathbf{q}_d - \ell_d^{\mathrm{D}}$, all on its own exchange D. The amount demanded should be equal to the amount supplied on each leg of the strategy, hence $\sum_{e\in\mathrm{D}\setminus\{d\}} \Delta\mathbf{q}_e^\ell = \ell_d^{\mathrm{D}}$ and $\sum_{e\in(\mathrm{D}\setminus\{d\})\cup\{ccp\}} \Delta\mathbf{q}_e^h = 0$, thus $\sum_{e\in(\mathrm{D}\setminus\{d\})\cup\{ccp\}} \Delta\mathbf{q}_e = \ell_d^{\mathrm{D}}$. As $\sum_{e\in\mathrm{D}\setminus\{d\}} \mathbf{q}_e = -\mathbf{q}_d$, the ensuing post-default clearing condition on D' is

$$\sum_{e\in\mathrm{D}'=(\mathrm{D}\setminus\{d\})\cup\{ccp\}} \mathbf{q}'_e = -h_d^{\mathrm{D}}, \tag{39}$$

We assume that both liquidation and hedging happen simultaneously at the same price $\mathbf{p}'^{\mathrm{D}}$. Hence each trading participant on the post-default market D' has a single member optimality condition (32) (with, in particular, $\mathcal{R}_{ccp} = (h_d^{\mathrm{D}})^\top(\mathcal{P} - \mathbf{p}^{\mathrm{D}})$). Then

$$\mathrm{MC} = \mathrm{MC}_{\mathrm{D}} = \underbrace{-(h_d^{\mathrm{D}})^\top(\mathbf{p}^{\mathrm{D}} - \mathbf{p}'^{\mathrm{D}})}_{\mathrm{LC}=\mathrm{LC}_{\mathrm{D}}} + \sum_{e\in\mathrm{D}'=(\mathrm{D}\setminus\{d\})\cup\{ccp\}} \Delta\boldsymbol{\rho}_e.$$

Remark 7. By change of variables $\mathbf{z}'_e = \mathbf{q}'_e + k_e h_d^E$ and $\mathcal{R}'_e = \mathcal{R}_e - k_e(h_d^E)^\top\mathcal{P}$, for reals k_e such that $\sum_{e\in\mathrm{D}'} k_e = 1$, the clearing condition (39) and the optimality conditions (32) relative to the post-default equilibrium $((\mathbf{q}'_e)_{e\in\mathrm{D}'}, \mathbf{p}'^{\mathrm{D}})$ respectively become $\sum_{e\in\mathrm{D}'} \mathbf{z}'_e = 0$ and

$$\rho_e(-\mathcal{R}'_e + (\mathbf{z}'_e)^\top(\mathbf{p}'^{\mathrm{D}} - \mathcal{P})) \le \rho_e(-\mathcal{R}'_e + z_e^\top(\mathbf{p}'^{\mathrm{D}} - \mathcal{P})),\ z_e \in \mathbb{R}^m.$$

C.5 Strategies also Involving Another Exchange

Table 33 provides the output of the computations, detailed in Bastide et al. (2025), for the four default resolution strategies similar to the ones considered earlier in this subsection, but implemented on another exchange E instead of the own exchange D of the CCP.

Table 32: Decomposition of the market costs of the four default resolution strategies considered in Subsubsections C.1-C.4.

	LC	$\sum_{e\in \mathrm{D}'} \Delta\boldsymbol{\rho}_e$
1.	0	$\sum_{e\in \mathrm{D}\setminus\{d\}} \Delta\boldsymbol{\rho}_e$
2.	$-\mathbf{q}_d^\top(\mathbf{p}^{\mathrm{D}}-\mathbf{p}'^{\mathrm{D}})$	$\sum_{e\in (\mathrm{D}\setminus\{d\})\cup\{ccp\}} \Delta\boldsymbol{\rho}_e$
3.	$-\mathbf{q}_d^\top(\mathbf{p}^{\mathrm{D}}-\mathbf{p}'^{\mathrm{D}})$	$\sum_{e\in (\mathrm{D}\setminus\{d\})\cup\{ccp\}} \Delta\boldsymbol{\rho}_e$ with $\Delta\boldsymbol{\rho}_{ccp}=\mathbf{q}_d^\top\left(\mathbf{p}^{\mathrm{D}}-\mathbf{p}'^{\mathrm{D}}\right)$
4.	$-(h_d^{\mathrm{D}})^\top(\mathbf{p}^{\mathrm{D}}-\mathbf{p}'^{\mathrm{D}})$	$\sum_{e\in (\mathrm{D}\setminus\{d\})\cup\{ccp\}} \Delta\boldsymbol{\rho}_e$

Table 33: Decomposition of the market costs on the four default resolution strategies similar to the ones of Table 32, but implemented on another exchange E, instead of the own exchange D of the CCP.

	LC	$\sum_E\sum_{e\in E'} \Delta\boldsymbol{\rho}_e$
1.	$-\mathbf{q}_d^\top(\mathbf{p}^{\mathrm{D}}-\mathbf{p}'^{\mathrm{D}})+\mathbf{q}_d^\top(\mathbf{p}^{\mathrm{E}}-\mathbf{p}'^{\mathrm{E}})$	$\sum_{e\in \mathrm{D}\setminus\{d\}} \Delta\boldsymbol{\rho}_e+\sum_{e\in\mathrm{E}}\Delta\boldsymbol{\rho}_e$
2.	$-\mathbf{q}_d^\top(\mathbf{p}^{\mathrm{D}}-\mathbf{p}'^{\mathrm{D}})$	$\sum_{e\in \mathrm{D}\setminus\{d\}} \Delta\boldsymbol{\rho}_e+\sum_{e\in\mathrm{E}\cup\{ccp\}}\Delta\boldsymbol{\rho}_e$
3.	$-\mathbf{q}_d^\top(\mathbf{p}^{\mathrm{D}}-\mathbf{p}'^{\mathrm{D}})$	$\sum_{e\in \mathrm{D}\setminus\{d\}} \Delta\boldsymbol{\rho}_e+\sum_{e\in\mathrm{E}\cup\{ccp\}}\Delta\boldsymbol{\rho}_e$ with $\Delta\boldsymbol{\rho}_{ccp}=\mathbf{q}_d^\top\left(\mathbf{p}^{\mathrm{D}}-\mathbf{p}'^{\mathrm{E}}\right)$
4.	$-\mathbf{q}_d^\top(\mathbf{p}^{\mathrm{D}}-\mathbf{p}'^{\mathrm{D}})+(\boldsymbol{h}_d^{\mathrm{E}})^\top(\mathbf{p}^{\mathrm{E}}-\mathbf{p}'^{\mathrm{E}})$	$\sum_{e\in \mathrm{D}\setminus\{d\}} \Delta\boldsymbol{\rho}_e+\sum_{e\in\mathrm{E}\cup\{ccp\}}\Delta\boldsymbol{\rho}_e$

§4 MARKET COST: THE CASE OF EXPECTED SHORTFALL

In this section, we provide numerical results regarding the market cost of default resolutions on D, hence (cf. (31) and (29))

$$\mathrm{LC}=\mathrm{LC}_{\mathrm{D}}=\sum_{e\in\mathrm{D}'}(\mathbf{q}'_e)^\top(\mathbf{p}^{\mathrm{D}}-\mathbf{p}'^{\mathrm{D}}),\qquad \mathrm{MC}=\mathrm{MC}_{\mathrm{D}}=\mathrm{LC}_{\mathrm{D}}+\sum_{e\in\mathrm{D}'}\Delta\boldsymbol{\rho}_e. \quad (40)$$

Since $\mathbf{q}_{ccp}=0$ and $\rho_{ccp}(0)=0$, we obtain by translation equivariance of the ρ_e, and (28), (9):

$$\Delta\boldsymbol{\rho}_e=r_e(\mathbf{q}'_e)-\mathbf{1}_{e\neq ccp}r_e(\mathbf{q}_e)+(\mathbf{q}'_e)^\top\mathbf{p}'^{\mathrm{D}}-\mathbf{q}_e^\top\mathbf{p}^{\mathrm{D}},\ e\in\mathrm{D}'. \quad (41)$$

Hence (40) yields

$$\mathrm{MC}=\mathrm{MC}_{\mathrm{D}}=\sum_{e\in\mathrm{D}'}(\Delta\mathbf{q}_e)^\top\mathbf{p}^{\mathrm{D}}+\sum_{e\in\mathrm{D}'}\Big(r_e(\mathbf{q}'_e)-\mathbf{1}_{e\neq ccp}r_e(\mathbf{q}_e)\Big). \quad (42)$$

We assume that the risk preferences of each market participant is an expected shortfall $\mathbb{ES}_{\alpha_e}$, each vector $(\mathcal{R}_e, \mathcal{P}) \sim \mathcal{E}_{m+1}(\mu_e, \Gamma_e, \psi)$ (or, sometimes, a more specific $\mathcal{T}_{m+1}(\mu_e, \Gamma_e, \nu_e)$), and the CCP only operates on its own exchange D. Hence MC = MC$_{\mathrm{D}}$ as in (42), with, by (21),

$$r_e(q_e) = -\mathbb{E}[\mathcal{R}_e] - q_e^\top \mu + \mathbb{ES}_{\alpha_e}(\varepsilon_e)\sqrt{\mathbb{V}\mathrm{ar}(\mathcal{R}_e) + 2q_e^\top \mathrm{cov}_e + q_e^\top \Gamma q_e} \\ \text{where } \varepsilon_e \sim \mathcal{E}_1(0,1,\psi) \text{ (or } \varepsilon_e \sim \mathcal{T}_1(0,1,\nu_e)), \ e \in \mathrm{D} \cup \mathrm{D}'. \tag{43}$$

We consider two different cases.

A Liquidation on D

Let us first consider the liquidation case of §3.C.1, but with possibly new participants in D′. If $(\mathcal{R}_e, \mathcal{P}) \sim \mathcal{E}_{m+1}(\mu_e, \Gamma_e, \psi)$ with Γ_e positive definite, $e \in \mathrm{D} \cup \mathrm{D}'$, then, by Proposition 1 and Remark 4, there exists a unique pre-default Radner equilibrium $((\mathbf{q}_e)_{e \in \mathrm{D}}, \mathbf{p}^{\mathrm{D}})$. Following (10) and (22), the pre-default member $e \in \mathrm{D}$ optimality condition yields

$$\mathbf{p}^{\mathrm{D}} = \mu - \frac{\mathbb{ES}_{\alpha_e}(\varepsilon)}{\sqrt{\mathbb{V}\mathrm{ar}(\mathcal{R}_e) + 2\mathbf{q}_e^\top \mathrm{cov}_e + \mathbf{q}_e^\top \Gamma \mathbf{q}_e}} (\mathrm{cov}_e + \Gamma \mathbf{q}_e), \quad e \in \mathrm{D}. \tag{44}$$

The zero clearing condition and (44) imply

$$\begin{cases} \sum_{e \in \mathrm{D}} \mathbf{q}_e = 0 \\ \frac{\mathbb{ES}_{\alpha_1}(\varepsilon)}{\mathbb{ES}_{\alpha_e}(\varepsilon)} \sqrt{\frac{\mathbb{V}\mathrm{ar}(\mathcal{R}_e) + 2\mathbf{q}_e^\top \mathrm{cov}_e + \mathbf{q}_e^\top \Gamma \mathbf{q}_e}{\mathbb{V}\mathrm{ar}(\mathcal{R}_1) + 2\mathbf{q}_1^\top \mathrm{cov}_1 + \mathbf{q}_1^\top \Gamma \mathbf{q}_1}} (\mathrm{cov}_1 + \Gamma \mathbf{q}_1) - \mathrm{cov}_e - \Gamma \mathbf{q}_e = 0, \quad e \in \mathrm{D}. \end{cases} \tag{45}$$

Therefore, computing the optimal position is equivalent to finding the root of a vector function from $\mathbb{R}^{m|\mathrm{D}|}$ into $\mathbb{R}^{m|\mathrm{D}|}$. Likewise, the post-default optimal positions solve

$$\begin{cases} \sum_{e \in \mathrm{D}'} \mathbf{q}'_e = 0 \\ \frac{\mathbb{ES}_{\alpha_1}(\varepsilon)}{\mathbb{ES}_{\alpha_e}(\varepsilon)} \sqrt{\frac{\mathbb{V}\mathrm{ar}(\mathcal{R}_e) + 2(\mathbf{q}'_e)^\top \mathrm{cov}_e + (\mathbf{q}'_e)^\top \Gamma \mathbf{q}'_e}{\mathbb{V}\mathrm{ar}(\mathcal{R}_1) + 2\mathbf{q}'^\top_1 \mathrm{cov}_1 + \mathbf{q}'^\top_1 \Gamma \mathbf{q}'_1}} (\mathrm{cov}_1 + \Gamma \mathbf{q}'_1) - \mathrm{cov}_e - \Gamma \mathbf{q}'_e = 0, \quad e \in \mathrm{D}' \end{cases} \tag{46}$$

and the post-default price is equal to

$$\mathbf{p}'^{\mathrm{D}} = \mu - \frac{\mathbb{ES}_{\alpha_e}(\varepsilon_e)}{\sqrt{\mathbb{V}\mathrm{ar}(\mathcal{R}_e) + 2(\mathbf{q}'_e)^\top \mathrm{cov}_e + (\mathbf{q}'_e)^\top \Gamma \mathbf{q}'_e}} \left(\mathrm{cov}_e + \Gamma \mathbf{q}'_e\right), \quad e \in \mathrm{D}'. \tag{47}$$

In the remainder of this subsection, we assume that $m = 1$, $\alpha_e = 0.975$ for each market participant, $\mu = 2$, and $\sigma = 0.2$.

Example 1 (Multivariate normal). Let $\mathrm{D} = \{1, \ldots, 15\}$, $d = \{15\}$, $\mathrm{D}' \setminus \mathrm{D} = \varnothing$, $\varrho_e = 1$, $(\mathcal{R}_e, \mathcal{P}) \sim \mathcal{N}_2(\mu_e, \Gamma_e)$, $\mathrm{cov}_e = c_e \sigma \sqrt{\mathbb{V}\mathrm{ar}(\mathcal{R}_e)}$ with $\sigma = 0.2, c_e = (-1)^{e+1} 0.8$ and $\mathbb{V}\mathrm{ar}(\mathcal{R}_e) = 0.09e^2$, $e \in \mathrm{D}$. Solving (45)-(46) yields the pre- and post-default optimal positions displayed in Table 34, which imply via (44), (47) and (34) that $\mathbf{p}^{\mathrm{D}} = 1.96, \mathbf{p}'^{\mathrm{D}} = 2.04$ and $\mathrm{MC}_{\mathrm{D}} = 0.69$ (with $\mathrm{LC} = 0$).

Table 34: Pre- and post-default optimal positions when the CCP liquidates on its own exchange D with $\mathrm{D}' \setminus \mathrm{D} = \varnothing$ in the expected shortfall case.

CM_e	1	2	3	4	5	6	7	8
cov_e	0.05	−0.10	0.14	−0.19	0.24	−0.29	0.34	−0.38
$\mathbf{q}_e$	−1.12	2.56	−3.36	5.12	−5.60	7.68	−7.84	10.24
$\mathbf{q}'_e$	−1.28	2.24	−3.84	4.48	−6.40	6.72	−8.96	8.96

CM_e	9	10	11	12	13	14	15
cov_e	0.43	−0.48	0.53	−0.58	0.62	−0.67	0.72
$\mathbf{q}_e$	−10.08	12.80	−12.32	15.36	−14.56	17.92	−16.80
$\mathbf{q}'_e$	−11.52	11.20	−14.08	13.44	−16.64	15.68	

Example 2 (Multivariate Student t). Reconsider Example 1 but for $(\mathcal{R}_e, \mathcal{P}) \sim \mathcal{T}_2(\mu_e, \Gamma_e, \nu_e)$ with $\nu_e = \nu = 2.5$, $e \in \mathrm{D}$. Then (43) yields

$$
\begin{aligned}
r_e(q_e) = &-\mathbb{E}[\mathcal{R}_e] - q_e^\top \mu + \mathbb{ES}_{\alpha_e}(\varepsilon) \sqrt{\mathbb{V}\mathrm{ar}(\mathcal{R}_e) + 2q_e^\top \mathrm{cov}_e + q_e^\top \Gamma q_e} \\
&\text{with } \varepsilon \sim \mathcal{T}_1(0, 1, \nu),\ e \in \mathrm{D}.
\end{aligned}
$$

By McNeil et al. (2015, Example 2.15, page 71), we obtain $\mathbb{ES}_{\alpha_e}(\varepsilon) = \sqrt{\frac{\nu-2}{\nu}} \frac{t_\nu\left(T_\nu^{-1}(\alpha_e)\right)\left[\nu + \left(T_\nu^{-1}(\alpha)\right)^2\right]}{(1-\alpha)(\nu-1)}$. By inspection, $\mathbf{q}$ and $\mathbf{q}'$ are the same as in Example 1, given by Table 34. The above result also yield $\mathbf{p}^{\mathrm{D}} = 1.94, \mathbf{p}'^{\mathrm{D}} = 2.06$ and $\mathrm{MC}_{\mathrm{D}} = 1.06$ (with $\mathrm{LC}_{\mathrm{D}} = 0$ as per the first line in Table 32).

B Hedging on D

We now turn to the "hedging on own exchange D" case of §3.C.2, but with possibly new participants beyond the CCP ccp in D'. Let $(\mathcal{R}_e, \mathcal{P}) \sim \mathcal{E}_{m+1}(\mu_e, \Gamma_e, \psi)$ with Γ_e positive definite, for each $e \in \mathrm{D} \cup (\mathrm{D}' \setminus \{ccp\})$. By Proposition 1, there exists a unique pre-default Radner equilibrium $((\mathbf{q}_e)_{e \in \mathrm{D}}, \mathbf{p}^{\mathrm{D}})$, which can be computed by (44) and (45). Since $\mathcal{R}_{ccp} = \mathbf{q}_d^\top (\mathcal{P} - \mathbf{p}^{\mathrm{D}})$, letting $z = (z_1, z_2, \ldots, z_{m+1}) = (z_1, \hat{z})$, we obtain

$$
\mathbb{E}\Big[\exp\left(\mathrm{i} z^\top (\mathcal{R}_{ccp}, \mathcal{P})\right)\Big] = \mathbb{E}\Big[\exp\Big(\mathrm{i}\big((z_1 \mathbf{q}_d + \hat{z})^\top \mathcal{P} - z_1 \mathbf{q}_d^\top \mathbf{p}^{\mathrm{D}}\big)\Big)\Big]
$$

$$= \exp(\mathrm{i}\, z^\top \mu_{ccp})\psi\left(\frac{1}{2}z^\top \Gamma_{ccp} z\right), \quad z \in \mathbb{R}^{m+1},$$

as $(z_1 \mathbf{q}_d + \hat{z})^\top \mathcal{P} - z_1 \mathbf{q}_d^\top \mathbf{p}^{\mathrm{D}} \sim \mathcal{E}_1\Big(z^\top \mu_{ccp}, z^\top \Gamma_{ccp} z, \psi\Big)$. Hence $(\mathcal{R}_{ccp}, \mathcal{P}) \sim \mathcal{E}_{m+1}(\mu_{ccp}, \Gamma_{ccp}, \psi)$. However, as $\mathcal{R}_{ccp}$ is in the span of $\mathcal{P}$, Γ_{ccp} is not positive definite (see Remark 3). To nevertheless ensure a unique post-default equilibrium (beyond the setup of Proposition 1), we assume that $\mathrm{D}' \setminus \{ccp\} \neq \varnothing$. By Theorem 2 page 259, we have a unique post-default price $\mathbf{p}'^{\mathrm{D}}$ and a unique post-default portfolio $\mathbf{q}'_e, e \in \mathrm{D}' \setminus \{ccp\}$. Hence by the post-default clearing condition, we also have a unique post-default position $\mathbf{q}'_{ccp}$.

Letting $\mathcal{R}'_e = \mathcal{R}_e + \mathbf{q}_e^\top \mathcal{P}$ play the role of $\mathcal{R}_e$ in (19), we obtain

$$\begin{aligned} r'_e(z_e) &:= \mathbb{ES}_{\alpha_e}(-\mathcal{R}'_e - z_e^\top \mathcal{P}) \\ &= -\mathbb{E}[\mathcal{R}_e] - (z_e + \mathbf{q}_e)^\top \mu + \mathbb{ES}_{\alpha_e}(\varepsilon)\sqrt{(1, z_e + \mathbf{q}_e)^\top \Gamma_e (1, z_e + \mathbf{q}_e)}, \quad e \in \mathrm{D}' \end{aligned}$$

with $\varepsilon \sim \mathcal{E}_1(0, 1, \psi)$. In view of Remark 5, by change of variable $\mathbf{z}'_e = \mathbf{q}'_e - \mathbf{q}_e$, the proof of Proposition 1 shows that, for $e \in \mathrm{D}' \setminus \{ccp\}$, r'_e is differentiable and strictly convex with

$$\begin{aligned} -\,\mathbf{p}'^{\mathrm{D}} &= \nabla r'_e(\mathbf{z}'_e) = -\mu + \frac{\mathbb{ES}_{\alpha_e}(\varepsilon)}{\sqrt{(1, \mathbf{z}'_e + \mathbf{q}_e)^\top \Gamma_e (1, \mathbf{z}'_e + \mathbf{q}_e)}} (\mathrm{cov}_e + \Gamma \mathbf{z}'_e + \Gamma \mathbf{q}_e) \\ &= -\mu + \frac{\mathbb{ES}_{\alpha_e}(\varepsilon)}{\sqrt{\mathbb{V}\mathrm{ar}(\mathcal{R}_e) + 2(\mathbf{q}'_e)^\top \mathrm{cov}_e + (\mathbf{q}'_e)^\top \Gamma \mathbf{q}'_e}} \left(\mathrm{cov}_e + \Gamma \mathbf{q}'_e\right). \end{aligned}$$

For the CCP, the member ccp optimality condition gives $-\mathbf{p}'^{\mathrm{D}} \in \partial r'_{ccp}(\mathbf{z}'_{ccp}) = \partial r'_{ccp}(\mathbf{q}'_{ccp})$, i.e. $(r'_{ccp})^*(-\mathbf{p}'^{\mathrm{D}}) = -{\mathbf{q}'_{ccp}}^\top \mathbf{p}'^{\mathrm{D}} - r'_{ccp}(\mathbf{q}'_{ccp})$. Hence computing the optimal post-default position reduces to the root-finding problem

$$\begin{cases} \sum_{e \in \mathrm{D}'} \mathbf{q}'_e + \mathbf{q}_d = 0 \\ \frac{\mathbb{ES}_{\alpha_1}(\varepsilon)}{\mathbb{ES}_{\alpha_e}(\varepsilon)} \sqrt{\frac{\mathbb{V}\mathrm{ar}(\mathcal{R}_e) + 2(\mathbf{q}'_e)^\top \mathrm{cov}_e + (\mathbf{q}'_e)^\top \Gamma \mathbf{q}'_e}{\mathbb{V}\mathrm{ar}(\mathcal{R}_1) + 2{\mathbf{q}'_1}^\top \mathrm{cov}_1 + {\mathbf{q}'_1}^\top \Gamma \mathbf{q}'_1}} (\mathrm{cov}_1 + \Gamma \mathbf{q}'_1) - \mathrm{cov}_e - \Gamma \mathbf{q}'_e = 0, \quad e \in \mathrm{D}' \setminus \{ccp\} \\ r'_{ccp}(\mathbf{q}'_{ccp}) + {\mathbf{q}'_{ccp}}^\top \mathbf{p}'^{\mathrm{D}} + (r'_{ccp})^*(-\mathbf{p}'^{\mathrm{D}}) = 0. \end{cases} \tag{48}$$

Example 3. We consider the same setup as in Example 1, but in the "hedging on own exchange D" case with $\mathrm{D}' \setminus \mathrm{D} = \{ccp\}$ as per §3.C.2. The corresponding pre- and post-default optimal positions computed by (45) and (48) are given in Table 35. For $\mu = 2$, we obtain $\mathbf{p}^{\mathrm{D}} = 1.96, \mathbf{p}'^{\mathrm{D}} = 2.04$ and $\mathrm{MC_D} = 0.69$ (with, by the third line in Table 32, $\mathrm{LC_D} = -1.39$).

Table 36 displays the impact of the default resolution of d in terms of LC_e and $\Delta \boldsymbol{\rho}_e, e \in \mathrm{D}'$. As can be seen from the table, the impact of the default resolution on the $\Delta \boldsymbol{\rho}_e$ is almost the same in the liquidation and hedging cases, whereas its impact on the LC_e is significantly different in both cases.

Table 35: Pre- and post-default optimal positions when the CCP hedges on its own exchange D with $D' \setminus D = \{ccp\}$ in the expected shortfall case.

CM_e	1	2	3	4	5	6	7	8
cov_e	0.05	−0.10	0.14	−0.19	0.24	−0.29	0.34	−0.38
$\mathbf{q}_e$	−1.12	2.56	−3.36	5.12	−5.60	7.68	−7.84	10.24
$\mathbf{q}'_e$	−1.28	2.24	−3.84	4.48	−6.40	6.72	−8.96	8.96

CM_e	9	10	11	12	13	14	15	*ccp*
cov_e	0.43	−0.48	0.53	−0.58	0.62	−0.67	0.72	−0.69
$\mathbf{q}_e$	−10.08	12.80	−12.32	15.36	−14.56	17.92	−16.80	
$\mathbf{q}'_e$	−11.52	11.20	−14.08	13.44	−16.64	15.68		16.80

Table 36: Impacts of the default resolution on the LC_e and $\Delta\boldsymbol{\rho}_e$ in the liquidation and hedging cases in the expected shortfall examples 1-3.

	CM_e	1	2	3	4	5	6	7	8
Liquidation	LC_e	0.11	−0.18	0.32	−0.37	0.53	−0.56	0.74	−0.74
	$\Delta\boldsymbol{\rho}_e$	−0.10	0.20	−0.30	0.40	−0.50	0.60	−0.70	0.80
Hedging	LC_e	0.09	−0.21	0.28	−0.42	0.46	−0.64	0.65	−0.85
	$\Delta\boldsymbol{\rho}_e$	−0.10	0.20	−0.30	0.40	−0.50	0.60	−0.70	0.80

	CM_e	9	10	11	12	13	14	*ccp*
Liquidation	LC_e	0.95	−0.93	1.167	−1.11	1.38	−1.30	
	$\Delta\boldsymbol{\rho}_e$	−0.89	0.99	−1.09	1.19	−1.29	1.39	
Hedging	LC_e	0.84	−1.06	1.02	−1.27	1.21	−1.48	0
	$\Delta\boldsymbol{\rho}_e$	−0.89	0.99	−1.09	1.19	−1.29	1.39	1.39

§5 CREDIT COST

The market cost (MC) term (29) only addresses the impact of the considered default resolution strategy in terms of mis-hedge of market risk. It remains to address its credit cost (CC). By this we need to account for counterparty credit risk in a broad sense including the implications of this risk in terms of capital and funding costs, which we quantify by the XVA metrics of Chapter VIII. The overall impact of a clearing member default's resolution strategy will then be assessed in terms of the following all-inclusive funds transfer price:

$$\mathrm{FTP} = \mathrm{MC} + \mathrm{CC}, \tag{49}$$

which should thus supersede MC in the default resolution approach of §3.A.

A Structure of the Exchanges

As displayed in Figure 48, clearing member bank i's trades with a CCP are divided into proprietary or house trades $\mathbf{q_i}$, which are in effect hedges of the bank's OTC bilateral trading exposures $\mathcal{R}_i = \sum_{b\in B} \mathcal{R}_i^b$ (where the non-cleared, end-clients b are "outside" of the exchange), and back-to-back hedges $\mathbf{q_c^i}$ of intermediated cleared client trades, through which non-member clients c (simple participants to the exchange) can access the CCP clearing services.

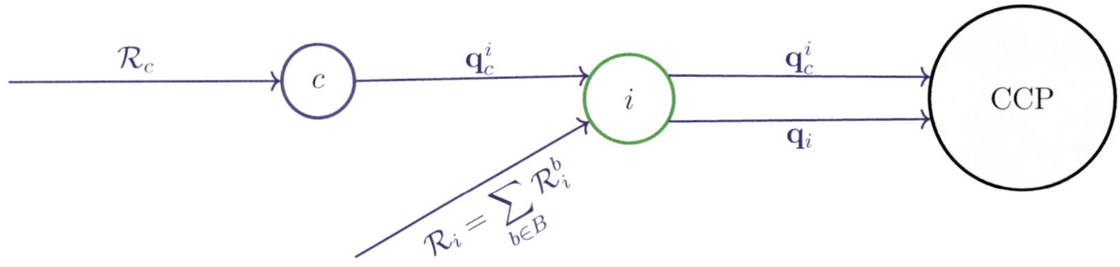

Figure 48: Client clearing ($\mathbf{q}_c^i$ transits from c to the CCP via i) versus bilateral hedging (i hedges $\mathcal{R}_i = \sum_{b\in B} \mathcal{R}_i^b$ by a proprietary trading position $\mathbf{q}_i$ with the CCP).

Our next task is to clarify the nature of the Radner equilibrium on an exchange E, accounting for this distinction between proprietary and cleared deals. Let $E = I \cup C$, with $I \cap C = \varnothing$, be the split between the set I of those participants i to the exchange that are also clearing members of the CCP and the set C of simple participants (non-clearing members) c, having recourse to the clearing members for intermediating trades with the CCP. Let q_i be the proprietary position of member i and q_c^i be the position of client (simple participant to the exchange) c cleared by member i. As depicted in Figure 48, the position q_c^i only transits from c to i and then passes from i to the CCP. Hence, even though the total position of i vis-à-vis the CCP is $q_i + \sum_{c\in C} q_c^i$, as i holds $-\sum_{c\in C} q_c^i$ vis-à-vis the clients it clears for, the respective positions of i and c involved in the Radner equilibrium on E are q_i and $q_c = \sum_{\imath\in I} q_c^\imath$. Once $((\mathbf{q}_e)_{e\in E=I\cup C}, \mathbf{p}^E)$ has been obtained as the solution of the corresponding Radner equilibrium (7)-(8), the splits $\mathbf{q}_c = \sum_{i\in I} \mathbf{q}_c^i$ (one for each $c \in C$) should follow in a second stage from pure credit risk considerations, quantifiable by suitable XVA metrics. Similar comments apply "with $\cdot'$ everywhere" to any post-default Radner equilibrium on $E' = I' \cup C'$.

B Credit Costs XVA Framework

Our XVA metrics are computed under the premise that the (random) loss triggered by the default of a market participant in the future is allocated between the surviving members of its CCP, pro-rata of their funded default fund contributions to the CCP (see after (60)). At time "0−", i.e. "right before" the instant default of the clearing member d at time 0, the participants e to the exchanges charge to their clients the expectation of their ensuing future default losses, in the form of their CVA_e, as well as their rehypothecated and segregated collateral funding costs FVA_e and MVA_e, and costs of capital KVA_e. These costs sum up to $\mathrm{XVA}_e = \mathrm{CVA}_e + \mathrm{FVA}_e + \mathrm{MVA}_e + \mathrm{KVA}_e$, computed for each participant e the way detailed for $e = 0$ in Subsection C below, based on the pre-default Radner equilibrium quantities and prices on each exchange. To quantify the XVA impact of a given default resolution procedure for a CCP portfolio of d, we also compute the time 0, post-default XVA'_e, for any participant $e \neq d$ to the exchanges. The credit cost of the settlement of the defaulted portfolio, coming on top of the already computed MC, is

$$\mathrm{CC} = \sum_{e \neq d} \underbrace{(\mathrm{XVA}'_e - \mathrm{XVA}_e)}_{\Delta \mathrm{XVA}_e} + \mathrm{AC}, \tag{50}$$

where the auctioning cost AC is another XVA incremental impact corresponding to the FTP in the sense of VIII.§7, which only involved credit costs, i.e. the XVA impact of auctioning any hedged (as opposed to liquidated) positions. Indeed, a CCP cannot keep the defaulted portfolio and the corresponding hedge on its book, it will look at auctioning them (Ferrara and Li, 2020; Oleschak et al., 2019).

In the end, the all-inclusive FTP (49) that emerges from the present chapter (where both market and credit costs are involved) is

$$\mathrm{FTP} = \underbrace{\mathrm{LC} + \sum_{E'} \sum_{e \in E'} \Delta \boldsymbol{\rho}_e}_{\mathrm{MC}} + \underbrace{\sum_{e \neq d} \Delta \mathrm{XVA}_e + \mathrm{AC}}_{\mathrm{CC}}. \tag{51}$$

Remark 8. Under a pure auctioning default resolution strategy (as for instance in VIII.§7), only the last AC term remains. But this AC term could be very expensive in the case of an unhedged portfolio.

C XVA Setup

The purpose of this part is to provide a bridge between the equilibrium setup of §2–§4 and the XVA setup of Chapter VIII, so that we are able to provide an overall FTP (51) quantifying the market but also credit costs of a given default resolution strategy. We leave for future research the extension of the approach of this chapter to a setup where not only the market costs, but also the credit costs, would be treated endogenously as part of a global equilibrium (or perhaps a two-stage equilibrium, accounting for the segregation between the market and credit spheres in banks (European Parliament, 2013, Article 92)), ideally in a dynamic setup.

We endorse a structure of the exchanges $E = I \cup C$ (with I and C implicitely parameterized by E) as per Subsection A and we also denote by B the set of all end-clients of the reference bank. Note that even for those clients clearing through a CCP member and also having OTC positions with that same member, the corresponding exposures must be treated separately, hence $C \cap B = \varnothing$.

Table 37: Some notation adaptation for the cash flows of the market participant 0 after the variation margin is subtracted, in the setup of Chapter VIII (with CCPs indexed by *ccp* in Chapter VIII and one CCP per exchange E in this chapter).

VIII.(18)-(19)	**This chapter**	**Description**
$\mathcal{P}_c^{ccp} - P_c^{ccp}$	$(\mathbf{q}_c^0)^\top(\mathbf{p}^E - \mathcal{P})$	Cash flows from a cleared client c of the CCP to the participant 0 (if the latter is not a clearing member of the CCP, these cash flows are simply zero).
$\mathcal{P}_i^{ccp} - P_i^{ccp}$	$\sum_{c\in C}(\mathbf{q}_c^i)^\top(\mathbf{p}^E - \mathcal{P})$	Client account cash flows from a clearing member i to the CCP.
$\overline{\mathcal{P}}_i^{ccp} - \overline{P}_i^{ccp}$	$\mathbf{q}_i^\top(\mathbf{p}^E - \mathcal{P})$	Proprietary account cash flows from a clearing member i to the CCP.
$\mathcal{P}_b - \mathrm{VM}_b$	$\mathcal{R}_0^b$	Cash flows from an end-user b to participant 0.

Let J_e be the survival indicator of participant e to the exchange E, i.e. $J_e = \mathbb{1}_{\{\tau_e > T\}}$, where τ_e is the default time of participant e, with probability $\gamma_e = \mathbb{Q}^*(J_e = 0)$ of default over $[0, T]$. We denote likewise $J_b = \mathbb{1}_{\{\tau_b > T\}}$ for any end-user $b \in B$. In what follows we focus on a participant $e = 0$ to the exchange, typically skipping the index 0 regarding this reference participant (except in the notation $\mathbf{q}_c^0$ and the likes, cf. Figure 48 page 274). A comparison of the present setup with Eqns. VIII.(18)-(19) yields the mapping of Table 37. This results in the following (pre-default equilibrium) credit loss profile $\mathcal{C}$ of the reference participant 0: $\mathcal{C} = \sum_{E=I\cup C}(\mathcal{C}^I + \mathcal{C}^C) + \mathcal{C}^B$, where (with I and C

are implicitely parameterized by E)

$$
\begin{aligned}
\mathcal{C}^I &= w^I \sum_{i\in I}(1-J_i)\Big[\Big(\Big(\sum_{c\in C}\mathbf{q}^i_c\Big)^\top(\mathbf{p}^E-\mathcal{P})-\mathrm{PIM}^I_i\Big)^+ + \\
&\qquad \Big(\mathbf{q}_i^\top(\mathbf{p}^E-\mathcal{P}\ )-\overline{\mathrm{PIM}}^I_i\Big)^+ - \mathrm{DFC}^I_i\Big]^+, \\
\mathcal{C}^C &= \sum_{c\in C}(1-J_c)\Big((\mathbf{q}^0_c)^\top(\mathbf{p}^E-\mathcal{P})-\mathrm{RIM}_c\Big)^+, \\
\mathcal{C}^B &= \sum_{b\in B}(1-J_b)(\mathcal{R}^b-\mathrm{RIM}_b)^+.
\end{aligned}
\tag{52}
$$

Here RIM_c is the initial margin (IM) requested by the reference participant $e=0$ to the simple participant $c\in C$ on the cleared position $\mathbf{q}^0_c$ (equal to 0 if $0\notin I$); w^I (equal to 0 if $0\notin I$) is the loss allocation coefficient of the participant 0 w.r.t. the CCP; PIM^I_i, $\overline{\mathrm{PIM}}^I_i$ and DFC^I_i are the initial margins for the cleared clients and proprietary accounts, as well as the funded default fund contribution requested by the CCP to the clearing member i; $\mathcal{R}^b$ is the exogenous receivable of the participant 0 from its OTC bilateral counterparty b, with corresponding initial margin RIM_b (which can be null under OTC agreement) requested by 0 to b (cf. Figure 48 page 274).

Likewise, the post-default equilibrium[4] default loss profile $\mathcal{C}'$ of the participant 0 is $\mathcal{C}'=\sum_{E'=I'\cup C'}(\mathcal{C}'^I+\mathcal{C}'^C)+\mathcal{C}^B$, for $\mathcal{C}^B$ as in (52) and (cf. Table 37 and (24))

$$
\begin{aligned}
\mathcal{C}'^I \underset{(27)}{=}\ & w'^I\sum_{i\in I'}(1-J_i)\times \\
&\Bigg(\Big(\sum_{c\in C'}(\mathbf{q}'^i_c)^\top(\mathbf{p}'^E-\mathcal{P})+\sum_{c\in C'}\Big(\mathbf{q}^i_c+(\Delta\mathbf{q}^i_c)^l\Big)^\top(\mathbf{p}^E-\mathbf{p}'^E)-\mathrm{PIM}'^I_i\Big)^+ + \\
&\Big((\mathbf{q}'_i)^\top(\mathbf{p}'^E-\mathcal{P})+(\mathbf{q}_i+\Delta\mathbf{q}^\ell_i)^\top(\mathbf{p}^E-\mathbf{p}'^E)-\overline{\mathrm{PIM}}'^I_i\Big)^+ - \mathrm{DFC}'^I_i\Bigg)^+, \\
\mathcal{C}'^C =\ & \sum_{c\in C'}(1-J_c)\Big((\mathbf{q}'^0_c)^\top(\mathbf{p}'^E-\mathcal{P})+\Big(\mathbf{q}^0_c+(\Delta\mathbf{q}^0_c)^l\Big)^\top(\mathbf{p}^E-\mathbf{p}'^E)-\mathrm{RIM}'^c\Big)^+,
\end{aligned}
\tag{53}
$$

where w'^I, PIM'^I_i, $\overline{\mathrm{PIM}}'^I_i$, DFC'^I_i and RIM'^c are the post-default analogs of w^I, PIM^I_i, $\overline{\mathrm{PIM}}^I_i$, DFC^I_i and RIM^c in (52).

By Theorem VIII.2, the pre- and post-default CVA of the participant 0 are given by

$$
\mathrm{CVA}=\mathbb{E}\Big[\mathcal{C}\Big|J=1\Big]=(1-\gamma)^{-1}\mathbb{E}\left[J\mathcal{C}\right],\ \mathrm{CVA}'=(1-\gamma)^{-1}\mathbb{E}\left[J\mathcal{C}'\right], \tag{54}
$$

[4]post-default referring in this chapter to the instant default at time 0 of a clearing member d, here assumed different from the reference clearing member 0.

Denoting by PIM_b the initial margin from the participant 0 to its OTC bilateral counterparty b, based on Remark 1 page 256, such margin remains constant in the post-default equilibrium, hence the pre- and post-default MVA of participant 0 are given by

$$\begin{aligned}\mathrm{MVA} &= \tilde{\gamma}\left(\sum_{b\in B}\mathrm{PIM}_b + \sum_{E=I\cup C}\left(\mathrm{PIM}^I + \overline{\mathrm{PIM}}^I + \mathrm{DFC}^I\right)\right),\\ \mathrm{MVA}' &= \tilde{\gamma}\left(\sum_{b\in B}\mathrm{PIM}_b + \sum_{I=I\cup C}\left(\mathrm{PIM}^{I'} + \overline{\mathrm{PIM}}^{I'} + \mathrm{DFC}^{I'}\right)\right),\end{aligned} \tag{55}$$

for some possibly blended funding rate $\tilde{\gamma} \le \gamma$ as detailed in III.§2.C. The pre-default KVA, defined for a hurdle rate h, is calculated based on an expected shortfall $\mathbb{ES}_{\tilde{\alpha}}$ of the credit loss $\mathcal{L} = \mathcal{C} - \mathrm{CVA}$ (on the event $J = 1$) of participant 0 under its own survival measure, $\mathbb{Q}^*(\cdot J)/(1-\gamma)$ (with $\tilde{\alpha} >$ the confidence level α introduced for the market cost computation in §4 when the risk measures used by the hedgers are expected shortfall risk measures[5]), as

$$\begin{aligned}\mathrm{KVA} &= \frac{h}{1+h}\mathbb{ES}_{\tilde{\alpha}}\left(\mathcal{C} - \mathrm{CVA}\right)\\ &= \frac{h}{1+h}\mathbb{E}\left[J\left(\mathcal{C} - \mathrm{CVA}\right)\middle|\mathcal{C} - \mathrm{CVA} \ge \mathbb{VaR}_{\tilde{\alpha}}\left(J(\mathcal{C} - \mathrm{CVA})\right), J = 1\right],\end{aligned} \tag{56}$$

by the last row of Table 19, where $\mathbb{VaR}_{\tilde{\alpha}}$ denotes the value-at-risk at the confidence level $\tilde{\alpha}$ under the measure $(1-\gamma)^{-1}\mathbb{Q}^*(\cdot J)$. The post-default KVA has a similar expression substituting $\mathcal{C}'$ to $\mathcal{C}$ and CVA' to CVA in (56):

$$\begin{aligned}\mathrm{KVA}' &= \frac{h}{1+h}\mathbb{ES}_{\tilde{\alpha}}\left(\mathcal{C}' - \mathrm{CVA}'\right)\\ &= \frac{h}{1+h}\mathbb{E}\left[J\left(\mathcal{C}' - \mathrm{CVA}\right)\middle|\mathcal{C}' - \mathrm{CVA} \ge \mathbb{VaR}_{\tilde{\alpha}}\left(J(\mathcal{C}' - \mathrm{CVA}')\right), J = 1\right].\end{aligned} \tag{57}$$

Finally, by the next-to-last row of Table 19, the pre- and post- FVA of the participant 0 is given by

$$\begin{aligned}\mathrm{FVA} &= \frac{\gamma}{1+\gamma}\left(\sum_{b\in B}\mathbb{E}\mathcal{R}^b - (\mathrm{CVA} + \mathrm{MVA}) - \mathbb{ES}_{\tilde{\alpha}}\left(\mathcal{C} - \mathrm{CVA}\right)\right)^+,\\ \mathrm{FVA}' &= \frac{\gamma}{1+\gamma}\left(\sum_{b\in B}\mathbb{E}\mathcal{R}^b - (\mathrm{CVA}' + \mathrm{MVA}') - \mathbb{ES}_{\tilde{\alpha}}\left(\mathcal{C}' - \mathrm{CVA}'\right)\right)^+.\end{aligned} \tag{58}$$

[5] regulatory and economic capital aim at capturing extreme losses that can occur once every 1000 years (Basel Committee on Banking Supervision, 2005, Paragraph 5.1), which leads to considering a much higher confidence level $\tilde{\alpha}$ for economic capital calculation, such as 0.9975, from which the KVA is defined, than the α used for market risk, set to 0.975, in line with (Bank for International Settlements, 2013, Section 1.4 (i)).

§6 EXAMPLES 1 AND 3 CONTINUED

We complete from the credit costs perspective Examples 1 and 3. In the case of default resolution procedures implemented on the own exchange D of the CCP, (51) boils down to

$$\mathrm{FTP} = \mathrm{LC}_{\mathrm{D}} + \underbrace{\sum_{e\in \mathrm{D}\setminus\{d\}} \Delta\boldsymbol{\rho}_e}_{\mathrm{MC}} + \underbrace{\sum_{e\in \mathrm{D}\setminus\{d\}} \Delta\mathrm{XVA}_e + \mathrm{AC}}_{\mathrm{CC}}\,. \tag{59}$$

We assume that only clearing members i participate to the only exchange of interest D. So all the participants e are clearing members i and there are no cleared clients c, i.e., in §5's notation, we only consider $E = \mathrm{D}$, with related $I = \mathrm{D}$ and $C = \varnothing$. Moreover, all the end-users (OTC bilateral counterparties) b are assumed to be default risk-free, Hence the pre-default credit loss (52) of member $0 \in \mathrm{D}$ reduces to

$$\mathcal{C} = w^{\mathrm{D}} \sum_{i\in \mathrm{D}} (1-J_i)\left((\mathbf{q}_i)^\top(\mathbf{p}^{\mathrm{D}} - \mathcal{P}) - \overline{\mathrm{PIM}}_i^{\mathrm{D}} - \mathrm{DFC}_i^{\mathrm{D}}\right)^+, \tag{60}$$

where $w^{\mathrm{D}} = \frac{\mathrm{DFC}^{\mathrm{D}} J}{\sum_{i\in\mathrm{D}} \mathrm{DFC}_i^{\mathrm{D}} J_i}$. Under the post-default equilibrium when the CCP fully liquidates on D (so that $\Delta\mathbf{q}_i^h = 0$), the post-default credit loss (53) of member $0 \in \mathrm{D}$ reduces to

$$\begin{aligned}\mathcal{C}' &= w'^{\mathrm{D}} \sum_{i\in \mathrm{D}'} (1-J_i)\left((\mathbf{q}_i')^\top(\mathbf{p}'^{\mathrm{D}} - \mathcal{P}) + (\mathbf{q}_i')^\top(\mathbf{p}^{\mathrm{D}} - \mathbf{p}'^{\mathrm{D}}) - \overline{\mathrm{PIM}}_i'^{\mathrm{D}} - \mathrm{DFC}_i'^{\mathrm{D}}\right)^+ \\ &= w'^{\mathrm{D}} \sum_{i\in \mathrm{D}'} (1-J_j)\left((\mathbf{q}_i')^\top(\mathbf{p}^{\mathrm{D}} - \mathcal{P}) - \overline{\mathrm{PIM}}_i'^{\mathrm{D}} - \mathrm{DFC}_i'^{\mathrm{D}}\right)^+,\end{aligned} \tag{61}$$

whereas, when the CCP fully hedges on D (so that $\Delta\mathbf{q}_i^l = 0$, $\mathbf{q}_j + \Delta\mathbf{q}_j^h = \mathbf{q}_j'$),

$$\mathcal{C}' = w'^{\mathrm{D}} \sum_{i\in \mathrm{D}'} (1-J_i)\left((\mathbf{q}_i')^\top(\mathbf{p}'^{\mathrm{D}} - \mathcal{P}) + (\mathbf{q}_i)^\top(\mathbf{p}^{\mathrm{D}} - \mathbf{p}'^{\mathrm{D}}) - \overline{\mathrm{PIM}}_i'^{\mathrm{D}} - \mathrm{DFC}_i'^{\mathrm{D}}\right)^+. \tag{62}$$

In both cases w'^{D}, $\overline{\mathrm{PIM}}_i'^{\mathrm{D}}$ and $\mathrm{DFC}_i'^{\mathrm{D}}$, $i \in \mathrm{D}'$, are the post-default analogs of w^{D}, $\overline{\mathrm{PIM}}_i^{\mathrm{D}}$ and $\mathrm{DFC}_i^{\mathrm{D}}$, $i \in \mathrm{D}$, based on the post-default updated portfolio positions. The pre- and post-default CVA, MVA and KVA of member $0 \in$

$D \setminus \{d\}$ are given by

$$\begin{aligned}
&\mathrm{CVA} = (1-\gamma)^{-1}\mathbb{E}\left[J\mathcal{C}\right], \ \mathrm{CVA}' = (1-\gamma)^{-1}\mathbb{E}\left[J\mathcal{C}'\right],\\
&\mathrm{MVA} = \tilde{\gamma}\left(\overline{\mathrm{PIM}}^{\mathrm{D}} + \mathrm{DFC}^{\mathrm{D}}\right), \ \mathrm{MVA}' = \tilde{\gamma}\left(\overline{\mathrm{PIM}}'^{\mathrm{D}} + \mathrm{DFC}'^{\mathrm{D}}\right),\\
&\mathrm{KVA} = \frac{h}{1+h}\mathbb{E}\left[J\left(\mathcal{C}-\mathrm{CVA}\right)\middle|\mathcal{C}-\mathrm{CVA} \geq \mathbb{V}\mathrm{a}\mathbb{R}_{\tilde{\alpha}}\left(J(\mathcal{C}-\mathrm{CVA})\right), J=1\right],\\
&\mathrm{KVA}' = \frac{h}{1+h}\mathbb{E}\left[J\left(\mathcal{C}'-\mathrm{CVA}'\right)\middle|\mathcal{C}'-\mathrm{CVA}' \geq \mathbb{V}\mathrm{a}\mathbb{R}_{\tilde{\alpha}}\left(J(\mathcal{C}'-\mathrm{CVA}')\right), J=1\right].
\end{aligned} \tag{63}$$

Moreover, in the present setup, $\mathbb{E}\mathcal{R}^b$ in (58) corresponds to $P_b - \mathrm{VM}_b$ in the setup of Chapter VIII. We assume, as it is most commonly the case in practice, that $\mathrm{VM}_b \approx P_b$. Hence we have $\mathbb{E}\mathcal{R}^b \approx 0$ and, in any case, dominated by $(\mathrm{CVA}+\mathrm{MVA}) - \mathbb{E}\mathbb{S}_{\tilde{\alpha}}\left(\mathcal{C}-\mathrm{CVA}\right)$ and $(\mathrm{CVA}'+\mathrm{MVA}') - \mathbb{E}\mathbb{S}_{\tilde{\alpha}}\left(\mathcal{C}'-\mathrm{CVA}'\right)$ in (58), leading to negligible FVA and FVA′, which we will therefore simply take as 0 (and do not report) in our numerics later below.

A Latent Factor Model

For default modeling purposes, we introduce for each member $i \in \mathrm{D}$ a latent variable $\mathcal{X}_i \sim \mathcal{N}_1(0,1)$ such that $\{J_i = 0\} \iff \left\{\mathcal{X}_i \leq \mathcal{N}^{-1}(\gamma_i)\right\}$, where $\mathcal{N}$ denotes the standard Gaussian cdf. These default latent variables are correlated as per $\mathcal{X}_i = \sqrt{\varrho^{cr}}\varepsilon + \sqrt{1-\varrho^{cr}}\varepsilon_i$, where ε and ε_i are i.i.d. $\mathcal{N}_1(0,1)$, while ϱ^{cr} is a positive credit/credit correlation coefficient. Writing $\mathcal{P} = \mu + \sigma\mathcal{Y}$ with $\mathcal{Y} \sim \mathcal{N}_1(0,1)$, the IM posted to the CCP by member i, based on the idea of a $\mathbf{q}_i(\mathbf{p}^{\mathrm{D}} - \mathcal{P})$ VM call not fulfilled over the time period T (versus Δ_s in Chapter VIII), is computed by a VaR metric $\mathbb{V}\mathrm{a}\mathbb{R}^i$ with respect to the survival measure of member i associated with $\mathbb{Q}^*$, at a confidence level $\alpha_{im} \in (1/2, 1)$, as

$$\overline{\mathrm{PIM}}_i^{\mathrm{D}} = \mathbb{V}\mathrm{a}\mathbb{R}^i_{\alpha_{im}}\left(\mathbf{q}_i(\mathbf{p}^{\mathrm{D}} - \mathcal{P})\right) = \mathbf{q}_i(\mathbf{p}^{\mathrm{D}} - \mu) + |\mathbf{q}_i|\,\sigma\mathcal{N}^{-1}(\alpha_{im}). \tag{64}$$

The liquidation time period Δ_l taken by the CCP in Chapter VIII to novate or liquidate the portfolio of d is also taken as T the present chapter.

Remark 9. In a one-period setup, the choice of Δ_l and Δ_s should reconcile with the hedging position time period Δ_l considered to be included in the expressions $\mathbf{q}_i^\top(\mathbf{p}^{\mathrm{D}} - \mathcal{P})$, $(\mathbf{q}_i')^\top(\mathbf{p}'^{\mathrm{D}} - \mathcal{P})$ and $(\mathbf{q}_i'))^\top(\mathbf{p}^{\mathrm{D}} - \mathcal{P})$ from (60), (61) and (62) respectively. The choice $\Delta_s = \Delta_l = T$ is made for consistent and comparable numerics with respect to the quantities MC and CC.

The funded default fund is calculated at the level of the considered CCP of d as the sum of the two highest stress losses over IM (VaRSLOIM), where VaRSLOIM is given for each member i as

$$\mathrm{VaRSLOIM}_i^{\mathrm{D}} = \mathbb{V}\mathrm{a}\mathbb{R}^i_{\alpha_{df}}\left(\mathbf{q}_i(\mathbf{p}^{\mathrm{D}} - \mathcal{P}) - \overline{\mathrm{PIM}}_i^{\mathrm{D}}\right) = |\mathbf{q}_i|\,\sigma\left(\mathcal{N}^{-1}(\alpha_{df}) - \mathcal{N}^{-1}(\alpha_{df})\right),$$

for some confidence level $\alpha_{df} > \alpha_{im}$. The default fund contribution of member i is given as

$$\mathrm{DFC}_i^{\mathrm{D}} = \frac{\mathrm{VaRSLOIM}_i^{\mathrm{D}}}{\sum_{a\in \mathrm{D}} \mathrm{VaRSLOIM}_a^{\mathrm{D}}} \left(\mathrm{VaRSLOIM}_{(0)}^{\mathrm{D}} + \mathrm{VaRSLOIM}_{(1)}^{\mathrm{D}}\right), \tag{65}$$

based on the Cover2 amount given as the sum of two largest stressed losses over IM ($\mathrm{VaRSLOIM}_i$) among its members, identified with subscripts (0) and (1).

Table 38: XVAs calculation configuration.

Portfolios maturity T	5 years
Liquidation period at default Δ_l	5 years
Credit factors correlation ϱ^{cr}	20%
IM covering period (margin period of risk) Δ_s	5 years
Default probabilities γ_i	39.3%
MVA funding blending ratio $\tilde{\gamma}_i/\gamma_i$	25%
Quantile levels $\tilde{\alpha}_i$ used for clearing members KVA	99.75%
Hurdle rate h used for KVA computations	10.0%
Number of Monte-Carlo simulations (for CVA and KVA computations)	10M
Number of batches (for KVA computations)	100

B Liquidation

In the "liquidation on own exchange D" case of Example 1 page 271, Table 39 provides, for each clearing member i successively envisioned in the role of the reference participant to the exchange labeled $e = 0$ before (so we now label the different metrics by lower indices i, while no index 0 was used for the sake of notational lightness in the above), the resulting "time 0−" XVA_i (with all members, d included) and time 0 XVA_i' (without d), using the allocated positions of Table 34 and the XVA specifications of Table 38 along with $\overline{\mathrm{PIM}}_i^{\mathrm{D}} = \mathrm{DFC}_i^{\mathrm{D}} = 0$, whilst Table 40 presents the same results for $\overline{\mathrm{PIM}}_i^{\mathrm{D}}$ and $\mathrm{DFC}_i^{\mathrm{D}}$ set at the 75% and 80% confidence levels. The chosen period length of $T = 5$ years covers the bulk (if not the final maturity) of most realistic CCP portfolios. Note that there is no auction in this (liquidation) case. The aggregated XVA cost (50) of liquidating the defaulted portfolio reduces to $\mathrm{CC} = \sum_{i\in \mathrm{D}\setminus\{d\}} \Delta\mathrm{XVA}_i$, namely -0.77 for the case $\overline{\mathrm{PIM}}_i^{\mathrm{D}} = \mathrm{DFC}_i^{\mathrm{D}} = 0$ and -0.25 with $\overline{\mathrm{PIM}}_i^{\mathrm{D}}$ and $\mathrm{DFC}_i^{\mathrm{D}}$ set at the 75% and 80% confidence level, coming on top of the market cost of $\mathrm{MC} = 0.70$ already obtained in Example 1.

Table 39: The pre- and post-default XVAs computed from (63) when the CCP liquidates the portfolio of the defaulted member $d = 15$ on its own exchange, D, in the expected shortfall case of Example 1 with $\overline{\mathrm{PIM}}_i^{\mathrm{D}}$ and $\mathrm{DFC}_i^{\mathrm{D}}$ set to 0.

CM_i	1	2	3	4	5	6	7	8
XVA_i	0.81	1.21	1.36	1.70	1.74	2.05	2.02	2.37
XVA_i'	0.92	1.17	1.47	1.61	1.82	1.93	2.07	2.19
$\Delta\mathrm{XVA}_i$	0.11	−0.04	0.12	−0.09	0.08	−0.12	0.05	−0.18

CM_i	9	10	11	12	13	14	15
XVA_i	2.25	2.62	2.42	2.84	2.57	3.04	2.69
XVA_i'	2.28	2.42	2.42	2.60	2.54	2.77	
$\Delta\mathrm{XVA}_i$	0.03	−0.20	−0.01	−0.24	−0.03	−0.27	

Table 40: The pre- and post-default XVAs computed from (63) when the CCP liquidates the portfolio of the defaulted member $d = 15$ on its own exchange D, in the expected shortfall case of Example 1 with $\overline{\mathrm{PIM}}_i^{\mathrm{D}}$ and $\mathrm{DFC}_i^{\mathrm{D}} = 0$ set at the 75% and 80% confidence levels.

CM_i	1	2	3	4	5	6	7	8
XVA_i	0.37	0.59	0.72	0.89	1.00	1.12	1.21	1.34
XVA_i'	0.42	0.56	0.80	0.84	1.07	1.05	1.28	1.22
$\Delta\mathrm{XVA}_i$	0.05	−0.03	0.08	−0.06	0.07	−0.07	0.07	−0.11

CM_i	9	10	11	12	13	14	15
XVA_i	1.42	1.50	1.57	1.64	1.72	1.77	1.85
XVA_i'	1.49	1.37	1.63	1.49	1.78	1.60	
$\Delta\mathrm{XVA}_i$	0.07	−0.12	0.06	−0.15	0.06	−0.17	

C Pure Auctioning

Instead of liquidation on the CCP's own exchange D, we now consider another default resolution strategy, in the form of an (idealized) auction inducing the taker giving rise to the least auction cost AC among all possible takers $i \in \mathrm{D} \setminus \{d\}$, the way developed in VIII.§7. The results are displayed in Table 41 for the case $\overline{\mathrm{PIM}}_i^{\mathrm{D}} = \mathrm{DFC}_i^{\mathrm{D}} = 0$ and in Table 42 for $\overline{\mathrm{PIM}}_i^{\mathrm{D}}$ and $\mathrm{DFC}_i^{\mathrm{D}}$ set at the 75% and 80% confidence levels. From Tables 41–42, participant 14 is the survivor taker leading to the smallest auctioning cost AC, namely −4.52

and −2.88, when taking over the defaulted portfolio of CM_{15} (and there are in this case no additional costs to consider, cf. Remark 8). Such take-over makes intuitive sense given the pre-default position $\mathbf{q}_{15} = -16.8$ of the defaulting member $d = 15$, compared with the position of member 14 at $\mathbf{q}_{14} = 17.92$, an almost offset of $\mathbf{q}_{15}$. Member 12 shows the second-best solution with an auctioning cost AC close to zero, also justifiable by his offsetting position at $\mathbf{q}_{12} = 15.36$.

Table 41: Auctioning costs AC corresponding to the different possible takers of the portfolio of the defaulted member $d = 15$, ranked by increasing value, for $\overline{\mathrm{PIM}}_i^{\mathrm{D}} = \mathrm{DFC}_i^{\mathrm{D}} = 0$. In parenthesis the contributions to AC of the considered possible taker itself.

CM_i	$\sum_{i\in\mathrm{D}\setminus\{d\}} (\mathrm{CVA}'_i - \mathrm{CVA}_i)$	$\sum_{i\in\mathrm{D}\setminus\{d\}} (\mathrm{KVA}'_i - \mathrm{KVA}_i)$	AC
14	-1.00 (-0.38)	-3.52 (-0.63)	-4.52 (-1.01)
12	-0.85 (-0.34)	-2.93 (-0.57)	-3.78 (-0.91)
10	-0.58 (-0.23)	-2.02 (-0.42)	-2.60 (-0.64)
8	-0.31 (-0.11)	-1.13 (-0.24)	-1.44 (-0.35)
6	-0.04 (0.02)	-0.30 (-0.04)	-0.34 (-0.02)
4	0.23 (0.15)	0.42 (0.20)	0.65 (0.35)
2	0.50 (0.29)	0.95 (0.52)	1.45 (0.81)
1	0.77 (0.43)	1.59 (0.86)	2.35 (1.29)
3	0.77 (0.39)	2.09 (0.62)	2.85 (1.01)
5	0.77 (0.36)	2.47 (0.47)	3.23 (0.82)
7	0.77 (0.32)	2.70 (0.37)	3.46 (0.70)
9	0.77 (0.29)	2.84 (0.30)	3.60 (0.60)
11	0.77 (0.26)	2.92 (0.27)	3.69 (0.53)
13	0.77 (0.23)	2.96 (0.25)	3.73 (0.48)

D Hedging Then Auctioning

Finally, we consider one more default resolution strategy, where the CCP hedges (as per §3.C.2) the defaulted portfolio before auctioning all its positions. In the hedging case, with the CCP *ccp* contributing to the post-default quantities and price discovery $((\mathbf{q}'_i)_{i\in\mathrm{D}'=(\mathrm{D}\setminus\{d\})\cup\{ccp\}}, \mathbf{p}'^{\mathrm{D}})$, resolving (32) (for $E = \mathrm{D}$) and (35)–(37) in the configuration of Example 3 page 272 leads to $\mathbf{q}'_{ccp} = -\mathbf{q}_d$. Coincidentally, this hedging resolution thus leads to a perfect replication as per §3.C.3. The corresponding $\sum_{i\neq d} \Delta\mathrm{XVA}_i$, detailed in Tables

Table 42: Auctioning costs AC corresponding to the different possible takers of the portfolio of the defaulted member $d = 15$, ranked by increasing value, for $\overline{\mathrm{PIM}}_i^{\mathrm{D}}$ and $\mathrm{DFC}_i^{\mathrm{D}}$ set at the 75% and 80% confidence levels. In parenthesis, the contributions to AC of the considered possible taker itself.

CM_i	$\sum_{i\in\mathrm{D}\setminus\{d\}}(\mathrm{MVA}_i' - \mathrm{MVA}_i)$	$\sum_{i\in\mathrm{D}\setminus\{d\}}(\mathrm{CVA}_i' - \mathrm{CVA}_i)$	$\sum_{i\in\mathrm{D}\setminus\{d\}}(\mathrm{KVA}_i' - \mathrm{KVA}_i)$	AC
14	-0.06 (-0.15)	-0.49 (-0.13)	-2.33 (-0.42)	-2.88 (-0.71)
12	-0.01 (-0.11)	-0.45 (-0.12)	-2.00 (-0.39)	-2.46 (-0.63)
10	0.06 (-0.04)	-0.36 (-0.08)	-1.47 (-0.27)	-1.77 (-0.40)
8	0.13 (0.03)	-0.27 (-0.04)	-0.93 (-0.14)	-1.07 (-0.16)
6	0.20 (0.10)	-0.17 (0.00)	-0.44 (0.01)	-0.41 (0.11)
4	0.26 (0.17)	-0.07 (0.04)	0.00 (0.19)	0.20 (0.40)
2	0.33 (0.23)	0.03 (0.09)	0.38 (0.41)	0.74 (0.74)
1	0.41 (0.31)	0.11 (0.14)	0.76 (0.64)	1.28 (1.09)
3	0.42 (0.31)	0.10 (0.13)	0.92 (0.51)	1.44 (0.95)
5	0.42 (0.31)	0.09 (0.11)	1.06 (0.42)	1.58 (0.85)
7	0.43 (0.32)	0.08 (0.10)	1.16 (0.37)	1.67 (0.78)
9	0.44 (0.32)	0.07 (0.09)	1.23 (0.31)	1.74 (0.72)
11	0.45 (0.32)	0.06 (0.08)	1.28 (0.28)	1.78 (0.68)
13	0.45 (0.33)	0.05 (0.06)	1.29 (0.26)	1.80 (0.65)

Table 43: The pre- and post-default XVAs computed from (63) when the CCP hedges the portfolio of the defaulted member $d = 15$ on its own exchange D, in the expected shortfall case of Example 3 with $\overline{\mathrm{PIM}}_i^{\mathrm{D}}$ and $\mathrm{DFC}_i^{\mathrm{D}}$ set to 0.

CM_i	1	2	3	4	5	6	7	8
XVA_i	0.81	1.21	1.36	1.7	1.74	2.05	2.02	2.37
XVA_i'	0.19	0.34	0.54	0.63	0.83	0.89	1.07	1.13
$\Delta\mathrm{XVA}_i$	−0.62	−0.87	−0.81	−1.06	−0.91	−1.15	−0.94	−1.24

CM_i	9	10	11	12	13	14	15	*ccp*
XVA_i	2.25	2.62	2.42	2.84	2.57	3.04	2.69	
XVA_i'	1.28	1.34	1.44	1.54	1.59	1.72		2.39
$\Delta\mathrm{XVA}_i$	−0.98	−1.27	−0.98	−1.3	−0.98	−1.32		

43 and 44, is -14.45 for $\overline{\mathrm{PIM}}_i^{\mathrm{D}} = \mathrm{DFC}_i^{\mathrm{D}} = 0$ and -6.80 for $\overline{\mathrm{PIM}}_i^{\mathrm{D}}$ and $\mathrm{DFC}_i^{\mathrm{D}}$ set at the 75% and 80% confidence levels. As the residual market risk is null in such a replication case, taking over the defaulted portfolio and its hedge does not generate any market risk. Hence in these cases no additional cost is generated by the auctioning process, which is why the corresponding AC vanish.

Table 44: The pre- and post-default XVAs computed from (63) when the CCP hedges the portfolio of the defaulted member $d = 15$ on its own exchange D, in the expected shortfall case of Example 3 with $\overline{\mathrm{PIM}}_i^{\mathrm{D}}$ and $\mathrm{DFC}_i^{\mathrm{D}}$ at the 75% and 80% confidence levels.

CM_i	1	2	3	4	5	6	7	8
XVA_i	0.37	0.59	0.72	0.89	1	1.12	1.21	1.34
XVA'_i	0.13	0.21	0.38	0.4	0.6	0.56	0.79	0.72
$\Delta\mathrm{XVA}_i$	−0.24	−0.38	−0.34	−0.5	−0.4	−0.56	−0.42	−0.62

CM_i	9	10	11	12	13	14	15	*ccp*
XVA_i	1.42	1.50	1.57	1.64	1.72	1.77	1.85	
XVA'_i	0.97	0.85	1.12	0.98	1.26	1.09		1.30
$\Delta\mathrm{XVA}_i$	−0.45	−0.64	−0.45	−0.67	−0.46	−0.68		

E Synthesis

We sum up in Tables 45 and 46 the FTP of each considered default management scheme without and with collateral (in the sense here of initial margins and default fund contributions), from the cheapest to the dearest (again, in this example, hedging then auctioning with $\mathrm{D}'\backslash\mathrm{D} = \{ccp\}$ happens to coincide with replicating then auctioning with $\mathrm{D}' \setminus \mathrm{D} = \varnothing$). The FTPs of the hedging then auctioning scheme provides much larger gains than the pure auctioning strategy, which itself provides more gains than the full liquidation strategy. However, our approach only endorses the point of view of the participants to the exchange. Indeed, our costs of settling the house portfolio of a defaulted clearing member ignore the damage of the default to the bilateral counterparties b (end-users external to the exchanges). From this last viewpoint, whenever available, centrally cleared trading is preferable to bilateral trading (but, as per today, centrally cleared trading can only concern the standardized half of the market).

Table 45: FTPs of different default management schemes on D split as per (59) for $\overline{\mathrm{PIM}}_i^{\mathrm{D}} = \mathrm{DFC}_i^{\mathrm{D}} = 0$ in the Gaussian setup of Examples 1-3.

	$\mathrm{LC_D}$	$\sum_{i\in\mathrm{D}'} \Delta\rho_i$	$\sum_{i\in\mathrm{D}\backslash\{d\}} \Delta\mathrm{XVA}_i$	AC	FTP
liquidating	0	0.70	-0.77	0	-0.08
auctioning	0	0	0	-4.52	-4.52
hedging then auctioning	-1.39	2.09	-14.45	0	-13.75

Table 46: FTPs of different default management schemes on D split as per (59) for $\overline{\mathrm{PIM}}_i^{\mathrm{D}}$ and $\mathrm{DFC}_i^{\mathrm{D}}$ set at the 75% and 80% confidence level in the Gaussian setup of Examples 1-3.

	LC_{D}	$\sum_{i \in \mathrm{D}'} \Delta \boldsymbol{\rho}_i$	$\sum_{i \in \mathrm{D} \setminus \{d\}} \Delta \mathrm{XVA}_i$	AC	FTP
liquidating	0	0.70	-0.25	0	0.45
auctioning	0	0	0	-2.88	-2.88
hedging then auctioning	-1.39	2.09	-6.80	0	-6.11

§7 CONCLUSION

The settlement of a CCP portfolio after a clearing member's default involves both market and counterparty credit risks, to some extent depending on the portfolio. A distinction is made in Table 47 between house and intermediated deals, with intermediated client transactions fully hedged by corresponding intermediary trades as depicted in Figure 49 (also recalling Figure 48 page 274). In case of a default, these transactions, along with their hedge, are transferred to a surviving clearing member, without market impact but involving a shift in counterparty credit risk. This transfer of risk is quantified by XVA costs, as explained in VIII.§7.

Table 47 also distinguishes between delta-one financial instruments, such as repo transactions and equity swaps, which are rolled over without upfront payments, and upfront derivatives like multi-leg swaps and options. While delta-one transactions have minimal counterparty credit risk due to their short reset periods, they can still involve significant liquidation or hedging costs. For swaps or derivatives portfolios, instead, XVA implications play a crucial role in the default resolution strategy.

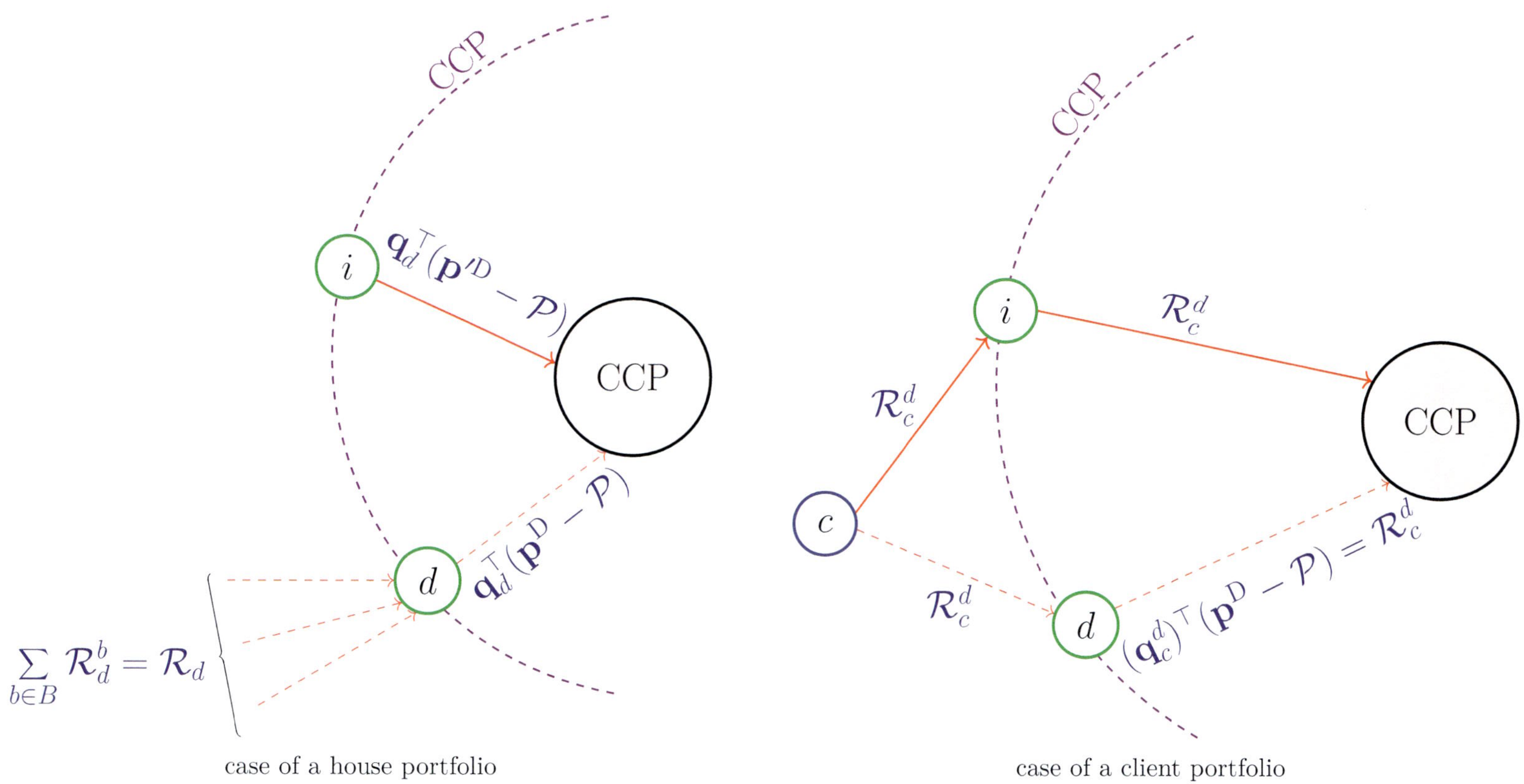

Figure 49: Default management of a CCP portfolio after clearing member d defaults with pre-default positions shown with dashed red lines and post-default positions with solid red lines: (*Left*) The default member d's receivable $\mathcal{R}_d$ from OTC bilateral counterparties is excluded from the default resolution, its house hedging portfolio $\mathbf{q}_d$ is transferred to surviving member i; (*Right*) Both d's client account $\mathbf{q}_d^c$ and the corresponding receivable $\mathcal{R}_d^c$ from client c are transferred as a package to the surviving clearing member i.

Table 47: Costs of the CCP for settling a netting set of deals of a defaulted clearing member, depending on the nature of the defaulted portfolio.
The transactions represented in the lower right cell would involve neither liquidity nor credit risk; however, this cell is actually empty because delta-one transactions do not require intermediation.

Account \ Assets	swaps and options	delta-one ($\subset$ house account)
house	liquidity (MC) and credit (CC $= \sum_{i \neq d} \Delta \mathrm{XVA}_i +$ AC)	liquidity (MC), no credit
client ($\subset$ pure auctioning)	no liquidity, credit (AC)	$\varnothing$

CHAPTER X

Quantitative Reverse Stress Testing, Bottom Up

This chapter outlines a dynamic framework for the analysis of extreme events based on forward-looking reverse stress testing. We perform a portfolio dynamic simulation and identify stress scenarios critical to the bank's solvency. Stress scenarios are determined based on their contribution to the capital cost, as expressed by KVA scenario differentials. We find that reverse stress testing can identify both the systemic and idiosyncratic weaknesses of the bank's portfolio without relying on historical events. Applications include solvency risk, extreme events hedging, liquidity risk management, trading and credit limits, model validation and model risk management.

§1 INTRODUCTION

In this chapter, we set out a dynamic framework and a computational methodology for analyzing extreme events in the form of forward-looking *reverse* stress testing (RST). The RST aims to gather the future scenarios that could impair a bank's solvency, using a reach forward-looking portfolio simulation sample as a scenario mine.

The interest in RST has been highlighted by the Bank of International Settlements and the Financial Services Authority, as regulators advocated the practice to share the responsibility of identifying potential future adverse states with the financial sector (see Chapter VIII and the introduction to the Risk part of the book). Although industry-wide preliminary RST approaches still mimic the spirit of historical scenario selection, we believe that there is merit in attempting a more systematic approach based on large-scale portfolio simulations of synthetic future events. This idea was already explored in a one-period setup in Chapter IX. In this chapter we want to extend this to a richer, dynamic setup.

DOI: 10.1201/9781003614944-15

RST is a developing field with scarce literature. Preliminary attempts have been made to cover portfolios with a handful of trades and risk-factors, as Grundke (2011) highlights. In the work of Breuer, Jandacka, Rheinberger and Summer (2009) constrained elliptical risk-factors distributions are used to analyze a portfolio of 100 loans and two risk-factors. Montesi and Papiro (2018) makes a step to model a bank's balance sheet, proposing a top-down approach. To simplify the computation, the bank's balance sheet is represented via a high-level accounting model, the bank's fragility is evaluated against simple regulatory capital thresholds, and future events' distributions are exogenously constrained to some variability range. Montesi, Papiro, Fazzini and Ronga (2020) extends Montesi and Papiro (2018), aiming at identifying extreme events via a Metropolis-Hastings variant. The scenario discovery method is applied to a common framework in which all the intricacies between the bank's portfolio composition, the risk-factors driving it, and the bank's accounting and regulatory representations collapse into a single explanatory variable. The current literature's common theme is to abstract the RST from the bank's view on the probability of future events and from the complexities of the portfolio held. Furthermore, because the proposed frameworks need exogenously marked parameters, their applicability would require a supra-entity delegated to run the RST for all the banks, a necessary level playground to ensure the absence of conflict of interests but with the risk of sharing parameters' marking errors.

Instead, in this chapter, we propose a bottom-up and multi-period RST framework, that robustly tights to

- the bank's portfolio composition,
- its risk-mitigation strategies, such as collateral exchange obligations specified in a CSA agreement,
- the bank's models that have been approved for internal and external use.

The RST framework we propose bases on simulating the future evolution of

- all the risk factors the bank is exposed to,
- the mark-to-market (MtM, i.e. counterparty-risk-free valuation) of all the claims collected in the bank's portfolio,
- the probability of default and the default losses of each one of the bank's counterparties and the counterparty credit and funding risk losses,
- the credit valuation adjustment (CVA) and funding valuation adjustment (FVA) since, as the 2007–09 global financial and the 2020 covid

(see XI.§4.A) crises showed, nowadays CVA and FVA volatilities have a significant impact on a bank's fragility.

After having projected a bank's future losses, we identify the scenarios critical to the bank's survival, and we data-mine them to surface the state of the economy and the counterparties driving the bank's future losses.

As compared to an RST top-down approach, a bottom-up approach faithfully captures the bespoke risk profile of the portfolio held and extends the set of actionable insights that an end-user would receive from the RST output by precisely pinpointing its weaknesses. The possibility of modelling the bank's portfolio with trade-level precision generates a broader set of actionable insights, which also comprehend the option to unwind, renegotiate or restructure specific trades or netting-sets (e.g. tightening the collateral agreement) or directly hedge, at the portfolio level, the systemic risk of future economic states. Contrarily, if a bank's portfolio is compressed into a handful of explanatory variables (see e.g. Montesi and Papiro (2018)), then the only actionable insight an end-user would receive is to alter the bank's regulatory capital buffers.

A Financial Modeling and Numerical Setup

We consider a filtered probability space $(\Omega, \mathcal{A}, \mathfrak{G}, \mathbb{Q})$, where $\mathfrak{G} = (\mathfrak{G}_t)$ is a filtration on the σ-algebra $\mathcal{A}$ and $\mathbb{Q}$ is the survival probability measure of the reference bank (conducting its XVA calculations) associated with a fininsurance probability measure $\mathbb{Q}^*$ (see the introduction to the Pricing part of the book). We denote by $\mathbb{E}_t$ and $\mathbb{Q}_t$ the $(\mathfrak{G}_t, \mathbb{Q})$ conditional expectation and probability (and for $t = 0$ we drop the index t).

To illustrate our methodology, we use a financial model diffusing interest rates, foreign rates, credit, and the conditional probability of default of each bank's counterparties: see Albanese and Trovato (2008), Albanese, Bellaj, Gimonet and Pietronero (2011) and Albanese and Vidler (2013) for the respective model's details and calibration. On this basis we sample the following processes at all nodes of an outer Monte Carlo simulation with time-grid (t_i):

- X^M, representing the set of market random variables, including as said above interest rates, foreign rates, credit and $\mathfrak{G}_{t_i}$ conditional probabilities of default of each bank's counterparties;

- Depending on the above, counterparty's specific random variables X^C, including the counterparty-risk-free value process P^b of all trades a bank has with each (assumed bilateral) counterparty b, the bank-counterparty's collateral obligations, the counterparty's exposure at default and its CVA;

- Depending on everything above, random variables are defined at the bank's whole portfolio level, e.g. the FVA, the trading loss and profit of the bank and economic capital.

Note that, by exception for simplicity in this chapter (see before IV.§1.A), we do not simulate the default events themselves, but only their probabilities.

We perform a simulation of both market and credit risk factors over 200-time points, covering 50 years and 20,000 primary scenarios for each risk-factor. Nested simulations branch off an additional 1,000 secondary scenarios at a yearly frequency, covering the portfolio's residual life. These nested simulations are used to account for portfolio CVA and FVA volatility risk. All the risk-factors realizations are coherently shared among the entire derivatives portfolio of the reference bank.

The chapter is organised as follows. Section §2 provides a simplification of the bilateral trading XVA setup in continuous time of III.§3, where the whole XVA suite can be computed by simply nested Monte Carlo simulations. §3 introduces our quantitative reverse stress testing framework in a finite specification of the latter. §4 introduces invalidation criteria for model choices based on RST scenarios breakdown. §5 is on credit and trading limits. §6 concludes.

§2 ECONOMICAL CAPITAL PROJECTIONS BASED KVA SETUP

For the sake of notation simplicity, we use the risk-free asset as a numéraire and we present the following formulas assuming no collateral exchanges between the bank and its (assumed all bilateral) counterparties, an assumption relaxed in our framework's implementation.

We start by defining the CVA, i.e. the bank's balance-sheet contra-asset (CA) entry that accounts for the counterparty default losses of the bank. Our CVA is defined as

$$\begin{aligned}\mathrm{CVA}_t &= \mathbb{E}_t\left[\sum_b \mathbb{1}_{\{t<\tau_b<\infty\}}(1-R_b)(P^b_{\tau_b})^+\right] = \mathbb{E}_t \int_t^\infty \underbrace{\left[\sum_b (1-R_b)(P^b_s)^+ \boldsymbol{\delta}_{\tau_b}(ds)\right]}_{dC_s}\\ &= \mathbb{E}_t \int_t^\infty \left[\sum_b \mathbb{1}_{\{t<\tau_b\}} e^{-\int_t^s \gamma^b_\zeta d\zeta}\gamma^b_s (1-R_b)(P^b_s)^+ ds\right],\end{aligned} \tag{1}$$

where sums run over each of the bank's netting-set b, P^b_s is the time-s cth netting set's default-free value, γ^b_s is the counterparty's default intensity, τ_b is the counterparty's default-time, and R_b is the counterparty's recovery rate.

Note that the formula in the second line of (1) assumes some conditional independence between τ_b and the remaining information in the model[1].

The FVA represents the bank's balance-sheet CA entry that reflects the shareholders' incurred funding costs to replenish the variation margin shortages in those future scenarios in which the bank is a net collateral poster. We follow the entity-level FVA definition initiated by Albanese, Andersen and Iabichino (2015) and revisited in Albanese, Crépey and Iabichino (2023), i.e.

$$\mathrm{FVA}_t = \mathbb{E}_t \int_t^{\infty} \underbrace{\left[\gamma_s \left(\sum_b P_s^b \mathbb{1}_{\{s<\tau_b\}} - \mathrm{CVA}_s - \mathrm{FVA}_s - \max(\mathrm{EC}_s, \mathrm{KVA}_s)\right)^+ ds\right]}_{dF_s}, \tag{2}$$

where γ_s is the unsecured funding spread of the bank at time s and EC and the KVA are defined below. Note that in Albanese et al. (2023), the equations are written under $\mathbb{Q}^*$ and there is a further factor $\mathbb{1}_{\{s<\tau\}}$ in the integrand in (2), where τ is the bank's default time, so that their FVA corresponds to the "cheaper FVA" discussed in II.§5.D. Here we stick to the form (2) under $\mathbb{Q}$ for consistency with the rest of the book.

We denote by L the bank's total accumulated loss process, i.e. the following mixture of default, funding risk and CA volatility losses (assuming the trading desks hedged but counterparty credit risk unhedged):

$$L_t = \int_0^t \left(dC_s + dF_s\right) + \mathrm{CVA}_t - \mathrm{CVA}_0 + \mathrm{FVA}_t - \mathrm{FVA}_0,\ t \geq 0. \tag{3}$$

The terms dC_s and dF_s are shorthand notation for the bracketed integrands in the second lines of (1) and in (2), i.e. the future losses realized by the bank due to defaults and funding expenses. The terms CVA_t and FVA_t are the conditional future CVA and FVA levels, computed by nested Monte Carlo simulations the way detailed in Albanese, Caenazzo and Crépey (2017, Section 5).

A spike in CVA gives rise to a capital requirement since CVA reserves which are set aside to cover counterparty default losses have to be increased. A spike in the FVA signals a liquidity risk event whereby the reserve capital set aside to cover cost of funding for the portfolio needs to be increased. The economic capital represents the capital buffer of the bank dedicated to absorbing extreme losses. Following the FRTB, an economic capital can be defined by the expected shortfall of the losses in the coming year with a confidence level 97.5%:

$$\mathrm{EC}_t = \mathbb{E}_t\left[L_{t+1} - L_t \mid L_{t+1} - L_t \geq \mathbb{V}\mathrm{a}\mathbb{R}_t\right], \tag{4}$$

[1] for the related technicalities, see e.g. Crépey, Bielecki and Brigo (2014, Section 13.7).

where

$$\mathbb{VaR}_t = \inf\{y; \mathbb{Q}_t(L_{t+1} - L_t \le y) \ge 97.5\%\} \tag{5}$$

is the corresponding value-at-risk (both conditional on $(\mathfrak{G}_t, \mathbb{Q})$).

Because a risk margin is also a loss-absorbing entry, the bank's capital at risk amounts to $\max(\mathrm{EC}_s, \mathrm{KVA}_s) = (\mathrm{EC}_s - \mathrm{KVA}_s)^+ + \mathrm{KVA}_s$, out of which only $(\mathrm{EC}_s - \mathrm{KVA}_s)^+$ corresponds to the *shareholder's* capital at risk (assuming the KVA is gathered from the client at trade inception). Accordingly and similarly to a Solvency II's risk margin, our KVA is defined as

$$\begin{aligned}\mathrm{KVA}_t &= \mathbb{E}_t \int_t^\infty h(\mathrm{EC}_s - \mathrm{KVA}_s)^+ ds = \mathbb{E}_t \int_t^\infty h\Big(\max(\mathrm{EC}_s, \mathrm{KVA}_s) - \mathrm{KVA}_s\Big) ds\\ &= \mathbb{E}_t \int_t^\infty h e^{-h(s-t)} \max(\mathrm{EC}_s, \mathrm{KVA}_s) ds,\ t \ge 0,\end{aligned}$$

for some constant hurdle rate term h. The coupling between FVA, EC (as the latter gauges the fluctuations of L, which include FVA fluctuations) and the KVA can be disentangled by Picard iterations as detailed in Algorithm 7 page 200 or by the explicit scheme of Abbas-Turki, Crépey, Li and Saadeddine (2025). But, for more simplicity in this chapter, we replace $\mathbb{E}_t$ and $\mathbb{Q}_t$ by $\mathbb{E}$ and $\mathbb{Q}$ in (4)-(5), which results in the following deterministic function of time, dubbed economic capital projection:

$$\mathrm{EC}(t) = \mathbb{E}\Big[L_{t+1} - L_t \mid L_{t+1} - L_t \ge \mathrm{VaR}(t)\Big], \tag{6}$$

where

$$\mathrm{VaR}(t) = \inf\{y; \mathbb{Q}(L_{t+1} - L_t \le y) \ge 97.5\%\}. \tag{7}$$

Solving the ensuing XVA equations (where, in particular, the KVA also becomes a deterministic function of time $\mathrm{KVA}(t)$) can then be done by simply nested Monte Carlo simulations, which we use for obtaining the CVA and FVA at all outer simulation nodes, in combination with some interpolations (cf. IV.§1.A).

Moreover, numerical experimentation shows that the inequality KVA $\le$ EC only stops holding when the hurdle rate h is very large and the term structure of EC starts out very small and has a sharp peak in a few years. This is quite unusual for a portfolio held on a run-off basis, as considered in XVA computations, which tends to amortize in time. So KVA $\le$ EC typically holds on realistic portfolios (cf. also Eqn. XI.(29)), i.e. $\max(\mathrm{KVA}, \mathrm{EC}) = \mathrm{EC}$. In practice, our KVA is based on an explicit simulation of the bank's loss process (3) that is used for computing economic capital projections as per (6)-(7) with $\max(\mathrm{EC}, \mathrm{KVA})$ replaced by KVA in (2). These economic capital projections are then integrated as

$$\begin{aligned}\mathrm{KVA}(t) &= \int_t^\infty h e^{-h(s-t)} \mathrm{EC}(s) ds,\ 0 \le t \le T,\\ \text{e.g. } \mathrm{KVA}(0) &= \int_0^\infty h e^{-hs} \mathrm{EC}(s) ds.\end{aligned} \tag{8}$$

Table 48: Toy portfolio of swaps (the nominal of each swap is $\$10^4$).

i	Mat.	Receiver Rate	Payer Rate
1	2y	Par – 6M	LIBOR – 3M
1	15y	Par – 6M	LIBOR – 3M
1	30y	LIBOR – 3M	Par – 6M
2	5y	Par – 6M	LIBOR – 3M
2	10y	LIBOR – 3M	Par – 6M
2	30y	Par – 6M	LIBOR – 3M
3	5y	LIBOR – 3M	Par – 6M
3	10y	Par – 6M	LIBOR – 3M
4	2y	LIBOR – 3M	Par – 6M
4	15y	LIBOR – 3M	Par – 6M

A Toy Portfolio Example

We first consider a portfolio of ten USD currency fixed-income swaps depicted in Table 48. The nominal of each swap is 10^4. The swaps are traded with four counterparties $i = 1, \ldots, 4$, with a 40% recovery rate and credit curves as in Figure 50. In this toy portfolio case the whole calculation takes roughly ten minutes to run, including two to three minutes for building the pre-calibrated market and credit models.

The corresponding XVA results are displayed in the left panel of Table 49. Since the portfolio is not collateralized, its CVA is quite high as compared with the nominal (10^4) of each swap. But its KVA is even higher. Again, given our deterministic term structure approximation (6) for economic capital, the computation of the KVA reduces to a deterministic time-integration, which is why there is no related standard error in Table 49. The number $\mathrm{FVA}_0^{(0)}$, accounting only for the re-hypothecability of variation margin amounts to \$73.87. However, if we consider the additional funding sources due to economic capital and reserve capital, we arrive at an FVA figure of $\mathrm{FVA}_0 = \$3.87$ only. We also show for comparison the "naively bilateral" FTDCVA and FTDDVA metrics, detailed in Albanese, Caenazzo and Crépey (2017, Proposition 4.1(ii)-(iii)), that ignore the wealth transfer issues related to the bank's own default (see I.§3.E.1 and II.§4). In particular, the FTDCVA is prone to decrease simply because the default risk of the bank increased, thus violating the monotonicity requirement of II.§1 (see also II.§5.A). As established in Albanese, Caenazzo and Crépey (2017, Proposition 4.1(iii)), under their technical assumption (18) that is satisfied in the basic immersion setup defined in Remark II.2 (which includes the market and credit model of the present chapter), it holds that FTDCVA – FTDDVA = FV (= CVA – DVA by Lemma II.8(ii)), the firm valuation of counterparty risk (valuation of

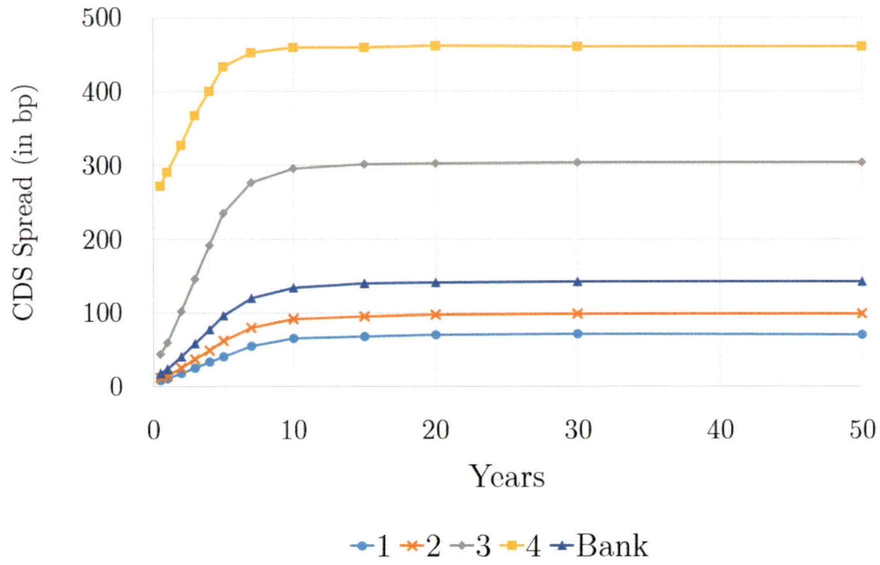

Figure 50: Credit curves of the bank and its four counterparties.

counterparty risk from the point of view of the bank as a whole). As one can see from the table, there is a complete disconnection between the XVA numbers accounting for, respectively ignoring, the related bank wealth transfer issues. Hence the latter must imperatively be accounted for in the analysis (the results ignoring them are so to say arbitrary numbers).

Table 49: Toy portfolio. *Left*: XVA values and standard relative errors (SE). *Right*: Respective impacts when Swaps 5 and 9 are added to the portfolio last.

	\$Value	SE
CVA_0	471.23	0.46%
$FVA_0^{(0)}$	73.87	1.06%
FVA_0	3.87	4.3%
KVA_0	668.83	N/A
$FTDCVA_0$	372.22	0.46%
$FTDDVA_0$	335.94	0.51%

	Swap 5	Swap 9
ΔCVA_0	155.46	−27.17
$\Delta FVA_0^{(0)}$	−85.28	−8.81
ΔFVA_0	−80.13	−5.80
ΔKVA_0	127.54	−52.85
$\Delta FTDCVA_0$	98.49	−23.83
$\Delta FTDDVA_0$	122.91	−80.13

The right panel of Table 49 shows the incremental XVA results when the fifth (resp. ninth) swap in Table 48 is added to the portfolio last. These results illustrate that, under the incremental run-off XVA methodology of I.§4.B, introducing financial contracts one after the other in one or the reverse order in the portfolio at time 0 results in the same aggregated incremental FTP amounts for the bank, equal to its "portfolio FTP", but in different FTPs for each given contract and counterparty.

Interestingly enough, all the incremental XVAs of Swap 9 (and also the incremental FVA of Swap 5) are negative, which is fine: the positivity and monotonicity requirements of II.§1 only apply at the bank portfolio level. Hence, Swap 9, when added last, is XVA profitable to the portfolio, meaning that a price maker should be ready to enter the swap for less than its counterparty-risk-free value, assuming it is already trading the rest of the portfolio: the corresponding FTP amounts to −\$85.82, versus \$202.87 in the case of Swap 5.

B Large Portfolio Example

Hereafter in this chapter our numerics are done in the context of a large portfolio of about 2,500 counterparties, 100,000 derivative trades including swaps, swaptions, FX options, inflation swaps, and CDS trades, with exposure to credit factors, all the G10 currencies' rates and their foreign-exchange rates. The results presented were gathered on a single server with a calculation time of around 3 hours.

Figure 51 represents the histograms of the corresponding bank's losses $(L_{t+1} - L_t)$ sampled at grid times t_i. The white dots in Figure 51 represent the economic capital projection at grid times t_i. Table 50 shows the time-0 XVA results. The FVA is much smaller than the KVA, especially after accounting for the economic and reserve capital funding sources (see Subsection A). The KVA amounts to \$275 M, which makes it the largest of the XVA numbers. Figure 52 shows the term structure of economic capital and the term structure of the KVA obtained by the deterministic term structure approximation $\mathrm{EC}(\cdot)$ of economic capital and by the corresponding KVA formula (8). Such a term structure of economic capital, with a starting hump followed by a slow decay after 2 or 3 years, is typical of an investment bank derivative portfolio assumed to be held on a run-off basis until its final maturity, where the bulk of the portfolio consists of trades with 3y to 5y maturity. In relation to the point made before the formula (8), note that the KVA computed by this formula based on the term structure of economic capital is below the latter at all times.

The funding needs reduction achieved by CR and RC = CA = CVA+FVA (in the absence of initial margin) is also shown in Figure 53 by the FVA blended curve. This is the FVA funding curve which, whenever applied to the FVA computed neglecting the impact of economic and reserve capital, gives rise to the same term structure for the forward FVA as the calculation carried out with the CDS curve of the bank as the funding curve but accounting for the economic and reserve capital funding sources. This blended curve is often inferred by consensus estimates based on the Markit XVA service.

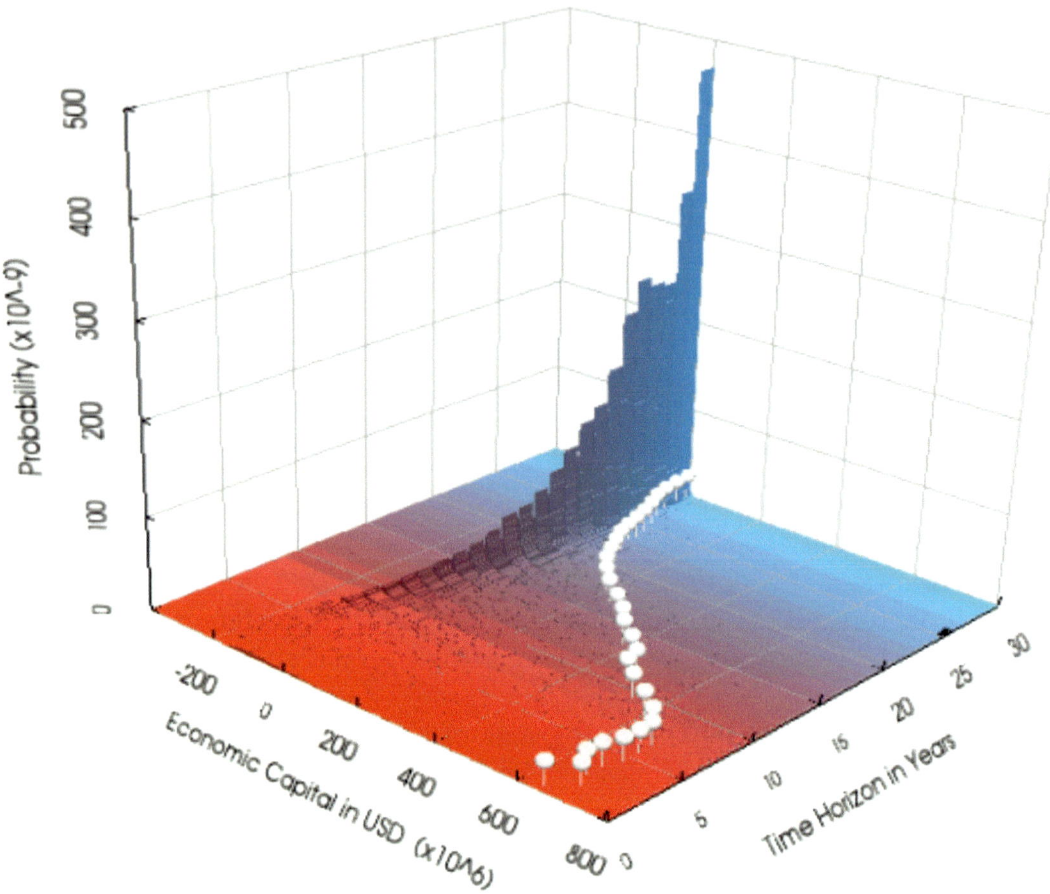

Figure 51: Histograms of the bank's losses ($L_{t_{i+1}} - L_{t_i}$) over time and economic capital projections.

Table 50: XVA values for the large portfolio.

XVA	\$Value
CVA_0	242 M
$\mathrm{FVA}_0^{(0)}$	126 M
FVA_0	62 M
KVA_0	275 M
FTDCVA	194 M
FTDDVA	166 M

However, here it is computed from the ground up based on full-fledged capital projections.

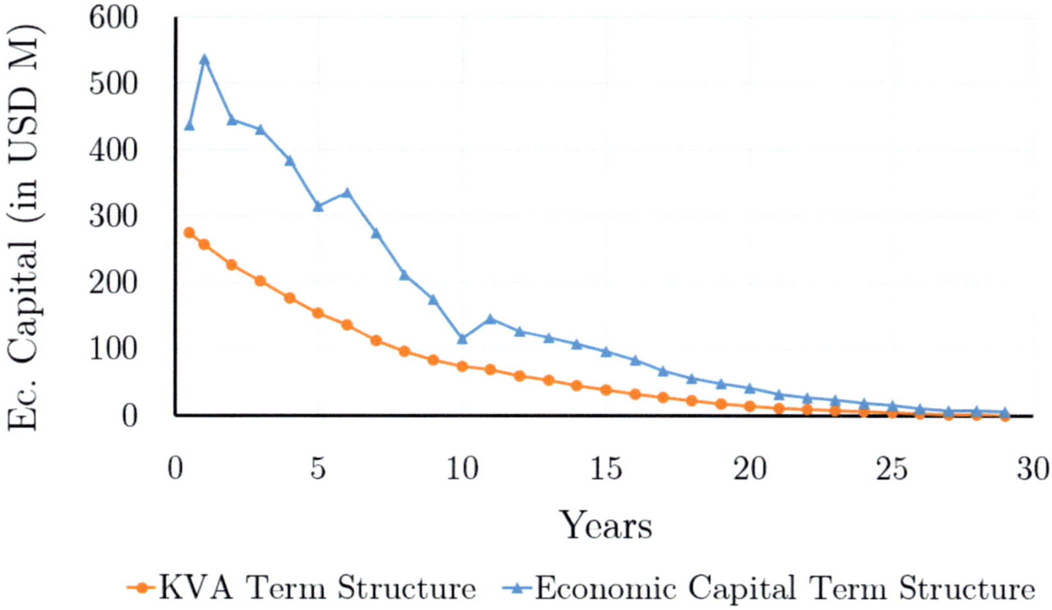

Figure 52: Term structure of economic capital compared with the term structure of KVA.

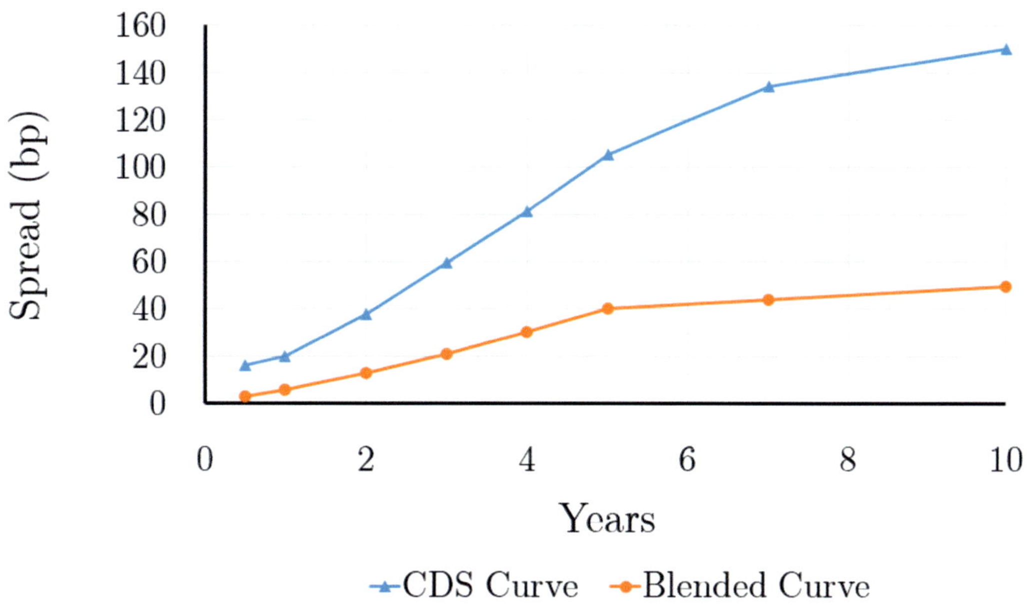

Figure 53: FVA blended funding curve computed from the ground up based on capital projections.

§3 THE REVERSE STRESS TESTING FRAMEWORK

In §2, for the notation simplicity's sake, we described our setup using a continuous-time and space formalism. However, we stress out that our RST approach relies on a finite specification. The RST framework we propose founds upon a multi-dimensional simulation engine and a landmark metric, i.e. the corresponding economic capital projections based valuation adjustment (KVA), synthesizing all the risks the bank is exposed to. In a nutshell, within a finite simulation framework, we compute the difference between the time-0 portfolio KVAs including and excluding the scenario ω (δ_ωKVA) and, by gathering the stress scenarios that impair the bank's resilience the most, we populate the stress scenario pool Ω^*. A stress scenario ($\omega^* \in \Omega^*$) is a scenario that sits between two pre-specified KVA differential's high quantile levels. For each $\omega^* \in \Omega^*$, we then identify the time-point t_{i^*} that maximizes capital consumption along the path ω^*. Having determined such stress scenario/stress time-point tuples, the RST is the function providing the corresponding states of the market and counterparty's losses random variables, i.e. the function

$$\Omega^* \ni \omega^* \mapsto (t_{i^*}, X^M(\omega^*, t_{i^*}), X^C(\omega^*, t_{i^*})), \tag{9}$$

A Scenario Differential KVA and Reverse Stress Testing

In principle, an RST framework could be designed with various scenario ranking methods (e.g. considering only the CVA or the FVA paths, or the time-integrated losses along the primary simulation paths), a choice that would isolate different risks (e.g. credit or liquidity risk). The choice of a scenario ranking method of course has a first-order impact on the RST's output.

Our scenario ranking proposal revolves around a bank's capital costs centric view, synthesized by the KVA. Therefore, the KVA represents the root vertex of our RST framework, a vertex that is sensitive to the future state of all the risk-factors that drive the bank's loss process L. The KVA naturally probes a way to rank and identify those future scenarios that drive the bank's shareholder's capital consumption, which could be led by the occurrence of a cluster of defaults, credit or liquidity risk losses. In our view, the KVA does not only provide a robust scenario ranking method. It could take the role of being the landmark metric on which a market-maker could harmonize and optimize its risk-reward decisions consistently.

In the context of a finite, pre-generated simulation framework, we denote the time-0 relative contribution of a scenario ω to the overall KVA as

$$\delta_\omega \text{KVA}. \tag{10}$$

The term δ_ωKVA expresses the difference between KVA(0) and KVA(0) recomputed omitting scenario ω. In view of (8), δ_ωKVA could be interpreted as

Figure 54: Bank's portfolio's losses over a stress scenario $\omega^* \in \Omega^*$ (the red rhomboid indicates the max loss, happening at t_{i^*}).

the dividend accrual stream paid by the bank to rise capital buffers along the scenario ω. Thus, the metric δ_ωKVA enables a scenario-ranking exercise, in which scenarios are ranked based on the shareholder's capital the bank would demand, or, in other words, on the scenario's total dividend liability.

Among the pool of future scenarios, we select the bank's most adverse ones in the confidence interval $[99\%, 99.9\%]$, identifying those 180 scenarios (out of the 20,000 primary ones) with the highest scenario differential KVA (10) (where the 20 most extreme are neglected). We denote the set of scenarios verifying the condition above as the stress scenario pool (Ω^*).

For each scenarios belonging to the stress scenario pool ($\omega^* \in \Omega^*$), we perform an event-based temporal analysis, in which we identify the point in time t_{i^*} in which the bank faces the most substantial losses (see Figure 54). For each extreme scenario/time-point tuple, we gather the vector collecting the risk-factors' actual future state ($X^M_{t_{i^*}}(\omega^*)$) through a data-mining exercise (see Figure 55).

Figure 56 depicts the USD univariate projection states data-mined on the back of all the identified stress scenario pool, while Figures 57–60 present the USD-GBP, the USD-CAD, the USD-JPY and the EUR-GBP bivariate projections. Due to established monetary policies, low rates environments occur at the downturn of credit cycles, as is reflected in the correlation estimates that have been used to perform the simulation. In our setup credit is

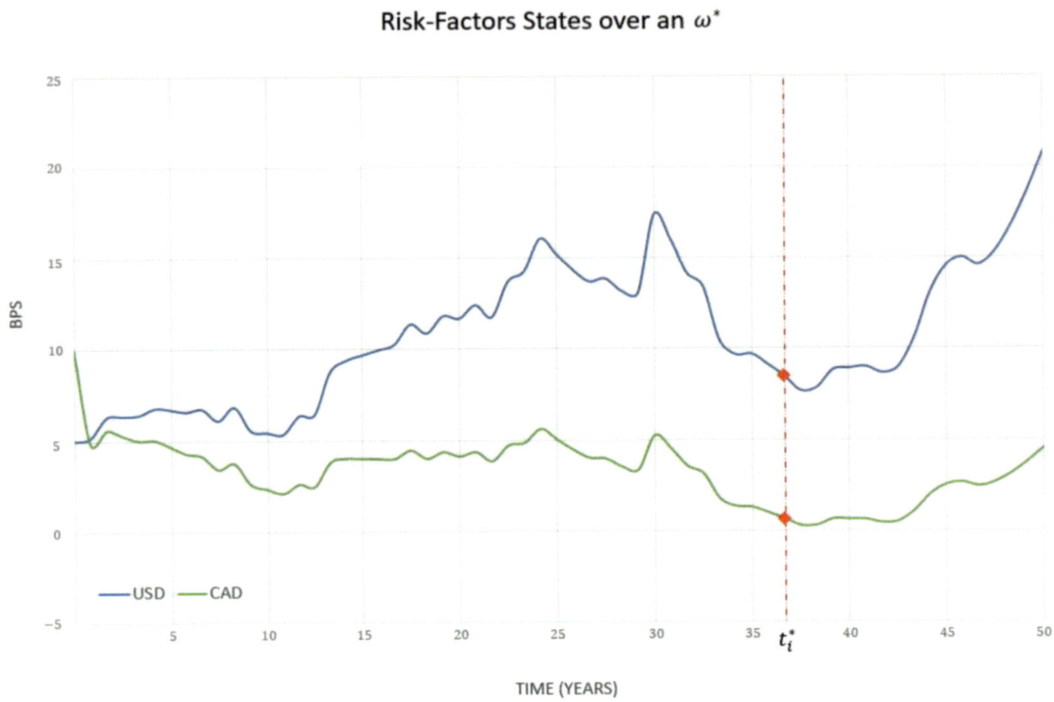

Figure 55: USD OIS and CAD CORRA stress state at (ω^*, t_{i^*}).

modeled by a distance-to-default model for each counterparty and implied spread returns are correlated with one another and with interest rates through a dynamic Gaussian copula (see Albanese and Vidler (2013) for details, including calibration). This choice implies that clusters of defaults materialize at the lowest points of the credit cycle. Consequently, default losses are more significant when interest rates are lower. For the chosen subset of risk factors in our portfolio, the correlation between stress scenario occurring during a downturn of the credit cycle is well visible from Figures 56 and 57-60. The analysis also reveals the presence of a second cluster which is idiosyncratic to the portfolio and reflects concentrations of cash-flows or convexity at certain points in time and under specific conditions in the future.

Having identified the most adverse economic states, we then extend our analysis to expose the counterparties that exacerbate the bank's losses in one of the scenarios belonging to the stress scenario pool: see Figure 61 for a visualization of the top 5 marginal contributors to an extreme stress scenario.

B Hedging Solvency Risk

The ex-ante identification of the risky time/risk-factors' states helps to identify candidate trades that have the potential of reducing the bank's losses in those states and, therefore, the bank's capital costs.

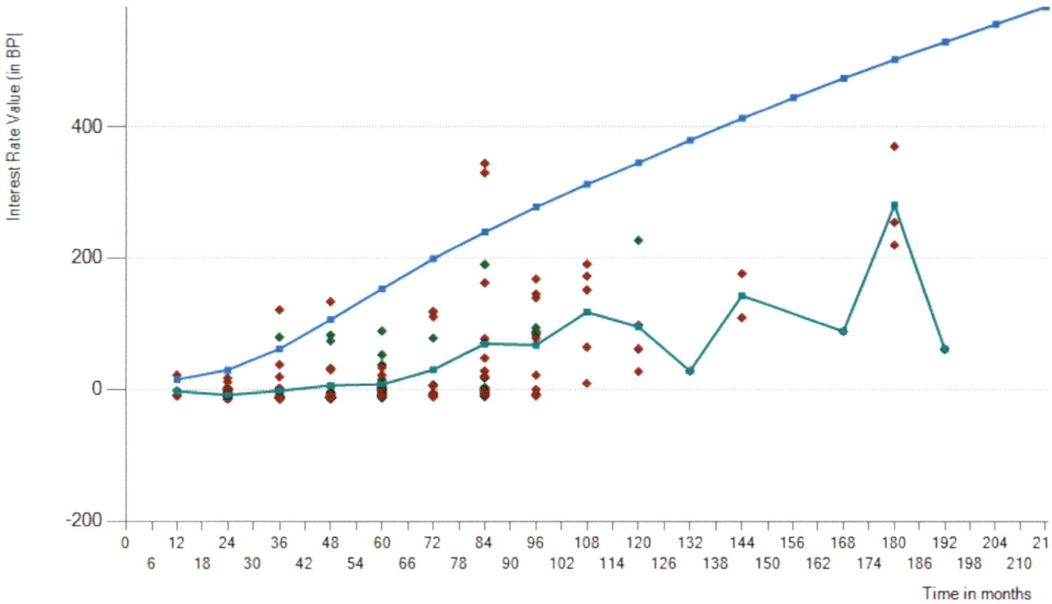

Figure 56: USD OIS overnight rate (in bps) for stress scenarios. The blue and green lines represent the forward curve and the forward curve conditional to stress scenarios (average of the dots), respectively.

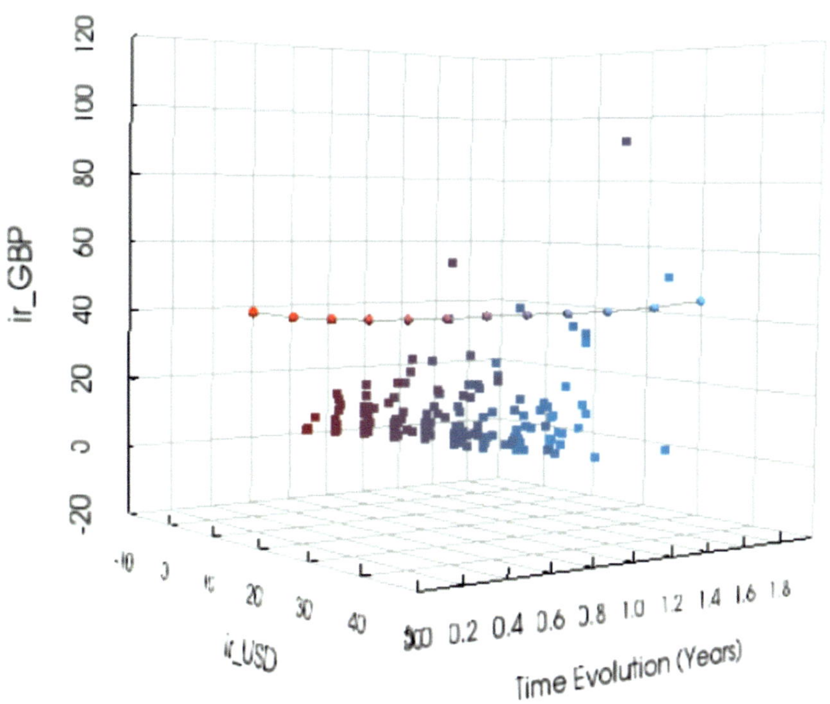

Figure 57: USD OIS vs. GBP SONIA overnight rates (in bps) for stress scenarios. The connected dots represent the USD-GBP forward curve.

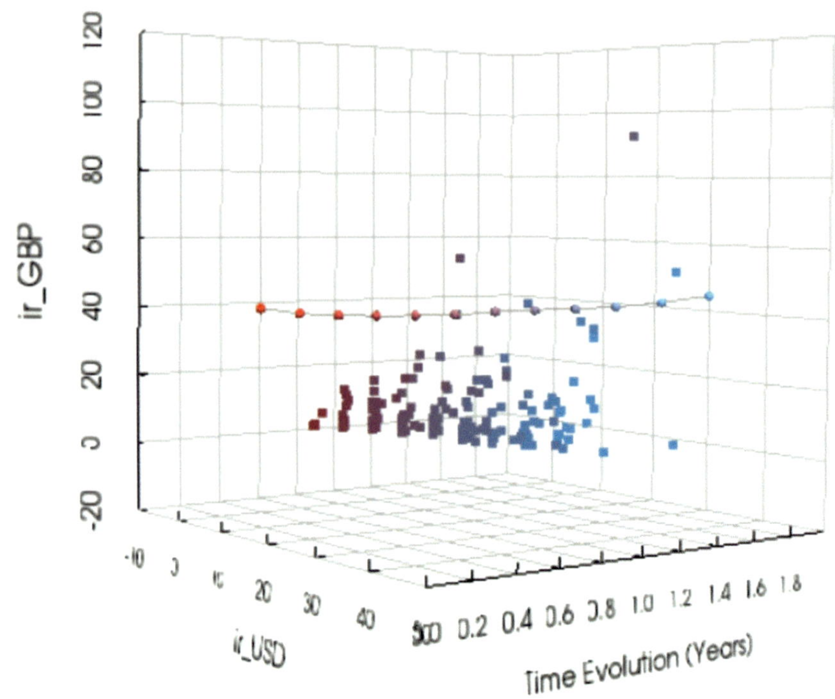

Figure 58: USD OIS vs. CAD CORRA overnight rates (in bps) for stress scenarios. The connected dots represent the USD-CAD forward curve.

After having identified a loss-driving counterparty, a similar scheme repeats to pinpoint those trades that exacerbate loss concentration on the back of the portfolio held. If more than one trade emerges to be loss-driving, then the bank identified a risk-concentration conditional to a fragility state. To hedge a risk-concentration, the bank could unwind, renegotiate or restructure specific transactions or CSA terms, or even build static hedges.

Assume that at time 0 the reference bank (labelled by 0) identifies a selling bank (b) for a trade (ϵ) that might hedge its losses against one of its states of fragility. The two banks will execute the transaction if they are able to fix a price p such that

$$\begin{aligned} P_0^\epsilon + (\Delta^b\mathrm{CVA}_0^{-\epsilon} + \Delta^b\mathrm{FVA}_0^{-\epsilon} + \Delta^b\mathrm{KVA}_0^{-\epsilon}) \le p \le \\ P_0^\epsilon - (\Delta^0\mathrm{CVA}_0^\epsilon + \Delta^0\mathrm{FVA}_0^\epsilon + \Delta^0\mathrm{KVA}_0^\epsilon), \end{aligned} \tag{11}$$

where P_0^ϵ is the risk-free valuation of the trade (seen from 0's viewpoint) and $\Delta^{0/b}\mathrm{XVA}_0^{\pm\epsilon}$ is the incremental impact of the trade on the XVA metrics of banks 0 and b. Or, equivalently to (11) in terms of the funds transfer price $\mathrm{FTP} := P_0^\epsilon - p$,

$$\begin{aligned} \Delta^0\mathrm{CVA}_0^\epsilon + \Delta^0\mathrm{FVA}_0^\epsilon + \Delta^0\mathrm{KVA}_0^\epsilon \le \mathrm{FTP} \le \\ -(\Delta^b\mathrm{CVA}_0^{-\epsilon} + \Delta^b\mathrm{FVA}_0^{-\epsilon} + \Delta^b\mathrm{KVA}_0^{-\epsilon}). \end{aligned} \tag{12}$$

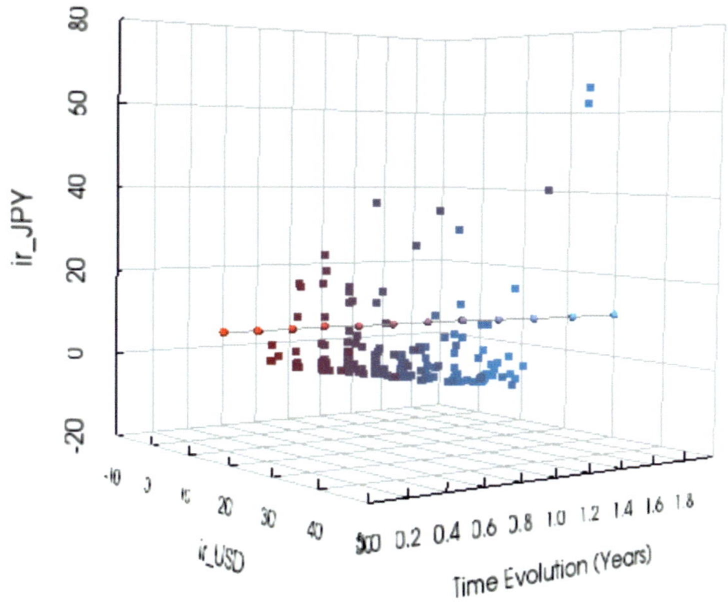

Figure 59: USD OIS vs. JPY JSCC overnight rates (in bps) for stress scenarios. The connected dots represent the USD-JPY forward curve.

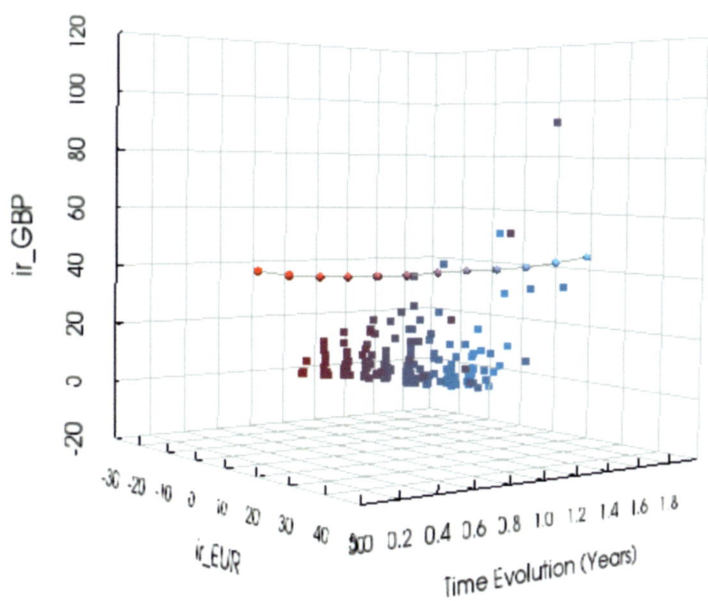

Figure 60: EUR EONIA vs. GBP SONIA overnight rates (in bps) for stress scenarios. The connected dots represent the EUR-GBP forward curve.

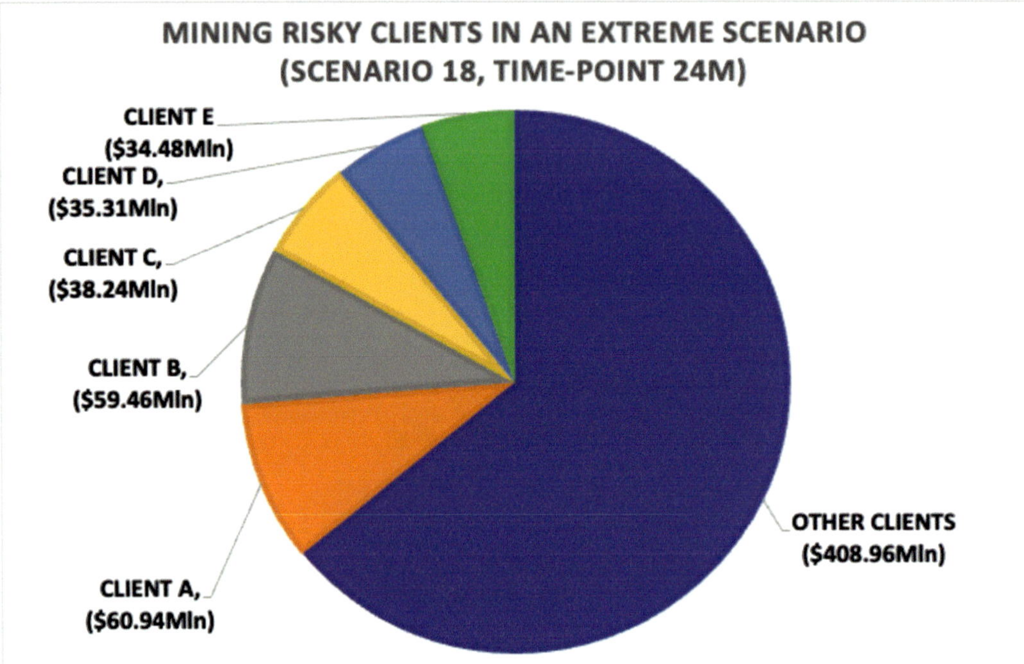

Figure 61: Top 5 marginal contributors to an extreme stress scenario (scenario number 18; time 24 months).

Assuming

$$\Delta^0\mathrm{CVA}_0^\epsilon + \Delta^0\mathrm{FVA}_0^\epsilon + \Delta^0\mathrm{KVA}_0^\epsilon + \Delta^b\mathrm{CVA}_0^{-\epsilon} + \Delta^b\mathrm{FVA}_0^{-\epsilon} + \Delta^b\mathrm{KVA}_0^{-\epsilon} \leq 0,$$

a win-win deal between the two banks can locate in any segment of the FTP price range defined by (12).

By singling out the path with the most significant marginal contribution to the KVA, a market-maker could restructure its portfolio or hedge the risk, reducing its capital costs. However, we don't know a priori a scenario's hedging costs. One could presumably expect that hedging an idiosyncratic scenario could be cheaper than hedging a systemic one. For instance, from Figure 56, we have identified two fragility clusters, an idiosyncratic and a systemic one. A bank could expect a wider FTP range (12) for an hedge that counterbalances an idiosyncratic fragility than a systemic one and higher chances to drive FTP closer to $\mathrm{FTP}^0 := \Delta^0\mathrm{CVA}_0^\epsilon + \Delta^0\mathrm{FVA}_0^\epsilon + \Delta^0\mathrm{KVA}_0^\epsilon$, given the lower market appetite. But the exact quantification of the trade-off between a scenario's hedging costs and the KVA gain can only be done by running capital optimization strategies.

§4 VALIDATION OR FALSIFIABILITY?

While in finance the common practice is to validate models, in the natural sciences experiments are aimed at invalidating them (Popper, 1934). The reverse stress testing method we are putting forward in this chapter gives us a framework for refining the model validation process embracing, to a greater degree, the scientific method by identifying scenarios of model breakdown. Our criteria to discriminate sound and unsound models is: a model to be deemed as fit for purpose must generate an econometrically meaningful set of stress scenarios.

In this section, we compare, from this perspective, the reverse stress testing properties of two different interest rate models, a "champion" and a "challenger", and we discuss individually their features in terms of their stress scenarios, XVA numbers, and the different actionable insight they suggest in terms of risk appetite and hedging.

A Champion and Challenger

As "champion" model, we choose the Hull and White (1990) 2-factor model (HW), a Gaussian model which, thanks to its analytical solution, is highly used in the XVA industry to speed up computation times. As a "challenger", we choose the 2-factor stochastic drift model (SD) of Albanese and Trovato (2008).

Both models are based on a mean-reverting process of the form

$$d\rho_t = \kappa(\theta_t - \rho_t)dt + \sigma(t)\rho_t^\beta dW^1, d\theta_t = k(a - \theta_t)dt + \nu dW^2, \tag{13}$$

where W^1 and W^2 are correlated Brownian motions. In the HW model, the short rate $r_t = \rho_t$ is an unbounded process, the parameter β equals 0, and the deterministic term structure for $\sigma(t)$ is calibrated to market observables. In the SD model, instead, the short rate process is defined by an equation of the form

$$r_t = \phi(t) + \gamma(t)\rho_t. \tag{14}$$

Here $\gamma(t)$ is a drift adjustment factor and

$$\phi(t) = \min(0, f(t) - \psi), \tag{15}$$

where $f(t)$ is the infinitesimal forward rate, and ψ (equal to 20 bp in our case study) is a bound calibrated econometrically. The process ρ_t is constrained to take non-negative values (apart for a handful of basis points expressed in ψ), the parameter β is adjustable, and the calibrated term structure $\sigma(t)$ is constrained to be of the form of a shifted exponential.

Unless stated otherwise, the numerics of this chapter use SD as interest rate model, but hereafter we also consider what happens with HW.

B Distorted Risk Appetite

Table 51 compares the time-0 XVA metrics computed with the champion and the challenger model. As could be expected, the KVA appears as the most model sensitive metric.

Table 51: Comparison of the time-0 XVA metrics (in million USD) between the SD and HW models. FVA (rho) accounts only for the re-hypothecation option (i.e. it ignores all the "−" terms in (2)), while FVA (rho + cva + fva + ec) deems reserve and economic capital as a fungible source of funding.

	HW model	SD model	model risk
CVA	242	248	+2.5%
FVA (rho)	126	109	−13%
FVA (rho + cva + fva + ec)	62	45	−27%
KVA	275	388	+41%

Concerning the models' RST properties, the two models exhibit a sensible discrepancy in the 97.5% quantile's location of future losses, resulting in significantly different economic capital projections and stress scenarios positioning. The different location of the USD stress scenarios clearly emerges if Figures 56 (SD model) and 62 (HW model) are analysed side-by-side. The comparison between Figures 56 and 62 surfaces the distorted signal that econometrically unrealistic models can provide to end-users. The SD model disperses the risk more homogeneously over the lifetime of the OTC book, while skewing only slightly the risk on shorter maturities. In contrast, the HW model clusters the bulk of the portfolio risk in the short term. In the SD model, the USD stress states are low but positive or nearly positive rates. In contrast, in the HW model, USD stress scenarios are deep, unrealistic, negative rates. The macroeconomic hedging ratios implied from the HW model do not just provide a false sense of security to the end-user; they put her at the bank's competitors mercy: external agents can easily exploit the potential distorted risk appetite that the HW model generates if the bank is willing to pay the HW price to purchase protection against unreachable economic states in the market (e.g. to hedge interest rates risk in the range between -100 and -350 bps in Figure 62).

It is straightforward to note how models that can not capture the risk-factor's econometric features can turn in a double-edged sword to the end-user. The model's elusive stress states can easily jeopardize both the suggested portfolio's risk hedging location and the hedging instruments' entry-price. Being willing to pay the price to gain exposure to future unreachable states is a form of arbitrage opportunity.

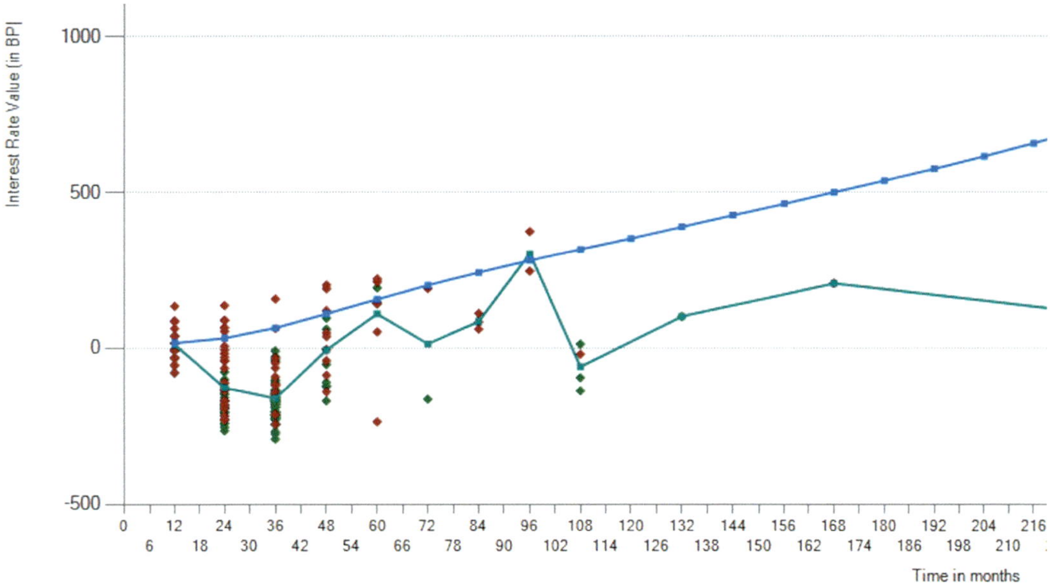

Figure 62: USD OIS overnight rate (in bps) for stress scenarios using the HW model. The blue and green lines represent the forward curve and the forward curve conditional to stress scenarios (average of the dots), respectively.

By surfacing the model's stressed states, the RST provides a model invalidity framework, as the effects of the most hidden assumptions are revealed and can be checked under the econometric realism's spotlight. As discussed in Albanese, Crépey and Iabichino (2021) and further quantified in HVA terms in Chapter XI, models that fail the reality check may exchange short-term profitability with long-term risk.

C Model Risk and Bayesian KVA

Market data do not uniquely determine valid models. Once the set of realistic models is identified, the residual model risk can be accounted for through Bayesian averages over the identified set, in the vein of Black and Litterman (1991, 1992), of economic capital projections and the ensuing KVA metric. These methods involve generating paths with several models to find aggregate metrics of interest, such as value-at-risk (see, e.g., Siu, Tong and Yang (2004)) or KVA (as in this chapter). In the KVA context, the Bayesian definition is the same as (8), except that the aggregation underlying (6)-(7) is computed over a composite set of econometrically realistic and calibrated models. Note that the Bayesian KVA mixing bears on the paths simulated in different co-calibrated models (cf. Hoeting, Madigan, Raftery and Volinsky (1999)), as opposed to simply mixing prices. Taking a Bayesian average among valid models by generating a composite scenario set based on multiple model specifications typically increases capital requirements. The difference derives from model risk and can be used as a model risk capital buffer (cf. Remark XI.12 page 344).

§5 CREDIT LIMITS

In addition to imposing validity constraints on models, RST also places restrictions on risk appetites. The possibility to identify the counterparties that exacerbate the capital demand (see Figure 61 page 306) sets the base to formulate a new credit limit metric based on counterparty incremental KVA.

The potential future exposure (PFE) is typically used in the industry for assessing the risk of a given counterparty based on potential future MtM losses. An extreme quantile (typically ranging between 95% and 99%) of the counterparty specific mark-to-market (P_t^b) distribution defines the PFE. The risk metric used for credit limit monitoring is typically the maximum PFE ($\max_{t\in[0,\Theta]}(\mathrm{PFE}(t))$, where Θ is either the residual life of the netting set or its first year of credit exposure. As observed in BCBS (2005) and in Canabarro and Duffie (2003), the prevailing PFE calculation method is to use the same simulation sample used to compute counterparty credit risk metrics. Therefore, the PFE generates through the interaction between the counterparty's portfolio composition, its (eventual) collateral, and the extreme states predicated by one (or more) model(s).

Credit risk officers use PFE-like metrics to assess the maximum total exposure a bank is willing to face over a counterparty's lifetime to prevent risk concentrations that might jeopardize the bank's solvency in extreme future states. However, PFE-like metrics are computed in counterparty-centric silos, in which each counterparty is evaluated in isolation; PFE-like metrics focus only on the default exposure, neglecting all the other risks a counterparty generates on the bank's entire portfolio held. In other words, individual counterparty credit losses are just one facet of the risk, while default contagion events or convexity concentrations for exotic derivatives have the potential to generate non-linear risks to the bank's solvency. Furthermore, fully collateralized derivative portfolios generate a zero PFE, although they are exposed to collateral funding risk.

In this direction, we find useful to alter our RST framework to perform, this time, a counterparty KVA incremental analysis. We find merit in performing this exercise as the KVA and the PFE share the following commonalities: both of them focus on extreme losses and assume a run-off basis portfolio holding. However, the KVA does not capture only counterparty's mark-to-market default loss. It also reflects all the potential sources of future losses, including the counterparty's CVA volatility risk and the counterparty's contribution to the bank's FVA volatility risk.

Taking the difference between the KVA of the entire portfolio and the KVA of the portfolio obtained by removing the counterparty b, we compute the time-0 incremental counterparty KVA:

$$\Delta_b \mathrm{KVA}. \tag{16}$$

The Δ_bKVA carries the intuitive meaning of representing the incremental cost of capital generated by a counterparty. Accounting for the full projections of economic capital, as opposed to just the current value, is in line with regulatory directives, such as CRR 292.7(b) (see Banking Association (2013)), which explicitly prescribes to monitor and potentially capitalize the concentration of exposure beyond the first year of credit exposure.

The benefits of using Δ_bKVA metrics in setting idiosyncratic risk appetites are visible in Figure 63, which depicts the comparison between the PFE and the Δ_bKVA for the whole set of counterparties of the analyzed portfolio. Figure 63 shows that the two metrics are virtually unrelated; the more risk-sensitive Δ_bKVA measure captures details related to wrong-way risk, portfolio effects and XVA volatility that the PFE misses. Furthermore, we note that the Δ_bKVA can be low in situations where the PFE is material. This situation highlights the presence of counterparties that are penalized from a PFE perspective, while they are in fact beneficial in reducing the overall risk of the portfolio held. On the opposite end of the spectrum, we find counterparties characterized by an immaterial PFE, but material Δ_bKVA. These counterparties are characterized by pronounced potential losses that go beyond mark-to-market losses, a risk that the PFE is not able to capture.

Hereafter, we analyze the way in which the PFE and the Δ_bKVA react to different model's assumptions, comparing results derived from the alternative usage of the HW and the SD model. From Figure 64 we observe that the PFE scatters roughly along the diagonal, indicating that the PFE is a fairly indifferent metric to the model of choice. Therefore, risk limited metrics such as the PFE could distort the bank's risk appetite on the counterparty selection process, in a similar way in which limited models' elusive future states could distort the bank's risk appetite around future economic conditions.

Figure 65 shows that the Δ_bKVA is sensitive to the radically different nature of stress scenarios generated by different models (see §4.B). The higher risk sensitivity of Δ_bKVA makes KVA-based credit limits a strong addition to PFE-based systems. Under the full implementation of the BCBS 261 directive on uncleared margin rules, where most client's trades will be collateralized, it may well be that the null PFE will be a too narrow metric to set risk appetites, as capital consumption may still be material.

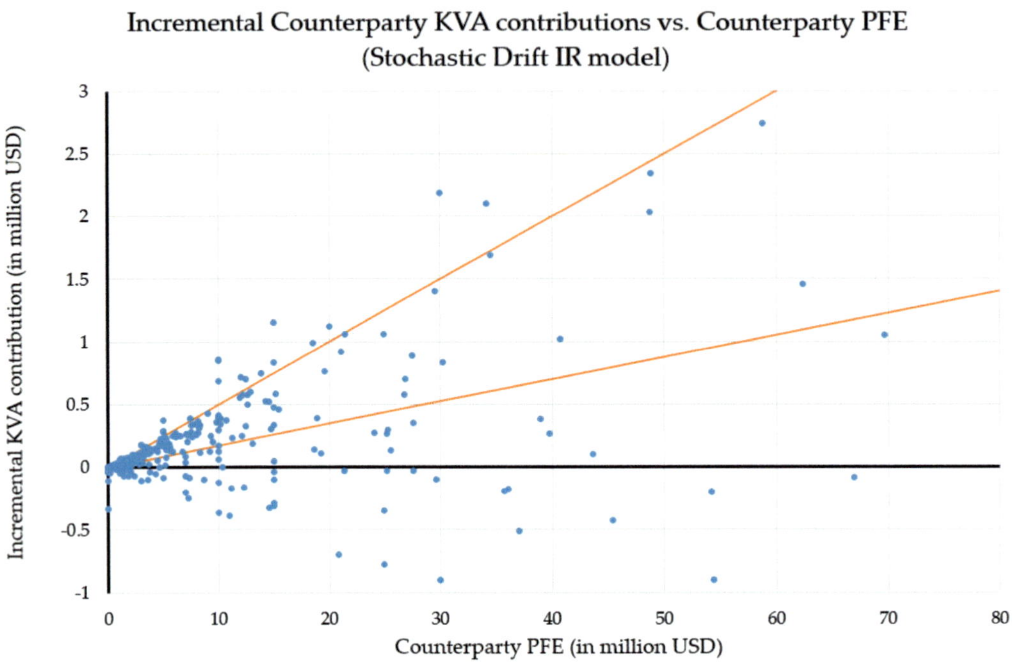

Figure 63: Δ_bKVA metrics and PFE using the SD model (see (15)). The orange lines denote the area within which the intermediary 50% of the data lie.

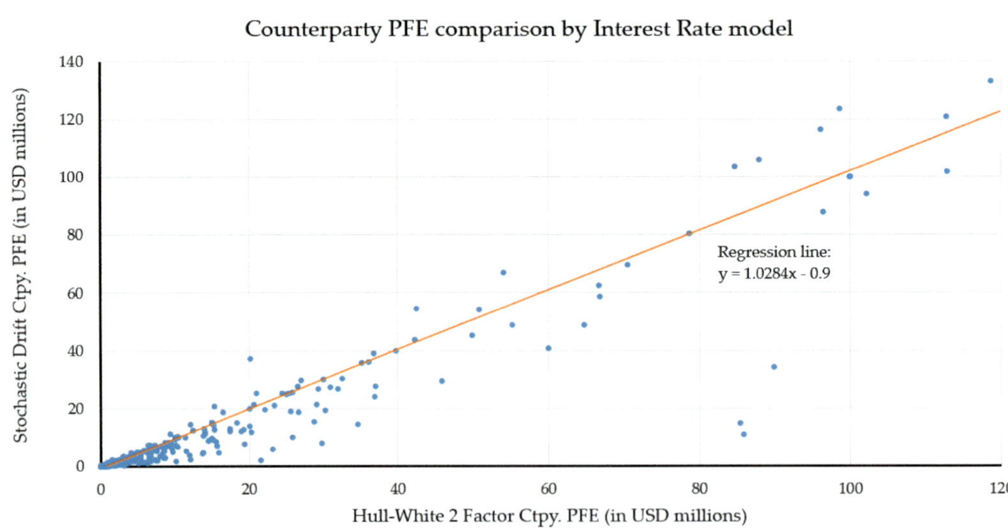

Figure 64: PFE metrics obtained with the HW model (x-axis) versus the same metric obtained with the SD model (y-axis). Note that the regression line (in orange) has an angular coefficient very close to 1, indicating a lack of sensitivity of the PFE metric.

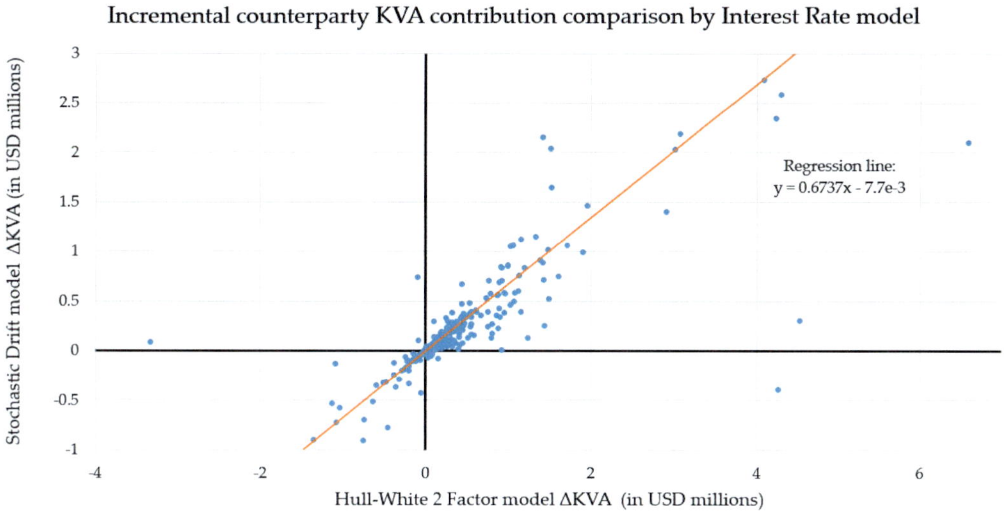

Figure 65: Δ_bKVA metrics obtained with the HW model (x-axis) versus the same metric obtained with the SD model (y-axis). Note that the regression line (in orange) now has an angular coefficient relatively far from 1, indicating a systematic bias due to model risk of the KVA metric.

§6 CONCLUSION

In this chapter, we introduced a consistent framework for reverse stress testing. We identified the future risk-factor's states that could impair the bank's resilience, mining them from a portfolio-wide simulation. Our reverse stress testing approach revealed that risky future conditions are linked to the macroeconomic risk-factors' states, connected to the credit cycle, the portfolio's idiosyncratic weaknesses and the bank's internal modelling choice. Once the stress states predicted by a model are surfaced, model users can compare the bank's modelling choice to their expectations. By adapting the same framework to risk appetites, we presented a credit limit's refinement, which captures a counterparty's risk on the back of the entire portfolio held.

A similar, more granular approach can be used to define trade incremental KVA (cf. I.§4.B). In particular, since the KVA is a portfolio metric, the trade incremental KVA is capable of distinguishing between trades, which are costly in terms of capital, from overall risk-reducing trades. The characteristics of the portfolio held by the bank at trade inception drive discrimination processes. Therefore, a trading desk may enter into a trade and then look for another (internal or external) desk whose portfolio is particularly suited to accommodate this trade. The two desks may then look for an agreement and split the advantages arising from reduced capital and XVA costs at the portfolio level (cf. §3.B and Albanese, Chataigner and Crépey (2020, Section 5)).

HVA IS WORTH A DETOUR

This concluding chapter is an excursion in the direction of model risk. We call it a detour because, as opposed to everything else before in the book, we do not claim that banks should do the corresponding HVA computations in practice, at least not systematically. Instead, the best practice would be that banks only rely on high-quality models, so that such computations are simply not needed. But this HVA detour is worth the trip, because we learn many things on our way: how market risk pops in back into XVA computations, by the entremise of model risk; how to paste local models together within the context of a global valuation framework. If, moreover, traders are too hard-headed, perhaps the regulator would still be well advised to take inspiration from this HVA framework to help banks handle their model risk.

CHAPTER XI

Hedging Valuation Adjustment and Model Risk

Burnett (2021) & Burnett and Williams (2021) introduced the notion of hedging valuation adjustment (HVA) in order to account for dynamic hedging frictions, e.g. when hedging the CVA (cf. VI.§6.A). In this chapter we revisit the notion of hedging valuation adjustment (HVA) in the direction of model risk. The resulting HVA can be seen as the bridge between a global fair valuation model and the local models used by the different desks of the bank. Moreover, model risk and dynamic hedging frictions indeed deserve a reserve, but a risk-adjusted one, so not only an HVA, but also a contribution to the KVA of the bank. Finally, we argue that the industry-standard XVA metrics are jeopardized by cash flows risk, which is in fact of the same mathematical nature than the one regarding pricing models, although at the higher level of aggregation characteristic of XVA metrics.

§1 INTRODUCTION

The 2008 global financial crisis triggered a shift from trade-specific pricing to netting-set CVA analytics. For tractability reasons, the market models used by banks for their CVA analytics are simpler than the ones that they use for individual deals. Given this coexistence of models, it is no surprise if FRTB emphasized the issue of model risk.

Typically (see e.g. European Parliament (2016, L 21/54, point (2))), model risk is accounted for by setting aside as a reserve the difference between the valuation computed with good models minus the valuation given by bad models. This indeed corresponds to the first layer of defense against model risk in this chapter. However, under current market practice, no model risk provision is linked to erroneous hedges. Model risk reserves that are predicated on differences in valuations have no effect on risk capital and risk exposures, which instead are a function of hedge ratios. Regulators have attempted to

DOI: 10.1201/9781003614944-17

remedy this shortcoming. We already mentioned FRTB, which insists that models used for risk management must be the exact same used to keep books and records. There is also Volker's prohibition of proprietary trading. But both requirements can be skirted by using low quality models.

In this chapter we propose to revise model risk reserves by adding an add-on sensitive to hedge ratios. Toward this aim, we put model risk in an XVA perspective. This leads us to propose a reserve for model risk and dynamic hedging frictions, in two parts: an HVA component, encapsulating Burnett (2021) & Burnett and Williams (2021) within a broader model risk perspective, restoring the correct prices that should have been used by the trader in the first place. This HVA can be seen as the bridge between a global fair valuation model and the local models used by the different desks of the bank. But the reserve should also be risk-adjusted, via a KVA component, i.e. a related contribution to the KVA of the bank.

Related literature Detering and Packham (2016) already insisted on "capital buffers necessary to sufficiently protect trading book positions against unexpected losses from model risk". However this was mostly envisioned at the level of individual deals. In this chapter we provide a global picture after aggregation throughout all deals and inclusion of hedging nonlinear imperfections at the hedging sets level, before capital at risk implications are eventually assessed at the balance-sheet level. We refer the reader to Detering and Packham (2016) for a discussion of the literature, including (Karoui, Jeanblanc-picqué and Shreve, 1998; Cont, 2006; Elices and Giménez, 2013), and for model risk regulatory guidelines, until 2014. More recent references include Barrieu and Scandolo (2015), who "introduce three quantitative measures of model risk when choosing a particular reference model within a given class", and Farkas, Fringuellotti and Tunaru (2020), who "propose a general method to account for model risk in capital requirements calculus related to market risk".

On the XVA side, Bichuch, Capponi and Sturm (2020) deal with XVAs that are conservative, or "robust" in the sense of superhedging with respect to the uncertainty of the credit spread of a client. Regarding our baseline cost-of-capital XVA approach, they write in their introduction:

> Despite the merits of this approach, in particular not having to rely on replication arguments for the value adjustments, it makes two critical assumptions. First, it assumes that the counterparty-free payoffs of the contract are perfectly replicated, rather than designing the replication strategy from first principles (and ignoring potential interaction of risk factors). Second, and most importantly, they assume that the historical and risk-neutral probability measure coincide. This, of course, exposes the calculation of the

> valuation adjustments to a substantial amount of model risk, which can be accounted for by the techniques proposed in this paper.

The present chapter provides an element of answer to their first point. As for their second point, the cost-of-capital XVA approach no longer assumes (as in early stages of the theory) that the historical and risk-neutral probability measure coincide: cf. the notion of fininsurance probability measure $\mathbb{Q}$ prevalent in this book. There is of course still an XVA model risk and uncertainty issue. But to address the latter, we prefer to a worst-case approach (see e.g. 0.§4-§5) the Bayesian-robust approach of X.§4.C, which we think is more scalable, i.e. it can more realistically be applied at the level of the derivative portfolio of a bank. It also does better justice to considerations of model realism, while a robust approach is typically unnecessarily over-conservative, hence unusable in practice, let aside the hopeless computational issues at large scale.

More fundamentally, at the economic level, a key difference between the XVA approaches underlying (Bichuch, Capponi and Sturm, 2020) and the cost-of-capital XVA approach of this book, even before model risk is introduced, is that they view XVAs as replication prices, whereas we say CVA and (especially) FVA cannot be replicated, hence the ensuing pnl deserves capital at risk, the KVA cost of which is in fact the mastermind of the theory and often the predominant XVA metric (see e.g. Figures 66-67 pages 336-337). This divergence also impacts the treatment of model risk. Regarding for instance model risk on the CVA, the approach of this chapter implies that, even if the correct CVA value is restored by boosting the default probability of the client (if too small in the first place), then the CVA hedge is still wrong and this should be reflected in the KVA. Regarding the FVA, which we say cannot seriously be hedged (see I.§3.E and II.§4.E.1), our emphasis is on cash flows risk, even prior to model risk. Cash flows risk is easily integrable to our HVA setup, but seems hardly amenable to a robust approach à la Bichuch, Capponi and Sturm (2020). Also note that their robust FVA only involves a single client and it is not clear how to scale it to several, unless distortive simplifications additive across counterparties are used, such as (from this viewpoint) the FCA and FBA discussed in §4.A.

Like Bichuch, Capponi and Sturm (2020), Silotto, Scaringi and Bianchetti (2021) is also about XVA valuation uncertainty linked to the existence of a range of different parameterizations and it also ignores the intrinsic non-replicability of counterparty default risk. The latter is manifested by their emphasis on the DVA metric, whereas, accounting for the counterparty credit risk incompleteness, the DVA cannot be monetized by the bank shareholders and should therefore not be considered in financial derivatives entry prices (see I.§3.E and II.§4.E.1).

Outline §2 casts the HVA and the related risk adjustment in a global valuation framework also encompassing mark-to-market (i.e. counterparty-risk-free) valuation (MtM), CVA, FVA and KVA. §3 and §4 illustrate the MtM and XVA sides of the HVA topic. §5 and §6 establish the pricing analytics in the Merton jump-to-ruin model used in §3, and derive the transaction costs component of the HVA in a delta-hedged jump-diffusion setup.

Standing notation All processes are adapted to the filtration $\mathfrak{G} = (\mathfrak{G}_t)_{t\in\mathbb{R}_+}$ of a reference stochastic basis. The risk-free asset chosen as a numéraire everywhere. All (cumulative) cash flows are finite variation processes (starting from 0) and all prices are special semimartingales in a càdlàg version. We denote by $\mathcal{N}$, the standard normal cumulative distribution function; T, a bound on the final maturity of the bank portfolio assessed on a run-off basis as relevant (see I.§4.B), also including the time (assumed bounded) of liquidating defaulted positions; Q_t and q_t, a reference fair valuation as per Definition 1 and a trader time-t price of a financial derivative of interest; $X_{(0)} = X - X_0$, for any process X.

We use italic capitals for cash flows and straight letters for prices, in respective upper and lower case for fair and trader valuations.

§2 THE GLOBAL VALUATION FRAMEWORK

We work throughout the chapter under the bank survival probability measure $\mathbb{Q}$ associated with a fininsurance measure $\mathbb{Q}^*$ (see the related notation in the introduction to the Pricing part of the book).

Definition 1. Given an optional, integrable process $\mathcal{Y}$ stopped at T interpreted as cumulative cash flow stream, its fair value process $Y = va(\mathcal{Y})$ (in the sense of bank shareholder value process everywhere in this chapter, cf. Definition II.1(ii) and Theorem II.1(i)) is the optional projection of $(\mathcal{Y}_T - \mathcal{Y})$, i.e.

$$Y_t = \mathbb{E}_t(\mathcal{Y}_T - \mathcal{Y}_t),\ t \leq T, \tag{1}$$

and Y vanishes on $[T, +\infty)$.

In particular, $(\mathcal{Y} + Y)$ is a martingale on $[0, T]$. But, as in X.§4-§5 already, our approach in this chapter considers a dual-model environment: on one side, the global fair valuation model (or reference model as advocated for model risk assessment in Barrieu and Scandolo (2015)), in which prices are fair value processes as per (1); on the other side, local trade-specific models used by traders, until at least the time of explosion of their strategy where their model no longer calibrates to fair valuation.

Remark 1. In this chapter we are not addressing model risk in the sense of misspecification of the risk-neutral and/or physical probability measures that underlie $\mathbb{Q}$. Our (Darwinian, as specified further in §3) model risk here refers to fair valuation in contrast with the local models that are effectively used by the traders of the bank.

The valuation approach developed in the Pricing part of this book can then be summarized as follows. $\mathrm{MtM} = va(\mathcal{M})$, $\mathrm{CVA} = va(\mathcal{C})$ and $\mathrm{FVA} = va(\mathcal{F})$ are the (shareholder) value processes (1) of the (other than dynamic hedging) cash flows impacting the eponymous desks of the bank, namely the cash flow processes $\mathcal{M}$ to the trading desks, $\mathcal{C}$ from the CVA desk and $\mathcal{F}$ from the funding desk. The ensuing loss process of the bank is then given by (recalling the notation $X_{(0)} = X - X_0$)

$$\mathcal{L} = -(\mathcal{M} + \mathrm{MtM}_{(0)}) + \mathcal{C} + \mathcal{F} + \mathcal{H} + \mathrm{CA}_{(0)}, \tag{2}$$

where

$$\mathrm{CA} = \mathrm{CVA} + \mathrm{FVA} \tag{3}$$

and the $\mathbb{Q}$ martingale $\mathcal{H}$ represents the dynamic hedging losses of the bank. ***In the extension of the theory provided by the present chapter, the no longer martingale process $\mathcal{H}$ additionally accounts for model risk and dynamic hedging frictions, and* (3) *is generalized as***

$$\mathrm{CA} = \mathrm{CVA} + \mathrm{FVA} + \mathrm{HVA}, \textit{ where } \mathrm{HVA} = va(\mathcal{H}). \tag{4}$$

Via compensation by the CA process (see after Definition 1), the above loss process $\mathcal{L}$, to be specified further in Lemma 2 page 328, is a $\mathbb{Q}$ martingale. The theory then proceeds as in Chapters I–III. To recap:

Definition 2. The economic capital (EC) of the bank is defined as the time-t conditional expected shortfall ($\mathbb{ES}_t$) of the random variable $(\mathcal{L}_{t'} - \mathcal{L}_t)$ at some confidence level $\alpha \in (\frac{1}{2}, 1)$, killed at the bank default time τ, where $\mathcal{L}$ is the loss process of the bank and $t' = (t+1) \wedge T$, i.e.[1] $\mathrm{EC}_t = \mathbb{1}_{\{t<\tau\}} \mathbb{ES}_t(\mathcal{L}_{t'} - \mathcal{L}_t)$, where

$$\mathbb{ES}_t(\mathcal{L}_{t'} - \mathcal{L}_t) = \frac{\mathbb{E}_t\Big((\mathcal{L}_{t'} - \mathcal{L}_t)\mathbb{1}_{\{\mathcal{L}_{t'} - \mathcal{L}_t \geq \mathbb{V}a\mathbb{R}_t(\mathcal{L}_{t'} - \mathcal{L}_t)\}}\Big)}{\mathbb{E}_t \mathbb{1}_{\{\mathcal{L}_{t'} - \mathcal{L}_t \geq \mathbb{V}a\mathbb{R}_t(\mathcal{L}_{t'} - \mathcal{L}_t)\}}}, \tag{5}$$

in which $\mathbb{V}a\mathbb{R}_t(\cdot)$ is the time-t conditional value-at-risk at the confidence level α.

As the process $\mathcal{L}$ is stopped at T, the $\cdot \wedge T$ in t' has no impact on the economic capital, we let it apparent only to emphasize that nothing happens beyond T.

[1] skipping here a minor measurability (predictability) issue that could be addressed as in Definition II.6, see Remark II.17 page 68.

Definition 3. We define the *shareholder* capital at risk (SCR), to be remunerated at a constant and nonnegative hurdle rate h, and the corresponding capital valuation adjustment (KVA) of the bank, as

$$\mathrm{SCR} = \max(\mathrm{EC}, \mathrm{KVA}) - \mathrm{KVA} = (\mathrm{EC} - \mathrm{KVA})^+, \tag{6}$$

where

$$\mathrm{KVA}_t = h\mathbb{E}_t \int_t^T \left(\mathrm{EC}_s - \mathrm{KVA}_s\right)^+ ds,\ t \le T. \tag{7}$$

So KVA is the value process (1) of $\int_0^{\cdot} h\mathrm{SCR}_s ds$. By Proposition II.4, (7) defines a unique square integrable KVA process, assuming EC square integrable.

The KVA process is thus designed so as to turn the dividend process

$$-(\mathcal{L} + \mathrm{KVA}_{(0)})$$

of the bank shareholders into a $\mathbb{Q}$ submartingale with drift coefficient hSCR, for some nonnegative constant hurdle rate h (e.g. 10%).

A HVA for Mark-to-Market

We index by "$\cdot$" the client originating deals of the bank, which may be hedged statically with other banks and/or dynamically through exchanges (see Chapters III and VIII). At the "model switch time" of explosion of the trader's strategy introduced after Definition 1, the valuation of the deal and its hedge in the trader's pnl respectively pass from local to fair valuation. This implies the following expression for the raw pnl of the trader, where "raw" is in reference to the fact that this pnl still ignores the to-be-defined HVA liability, as well as extra contributions, to be introduced later, that can only be addressed at the portfolio level:

Definition 4. The raw pnl of the client originating deal "$\cdot$" of the bank, with maturity $T^{\cdot}$, is given by

$$pnl^{\cdot} = \left(\mathcal{Q}^{\cdot} + J^{s,\cdot} q^{\cdot} + (1 - J^{s,\cdot})Q^{\cdot}\right)_{(0)}^{\tau^{\cdot}} - \left(\mathcal{P}^{\cdot} + J^{s,\cdot} p^{\cdot} + (1 - J^{s,\cdot})P^{\cdot}\right)_{(0)}^{\tau^{\cdot}} - h^{\cdot}, \tag{8}$$

where $\mathcal{Q}^{\cdot}$ represents the cash flows contractually promised to the bank through the deal, $Q^{\cdot} = va(\mathcal{Q}^{\cdot})$ as per Definition 1, $\mathcal{P}^{\cdot}$ and $P^{\cdot}$ are the analogous quantities regarding a static hedge component of the deal, $\tau^{\cdot} = \tau_d^{\cdot} \wedge \tau_e^{\cdot}$ in which $0 < \tau_e^{\cdot} \le T^{\cdot}$ is a deactivation (knock-out, as in autocalls) time, and $\tau_d^{\cdot} > 0$ is the time of default of the client of the deal, $q^{\cdot}$ and $p^{\cdot}$ are the prices of the product and of its static hedge in the local model of the trader, $h^{\cdot}$ is the dynamic hedging loss of the trader, and $J^{s,\cdot} = \mathbb{1}_{[\![0,\tau_s^{\cdot}[\![}$, where $\tau_s^{\cdot} > 0$ is the time of model switch.

Remark 2. The bank may consider liquidating the deal at $\tau_s^{\cdot}$ when earlier than $\tau^{\cdot}$. To render this case, one just needs to redefine $\tau^{\cdot}$ as $\tau_d^{\cdot} \wedge \tau_e^{\cdot} \wedge \tau_s^{\cdot}$ instead of $\tau_d^{\cdot} \wedge \tau_e^{\cdot}$.

Remark 3. We could also consider American claims with exercise times possibly $< T^{\cdot}$ under the control of the bank and/or client, in which case $\tau_e^{\cdot}$ in $\tau^{\cdot} = \tau_d^{\cdot} \wedge \tau_e^{\cdot}$ (or $\tau_d^{\cdot} \wedge \tau_e^{\cdot} \wedge \tau_s^{\cdot}$ as above) should be understood as the corresponding exercise time. Further adjustments are then required to deal with possibly unoptimal stopping by the bank (unoptimal stopping by the client can be conservatively ignored in the modeling). Callability by the bank is actually the source of model risk in Albanese, Crépey and Iabichino (2021) and Bénézet, Crépey and Essaket (2022). In this book American early exercise features are ignored for simplicity.

The raw pnl ignores the to-be-defined HVA liability, starting with:

Definition 5. $\mathrm{HVA}^{\cdot} = -va(pnl^{\cdot})$.

Lemma 1. *We have*

$$\mathrm{HVA}^{\cdot} = va(\mathcal{H}^{\cdot}), \tag{9}$$

where

$$\mathcal{H}^{\cdot} = (\mathcal{Q}^{\cdot} + Q^{\cdot})^{\tau^{\cdot}}_{(0)} - (\mathcal{P}^{\cdot} + P^{\cdot})^{\tau^{\cdot}}_{(0)} - pnl^{\cdot} \tag{10}$$

$$= \Big(J^{s,\cdot}\big(Q^{\cdot} - q^{\cdot} - (P^{\cdot} - p^{\cdot})\big)\Big)^{\tau^{\cdot}}_{(0)} + h^{\cdot}. \tag{11}$$

Proof. In view of (8) and (10), we have (11). Moreover, by the observation following Definition 1, $(\mathcal{Q}^{\cdot} + Q^{\cdot})^{\tau^{\cdot}}_{(0)}$ and $(\mathcal{P}^{\cdot} + P^{\cdot})^{\tau^{\cdot}}_{(0)}$ are martingales. Hence (9) proceeds from Definition 5. □

Remark 4. In view of (11), in the "continuously recalibrated case" $p^{\cdot} = P^{\cdot}$, the static side of the hedge does not impact $\mathcal{H}^{\cdot}$ nor $\mathrm{HVA}^{\cdot}$.

The process $h^{\cdot}$ is meant for standard dynamic hedging cash flows ignoring nonlinear frictions such as transaction costs, which can only be assessed at the portfolio level and will therefore be added later, in §2.C. Hereafter we postulate that, as natural in view of their financial interpretation:

Assumption 1. All prices $Q^{\cdot}, q^{\cdot}, P^{\cdot}, p^{\cdot}$ share a nil terminal condition at time $T^{\cdot}$ and $h^{\cdot}$ is a martingale stopped at $\tau^{\cdot}$.

Proposition 1. *We have*

$$\begin{aligned} \mathrm{HVA}^{\cdot} &= \Big(J^{s,\cdot}\big(P^{\cdot} - p^{\cdot} - (Q^{\cdot} - q^{\cdot})\big)\Big)^{\tau^{\cdot}} - V^{\cdot}, \\ \mathcal{H}^{\cdot} + \mathrm{HVA}^{\cdot}_{(0)} &= h^{\cdot} - V^{\cdot}_{(0)}, \end{aligned} \tag{12}$$

where

$$V^{\cdot}_t = \mathbb{E}_t\Big(\mathbb{1}_{\{\tau^{\cdot}<\tau^{\cdot}_s\wedge T^{\cdot}\}}\big(P^{\cdot}-p^{\cdot}-(Q^{\cdot}-q^{\cdot})\big)_{\tau^{\cdot}}\Big),\ t\geq 0. \tag{13}$$

In particular,

$$\mathrm{HVA}^{\cdot}_0 = \big(P^{\cdot}-p^{\cdot}-(Q^{\cdot}-q^{\cdot})\big)_0 - \mathbb{E}\Big(\mathbb{1}_{\{\tau^{\cdot}<\tau^{\cdot}_s\wedge T^{\cdot}\}}\big(P^{\cdot}-p^{\cdot}-(Q^{\cdot}-q^{\cdot})\big)_{\tau^{\cdot}}\Big). \tag{14}$$

Proof. Under Assumption 1, (9)–(11) yield (12)–(14). □

Remark 5. $\mathrm{HVA}^{\cdot}$ corresponds essentially to the current market practice for handling model risk, in the form of a reserve put aside at initial time. In actual practice, rather than paying $q^{\cdot}_0$ to the client as implied by (8) while the client would provide $\mathrm{HVA}^{\cdot}_0$ as reserve capital to the bank, the trader pays $Q^{\cdot}_0$ to the client and puts by himself $\mathrm{HVA}^{\cdot}_0$ in the reserve capital account, which is equivalent (at least if $\tau^{\cdot}=+\infty$ and $P^{\cdot}=p^{\cdot}$ a.s., so that $V^{\cdot}_0=0$ and $\mathrm{HVA}^{\cdot}_0=q^{\cdot}_0-Q^{\cdot}_0$ a.s.).

Note that, in the pure static hedging case where $h^{\cdot}=0$, (8)–(14) reduce to

$$\begin{aligned}
pnl^{\cdot} &= \big(\mathcal{Q}^{\cdot}+J^{s,\cdot}q^{\cdot}+(1-J^{s,\cdot})Q^{\cdot}\big)^{\tau^{\cdot}}_{(0)} - \big(\mathcal{P}^{\cdot}+J^{s,\cdot}p^{\cdot}+(1-J^{s,\cdot})P^{\cdot}\big)^{\tau^{\cdot}}_{(0)},\\
\mathcal{H}^{\cdot} &= \Big(J^{s,\cdot}\big(Q^{\cdot}-q^{\cdot}-(P^{\cdot}-p^{\cdot})\big)\Big)^{\tau^{\cdot}}_{(0)}\\
\mathrm{HVA}^{\cdot} &= \Big(J^{s,\cdot}\big(P^{\cdot}-p^{\cdot}-(Q^{\cdot}-q^{\cdot})\big)\Big)^{\tau^{\cdot}} - V^{\cdot},\\
\mathrm{HVA}^{\cdot}_0 &= \big(P^{\cdot}-p^{\cdot}-(Q^{\cdot}-q^{\cdot})\big)_0 - \mathbb{E}\Big(\mathbb{1}_{\{\tau^{\cdot}<\tau^{\cdot}_s\wedge T^{\cdot}\}}\big(P^{\cdot}-p^{\cdot}-(Q^{\cdot}-q^{\cdot})\big)_{\tau^{\cdot}}\Big),
\end{aligned} \tag{15}$$

whereas, in the pure dynamic hedging case where $P^{\cdot}=p^{\cdot}=0$,

$$\begin{aligned}
pnl^{\cdot} &= \big(\mathcal{Q}^{\cdot}+J^{s,\cdot}q^{\cdot}+(1-J^{s,\cdot})Q^{\cdot}\big)^{\tau^{\cdot}}_{(0)} - h^{\cdot},\\
\mathcal{H}^{\cdot} &= \big(J^{s,\cdot}(Q^{\cdot}-q^{\cdot})\big)^{\tau^{\cdot}}_{(0)} + h^{\cdot},\\
\mathrm{HVA}^{\cdot} &= \big(J^{s,\cdot}(q^{\cdot}-Q^{\cdot})\big)^{\tau^{\cdot}} - V^{\cdot}\ ,\ \text{where } V^{\cdot}_t = \mathbb{E}_t\Big(\mathbb{1}_{\{\tau^{\cdot}<\tau^{\cdot}_s\wedge T^{\cdot}\}}(q^{\cdot}-Q^{\cdot})_{\tau^{\cdot}}\Big),\ t\geq 0,\\
\mathrm{HVA}^{\cdot}_0 &= (q^{\cdot}-Q^{\cdot})_0 - \mathbb{E}\Big(\mathbb{1}_{\{\tau^{\cdot}<\tau^{\cdot}_s\wedge T^{\cdot}\}}(q^{\cdot}-Q^{\cdot})_{\tau^{\cdot}}\Big).
\end{aligned} \tag{16}$$

B HVA for CVA and FVA

If accounting methods are misaligned with economics they cause a misalignment of interests between executives and shareholders. Accounting methods are also models in the sense of SR-11-7 because they produce numbers, are based on assumptions, and have an impact on strategies: cf. https:

//www.federalreserve.gov/supervisionreg/srletters/sr1107.htm. Hence, model risk is a concept that does not apply only to pricing models, but should be extended to accounting principles for dealer banks, including the specification of their CVA and FVA metrics (as these are liabilities to the bank: see Figure 3 page 28).

In the case of the FVA, which is not marketed and can hardly be hedged (see I.§3.E and II.§4.E.1), the model risk game is in fact simpler than for client deals and their hedges, even though at the higher level of aggregation characteristic of XVA metrics. It is just the temptation for the bank to underestimate its FVA for being more competitive with clients, at the long-term risk of seeing the FVA spiking and triggering huge losses to the bank as happened for instance during the Covid financial crisis of Q1 2020 (see §4.A). The CVA, which can only partially be hedged (only its market risk and only to some extent, fault of liquid enough CDS markets, cf. VI.§5.A and VI.§6.A), is from this point of view in an intermediate situation between mark-to-market (MtM) considered in Subsection A and the FVA. Accordingly:

Assumption 2. The CVA is purely dynamically hedged and the FVA is unhedged.

We denote by $\mathcal{C}$ and $\mathcal{F}$ the counterparty default and risky funding cash flows (cf. the Pricing part of the book), with fair vs. trader prices CVA/FVA vs. cva/fva. Developments similar to (8)–(16) with analogous notation, detailed in Table 52 page 330, yield under Assumption 2 (see Remark 6 below):

$$\begin{aligned}
-\,pnl^{cva} &= \Big(\mathcal{C} + J^{s,cva}\mathrm{cva} + (1-J^{s,cva})\mathrm{CVA}\Big)_{(0)} - h^{\mathrm{cva}},\\
\mathcal{H}^{cva} &= \Big(J^{s,cva}(\mathrm{CVA}-\mathrm{cva})\Big)_{(0)} + h^{\mathrm{cva}},\\
\mathrm{HVA}^{cva} &= va(-\mathcal{H}^{cva}) = va(-pnl^{cva}) = J^{s,cva}\Big(\mathrm{CVA}-\mathrm{cva}\Big),\\
\mathrm{HVA}^{cva}_0 &= \Big(\mathrm{CVA}-\mathrm{cva}\Big)_0,\\
\mathcal{H}^{cva} - \mathrm{HVA}^{cva}_{(0)} &= h^{cva},
\end{aligned} \tag{17}$$

and

$$\begin{aligned}
-\,pnl^{fva} &= \Big(\mathcal{F} + J^{s,fva}\mathrm{fva} + (1-J^{s,fva})\mathrm{FVA}\Big)_{(0)},\\
\mathcal{H}^{fva} &= \Big(J^{s,fva}(\mathrm{FVA}-\mathrm{fva})\Big)_{(0)},\\
\mathrm{HVA}^{fva} &= va(-\mathcal{H}^{fva}) = va(-pnl^{fva}) = J^{s,fva}\Big(\mathrm{FVA}-\mathrm{fva}\Big),\\
\mathrm{HVA}^{fva}_0 &= \Big(\mathrm{FVA}-\mathrm{fva}\Big)_0,\\
\mathcal{H}^{fva} - \mathrm{HVA}^{fva}_{(0)} &= 0.
\end{aligned} \tag{18}$$

C HVA for Dynamic Hedging Frictions

The above processes $h^{\cdot}$ and h^{cva} are meant for standard dynamic hedging cash flows ignoring nonlinear frictions such as transaction costs, i.e. ignoring HVA originating cash flows à la Burnett (2021) & Burnett and Williams (2021). Indeed, as the latter are nonlinear, they can only be addressed at the level of each book of contracts or exposures that are hedged together, or hedging sets "$\star$". Let $proc^{mtm} = \sum_{\cdot} proc^{\cdot}$, for each process $proc = pnl, h, \mathcal{H}, \mathrm{HVA}, V$, and let

$$\begin{aligned}
&\mathcal{M} + \mathrm{MtM}_{(0)} = \sum_{\cdot}\Big((\mathcal{Q}^{\cdot} + Q^{\cdot})^{\tau^{\cdot}}_{(0)} - (\mathcal{P}^{\cdot} + P^{\cdot})^{\tau^{\cdot}}_{(0)}\Big),\\
&\mathcal{H} = \mathcal{H}^{mtm} - \mathcal{H}^{cva} - \mathcal{H}^{fva} + f, \qquad (19)\\
&\mathrm{HVA} = \mathrm{HVA}^{mtm} + \mathrm{HVA}^{cva} + \mathrm{HVA}^{fva} + \mathrm{HVA}^{f},
\end{aligned}$$

where $f = \sum_{\star} f^{\star}$ is the sum of the hedging friction $f^{\star}$ on each hedging set $\star$, valued by $\mathrm{HVA}^{f} = va(f)$.

Remark 6. As local XVA pricing models typically underestimate CVA and FVA whereas local market models overestimate counterparty-risk-free values, our sign convention for the XVA-related HVA cash flows in (17)–(18) is opposite to the one regarding the MtM component. Hence we arrive at a final HVA formula as per the last line in (19), with all terms nonnegative there. Also note that the first line in (19) is actually not a definition but the identity established as III.(15).

Accounting for raw pnls, hedging frictions, and HVA compensators for all, we obtain the overall trading loss of the bank

$$\mathcal{L} = -pnl^{mtm} + \mathrm{HVA}^{mtm}_{(0)} - pnl^{cva} + \mathrm{HVA}^{cva}_{(0)} - pnl^{fva} + \mathrm{HVA}^{fva}_{(0)} + f + \mathrm{HVA}^{f}_{(0)}. \qquad (20)$$

Lemma 2. **(i)** *The process $\mathcal{L}$ as per* (20) *is a $\mathbb{Q}$ martingale of the form* (2)-(4), *with the different terms specified as in* (19).
(ii) *It holds:*

$$\begin{aligned}
\mathcal{H} + \mathrm{HVA}_{(0)} &= h^{mtm} - V^{mtm}_{(0)} - h^{cva} + f + \mathrm{HVA}^{f}_{(0)},\\
\mathcal{L} &= -\sum_{\cdot}\Big((\mathcal{Q}^{\cdot} + Q^{\cdot})^{\tau^{\cdot}}_{(0)} - (\mathcal{P}^{\cdot} + P^{\cdot})^{\tau^{\cdot}}_{(0)} - h^{\cdot} + V^{\cdot}_{(0)}\Big) \qquad (21)\\
&\quad + \mathcal{C} + \mathrm{CVA}_{(0)} - h^{cva} + \mathcal{F} + \mathrm{FVA}_{(0)} + f + \mathrm{HVA}^{f}_{(0)}.
\end{aligned}$$

Proof. **(i)** follows from (10) and its credit and funding analogs

$$-pnl^{cva} + \mathcal{H}^{cva} = \mathcal{C} + \mathrm{CVA}_{(0)} \text{ and } -pnl^{fva} + \mathcal{H}^{fva} = \mathcal{F} + \mathrm{FVA}_{(0)}$$

that stem from (17) and (18).

(ii) The expression for $\mathcal{H} + \mathrm{HVA}_{(0)}$ in (21) proceeds from the last lines in (12), (17) and (18). Substituting this expression for $\mathcal{H}+\mathrm{HVA}_{(0)}$ in the identity $\mathcal{L} = -(\mathcal{M} + \mathrm{MtM}_{(0)}) + \mathcal{C} + \mathcal{F} + \mathrm{CVA}_{(0)} + \mathrm{FVA}_{(0)} + \mathcal{H} + \mathrm{HVA}_{(0)}$ that stems from (2)-(4) yields

$$\mathcal{L} = -(\mathcal{M}+\mathrm{MtM}_{(0)})+h^{mtm}-V^{mtm}_{(0)}+\mathcal{C}+\mathrm{CVA}_{(0)}-h^{cva}+\mathcal{F}+\mathrm{FVA}_{(0)}+f+\mathrm{HVA}^{f}_{(0)},$$

where, by (19) and definition of h^{mtm} and V^{mtm},

$$-(\mathcal{M}+\mathrm{MtM}_{(0)})+h^{mtm}-V^{mtm}_{(0)} = -\sum_{\cdot}\Big((\mathcal{Q}^{\cdot}+Q^{\cdot})^{\tau^{\cdot}}_{(0)}-(\mathcal{P}^{\cdot}+P^{\cdot})^{\tau^{\cdot}}_{(0)}-h^{\cdot}+V^{\cdot}_{(0)}\Big).$$

This yields the second line in (21). □

We then need to specify the friction processes $f^{\star}$. In §6 this is achieved by passage to the continuous-time limit starting from a classical discrete-time specification. This sheds more rigor in the seminal contribution of Burnett (2021), who derives a PDE for the transaction costs at the limit, while only rebalancing when the delta of the underlying portfolio is shifted by some positive and constant threshold (so it seems that Burnett (2021)'s limiting HVA should increase at discrete rebalancing times only, rather than being given by a PDE). Our approach also allows computing $\mathrm{HVA}^{\star} = va(f^{\star})$ numerically in a model risk setup accounting for the impact of recalibration on transaction costs, which is not considered in Burnett (2021).

Table 52 recapitulates the HVA-related data of the global valuation problem of the bank. From an organizational viewpoint, the computation of the $\mathrm{HVA}^{\cdot}$ (summing up to HVA^{mtm}), HVA^{cva} and HVA^{fva} components could be delegated to each related trader (under regulatory control). The HVA^{f} component(s) calculations would typically require a dedicated (regulated) HVA desk, as such computations typically need a mix of data from the different trading and/or CVA/funding desks of the bank.

D KVA Also Accounting for Dynamic Frictions and Model Risk

After compensation by the HVA, the price is right (cf. Remark 5 page 326), but the hedge is still wrong (as it is not the hedging strategy corresponding to the corrected price). Under a cost-of-capital valuation approach, a reserve for model risk and dynamic hedging frictions does not reduce to HVA terms summed over all the pnl centers of the banks. This reserve is also be risk-adjusted, via the impact of model risk and dynamic hedging frictions on $\mathcal{L}$ as per (2)-(20)-(21). Namely, going by Definitions 2 and 3, the ensuing volatile swings of $\mathcal{L}$ are reflected in the economic capital and in the KVA of the bank.

Table 52: HVA related data.

$\cdot$	a generic client deal of the bank
$\mathcal{Q}^{\cdot}, \mathcal{P}^{\cdot}$	cash flows promised to the bank on the deal and its static hedge components
$Q^{\cdot}, P^{\cdot}$	corresponding fair value processes (1)
$q^{\cdot}, p^{\cdot}$	trader's local model price for the deal "·" and its static hedge component (if any)
$\tau^{\cdot}_{e/s/d}$	related deactivation (barrier) time, model switch time, default time
$h^{\cdot}$	dynamic (martingale) hedging cash flows related to the deal "·"
$\mathcal{C}, \mathcal{P}^{cva} = 0, \mathrm{CVA}, P^{cva} = 0$, cva, $p^{cva} = 0$ $\tau^{cva}_{e/d} = +\infty, \tau^{cva}_{s}, h^{cva}$	similar data regarding the CVA of the bank
$\mathcal{F}, \mathcal{P}^{fva} = 0, \mathrm{FVA}, P^{fva} = 0$, fva, $p^{fva} = 0$ $\tau^{fva}_{e/d} = +\infty, \tau^{fva}_{s}, h^{fva} = 0$	similar data regarding the FVA of the bank
f	dynamic hedging friction costs at the aggregated bank level

Remark 7. We saw in Remark 4 that, in the continuously recalibrated case, the static side of the hedge does not impact $\mathcal{H}^{\cdot}$ nor $\mathrm{HVA}^{\cdot}$. But it does modify the raw $pnl^{\cdot}$ of the trader (8), hence the trading loss of the bank (20), and therefore the related contribution to the KVA.

§3 MTM EXAMPLE

In the context of structured products, Albanese, Crépey and Iabichino (2021) introduced the notion of Darwinian model risk, whereby the trader of a bank prefers to a reference fair valuation model to an alternative pricing model, which renders a trade more competitive (more attractive for the client of the bank) in valuation terms (*first Darwinian principle*). But the recalibration of such a model introduces alpha leakage on the asset valuation side, which has to be compensated on the hedging side so that the model stands a chance to be accepted by the management of the bank (*second Darwinian principle*). However, even in this case, the systematic gains on the hedging side of the position could be a short-to-medium viewpoint: in the long run, large losses may be incurred by the bank when market conditions reveal the unsoundness of the trader's pricing and hedging model (*third Darwinian principle*).

The following example is devised for the sake of analytical tractability. But the Darwinian model risk mechanism at hand here is essentially the same as the one affecting huge amounts of structured derivative products, including range accruals in the fixed-income world, autocallables and cliquets on equities, or power-reversal dual currency options and target redemption forwards on foreign-exchange. It was thus reported on https://www.risk.net/derivatives/6556166/remembering-the-range-accrual-bloodbath (11 April 2019) that Q4 of 2019, a \$70bn notional of range accrual had to be unwound at very large losses by the industry. Albanese, Crépey and Iabichino (2021) illustrate this episod by a case study regarding a callable range accrual hedged by digital swaption streams. Bénézet, Crépey and Essaket (2022) quantify it in HVA terms. In these papers, Darwinian model risk enters the picture through the callability of the asset (see Remark 3 page 325): a trader long an extreme (corridor exit) event ends-up being short in the event, hence "gamma negative", by shorting excessive vanilla option (digital swaption) positions as his hedge, as these are computed with a model overestimating the probability of the extreme event and therefore the optimal call time by the bank. The trader makes systematic profits in the short-to-medium term followed by a large loss when a related extreme event occurs. In our example below, a trader is short an extreme (default) event but pretends he does not see it, only hedging market risk. Hence the hedged position is still short the default event, which can be seen as an extreme case of "gamma negative" type position. If the extreme event happens, the mis-hedged position blows up (see Remark 8 page 333).

We denote by

$$Q^{mo}(\text{t, market risk parameters[; model parameters, whenever relevant]}),$$

or simply Q^{mo}_t when the parameters are clear enough from the context, the value process (1) of a given product (or cumulative cash flow stream) in the fair valuation model mo, and we use analogous conventions with P instead of Q regarding a static hedge component of the claim.

We consider financial derivatives of maturity T on a stock S, with dividend yields on S and interest rates in the economy set to 0 for notational simplicity. The role of the fair valuation model is played by the jump-to-ruin (jr) model

$$dS_t = \lambda S_t dt + \sigma S_t dW_t - S_{t-} dN_t = \sigma S_t dW_t - S_{t-} dM_t, \tag{22}$$

where W is a standard Brownian motion, $\sigma > 0$ is a constant volatility parameter and N is a Poisson process with intensity $\lambda > 0$ and compensated martingale $M = N - \lambda t$, with W and N independent (as in fact always the case for a Brownian motion and a Poisson process with respect to a common stochastic basis, see e.g. He, Wang and Yan (1992, Theorem 11.43 page 316)). Hence the stock S jumps to 0 at the first jump time θ of the driving Poisson process N.

The role of a local model is played by a Black-Scholes model bs, with volatility Σ continuously recalibrated to the jump-to-ruin price P^{jr} of a European vanilla put with maturity T and strike K, which can be used by the trader as an hedging asset along with S and the risk-free constant asset. The claim of interest consists of an exotic variation on this put including an additional barrier clause. At each time t the option is valued and hedged by the trader in the Black-Scholes model with implied volatility Σ_t (see Definition 6) of the vanilla put. At least this holds for $t < \theta \wedge T$. Indeed, an application of the formula (53) for $S = 0$ and $-d_\pm = +\infty$ shows that $P^{jr} = K$ on $[\theta, \theta \wedge T[$. As detailed in Remark 13 page 350, at time θ (if $< T$), the implied volatility of the vanilla put ceases to be well-defined, hence the falsity of the local model is revealed, and the position must be unwound by the trader at exit prices dictated by the fair valuation (jr) model. Hence in this example $\tau_s^{\cdot} = \theta$.

A Static Hedging of a Vulnerable Put

We consider the limiting case in the above, where the option bought (at $t = 0$) and hedged by the trader is a put option of maturity T and strike K with a deactivating barrier at the level $S = 0$.

In the jump-to-ruin model, this barrier is attainable and absorbing and the above option is equivalent to the vulnerable put of Proposition 6, so that, by (56), $Q^{jr} = 0$ from time θ onward. But the trader values the vulnerable put at its bs price with implied volatility Σ_t of the vanilla put. As the barrier is immaterial in bs, for $t < \theta \wedge T$,

$$\begin{aligned} q_t &= Q^{bs}(t, S_t; \Sigma_t) = P^{bs}(t, S_t; \Sigma_t) = P^{jr}(t, S_t) \\ &= Q^{jr}(t, S_t) + K(1 - e^{-\lambda(T-t)}) > Q^{jr}(t, S_t) = Q_t, \end{aligned} \tag{23}$$

where the first equality in the second line follows from (57). The inequality in (23) is in line with the first Darwinian principle stated in the beginning of this section. We assume the portfolio of the bank is restricted to the vulnerable put and its hedge, as well as a risk-free bank, client and hedge counterparties. In the notation of §2 summarized in Table 52 page 330, we thus have $\tau_s^{\cdot} = \theta$ as already seen above and

$$\begin{aligned} &\mathcal{Q}^{\cdot} = \mathbb{1}_{\{\theta > T\}}(K - S_T)^+ \mathbb{1}_{[T,+\infty)},\ Q^{\cdot} = Q^{jr},\ \mathcal{P}^{\cdot} = (K - S_T)^+ \mathbb{1}_{[T,+\infty)} \\ &q^{\cdot} = p^{\cdot} = P^{\cdot} = P^{jr},\ \tau^{\cdot} = +\infty,\ V^{\cdot} = 0 \end{aligned} \tag{24}$$

(cf. (13)).

In view of (24), (57), (15) and (20) (cf. also (19)), the HVA equations reduce to

$$\begin{aligned}
pnl^{\cdot} &= -\mathbb{1}_{\{\theta\leq T\}}K\mathbb{1}_{[T,+\infty)} + \mathbb{1}_{[\theta,+\infty)}(Q^{jr}-P^{jr})\\
&= -\mathbb{1}_{\{\theta\leq T\}}K\mathbb{1}_{[T,+\infty)} - \mathbb{1}_{[\theta,+\infty)}K\mathbb{1}_{[0,T)}\\
&= -\mathbb{1}_{\{\theta\leq T\}}K\mathbb{1}_{[T,+\infty)}\mathbb{1}_{[\theta,+\infty)} - \mathbb{1}_{[\theta,+\infty)}K\mathbb{1}_{[0,T)}\mathbb{1}_{\{\theta\leq T\}} = -\mathbb{1}_{[\theta,+\infty)}K\mathbb{1}_{\{\theta\leq T\}}\\
\mathcal{H} &= \mathcal{H}^{\cdot} = \mathbb{1}_{[0,\theta)}\Big(Q^{jr}-P^{jr}\Big) - \Big(Q^{jr}-P^{jr}\Big)_0 = -\mathbb{1}_{[0,\theta)}K(1-e^{-\lambda(T-\cdot)}) + K(1-e^{-\lambda T})\\
\mathrm{HVA} &= \mathrm{HVA}^{\cdot} = \mathbb{1}_{[0,\theta)}\Big(P^{jr}-Q^{jr}\Big) = \mathbb{1}_{[0,\theta)}K(1-e^{-\lambda(T-\cdot)}),\ \mathrm{HVA}_0 = K(1-e^{-\lambda T}),\\
\mathcal{L} &= -pnl^{\cdot} + \mathrm{HVA}^{\cdot}_{(0)}.
\end{aligned} \tag{25}$$

Remark 8. The raw pnl process in (25) and the corresponding pnl net of $\mathrm{HVA}^{\cdot}_{(0)}$ satisfy (starting from 0)

$$\begin{aligned}
dpnl^{\cdot}_t &= -K\mathbb{1}_{\{t\leq T\}}\boldsymbol{\delta}_\theta(dt) = \mathbb{1}_{\{t\leq\theta\wedge T\}}\Big(-\lambda K dt - (KdN_t - \lambda K dt)\Big)\\
dpnl^{\cdot}_t - d\mathrm{HVA}^{\cdot}_t &= K\mathbb{1}_{\{t\leq\theta\wedge T\}}e^{-\lambda(T-t)}\Big(\lambda dt - \boldsymbol{\delta}_\theta(dt)\Big).
\end{aligned} \tag{26}$$

Consistently with the Darwinian model risk pattern featured in the beginning of this section, a seemingly positive drift

$$\mathbb{1}_{\{t\leq\theta\wedge T\}}\lambda K e^{-\lambda(T-t)}dt$$

in the second line is only the compensator of the loss

$$-\mathbb{1}_{\{t\leq\theta\wedge T\}}K e^{-\lambda(T-t)}dN_t$$

that hits the bank in case the extreme event materializes.

Recall Definitions 2 and 3.

Proposition 2. *Denoting* $\Theta = (T + \frac{\ln(\alpha)}{\lambda})^+ \leq T$, *we have* $\mathrm{EC} = \mathbb{1}_{[0,\theta)}\widetilde{\mathrm{EC}}$ *and* $\mathrm{KVA} = \mathbb{1}_{[0,\theta)}\widetilde{\mathrm{KVA}}$, *where*

$$\begin{aligned}
\widetilde{\mathrm{EC}} &= \mathbb{1}_{\lambda>-\ln(\alpha)}\mathbb{1}_{[0,\Theta)}Ke^{-\lambda(T-\cdot)},\\
\widetilde{\mathrm{KVA}} &= \mathbb{1}_{\lambda>-\ln(\alpha)}Ke^{-\lambda(T-\cdot)}\mathbb{1}_{[0,\Theta)}(1-e^{-h(\Theta-\cdot)}),\\
\mathrm{KVA}_0 &= \mathbb{1}_{\lambda>-\ln(\alpha)}Ke^{-\lambda T}\mathbb{1}_{\Theta>0}(1-e^{-h\Theta}).
\end{aligned} \tag{27}$$

Proof. For $t < t' \leq T$, the last line in (25) yields

$$\begin{aligned}
\mathcal{L}_{t'} - \mathcal{L}_t &= (-pnl^{\cdot} + \mathrm{HVA}^{\cdot})_{t'} - (-pnl^{\cdot} + \mathrm{HVA}^{\cdot})_t\\
&= \mathbb{1}_{\{t'\geq\theta>t\}}K + \mathbb{1}_{\{t'<\theta\}}K(1-e^{-\lambda(T-t')}) - \mathbb{1}_{\{t<\theta\}}K(1-e^{-\lambda(T-t)})\\
&= \mathbb{1}_{\{t<\theta\}}\Big(\mathbb{1}_{\{t'\geq\theta\}}(K - K(1-e^{-\lambda(T-t')})) + K(1-e^{-\lambda(T-t')}) - K(1-e^{-\lambda(T-t)})\Big)\\
&= \mathbb{1}_{\{t<\theta\}}B^t_{t'},\ \text{where } B^t_{t'} = \mathbb{1}_{\{t'\geq\theta\}}Ke^{-\lambda(T-t')} + K(e^{-\lambda(T-t)} - e^{-\lambda(T-t')}).
\end{aligned}$$

On $\{t < \theta\}$, the Bernoulli random variable $\mathbb{1}_{\{t'\geq\theta\}}$ satisfies $\mathbb{E}_t\Big[\mathbb{1}_{\{t'\geq\theta\}}=0\Big]=e^{-\lambda(t'-t)}$ and, for any confidence level $\alpha > e^{-\lambda(t'-t)}$, i.e. such that $t'-t>\frac{-\ln(\alpha)}{\lambda}$, $\mathbb{V}a\mathbb{R}_t(\mathcal{L}_{t'}-\mathcal{L}_t)$ is the largest of the two possible values of $(\mathcal{L}_{t'}-\mathcal{L}_t)$, so that the latter never exceeds $\mathbb{V}a\mathbb{R}_t(\mathcal{L}_{t'}-\mathcal{L}_t)$. As a consequence, for $t'-t>\frac{-\ln(\alpha)}{\lambda}$, we have:

$$\begin{aligned}&\mathbb{ES}_t(\mathcal{L}_{t'}-\mathcal{L}_t)=\mathbb{V}a\mathbb{R}_t(\mathcal{L}_{t'}-\mathcal{L}_t)=\\&\qquad\mathbb{1}_{\{t<\theta\}}\Big(Ke^{-\lambda(T-t')}+K(e^{\lambda(T-t)}-e^{-\lambda(T-t')})\Big)=\mathbb{1}_{\{t<\theta\}}Ke^{-\lambda(T-t)}.\end{aligned}$$

For $t'-t\leq\frac{-\ln(\alpha)}{\lambda}$, we have

$$(\mathcal{L}_{t'}-\mathcal{L}_t)\mathbb{1}_{\{\mathcal{L}_{t'}-\mathcal{L}_t\geq\mathbb{V}a\mathbb{R}_t(\mathcal{L}_{t'}-\mathcal{L}_t)\}}=\mathcal{L}_{t'}-\mathcal{L}_t,$$

which is a time-t conditionally centered random variable as the increment of the martingale $\mathcal{L}$. Hence

$$0=\mathbb{E}_t(\mathcal{L}_{t'}-\mathcal{L}_t)=\mathbb{E}_t\Big((\mathcal{L}_{t'}-\mathcal{L}_t)\mathbb{1}_{\{\mathcal{L}_{t'}-\mathcal{L}_t\geq\mathbb{V}a\mathbb{R}_t(\mathcal{L}_{t'}-\mathcal{L}_t)\}}\Big)=\mathbb{ES}_t(\mathcal{L}_{t'}-\mathcal{L}_t).$$

Setting $t'=(t+1)\wedge T$ so that $t'-t>\frac{-\ln(\alpha)}{\lambda}\Leftrightarrow t<\Theta$, we obtain by Definition 2:

$$\mathrm{EC}_t=\mathbb{ES}_t(\mathcal{L}_{t'}-\mathcal{L}_t)=\mathbb{1}_{\{t<\theta\}}\mathbb{1}_{\lambda>-\ln(\alpha)}\mathbb{1}_{t<\Theta}Ke^{-\lambda(T-t)},$$

which is the first line in (27).

Assuming $\lambda>-\ln(\alpha)$ (otherwise $\mathrm{EC}=\mathrm{KVA}=0$), let us define the process

$$\mathrm{KVA}^{\dagger}_t:=h\mathbb{E}_t\int_t^T e^{-h(u-t)}\mathrm{EC}_u du=h\mathbb{E}_t\int_t^T\Big(\mathrm{EC}_s-\mathrm{KVA}^{\dagger}_s\Big)ds,\ t\leq T. \tag{28}$$

We have

$$\begin{aligned}\mathrm{KVA}^{\dagger}_t&=hK\mathbb{E}_t\mathbb{1}_{\{t<\Theta\}}\int_t^{\Theta}e^{-h(u-t)}\mathbb{1}_{\{u<\theta\}}e^{-\lambda(T-u)}du\\&=hKe^{-\lambda(T-\Theta)}\mathbb{1}_{\{t<\Theta\}}\mathbb{1}_{\{t<\theta\}}\int_t^{\Theta}e^{-h(u-t)}e^{-\lambda(u-t)}e^{-\lambda(\Theta-u)}du\\&=\mathbb{1}_{\{t<\theta\}}hKe^{-\lambda(T-\Theta)}e^{-\lambda(\Theta-t)}\mathbb{1}_{\{t<\Theta\}}\int_t^{\Theta}e^{-h(u-t)}du\\&=\mathbb{1}_{\{t<\theta\}}Ke^{-\lambda(T-t)}\mathbb{1}_{t<\Theta}(1-e^{-h(\Theta-t)})\leq\mathbb{1}_{\{t<\theta\}}K\mathbb{1}_{t<\Theta}e^{-\lambda(T-t)}=\mathrm{EC}_t.\end{aligned} \tag{29}$$

Back to the right-hand side in (28), the process $\mathrm{KVA}^{\dagger}$ therefore satisfies

$$\mathrm{KVA}^{\dagger}_t=\mathbb{E}_t\int_t^T h\Big(\mathrm{EC}_s-\mathrm{KVA}^{\dagger}_s\Big)ds=\mathbb{E}_t\int_t^T h\Big(\mathrm{EC}_s-\mathrm{KVA}^{\dagger}_s\Big)^+ds,\ t\leq T, \tag{30}$$

which is the KVA equation (7). As EC and $\mathrm{KVA}^{\dagger}$ are bounded processes, the property recalled after Definition 3 implies that $\mathrm{KVA}^{\dagger}$ is the unique bounded (or even square integrable) solution to this equation, i.e. $\mathrm{KVA}^{\dagger}=\mathrm{KVA}$. The first identity in the last line of (29) then yields the second line in (27). □

S. Crépey

A.1 Numerical Results

For $\lambda = 1\%$, $T = 10y$ and $h = 10\%$, (26) and (27) yield as $\alpha \downarrow e^{-0.01} \approx 99\%$:

$$\begin{aligned}&\mathrm{HVA}_0 = K(1-e^{-0.1}) \approx 0.095K,\ \mathrm{KVA}_0 \downarrow Ke^{-0.1}(1-e^{-1+0.1}) \approx 0.54K\\ &\frac{\mathrm{KVA}_0}{\mathrm{HVA}_0} \downarrow \frac{(1-e^{-0.9})}{(e^{0.1}-1)} \approx 5.64.\end{aligned} \tag{31}$$

In the present case, $f = 0$ and a pure frictions HVA^f à la Burnett (2021) & Burnett and Williams (2021) vanishes. But we see from Figure 66 that the Darwinian model risk HVA can be extreme. As visible on Figure 67, where we also denote HVA + KVA =: AVA for "additional valuation adjustment" in the spirit of European Parliament (2013, 2016) (see also https://www.eba.europa.eu/regulation-and-policy/market-risk/draft-regulatory-technical-standards-on-prudent-valuation, accessed on 25 February 2025), the corresponding KVA adjustment can be even several times larger. The latter holds for $\alpha > e^{-\lambda}$. For $\alpha \le e^{-\lambda}$, instead, there is no tail risk at the envisioned confidence level, hence EC = KVA = 0.

B Delta Hedging of a Vulnerable Put

Let $\mathcal{N}$ be the standard normal cumulative distribution function. We denote by

$$\Delta_t^{bs} := \partial_S Q^{bs}(t, S_t; \Sigma_t) = \partial_S P^{bs}(t, S_t; \Sigma_t) = -\mathcal{N}\Big(-d_-(t, S_t; 0, \Sigma_t)\Big)$$

the delta of the vulnerable put computed in the local pricing model: cf. the Black-Scholes formula for puts and (47), and note that Δ^{bs} is negative before θ. Instead of the previous static hedge, we now assume a dynamic delta hedging scheme, whereby the trader uses, on top of the Black-Scholes implied price[2]

$$q_t = Q^{bs}(t, S_t; \Sigma_t) = Q^{bs}(t, S_t; \Sigma_t) = P^{jr}(t, S_t) > Q_t^{jr},$$

the delta $\delta_t = \Delta_{t-}^{bs}$ in S, so that $h^{\cdot} = \int_0^{\cdot} \Delta_{t-}^{bs} dS_t$ (a martingale).

Using (24)[3], (56)-(57), (16) and (20), the HVA$^{\cdot}$ equations reduce to

$$\begin{aligned}&pnl^{\cdot} = \mathbb{1}_{\{\theta>T\}}(K-S_T)^+\mathbb{1}_{[T,+\infty)} + \mathbb{1}_{[0,\theta)}P^{jr} - P_0^{jr} - h^{\cdot},\\ &\mathcal{H}^{\cdot} = \mathbb{1}_{[0,\theta)}(Q^{jr} - P^{jr}) - (Q^{jr} - P^{jr})_0 + h^{\cdot} =\\ &\qquad - \mathbb{1}_{[0,\theta)}K(1-e^{-\lambda(T-\cdot)})\mathbb{1}_{[0,T)} + K(1-e^{-\lambda T}) + h^{\cdot},\\ &\mathrm{HVA}^{\cdot} = \mathbb{1}_{[0,\theta)}(P^{jr} - Q^{jr}) = \mathbb{1}_{[0,\theta)}K(1-e^{-\lambda(T-\cdot)}),\ \mathrm{HVA}_0^{\cdot} = K(1-e^{-\lambda T}).\end{aligned} \tag{32}$$

[2]cf. (23).

[3]but with $\mathcal{P}^{\cdot} = p^{\cdot} = P^{\cdot} = 0$ in the present purely dynamic hedging setup.

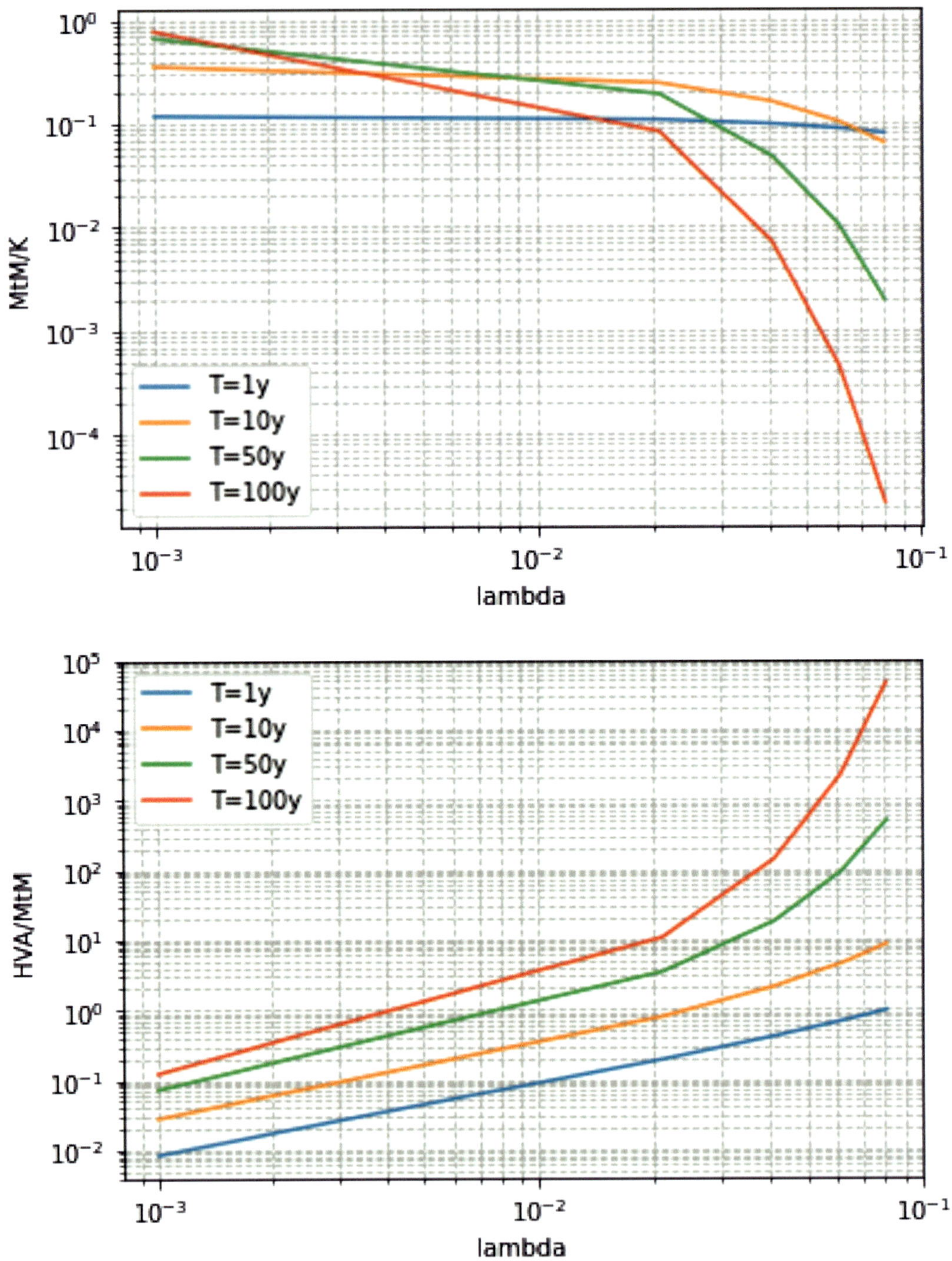

Figure 66: HVA computations in the static hedging case. At-the-money $S_0 = K$, denoting $\mathrm{MtM}_0 = Q_0^{jr}$ and assuming $\alpha \downarrow e^{-\lambda}$ everywhere in the bottom panels (where the limiting value of the confidence level α that underlies the KVA therefore depends of the abscissa λ): *(Top)* $\frac{\mathrm{MtM}_0}{K}$; *(Bottom)* $\frac{\mathrm{HVA}_0}{\mathrm{MtM}_0}$.

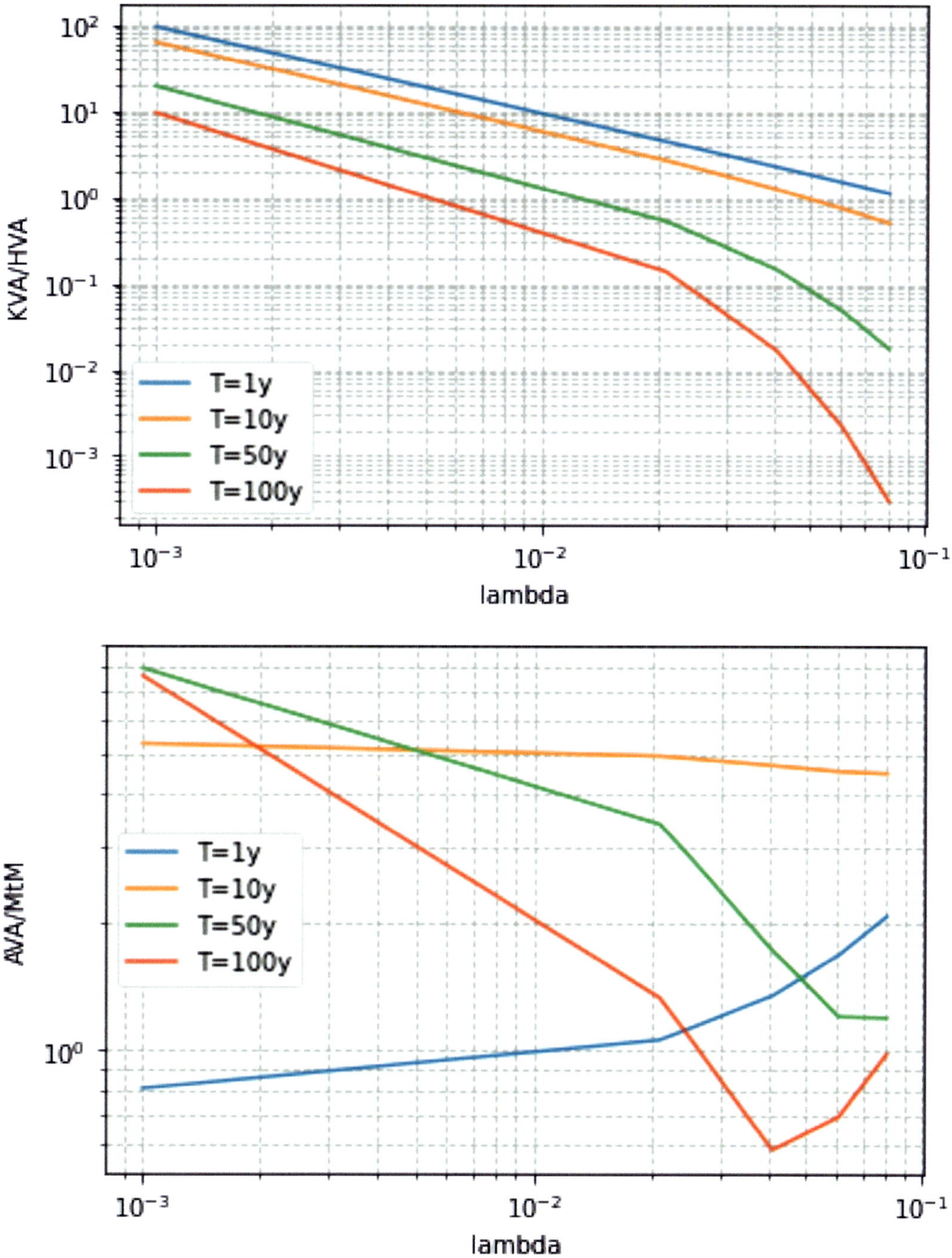

Figure 67: KVA computations in the static hedging case. At-the-money $S_0 = K$, denoting $\mathrm{MtM}_0 = Q_0^{jr}$ and assuming $\alpha \downarrow e^{-\lambda}$ everywhere in the bottom panels (where the limiting value of the confidence level α that underlies the KVA therefore depends of the abscissa λ): *(Top)* $\frac{\mathrm{KVA}_0}{\mathrm{HVA}_0}$; *(Bottom)* $\frac{\mathrm{AVA}_0}{\mathrm{MtM}_0}$, where $\mathrm{AVA} := \mathrm{HVA} + \mathrm{KVA}$.

Remark 9. The raw $pnl^{\cdot}$ in (32), satisfies, for $t < \theta$,

$$dpnl^{\cdot}_t = \boldsymbol{\delta}_T(dt)(K - S_T)^+ + dP^{jr}_t - \delta_t dS_t,$$

whereas at θ (if $\leq T$) the bank incurs a loss

$$\begin{aligned} pnl^{\cdot}_\theta - pnl^{\cdot}_{\theta-} &= -P^{jr}_{\theta-} + h^{\cdot}_{\theta-} - h^{\cdot}_\theta = -P^{bs}_{\theta-} + \Delta^{bs}_{\tau_s-}(S_{\theta-} - S_\theta) \\ &= -P^{bs}_{\theta-} + \Delta^{bs}_{\theta-} S_{\theta-} = -K\mathcal{N}(-d_-(\theta, S_{\theta-}; 0, \Sigma_{\theta-})) < 0 \end{aligned} \tag{33}$$

(cf. the Black-Scholes formula for puts), consistent with the blow-up pattern of the Darwinian model risk described at the beginning of this section.

Remark 10. While the static hedge of Subsection A is perfect before θ, and the continuous-time delta hedge is not (due to the continuous recalibration of the pricing model), one observes a smaller loss at $\theta < T$ in the delta hedge case:

$$K\mathcal{N}(-d_-(\theta, S_{\theta-}; 0, \Sigma_{\theta-})) \leq K = P^{bs}_\theta,$$

cf. (33), (26) and see Figure 68.

Following up on (32) and (20), we obtain

$$\mathcal{L} = -pnl^{\cdot} + \mathrm{HVA}^{\cdot}_{(0)} + f + \mathrm{HVA}^{f}_{(0)}, \tag{34}$$

where, for $t < t' \leq T$,

$$\begin{aligned} &(-pnl^{\cdot} + \mathrm{HVA}^{\cdot}_{(0)})_{t'} - (-pnl^{\cdot} + \mathrm{HVA}^{\cdot}_{(0)})_t \\ &\quad = -\Big((K - S_T)^+ \mathbb{1}_{[T,+\infty)} \mathbb{1}_{\{\theta > T\}} + \mathbb{1}_{[0,\theta)} Q^{jr}\Big)_{t'} + \\ &\qquad \Big((K - S_T)^+ \mathbb{1}_{[T,+\infty)} \mathbb{1}_{\{\theta > T\}} + \mathbb{1}_{[0,\theta)} Q^{jr}\Big)_t + (h_{t'} - h_t) \\ &\quad = -(K - S_T)^+ \mathbb{1}_{t < T \leq t'} \mathbb{1}_{\{\theta > T\}} - \Big(\mathbb{1}_{[0,\theta)} Q^{jr}\Big)_{t'} + \Big(\mathbb{1}_{[0,\theta)} Q^{jr}\Big)_t + (h_{t'} - h_t). \end{aligned} \tag{35}$$

B.1 HVA^f *for the Vulnerable Put Under the Delta Hedging Scheme*

Regarding frictions, we assume (unrealistically but with some genericity as explained in the beginning of this section) the bank portfolio reduced to the vulnerable put and its dynamic delta-hedge in S. We are thus in the setup of §6 with $T^\star = T, \mathcal{X} = (\tilde{S}, \Sigma; \mathbb{1}_{[0,\theta)})$, and $q^\star_t = P^{bs}(t, \tilde{S}_t; \Sigma_t)$. Here $\tilde{S}$ is the auxiliary Black-Scholes model (45) (which satisfies $\tilde{S}_t = S_{t-}$ on $\{t \leq \theta\}$), Σ_t is the Black-Scholes implied volatility of the vanilla put as per Definition 6, and $P^{bs}(t, \tilde{S}_t; \Sigma)$ is the pricing function of the vanilla put with strike K and maturity T in the Black-Scholes model with zero interest rates and volatility Σ_t, with dynamic hedging ratio $\mathbb{1}_{\{t \leq \theta\}} \Delta^{bs}_{t-}$ for the vulnerable put, where $\Delta^{bs}_t = \partial_S P^{bs}(t, \tilde{S}_t; \Sigma_t)$.

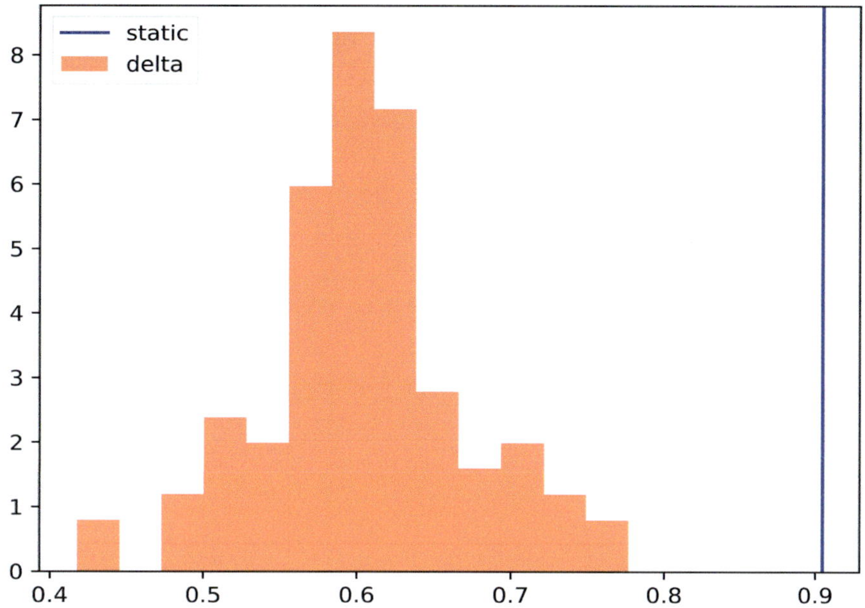

Figure 68: The red histogram is the density of $-pnl^{\cdot}_1 + \mathrm{HVA}^{\cdot}_1 - \mathrm{HVA}^{\cdot}_0$ conditional on model switch occurring before time 1, i.e. on $\{0 < \theta \leq 1\}$, for delta hedging without frictions. The vertical blue line in the right panel corresponds to the deterministic loss $-pnl^{\cdot}_1 + \mathrm{HVA}^{\cdot}_1 - \mathrm{HVA}^{\cdot}_0 = K + \mathrm{HVA}^{\cdot}_1 - \mathrm{HVA}^{\cdot}_0$ for static hedging, also conditional on $\{0 < \theta \leq 1\}$. Note that, in both cases, $\mathrm{HVA}^{\cdot}_1 - \mathrm{HVA}^{\cdot}_0 = 0 - K(1-e^{-\lambda T}) \simeq -0.095$ on $\{0 < \theta \leq 1\}$.

Proposition 3. *Assume that trading is permitted only at the discrete dates* $i\mathrm{h}$, $1 \leq i \leq n$, *with* $\mathrm{h} = \frac{T}{n}$ *(for any* $n \geq 1$*). Assume further that implementing the delta hedging strategy triggers a cumulative cost at time* $i\mathrm{h}$ *and a discrete-time hedging valuation adjustment for frictions as per Definition 7 page 353, i.e. in the present setup, for* $0 \leq i \leq n$*,*

$$
\begin{aligned}
f^{\mathrm{h}}_{i\mathrm{h}} &:= \sum_{j=1}^{i} \mathrm{k}\frac{\sqrt{\mathrm{h}}}{2} S_{j\mathrm{h}} \left| \mathbb{1}_{\{j\mathrm{h}\leq\theta\}} \Delta^{bs}_{j\mathrm{h}-} - \mathbb{1}_{\{(j-1)\mathrm{h}\leq\theta\}} \Delta^{bs}_{(j-1)\mathrm{h}-} \right|, \\
\mathrm{HVA}^{\mathrm{h}}_{i\mathrm{h}} &:= \mathbb{E}_{i\mathrm{h}} \left[f^{\mathrm{h}}_{n\mathrm{h}} - f^{\mathrm{h}}_{i\mathrm{h}} \right].
\end{aligned}
\tag{36}
$$

Then, as h *goes to* 0*, the discrete HVA for frictions* $\mathrm{HVA}^{\mathrm{h}}_{\lfloor \frac{t}{\mathrm{h}} \rfloor \mathrm{h}}$ *converges almost surely to* $\mathrm{HVA}^f = va(f)$ *on* $[0,T]$*, for the process* f *such that*

$$
df_t = \mathbb{1}_{\{t\leq\theta\}} \frac{\mathrm{k}}{\sqrt{2\pi}} S_t \left| \sigma \Gamma^{bs}_{t-} + \varsigma_t \partial^2_{\Sigma,S} P^{bs}(t, S_{t-}; \Sigma_{t-}) \right| dt, \tag{37}
$$

where $\Gamma_t^{bs} = \partial^2_{S^2} P^{bs}(t, S_t; \Sigma_t) \geq 0$, *while* ς *is the diffusion coefficient of the implied volatility process* Σ.

Proof. This follows by an application of Theorem 1 (cf. (64)) with $\mathcal{X} = (\tilde{S}, \Sigma; \mathbb{1}_{[0,\theta)})$ and $\mathrm{k}_1 = \mathrm{k} \geq 0$, $\mathrm{k}_2 = 0$ (as the position is not hedged against the fluctiations of Σ_t in the *jr* model, see Remark 14). □

Interestingly, the cost of delta-hedging in the *bs* model computed within the *jr* model also depends on the derivative of the delta with respect to the implicit volatility, or implied "vanna", $\partial^2_{\Sigma,S} P^{bs}(t, S_{t-}; \Sigma_{t-})$. This comes from the continuous recalibration of the trader's model to the fair valuation of the vanilla put. Because of this impact of recalibration into transaction costs, (37) would be quite demanding to implement directly, but its discrete counterpart (36) is rather straightforward (the consistency between the two being insured by Proposition 3).

B.2 Numerical Results

The numerical parameters are the same as in Subsubsection A.1 above, along with $S_0 = K = 1$ and $\sigma = 0.3$, and with $\mathrm{k} = 0.1$ in (37).

HVA for frictions We perform Monte-Carlo simulations with $M = 50,000$ paths to estimate $\mathrm{HVA}_0^{\mathrm{h}} = \mathbb{E}\left[f^{\mathrm{h}}_{n\mathrm{h}}\right]$ as per (36), for a monthly time-discretization, i.e. $n = 120$ and $\mathrm{h} = \frac{\mathrm{T}}{\mathrm{n}} = \frac{1}{12}$. As a sanity check, we also price by Monte Carlo $\mathrm{HVA}_0^{\cdot}$ already known from (32) and (31). We can see from the right panel in Figure 68, where the horizontal red line corresponds to $\mathrm{HVA}_0^{\cdot} = 1 - e^{-0.1}$, that HVA_0^f is dominated by $\mathrm{HVA}_0^{\cdot}$.

Economic capital and KVA In the present dynamic hedging setup, we rely on numerical approximations to estimate the economic capital and the KVA of the bank at a quantile level α set in the numerics to 99%. In fact, in this Markovian framework, each process $Z = \mathrm{HVA}^f, \mathrm{EC}, \mathbb{V}a\mathbb{R}.(\mathcal{L}_{\cdot'} - \mathcal{L}_{\cdot})$ and KVA satisfies, for all $t \geq 0$,

$$Z_t = \widetilde{Z}(t, S_t) = \mathbb{1}_{\{t<\theta\}} \widetilde{Z}(t, \widetilde{S}_t),$$

where $\tilde{S}$ is the auxiliary Black-Scholes model (45) and $\widetilde{\mathrm{HVA}^f}(t, 0) = \widetilde{\mathbb{V}a\mathbb{R}}(t, 0) = \widetilde{\mathrm{EC}}(t, 0) = \widetilde{\mathrm{KVA}}(t, 0) = 0$, while, for all $(t, S) \in [0, T] \times (0, \infty)$, setting $t' = (t+1) \wedge T$,

$$\begin{aligned}
\widetilde{\mathrm{HVA}}^f(t,S) &= \mathbb{E}\left[f_T - f_t \,|\, S_t = S\right], \\
\widetilde{\mathbb{V}a\mathbb{R}}(t,S) &= \mathbb{V}a\mathbb{R}\left[\mathcal{L}_{t'} - \mathcal{L}_t \,|\, S_t = S\right], \\
\widetilde{\mathrm{EC}}(t,S) &= \mathbb{ES}\left[\mathcal{L}_{t'} - \mathcal{L}_t \,|\, S_t = S\right], \\
\widetilde{\mathrm{KVA}}(t,S) &= h\,\mathbb{E}\left[\int_t^T (\mathrm{EC}_u - \mathrm{KVA}_u)^+ \,du \,\middle|\, S_t = S\right].
\end{aligned} \tag{38}$$

On this basis, one can obtain approximations $\widehat{\mathrm{HVA}^f}$, $\widehat{\mathrm{EC}}$ and $\widehat{\mathrm{KVA}}$ of the HVA^f, EC and KVA processes at all nodes of a forward simulated grid $(S^m_{t_k})^{1\le m\le M}_{0\le k\le 10}$ of S, by neural net regressions and quantile regressions that are used backward in time for solving the above equations numerically (cf. IV.§2-§3), the way detailed in §7.

We plotted on Figures 69 and 70 the processes $\widehat{\mathrm{EC}}(\cdot,\widetilde{S}_{\cdot})$ and $\widehat{\mathrm{KVA}}(\cdot,\widetilde{S}_{\cdot})$, represented by the term structures of their means (in green) and quantiles of levels 10%, 90% (in blue) and 2.5% and 97.5% (in red), both with and without frictions f. We also represented the deterministic counterparts of these processes in the static hedging case (27).

For the same numerical parameters as the ones used in Subsubsection B.1, a quantile level α for the EC computations set at 99%, and a hurdle rate r for the KVA computations set at 10%, we obtain in the dynamic hedging case :

$$\widehat{\mathrm{HVA}^{\cdot}}_0 \simeq 0.095 \text{ and } \widehat{\mathrm{HVA}^f}_0 \simeq 0.046, \text{ hence } \widehat{\mathrm{HVA}}_0 = \widehat{\mathrm{HVA}^{\cdot}}_0 + \widehat{\mathrm{HVA}^f}_0 \simeq 0.141,$$
$$\widehat{\mathrm{KVA}}_0 \simeq 0.407, \frac{\widehat{\mathrm{KVA}}_0}{\widehat{\mathrm{HVA}}_0} \simeq 2.881. \tag{39}$$

As could be expected from Remark 10 page 338, there is ultimately less risk (as assessed by economic capital and KVA, cf. (31) and Figures 69-70) with the delta hedge than with the static hedge.

In the frictionless case $f = 0$, we obtain by the same methodology

$$\begin{aligned}
\widehat{\mathrm{HVA}}_0 &= \widehat{\mathrm{HVA}^{\cdot}}_0 \simeq 0.095, \\
\widehat{\mathrm{KVA}}_0 &\simeq 0.433, \frac{\widehat{\mathrm{KVA}}_0}{\widehat{\mathrm{HVA}}_0} \simeq 4.550.
\end{aligned} \tag{40}$$

In view of (39) (see also Figures 69-70), the dynamic hedging frictions are risk-reducing in this case, meaning that the components $-pnl^{mtm} + \mathrm{HVA}^{mtm} - \mathrm{HVA}^{mtm}_0 = -pnl^{\cdot} + \mathrm{HVA}^{\cdot} - \mathrm{HVA}^{\cdot}_0$ and $f + \mathrm{HVA}^f - \mathrm{HVA}^f_0$ of (20) happen to be negatively correlated.

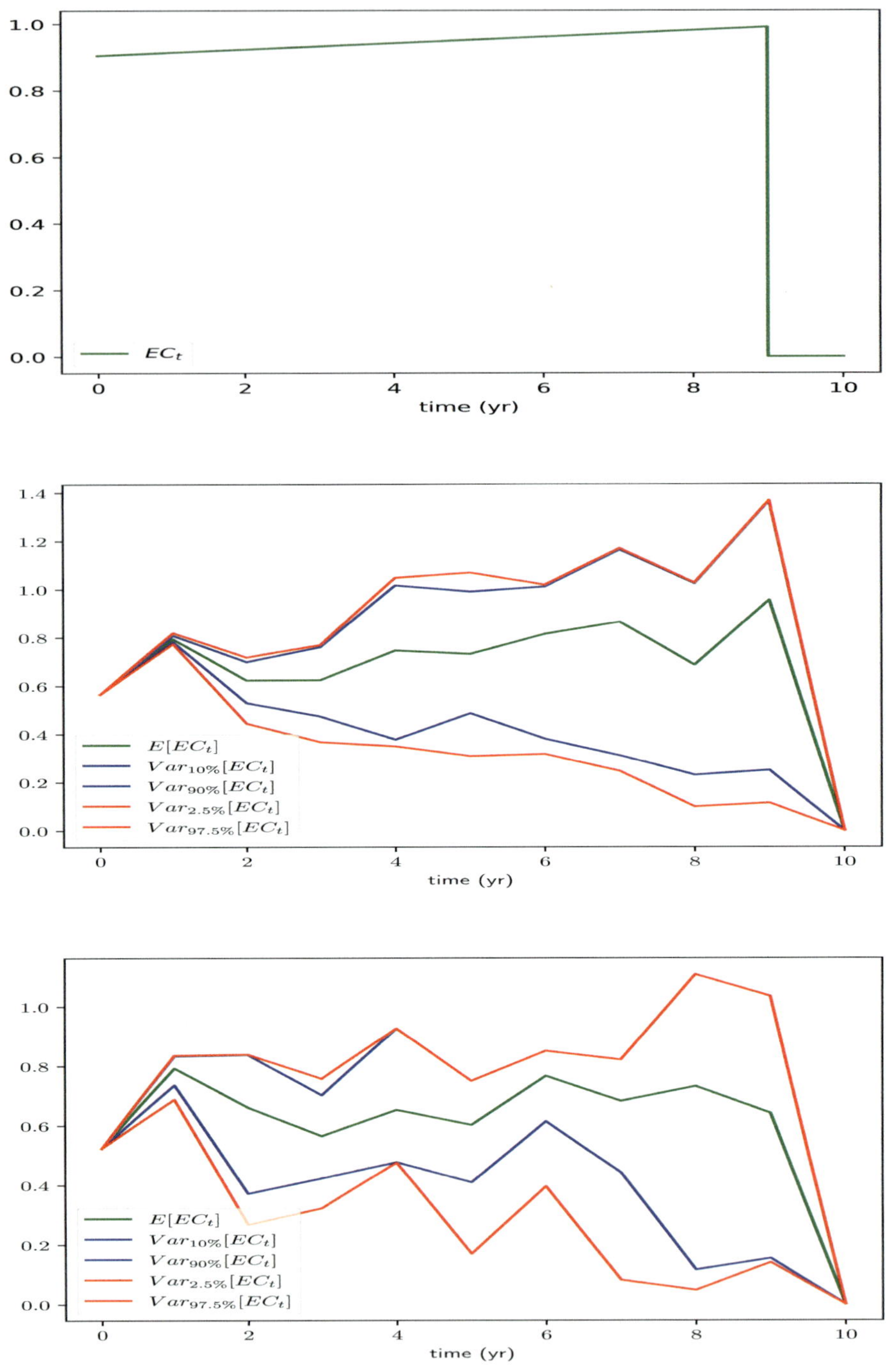

Figure 69: Plot of the deterministic map $t \mapsto \widetilde{\mathrm{EC}}(t)$ in the static hedging case (27) (*top*). Plots of mean (in green) and quantiles at levels 10% and 90% (in blue) and 2.5% and 97.5% (in red) of $\widehat{\mathrm{EC}}(t, \widetilde{S}_t)$ in the delta hedging case without friction (*middle*) and in the delta hedging case with frictions (*bottom*).

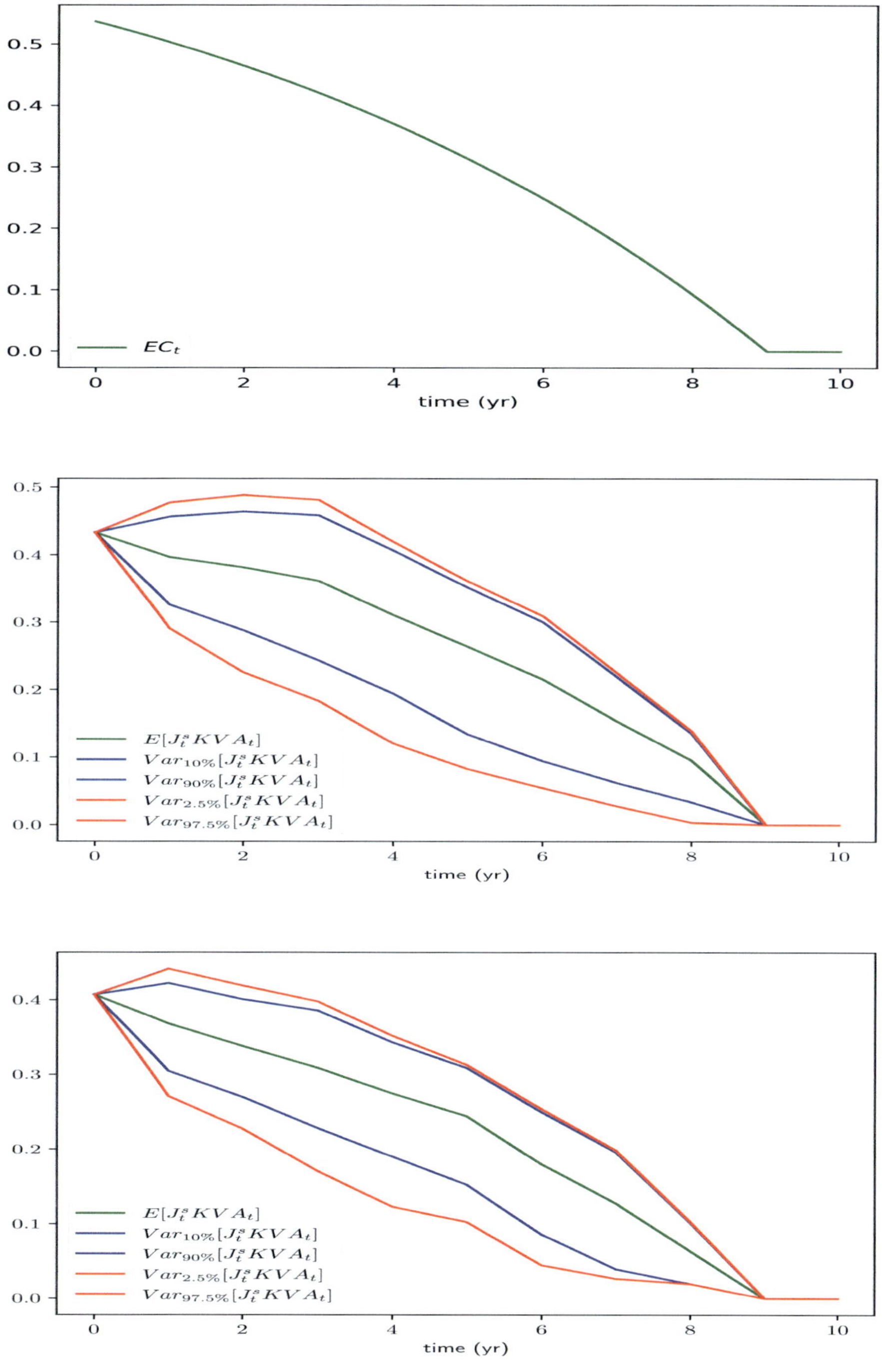

Figure 70: Plot of the deterministic map $t \mapsto \widetilde{\mathrm{KVA}}(t)$ in the static hedging case (27) (*top*). Plots of mean (in green) and quantiles at levels 10% and 90% (in blue) and 2.5% and 97.5% (in red) of $\widehat{\mathrm{KVA}}(t, \widetilde{S}_t)$ in the delta hedging case without friction (*middle*) and in the delta hedging case with frictions (*bottom*).

Remark 11. The statically hedged position is delta and vega neutral. Hence our vulnerable put example yields a case where delta-vega hedging the option actually increases model risk with respect to delta-hedging it only.

§4 XVA MODEL RISK EXAMPLES

To build a bad CVA model, one can for instance neglect credit vegas. It is well known that credit spread volatilities spike up whenever spreads jump up. If one uses a credit model where hazard rates for default are either deterministic or follow a diffusion process, credit hedge ratios are under-stated and the CVA is under-valued, giving rise to more competitive pricing possibly masked by advantageous hedge in the short-to medium-term but leading to blow up in the long-term, in line with the Darwinian model risk pattern depicted in the beginning of §3.

Another route to build a bad CVA model is to ignore wrong way risk (regarding the latter see e.g. Li and Mercurio (2015), Crépey and Song (2016)). In particular, upon the occurence of default of a large and systemically important entity, a waterfall of consequences and market disruptions ensue. Accordingly, defaults by major counterparties should be modeled as binary occurrences, not as a probabilities of default. By modeling defaults only through probability distributions and not as binary events, one reduces the CVA and benefits out of selling uncovered puts on the default of those names.

Regarding funding valuation adjustments, as illustrated in Subsection A below, a bad FVA model widely used for its simplicity by banks for their pricing purposes is the symmetric FVA, i.e. fva $=$ FCA$-$FBA as per (41)-(42), while the fair valuation model should be an asymmetric FVA (see II.§5.D). The stochastic models to evaluate fva and FVA may or may not be the same, there is in any case, a cash flow risk in the first place. The situation is even worse than that as, for their accounting purposes, banks just use the too conservative model FCA, instead of what should be the asymmetric FVA again, then compensating for the lack of modeling of the rehypothecation option by slashing the funding rate by a factor 3 to 5 (cf. Figure 53 page 299). This is commonly achieved by pulling data from peers and adopting a similar funding spread. Although in tranquil times this method yields a number roughly in the right FVA ballpark, it leads to an unavoidable blow-up at times of stress.

Remark 12. As for KVA computations, to enhance its competitiveness in the short term, a bank might be tempted to use a model understating the risk and economic capital of the bank. A sound practice in this regard is to combine different, equally valid (realistic and co-calibrated) models for simulating the set of trajectories underlying the economic capital and KVA computations (see X.§4.C). Such a Bayesian KVA approach typically fattens the tails of the simulated distributions and avoids under-stated risk estimates.

A FVA Proxies at the Test of the Covid-19

The 2013-2016 XVA debate revolved around the definition of suitable FVA (cost of funding) and KVA (cost of capital) metrics. The FVA number supposedly captures the present value of funding costs by projecting out into the future requirements and credit spreads. The KVA number is the cost of capital metric designed as an overall proxy for shareholder return and for general guidance on a broad spectrum of strategic actions from hedging to executive compensation, credit limits and dividend policy. Accordingly, FVA and KVA require computations at an even broader level than the netting-set CVA, i.e. at the funding-set level for the FVA and at the overall derivative portfolio (i.e. balance-sheet) level of the bank for the KVA: see the earlier chapters of the book.

Under current market practice, however, banks are calculating their funding costs by aggregation of client (or netting-set) specific numbers, which does not reflect the economics of collateral management (see II.§5.D). This arises from the desire of the banks to arrive at the FVA numbers by simply retrofitting their CVA calculators, which are based on distributed computing and are performed netting set by netting set, often with netting set-specific approximations.

In this subsection, we provide a qualitative analysis of the limits and dangers of this strategy revealed by the COVID-19 financial crisis of Q1 2020. Quoting a statement on the website of the Bank of International Settlements (see https://www.bis.org/topic/coronavirus.htm, last accessed on December 23, 2021):

> The coronavirus (Covid-19) pandemic is a major disruptive event for the global economy. *It is revealing financial vulnerabilities and testing the post-financial crisis economic system.*

As we detail in what follows, the Covid-19 Q1 2020 financial crisis demonstrates the inappropriateness of using netting-set aggregation typical to CVA analytics for FVA and KVA computations.

Let γ and $D_b(\omega)$ respectively denote the bank credit spread and the (possibly negative) debt of the (assumed all bilateral) counterparty b to the bank in the scenario ω (skipped in the notation hereafter). Assuming for simplicity no collateral on the derivative portfolios of the bank, but perfect variation margining on their hedges (and no initial margins), the amount that needs be borrowed by the bank to be posted as collateral on its hedge is $(\sum_b D_b)^+$. The reason why the positive part sits above the sum is that variation margin is rehypothecable, i.e. fungible across netting sets (see I.§2). In view of this, the economically correct cost of funding formula, which should be used both for decision-making and as a capital deduction by the bank (as the cost of its future funding expenses is a liability to the bank), is the asymmetric FVA $= \gamma \mathbb{E}[(\sum_b D_b)^+]$. Instead of this:

- for all their decision-making purposes, such as pricing and executives compensation, banks use

$$\gamma \mathbb{E} \sum_b D_b = \underbrace{\gamma \mathbb{E} \sum_b D_b^+}_{=:\mathrm{FCA}} - \underbrace{\gamma \mathbb{E} \sum_b D_b^-}_{=:\mathrm{FBA}}; \tag{41}$$

- as a capital deduction, they use

$$\mathrm{FCA} = \gamma \mathbb{E} \sum_b D_b^+. \tag{42}$$

Indeed, regulators insist, rightly so, that only asymmetric, nonnegative numbers should be used for the purpose of calculating a capital deduction. They do not specify the aggregation level, which could be at the netting set or funding set level. They are indifferent, which from their point of view is understandable, as the smaller is the level of aggregation, the larger and more conservative is the size of capital deduction.

Now, in calm times:

- equity capital buffers are large enough to absorb the conservative capital deduction;

- banks' balance sheets are dominated by assets, i.e.

$$0 < \sum_b D_b = (\sum_b D_b)^+ \text{ holds in most scenarios,} \tag{43}$$

so

$$\gamma \mathbb{E} \sum_b D_b \approx \gamma \mathbb{E}[(\sum_b D_b)^+] \text{ , i.e. FCA} - \text{FBA} \approx \text{FVA}. \tag{44}$$

The reason for (43) is that, if a corporate holds a bank payable, it typically has a desire to close it, receive cash, and restructure the trade with a par contract (the bank would agree to close the deal as a market maker, charging fees for the new trade). Because of this natural selection, in normal times, a bank is mostly in the receivables in its derivative business with corporates.

However, during the Covid-19 financial crisis:

- markdowns swinged bank balance sheets towards liabilities, invalidating (43) and the ensuing approximation (44);

- we saw an 8-fold credit spread widening;
- increasing default rates put pressure on bank capital.

As a result:

- the FCA – FBA number (41) used for decision-making by banks went further and further from the correct FVA number, implying erroneous hedges and executive compensation;
- the FCA number (42) exploded and the corresponding capital reduction became needlessly punitive for banks, at the precise time when capital was becoming scarce for banks;
- the discrepancy between the (both wrong) FCA – FBA and FCA numbers increased, enhancing the corresponding misalignment of interest between the executives and the shareholders of the bank.

As FVA fluctuations contribute to economic capital (cf. (2)-(5)), the cost of which is the KVA (7), wrong FVA computations then compromise the KVA computations. This is a perfect storm weather, through which only a mathematically and numerically rigorous treatment of accounting numbers, capital models and funding strategies can be of guidance. The intent of the cost-of-capital XVA approach of this book is precisely to define economically and mathematically correct accounting principles for dealer banks. Until banks adopt a correct accounting framework, they are exposed to major cash flows risk, like with funding desks pricing through their fva metric a wrong $\mathcal{F}$, say fva $= \widetilde{va}(\tilde{\mathcal{F}})$, with not only $\widetilde{va} \neq va$ (wrong FVA pricing model), but $\tilde{\mathcal{F}} \neq \mathcal{F}$ (erroneous FVA originating cash flows in the first place), implying a very significant HVA^{fva} as per (18). The above even shows a usage by the bank of two different (and both wrong) specifications (or "models") $\sum_b D_b$ and $\sum_b D_b^+$ of the same cash flow $(\sum_b D_b)^+$ for serving two different purposes within the bank: pricing and capital reduction.

Conclusion

In this chapter we revisited Burnett (2021) & Burnett and Williams (2021)'s notion of hedging valuation adjustment (HVA) in the direction of model risk. The fact evidenced by Remark 11 that vega hedging may actually increase model risk illustrates well that model risk cannot be hedged. For instance, "CVA hedging" industry-practices such as the one depicted in VI.§6.A should not really be seen as hedging model risk, but only hedging model shifts. In fact model risk can only be provisioned against or, preferably, compressed by improving the quality of the models used by traders.

In any case, a provision for model risk should be risk-adjusted. But, as this chapter illustrates, a risk-adjusted reserve would be much greater than the "HVA uptick" (price difference) currently used in banks, by a factor 3 to 5 in our experiments (cf. Remark 5 page 326 and (39)-(40)), and it could be even more if one accounted for the price impact of a liquidation in extreme market conditions (cf. https://www.risk.net/derivatives/6556166/remembering-the-range-accrual-bloodbath and Chapter IX).

Risk-adjusted HVA computations are also very demanding. In particular, beyond analytical toy examples such as that one of §3.A, and already in the case of §3.B, HVA risk-adjusted KVA computations require dynamic recalibration in a simulation setup, for assessing the hedging ratios used by the traders at future time points as well as the time of explosion of the trader's strategy (time of model switch $\tau_s^{\cdot}$). Dynamic recalibration in a simulation setup at a banking portfolio level is unreasonable from the computational workload viewpoint. Hence, from this viewpoint too, the best practice would be that banks only rely on high-quality models, so that such computations are simply not needed.

In the end, the orders of magnitude of the corrections that would be required for duly compensating model risk (accounting not only for misvaluation but also for the associated mishedge), as well as the corresponding computational burden for a precise assessment of the latter, suggest that bad models should not so much be managed via reserves, as excluded altogether (cf. X.§4).

§5 PRICING EQUATIONS IN THE JUMP-TO-RUIN MODEL (∗)

In this section we provide pricing analytics in the *jr* model (22) for S. We also consider the auxiliary Black-Scholes model

$$d\tilde{S}_t = \lambda \tilde{S}_t dt + \sigma \tilde{S}_t dW_t, \tag{45}$$

starting from $\tilde{S}_0 = S_0$, where λ and σ (omitted in the notation for $d_\pm$ below when clear from the context) were introduced after (22). Hence

$$S_t = \mathbb{1}_{\{N_t=0\}} \tilde{S}_t,\ t \geq 0. \tag{46}$$

Given the maturity $T = T$ and strike $K > 0$ of an option, let, for every pricing time t and stock value S,

$$d_\pm(t, S; \lambda, \sigma) = \frac{\ln(\frac{S}{K}) + \lambda(T-t)}{\sigma\sqrt{T-t}} \pm \frac{1}{2}\sigma\sqrt{T-t}. \tag{47}$$

We first consider the pricing of a vanilla call option.

Proposition 4. *The jr value process* (1) *of the call option with payoff* $(S_T - K)^+$ *at time* T *can be represented as*

$$C_t^{jr} = u(t, S_t)\mathbb{1}_{[0,T)},\ t \in [0, T],$$

where the pricing function $u = u(t, S) := \mathbb{E}\Big((S_T - K)^+\Big|S_t = S\Big)$ *is the unique classical solution*[4] *with linear growth in* S *to the PDE*

$$\begin{cases} u(T, S) = (S - K)^+,\ S \geq 0 \\ \partial_t u(t, S) + \lambda S \partial_S u(t, S) + \frac{\sigma^2 S^2}{2}\partial^2_{S^2} u(t, S) \\ \qquad\qquad -\lambda u(t, S) = 0,\ t < T,\ S \geq 0. \end{cases} \tag{48}$$

For $t < T$,

$$C_t^{jr} = S_t \mathcal{N}(d_+(t, S_t)) - K e^{-\lambda(T-t)}\mathcal{N}(d_-(t, S_t)). \tag{49}$$

Proof. We have $S_T = \mathbb{1}_{\{\theta>T\}}\tilde{S}_T = \mathbb{1}_{\{\theta>T\}}S_0 \exp\Big(\sigma W_T + (\lambda - \frac{\sigma^2}{2})T\Big)$. Since $(S_T - K)^+ = 0$ on $\theta \leq T$ and $S_T = \tilde{S}_T$ on $\theta > T$, it follows that, on $\{t < \theta\}$,

$$\begin{aligned} \mathbb{E}_t\Big[(S_T - K)^+\Big] &= \mathbb{E}_t\Big[\mathbb{1}_{\{\theta>T\}}(S_T - K)^+\Big] = \\ &= \mathbb{E}_t\Big[\mathbb{1}_{\{\theta>T\}}(\tilde{S}_T - K)^+\Big] = \mathbb{E}_t\Big[e^{-\lambda(T-\tau)}(\tilde{S}_T - K)^+\Big], \end{aligned} \tag{50}$$

by independence between W and N. One recognizes the probabilistic expression for the time-t price of the vanilla call option in the auxiliary Black-Scholes model (45), hence the proposition follows from standard Black-Scholes results. □

We now consider the pricing of a put option in the *jr* model, in two forms: either a vanilla put with payoff $(K - S_T)^+$, or a vulnerable put with payoff $\mathbb{1}_{\{\theta>T\}}(K - S_T)^+$. Note that for a call option, vulnerable or not makes no difference in the *jr* model, where $S_T = (S_T - K)^+ = 0$ holds on $\{\theta \leq T\}$.

Proposition 5. *The jr value process* (1) *of the vanilla put can be represented as*

$$P_t^{jr} = v(t, S_t)\mathbb{1}_{[0,T)},\ t \in [0, T], \tag{51}$$

where the vanilla put pricing function $v = v(t, S) := \mathbb{E}\Big((K - S_T)^+\Big|S_t = S\Big)$ *is the unique bounded classical solution to the PDE*

$$\begin{cases} v(T, S) = (K - S)^+,\ S \geq 0 \\ \partial_t v(t, S) + \lambda S \partial_S v(t, S) + \frac{\sigma^2 S^2}{2}\partial^2_{S^2} v(t, S) \\ \qquad\qquad -\lambda v(t, S) + \lambda K = 0,\ t < T,\ S \geq 0. \end{cases} \tag{52}$$

For $t < T$,

$$P_t^{jr} = K e^{-\lambda(T-t)}\mathcal{N}(-d_-(t, S_t)) - S_t \mathcal{N}(-d_+(t, S_t)) + K(1 - e^{-\lambda(T-t)}). \tag{53}$$

[4] of class $\mathcal{C}^{1,2}\big([0, T) \times [0, +\infty)\big) \cap \mathcal{C}^0\big([0, T] \times [0, +\infty)\big)$.

Proof. Taking expectation in the decomposition $S_T - K = (S_T - K)^+ - (S_T - K)^-$ yields the (model-free) call-put parity relationship

$$S_t - K = u(t, S_t) - v(t, S_t),\ t \leq T, \tag{54}$$

hence $v = u - (S - K)$, from which the PDE characterization based on (52) for v results from the PDE characterization based on (48) for u. Moreover, we deduce from (49) that, for $t < T$,

$$\begin{aligned} P_t^{jr} &= C_t^{jr} - (S_t - K) = S_t\left(\mathcal{N}(d_+(t, S_t)) - 1\right) - K\left(e^{-\lambda(T-t)}\mathcal{N}(d_-(t, S_t)) - 1\right) \\ &= Ke^{-\lambda(T-t)}\mathcal{N}(-d_-(t, S_t)) - S_t\mathcal{N}(-d_+(t, S_t)) + K(1 - e^{-\lambda(T-t)}), \end{aligned}$$

which is (53). □

In accordance with (53):

Definition 6. For $t < \theta \wedge T$, given the observed spot price $S_t = S > 0$, the Black-Scholes implied volatility $\Sigma_t = \Sigma(t, S)$ of the vanilla put in the jr model is the unique solution to

$$\begin{aligned} &Ke^{-\lambda(T-t)}\mathcal{N}(-d_-(t, S; \lambda, \sigma)) - S\mathcal{N}(-d_+(t, S; \lambda, \sigma)) + K(1 - e^{-\lambda(T-t)}) \\ &\qquad = K\mathcal{N}(-d_-(t, S; 0, \Sigma_t)) - S\mathcal{N}(-d_+(t, S; 0, \Sigma_t)). \end{aligned} \tag{55}$$

We also set $\Sigma(t, 0) = 0$.

Remark 13. For $S = 0$, any $\Sigma \geq 0$ solves (55): for any Σ, $d_\pm = -\infty$ as $\ln(\frac{0}{K}) = -\infty$, so $K\mathcal{N}(-d_-) - S\mathcal{N}(-d_+) = K - S = K$ (for $S = 0$).

Proposition 6. *The value process* (1) *of the vulnerable put is given by*

$$\begin{aligned} Q_t^{jr} &= \mathbb{1}_{t<\theta\wedge T}\left(P_t^{jr} - (1 - e^{-\lambda(T-t)})K\right) = \\ &\quad \mathbb{1}_{t<\theta\wedge T}\left(Ke^{-\lambda(T-t)}\mathcal{N}\big(-d_-(t, S_t)\big) - S_t\mathcal{N}\big(-d_+(t, S_t)\big)\right). \end{aligned} \tag{56}$$

For $t < T$,

$$P_t^{jr} - Q_t^{jr} = \mathbb{1}_{t<\theta}K(1 - e^{-\lambda(T-t)}) + \mathbb{1}_{t\geq\theta}K. \tag{57}$$

Proof. We have

$$\mathbb{1}_{\{\theta>T\}}(S_T - K) = \mathbb{1}_{\{\theta>T\}}\left((S_T - K)^+ - (S_T - K)^-\right),$$

which in jr reduces to

$$S_T - \mathbb{1}_{\{\theta>T\}}K = (S_T - K)^+ - \mathbb{1}_{\{\theta>T\}}(S_T - K)^-.$$

By taking time-t conditional expectations, we have, on $\{t < \theta \wedge T\}$, that $S_t - Ke^{-\lambda(T-t)} = C_t^{jr} - Q_t^{jr}$, which yields

$$Q_t^{jr} = C_t^{jr} - S_t + Ke^{-\lambda(T-t)},$$

out of which (still on $\{t < \theta \wedge T\}$) the first identity in (56) follows from (54) and the second identity in turn follows from (53). Besides, on $\{t \geq \theta\}$, we have $Q^{jr} = 0$ and $P^{jr} = K$, whereas on $\{t \geq T\}$ we have $Q^{jr} = 0$, which completes the proof of (56) and (57). □

Proposition 7. *Setting* $w(t,S) = v(t,S) - K(1 - e^{-\lambda(T-t)})$[5], *the vulnerable put is replicable on* $[0, \theta \wedge T]$ *in the jr model (in the absence of model risk and hedging frictions), by the dynamic strategy* ζ *in* S *and* η *in the vanilla put given by*

$$\zeta_t = -\frac{\mathcal{N}\big(-d_+(t,S_t)\big)}{1-\mathcal{N}\big(-d_-(t,S_t)\big)},\ \eta_t = -\frac{\mathcal{N}\big(-d_-(t,S_t)\big)}{1-\mathcal{N}\big(-d_-(t,S_t)\big)},\ t < \theta \wedge T, \qquad (58)$$

and the number of constant riskless assets deduced from the budget condition $w(t, S_t)$ *on the strategy.*

Proof. In the absence of model risk and hedging frictions, the position being assumed to be funded by the constant riskless asset and unwound at θ, the profit-and-loss associated with the hedging strategy ζ in S and η in the vanilla put, both assumed càglàd processes, evolve following

$$dpnl_t = \mathbb{1}_{\{t \leq \theta\}}(dQ_t^{jr} - \zeta_t dS_t - \eta_t dP_t^{jr})$$

(with $pnl_0 = 0$). Itô formulas with (elementary) jump exploiting the results of Propositions 5 and 6 yield[6]

$$dpnl_t = \mathbb{1}_{\{t \leq \theta\}}(\alpha_t dW_t + \beta_t dM_t),$$

where $\alpha_t = \sigma S_t\Big(\partial_S w(t, S_{t-}) - \zeta_t - \eta_t \partial_S v(t, S_{t-})\Big)$ and $\beta_t = -w(t, S_{t-}) + \zeta_t S_{t-} + \eta_t\Big(v(t, S_{t-}) - K\Big)$ (noting from the Itô isometry that $\int_0^{\cdot\wedge\theta}(\zeta_t\sigma S_t - \zeta_t\sigma S_{t-})dW_t = \int_0^{\cdot\wedge\theta}\Big(\eta_t\sigma S_t\partial_S v(t,S_t) - \eta_t\sigma S_{t-}\partial_S v(t,S_{t-})\Big)dW_t = 0$). Hence the replication condition $\alpha = \beta = 0$ reduces to the linear systems

$$\begin{aligned} &\partial_S w(t, S_{t-}) - \zeta_t - \eta_t \partial_S v(t, S_{t-}) = \\ &\quad - w(t, S_{t-}) + \zeta_t S_{t-} + \eta_t\Big(v(t, S_{t-}) - K\Big) = 0 \end{aligned} \qquad (59)$$

in (ζ_t, η_t) (one system for each $t < \theta \wedge T$). Using (53) for the first line and (56) and (57) for the second line, one verifies that (58) solves (59). □

[5]see Proposition 5 and (57).
[6]cf. (22).

§6 THE LIMITING FRICTION PROCESS FOR PROPORTIONAL TRANSACTION COSTS (∗)

Our HVA encompasses nonlinear transaction costs à la Burnett (2021) & Burnett and Williams (2021). In this section we derive a specification for the corresponding cumulative friction costs f of §2.C, by passage to the continuous-time limit starting from a classical discrete-time specification.

A Fair Valuation Setup

This is achieved in the setup of the following fair valuation model stated under the probability measure $\mathbb{Q}$ with the risk-free asset as a numéraire:

$$\begin{aligned} dX_t &= \mu(t, \mathcal{X}_t)dt + \sigma(t, \mathcal{X}_t)dW_t, \\ dJ_t &= \sum_{\jmath=1}^{L}(\jmath - J_{t-})d\nu_t^{\jmath},\ \lambda_t^{\jmath} = \lambda_{\jmath}(t, \mathcal{X}_{t-}), \end{aligned} \tag{60}$$

where W is a multivariate Brownian motion and $\nu^{\jmath}$ is a random measure counting the number of transitions of the "Markov chain like" component[7] J to the state $\jmath$ on $(0, \cdot]$, with compensated martingale $d\nu_t^{\jmath} - \lambda_t^{\jmath}dt$ of $\nu^{\jmath}$. Jumps could also be introduced in X, but we refrain from doing so for notational simplicity. This setup encompasses the *jr* fair valuation model in our vulnerable put example. It also includes XVA models, with room for a client default indicator processes in the J components of $\mathcal{X}$.

We assume that the function-coefficients μ, σ, λ are continuous maps such that the above-model is well-posed, referring to Crépey (2013, Proposition 12.3.7) for a set of explicit assumptions ensuring it. In particular:

Assumption 3. **(i)** The maps $\lambda_{\jmath}$, $1 \le \jmath \le L$, are bounded by a constant $\Lambda \ge 0$.
(ii) The map $(t, x, \jmath) \mapsto (\mu, \sigma)(t, x, \jmath)$ is Lipschitz in $x \in \mathbb{R}^d$, uniformly in $(t, \jmath)$, and the map $(t, \jmath) \mapsto (\mu, \sigma)(t, 0, \jmath)$ is bounded.

Hence (see e.g. Élie (2006, Eqn. (II.83) page 123)), there exists a constant $C_1 \ge 0$ such that

$$\mathbb{E}\left[|X_t - X_s|^2\right]^{\frac{1}{2}} \le C_1(t-s)^{\frac{1}{2}}. \tag{61}$$

In addition, for all $1 \le l \le d$,

$$C^l := \sup_{t\in[0,T]} \mathbb{E}\left[(X_t^l)^2\right]^{\frac{1}{2}} < +\infty. \tag{62}$$

[7]but with transition probabilities modulated by X.

B Transaction costs for Discrete Rebalancing

We assume that a trader values a hedging set "$\star$" as $q^\star_t = q^\star(t, \mathcal{X}_t)$ for some smooth map $q^\star$, and that the trader delta-hedges its position with respect to the d-dimensional risky asset X, discretely at the times of the uniform grid $(i\mathrm{h})_{0\le i\le n}$ with $\mathrm{h} := \frac{T^\star}{n}$ for some $n \ge 1$, where $T^\star \le T$ is the final maturity of this hedging set.

Remark 14. More generally, one can consider delta-hedging only with respect to some components of X. It is actually what we do in §3.B.1 while delta-hedging in the Black-Scholes model with respect to $\tilde{S}$ only in $\mathcal{X} = (\tilde{S}, \Sigma; \mathbb{1}_{[0,\theta)})$ there (see (45) and (37)). The extension is straightforward, as it is (at least for our purpose) equivalent to considering no transaction costs for those non-delta-hedged assets, i.e. setting the corresponding diagonal entries of $\mathbf{k}$ to 0 below.

We work in a setting similar to Kabanov and Safarian (2009, Chapter 1, Section 2), with proportional transaction costs scaled to the rebalancing time by a factor $\sqrt{\mathrm{h}}$, where h is the time interval between two rebalancing dates.

Remark 15. In their case, scaling proportional transactions costs by h^α, with $\alpha \in (0, \frac{1}{2}]$, allows showing, in the Black-Scholes model, that perfect replication of a vanilla call can be achieved in the limit as the number of rebalancing dates goes to infinity by delta-hedging the portfolio's value computed with a modified volatility. In our case, scaling the transaction costs by $\sqrt{\mathrm{h}}$ allows passing to the continuous time limit and deriving the dynamics of the transaction costs and the PDE for the HVA* with trading indeed occurring continuously, and not only along a sequence of stopping times as in Burnett (2021).

Abbreviating ∂_{x_l} into ∂_l, let $a_t := \left(a^l_t\right)_{1\le l\le d}$ with $a^l_t := \partial_l q^\star(t, \mathcal{X}_t)$, $0 \le t \le T^\star$, $1 \le l \le d$. We also denote $(\delta a)^{\mathrm{abs}}_{i\mathrm{h}} := (|a^l_{i\mathrm{h}} - a^l_{(i-1)\mathrm{h}}|, 1 \le l \le d)$, and

$$\varphi^{\mathrm{h}}_{i\mathrm{h}} := \frac{\sqrt{\mathrm{h}}}{2} X^\top_{i\mathrm{h}} \mathbf{k} (\delta a)^{\mathrm{abs}}_{i\mathrm{h}}, \quad 0 < i < n, \quad \varphi^{\mathrm{h}}_0 = \varphi^{\mathrm{h}}_{t^{\mathrm{h}}_n} := 0,$$

in which $\mathbf{k} := diag(\mathrm{k}_l, 1 \le l \le d)$ for some constants $\mathrm{k}_l \ge 0$, $1 \le l \le d$

Remark 16. Unless there is a Markov-chain-like component J involved in $\mathcal{X}$ (i.e. for $L = 0$ in (60)), the replication hedging ratios in such setups also involve finite differences (as opposed to partial derivatives only in the above): see e.g. Proposition 7. However practitioners typically only use partial derivatives as their hedging ratios, motivating the present framework, which encompasses in particular the use-case of §3.B.1.

Assumption 4. The transaction cost to go from portfolio $a = \left(a^l\right)_{1\le l\le d}$ at time $t = i\mathrm{h}$ to portfolio $a + \delta a = (a^l + \delta a^l)_{1\le l\le d}$ at time $t + \mathrm{h}$ is given by $X^\top_{t+\mathrm{h}} \delta a + \varphi^{\mathrm{h}}_{i\mathrm{h}}$.

The transaction costs are thus proportional to the risky assets price (measured in units of the risk-free asset price). In the context of proportional transaction costs, Assumption 4 is classical: see e.g. Kabanov and Safarian (2009, page 8).

Remark 17. We neglect the transaction costs at time $t=0$, given by (assuming $d=1$ for simplicity) $\sqrt{\mathrm{h}}\frac{\mathrm{k}}{2}X_0|a_0-a_{0-}|$ (where a_{0-} is the initial quantity of risky asset possessed before entering the deal), and at time $t=T^\star=t_n^{\mathrm{h}}$, given by $\sqrt{\mathrm{h}}\frac{\mathrm{k}}{2}X_{T^\star}|a_{(n-1)\mathrm{h}}|$ (to liquidate the hedging portfolio). In view of the numerical results of Burnett and Williams (2021, Figure 2, Table 2), in which their closeout HVA is negligible, so we also neglect the liquidation costs occurring at the default time of a counterparty.

The discrete-time hedging valuation adjustment for frictions ($\mathrm{HVA}^{\mathrm{h}}$) is then a process that aims to compensate the bank for these transaction costs.

Definition 7. The hedging frictions and the HVA for frictions associated to discrete hedging along the time-grid $(i\mathrm{h}:=i\mathrm{h})_{0\le i\le n}$ are defined as the nonnegative processes $f^{\mathrm{h}}_{i\mathrm{h}}:=\sum_{u=0}^{i}\varphi^{\mathrm{h}}_{u\mathrm{h}}$ and $\mathrm{HVA}^{\mathrm{h}}$ such that $\mathrm{HVA}^{\mathrm{h}}_{t_n^{\mathrm{h}}}=0$ and, for $0\le i<n$,

$$\begin{aligned}\mathrm{HVA}^{\mathrm{h}}_{i\mathrm{h}}&=\mathbb{E}_{i\mathrm{h}}\left[f^{\mathrm{h}}_{t_n^{\mathrm{h}}}-f^{\mathrm{h}}_{i\mathrm{h}}\right]\\&=\mathbb{E}_{i\mathrm{h}}\left[f^{\mathrm{h}}_{(i+1)\mathrm{h}}-f^{\mathrm{h}}_{i\mathrm{h}}+\mathrm{HVA}^{\mathrm{h}}_{(i+1)\mathrm{h}}\right]=\mathbb{E}_{i\mathrm{h}}\left[\varphi^{\mathrm{h}}_{(i+1)\mathrm{h}}+\mathrm{HVA}^{\mathrm{h}}_{(i+1)\mathrm{h}}\right].\end{aligned}\tag{63}$$

Note that the second line in (63) yields a numerical scheme to recursively compute the discrete HVA process, backward in time starting from $\mathrm{HVA}^{\mathrm{h}}_{t_n^{\mathrm{h}}}=0$.

C Transaction Costs in the Continuous-Time Rebalancing Limit

The subsection specifies the cumulative friction costs $f^\star$ of §2.C and the ensuing $\mathrm{HVA}^\star$ that arises in the above setup when the rebalancing frequency of the hedge goes to infinity, i.e. when $\mathrm{h}\to 0$.

Definition 8. For all $t\in[0,T^\star]$, let $\varphi_t:=\varphi(t,\mathcal{X}_t)$ with, for all $(t,x,\jmath)\in[0,T^\star]\times\mathbb{R}^d\times\{1,\cdots,L\}$,

$$\varphi(t,x,\jmath)=\frac{1}{\sqrt{2\pi}}x^\top\mathbf{k}(\Gamma\sigma)^{\mathrm{abs}}(t,x,\jmath),\tag{64}$$

where $(\Gamma\sigma)^{\mathrm{abs}}:=(|\partial_x(\partial_l q^\star)\sigma|,1\le l\le d)$; let then $f^\star_t=\int_0^t\varphi_s ds$ and

$$\mathrm{HVA}^\star_t=va(f^\star)_t=\mathbb{E}_t\left[\int_t^{T^\star}\varphi_s ds\right].\tag{65}$$

For any smooth map $u=u(t,x,\jmath)$, let $\mathcal{F}u:=\partial_t u+(\partial_x u)\mu+\frac{1}{2}\mathrm{tr}\left[\sigma\sigma^\top\partial^2_{x^2}u\right]$, where ∂_x is the row-gradient with respect to x, $\partial^2_{x^2}$ the Hessian matrix with respect to x, and tr is the trace operator, and let $\mathcal{G}u=\mathcal{F}u+\sum_{\jmath=1}^{L}(u(\cdot,\jmath)-u)\lambda_\jmath$.

Note that the map HVA* defined by $\mathrm{HVA}^\star(t,x,\jmath) := \mathbb{E}\left[\mathrm{HVA}^f_t \,\middle|\, \mathcal{X}_t = (x,\jmath)\right]$ solves the PDE

$$\begin{aligned} &\mathrm{HVA}^\star(T^\star,\cdot) = 0 \text{ on } \mathbb{R}\times\{1,\ldots,L\},\\ &(\partial_t + \mathcal{G})\mathrm{HVA}^\star + \varphi = 0 \text{ on } [0,T^\star)\times\mathbb{R}\times\{1,\ldots,L\}. \end{aligned} \tag{66}$$

We make the following technical hypothesis on the model coefficients:

Assumption 5. (i) There exists $0<\alpha<\frac{1}{2}$ such that, for all $1\le l\le d$ and $1\le \jmath\le L$, the maps $(t,x)\mapsto \partial_l q^\star(t,x,\jmath)$ and $(t,x)\mapsto(\partial_x(\partial_l q^\star)\sigma(t,x,\jmath)$ is α-Hölder continuous in t and Lipschitz continuous in x.
(ii) There exists $C_2>0$ such that, for any $u\in\{\partial_l q^\star,(\partial_x(\partial_l q^\star)\sigma\,|\,1\le l\le d\}$,

$$\sup_{(t,x,\imath,\jmath)} |u(t,x,\imath)-u(t,x,\jmath)| \le C_2 < \infty.$$

(iii) $\sup_{t\in[0,T^\star]}\mathbb{E}\left[|(\partial_t+\mathcal{F})(\partial_l q^\star)(t,\mathcal{X}_t)|^2\right]^{\frac{1}{2}} \le C_2<\infty,\ 1\le l\le d.$

These assumptions, which are not minimal, are satisfied for regularized payoffs of vanilla European claims in the jr setup (22).

Theorem 1. *We set* $\mathrm{HVA}^{\mathrm{h}}_t := \mathrm{HVA}^{\mathrm{h}}_{\lfloor\frac{t}{\mathrm{h}}\rfloor\mathrm{h}}, 0\le t\le T^\star$. *Under Assumptions 3, 4 and 5, we have, for all* $t\in[0,T^\star]$,

$$\mathrm{HVA}^{\mathrm{h}}_t \to_{\mathrm{h}\to 0} \mathrm{HVA}^f_t.$$

Note that (65) would be virtually impossible to implement without the connection to HVA^{h} provided by the underlying discrete setup: transaction costs with model risk are a case where the approximation to a limiting problem in continuous time is problematic unless one knows where the limiting problem is coming from the first (discrete) place. But Theorem 1 is interesting from a theoretical viewpoint, and also important in practice, to guarantee the meaningfulness (stability for small h) of the numbers $\mathrm{HVA}^{\mathrm{h}}_t$ to be computed numerically based on (63).

C.1 Proof of Theorem 1

Lemma 3. *Under Assumptions 3 and 5, there exists* $C_3>0$ *such that, for all* $\mathrm{h}>0$ *and* $u\in\{\partial_l q^\star,(\partial_x(\partial_l q^\star)\sigma\,|\,1\le l\le d\}$,

$$\sup_{0<t-s<\mathrm{h}} \mathbb{E}\left[|u(t,\mathcal{X}_t)-u(s,\mathcal{X}_s)|^2\right]^{\frac{1}{2}} \le C_3\mathrm{h}^\alpha.$$

Proof. Since

$$1-\prod_{\jmath=1}^{L}\mathbb{1}_{\{\nu^\jmath_t=\nu^\jmath_s\}} \le \sum_{\jmath=1}^{L}(\nu^\jmath_t-\nu^\jmath_s),$$

we have, for some constant $C \geq 0$ varying from line to line,

$$
\begin{aligned}
&\mathbb{E}\left[|u(t,\mathcal{X}_t)-u(s,\mathcal{X}_s)|^2\right]^{\frac{1}{2}}\\
&\leq \sqrt{2}\mathbb{E}\left[|u(t,X_t,J_t)-u(t,X_t,J_s)|^2\right]^{\frac{1}{2}}+\sqrt{2}\mathbb{E}\left[|u(t,X_t,J_s)-u(s,X_s,J_s)|^2\right]^{\frac{1}{2}}\\
&\leq C\left(\mathbb{E}\left[|u(t,X_t,J_t)-u(t,X_t,J_s)|^2\sum_{\jmath=1}^{L}(\nu_t^\jmath-\nu_s^\jmath)\right]^{\frac{1}{2}}+(t-s)^\alpha+(t-s)^{\frac{1}{2}}\right)\\
&\leq C\left(\sum_{\jmath=1}^{L}\mathbb{E}\left[\int_s^t\lambda_r^\jmath dr\right]^{\frac{1}{2}}+(t-s)^\alpha\right)\\
&\leq C\left(L\sqrt{\Lambda(t-s)}+(t-s)^\alpha\right)\leq C_3\mathrm{h}^\alpha,
\end{aligned}
$$

Where we used equation (61) and the bound on the maps $\lambda^\jmath$. □

Coming to the proof of the theorem, we have, for $t=0$, for notational simplicity,

$$
\begin{aligned}
&\left|\mathrm{HVA}_0^{\mathrm{h}}-\mathrm{HVA}_0^{\star}\right|=\left|\mathbb{E}\left[\sum_{i=1}^{n}\varphi_{i\mathrm{h}}^{\mathrm{h}}\right]-\mathbb{E}\left[\int_0^{T^\star}\varphi_t dt\right]\right|\\
&=\left|\frac{\sqrt{\mathrm{h}}}{2}\mathbb{E}\left[\sum_{i=1}^{n}X_{i\mathrm{h}}^\top\mathbf{k}(\delta a)_{i\mathrm{h}}^{\mathrm{abs}}\right]-\frac{1}{\sqrt{2\pi}}\mathbb{E}\left[\sum_{i=1}^{n}\int_{(i-1)\mathrm{h}}^{i\mathrm{h}}X_t^\top\mathbf{k}(\Gamma\sigma)^{\mathrm{abs}}(t,\mathcal{X}_t)dt\right]\right|\\
&\leq\sum_{l=1}^{d}\mathrm{k}_l\left|\frac{\sqrt{\mathrm{h}}}{2}\mathbb{E}\left[\sum_{i=1}^{n}X_{i\mathrm{h}}^l\left|a_{i\mathrm{h}}^l-a_{(i-1)\mathrm{h}}^l\right|\right]\right.\\
&\qquad\left.-\frac{1}{\sqrt{2\pi}}\mathbb{E}\left[\sum_{i=1}^{n}\int_{(i-1)\mathrm{h}}^{i\mathrm{h}}X_t^l|\partial_x(\partial_l q^\star)\sigma|(t,\mathcal{X}_t)dt\right]\right|\\
&=\mathrm{h}\sum_{l=1}^{d}\sum_{i=1}^{n}\mathrm{k}_l\times\\
&\qquad\left|\mathbb{E}\left[\frac{1}{2\sqrt{\mathrm{h}}}X_{i\mathrm{h}}^l\left|a_{i\mathrm{h}}^l-a_{(i-1)\mathrm{h}}^l\right|-\frac{1}{\mathrm{h}\sqrt{2\pi}}\int_{(i-1)\mathrm{h}}^{i\mathrm{h}}X_t^l|\partial_x(\partial_l q^\star)\sigma|(t,\mathcal{X}_t)dt\right]\right|\\
&\leq T^\star\sum_{l=1}^{d}\mathrm{k}_l\sup_{0\leq s<t\leq T^\star,t-s=\mathrm{h}}\\
&\qquad\left|\mathbb{E}\left[\frac{1}{2\sqrt{\mathrm{h}}}X_t^l\left|a_t^l-a_s^l\right|-\frac{1}{\mathrm{h}\sqrt{2\pi}}\int_s^t X_u^l|\partial_x(\partial_l q^\star)\sigma|(u,\mathcal{X}_u)du\right]\right|.
\end{aligned}
$$

We fix $1 \leq l \leq d$ and we show that

$$\sup_{t-s=\mathrm{h}} \left| \mathbb{E}\left[\frac{1}{2\sqrt{\mathrm{h}}} X^l_t \left| a^l_t - a^l_s \right| - \frac{1}{\mathrm{h}\sqrt{2\pi}} \int_s^t X^l_u |\partial_x (\partial_l q^\star)\sigma|(u, \mathcal{X}_u) du \right] \right| \to_{\mathrm{h}\to 0} 0. \tag{67}$$

In fact, for all $0 \leq s < t \leq T^\star$ such that $t - s = \mathrm{h}$,

$$\begin{aligned}
&\left| \mathbb{E}\left[\frac{1}{2\sqrt{\mathrm{h}}} X^l_t \left| a^l_t - a^l_s \right| - \frac{1}{\mathrm{h}\sqrt{2\pi}} \int_s^t X^l_u |\partial_x (\partial_l q^\star)\sigma|(u, \mathcal{X}_u) du \right] \right| \\
&\leq \frac{1}{2\sqrt{\mathrm{h}}} \left| \mathbb{E}\left[(X^l_t - X^l_s) \left| a^l_t - a^l_s \right| \right] \right| \\
&+ \frac{1}{2\sqrt{\mathrm{h}}} \left| \mathbb{E}\left[X^l_s \left(\left| a^l_t - a^l_s \right| - |\partial_x (\partial_l q^\star)\sigma(s, \mathcal{X}_s)(W_t - W_s)| \right) \right] \right| + \\
&\left| \mathbb{E}\left[\frac{1}{2\sqrt{\mathrm{h}}} X^l_s \left| \partial_x (\partial_l q^\star)\sigma(s, \mathcal{X}_s)(W_t - W_s) \right| - \frac{1}{\mathrm{h}\sqrt{2\pi}} \int_s^t X^l_u |\partial_x (\partial_l q^\star)\sigma|(u, \mathcal{X}_u) du \right] \right|.
\end{aligned} \tag{68}$$

Regarding the first term in the r.h.s. of (68), we have, by Assumption 3 and Lemma 3,

$$\begin{aligned}
&\frac{1}{2\sqrt{\mathrm{h}}} \left| \mathbb{E}\left[\left(X^l_t - X^l_s \right) \left| a^l_t - a^l_s \right| \right] \right| \\
&\qquad \leq \frac{1}{2\sqrt{\mathrm{h}}} \mathbb{E}\left[\left| X^l_t - X^l_s \right|^2 \right]^{\frac{1}{2}} \mathbb{E}\left[\left| a^l_t - a^l_s \right|^2 \right]^{\frac{1}{2}} \\
&\qquad \leq \frac{C_1}{2} \mathrm{h}^{-\frac{1}{2}+\frac{1}{2}+\alpha} = \frac{C_1}{2} \mathrm{h}^\alpha.
\end{aligned} \tag{69}$$

We now consider the second term in the r.h.s. of (68). With $\delta\partial_l q^\star(t, x, \imath, \jmath) := \partial_l q^\star(t, x, \imath) - \partial_l q^\star(t, x, \jmath)$ and C^l defined in (62), recalling that $|\delta\partial_l q^\star(t,x,\imath,\jmath)| \le C_2$ by Assumption 5, we compute by Itô's formula:

$$\begin{aligned}
&\frac{1}{2\sqrt{\mathrm{h}}}\left|\mathbb{E}\left[X^l_s\left(\left|a^l_t-a^l_s\right|-|\partial_x(\partial_l q^\star)\sigma(s,\mathcal{X}_s)(W_t-W_s)|\right)\right]\right|\\
&\le \frac{1}{2\sqrt{\mathrm{h}}}\mathbb{E}\left[X^l_s\int_s^t |(\partial_t+\mathcal{F})\partial_l q^\star(u,\mathcal{X}_u)|\,du\right]\\
&\qquad+\sum_{\jmath=1}^{L}\frac{1}{2\sqrt{\mathrm{h}}}\mathbb{E}\left[X^l_s\left|\int_s^t \delta\partial_l q^\star(u,\mathcal{X}_u,\jmath)d\nu^\jmath_u\right|\right]\\
&\qquad+\frac{1}{2\sqrt{\mathrm{h}}}\mathbb{E}\left[X^l_s\left|\int_s^t \left(\partial_x(\partial_l q^\star)\sigma(u,\mathcal{X}_u)-\partial_x(\partial_l q^\star)\sigma(s,\mathcal{X}_s)\right)dW_u\right|\right] \qquad (70)\\
&\le \frac{C^l}{2}\mathbb{E}\left[\int_s^t |(\partial_t+\mathcal{F})\partial_l q^\star(u,\mathcal{X}_u)|^2\,du\right]^{\frac12}+\frac{C_2}{2\sqrt{h}}\sum_{\jmath=1}^{L}\mathbb{E}\left[X^l_s(\nu^\jmath_t-\nu^\jmath_s)\right]\\
&\qquad+\frac{C^l}{2\sqrt{\mathrm{h}}}\mathbb{E}\left[\int_s^t |\partial_x(\partial_l q^\star)\sigma(u,\mathcal{X}_u)-\partial_x(\partial_l q^\star)\sigma(s,\mathcal{X}_s)|^2\,du\right]^{\frac12}\\
&\le \frac{C^lC_2}{2}\sqrt{\mathrm{h}}+\frac{C^lC_2\Lambda K\sqrt{\mathrm{h}}}{2}+\frac{C^lC_3}{2}\mathrm{h}^\alpha \le C\mathrm{h}^\alpha,
\end{aligned}$$

where we used

$$\begin{aligned}
\mathbb{E}\left[X^l_s(\nu^\jmath_t-\nu^\jmath_s)\right] &= \mathbb{E}\left[X^l_s\mathbb{E}_s\left[\nu^\jmath_t-\nu^\jmath_s\right]\right]=\mathbb{E}\left[X^l_s\int_s^t\lambda^\jmath_u du\right]\\
&\le \mathbb{E}\left[\left(X^l_s\right)^2\right]^{\frac12}\mathbb{E}\left[\left(\int_s^t\lambda^\jmath_u du\right)^2\right]^{\frac12}\le C^l\Lambda\mathrm{h},
\end{aligned}$$

$$\begin{aligned}
\mathbb{E}\left[\int_s^t |(\partial_t+\mathcal{F})\partial_l q^\star(u,\mathcal{X}_u)|^2\,du\right]^{\frac12} &= \left(\int_s^t \mathbb{E}\left[|(\partial_t+\mathcal{F})\partial_l q^\star(u,\mathcal{X}_u)|^2\right]du\right)^{\frac12}\\
&\le \sqrt{\mathrm{h}}\sup_{t\in[0,T^\star]}\mathbb{E}\left[|(\partial_t+\mathcal{F})\partial_l q^\star(u,\mathcal{X}_u)|^2\right]^{\frac12}\le C_2\sqrt{\mathrm{h}},
\end{aligned}$$

and Lemma 3.

We finally deal with the last term in the r.h.s. of (68). As $\partial_x(\partial_l q^\star)\sigma(s,\mathcal{X}_s)\,(W_t-W_s)$ has, conditionally on $\mathfrak{G}_s$, the law $\mathcal{N}\left(0,\mathrm{h}\,|\partial_x(\partial_l q^\star)\sigma(s,\mathcal{X}_s)|^2\right)$, we have

$$\frac{1}{2\sqrt{\mathrm{h}}}\mathbb{E}\left[X^l_s\,|\partial_x(\partial_l q^\star)\sigma(s,\mathcal{X}_s)\,(W_t-W_s)|\right]=\frac{1}{\sqrt{2\pi}}\mathbb{E}\left[X^l_s\,|\partial_x(\partial_l q^\star)\sigma(s,\mathcal{X}_s)|\right].$$

We then obtain for this last term

$$\left|\mathbb{E}\left[\frac{1}{2\sqrt{\mathrm{h}}}X^l_s\,|\partial_x(\partial_l q^\star)\sigma(s,\mathcal{X}_s)\,(W_t-W_s)|-\frac{1}{\mathrm{h}\sqrt{2\pi}}\int_s^t X^l_u|\partial_x(\partial_l q^\star)\sigma|(u,\mathcal{X}_u)du\right]\right|$$

$$
\begin{aligned}
&= \frac{1}{\sqrt{2\pi}} \left| \mathbb{E}\left[X^l_s \left|\partial_x(\partial_l q^\star)\sigma(s,\mathcal{X}_s)\right| - \frac{1}{\mathrm{h}}\int_s^t X^l_u |\partial_x(\partial_l q^\star)\sigma|(u,\mathcal{X}_u)du \right] \right| \\
&\leq \frac{1}{\sqrt{2\pi}} \left| \mathbb{E}\left[X^l_s \left|\partial_x(\partial_l q^\star)\sigma(s,\mathcal{X}_s)\right| - \frac{1}{\mathrm{h}}\int_s^t X^l_u |\partial_x(\partial_l q^\star)\sigma|(u,X_u,J_s)du \right] \right| \\
&\quad + \frac{1}{\mathrm{h}\sqrt{2\pi}} \left| \mathbb{E}\left[\int_s^t X^l_u \left(|\partial_x(\partial_l q^\star)\sigma|(u,X_u,J_s) - |\partial_x(\partial_l q^\star)\sigma|(u,X_u,J_u) \right) du \right] \right| \\
&\leq \frac{1}{\sqrt{2\pi}} \left| \mathbb{E}\left[X^l_s |\partial_x(\partial_l q^\star)\sigma(s,\mathcal{X}_s)| - X^l_r |\partial_x(\partial_l q^\star)\sigma(r,X_r,J_s)| \right] \right| + \\
&\quad \frac{C^l}{\sqrt{\mathrm{h}2\pi}} \mathbb{E}\left[\int_s^t \left(\partial_x(\partial_l q^\star)\sigma(u,X_u,J_s) - \partial_x(\partial_l q^\star)\sigma(u,X_u,J_u) \right)^2 du \sum_{\jmath=1}^{L} (\nu^\jmath_t - \nu^\jmath_s) \right]^{\frac{1}{2}},
\end{aligned}
$$

where the (random) $r \in (s,t)$ in the first term is obtained via the mean value theorem. For this first term, we have, for a constant C changing from term to term,

$$
\begin{aligned}
&\left| \mathbb{E}\left[X^l_s |\partial_x(\partial_l q^\star)\sigma(s,\mathcal{X}_s)| - X^l_r |\partial_x(\partial_l q^\star)\sigma(r,X_r,J_s)| \right] \right| \\
&\leq \mathbb{E}\left[|X^l_s - X^l_r| |\partial_x(\partial_l q^\star)\sigma(s,\mathcal{X}_s)| \right] \\
&\qquad + \mathbb{E}\left[X^l_r |\partial_x(\partial_l q^\star)\sigma(s,\mathcal{X}_s)| - \partial_x(\partial_l q^\star)\sigma(r,X_r,J_s)| \right] \\
&\leq C_1 \mathrm{h}^{\frac{1}{2}} \sup_{t\in[0,T^\star]} \mathbb{E}\left[|\partial_x(\partial_l q^\star)\sigma(t,\mathcal{X}_t)|^2 \right]^{\frac{1}{2}} + C\mathrm{h}^\alpha \mathbb{E}\left[X^l_r \right] + C\mathbb{E}\left[X^l_r |X_r - X_s| \right] \\
&\leq C\mathrm{h}^\alpha,
\end{aligned}
\tag{71}
$$

as

$$
\sup_{t\in[0,T^\star]} \mathbb{E}\left[|\partial_x(\partial_l q^\star)\sigma(t,\mathcal{X}_t)|^2 \right]^{\frac{1}{2}} \leq CT^{\star\alpha} + C^l + C \max_{1\leq \jmath \leq L} |\partial_x(\partial_l q^\star)\sigma(0,0,\jmath)| < \infty.
$$

Finally,

$$
\begin{aligned}
&\frac{C^l}{\sqrt{\mathrm{h}2\pi}} \mathbb{E}\left[\int_s^t \left(\partial_x(\partial_l q^\star)\sigma(u,X_u,J_s) - \partial_x(\partial_l q^\star)\sigma(u,X_u,J_u) \right)^2 du \sum_{\jmath=1}^{L} (\nu^\jmath_t - \nu^\jmath_s) \right]^{\frac{1}{2}} \\
&\qquad \leq \frac{C^l C_2}{\sqrt{2\pi}} \mathbb{E}\left[\sum_{\jmath=1}^{L} \int_s^t \lambda^\jmath_u du \right]^{\frac{1}{2}} \leq \frac{C^l C_2}{\sqrt{2\pi}} \sqrt{L\Lambda \mathrm{h}}.
\end{aligned}
\tag{72}
$$

Using (69)–(72), we obtain, for some constant $C \geq 0$,

$$
\sup_{t-s=\mathrm{h}} \left| \mathbb{E}\left[\frac{1}{2\sqrt{\mathrm{h}}} X^l_t \left| a^l_t - a^l_s \right| - \frac{1}{\mathrm{h}\sqrt{2\pi}} \int_s^t X^l_u |\partial_x(\partial_l q^\star)\sigma|(u,\mathcal{X}_u)du \right] \right| \leq C\mathrm{h}^\alpha,
$$

which proves (67) and therefore Theorem 1.

§7 NEURAL NET REGRESSIONS AND QUANTILE REGRESSIONS FOR THE PATHWISE HVA^F, EC AND KVA OF §3.B.2

The setup and notation are those of §3.B.2.

A HVA^f Computations

The function $\widetilde{\mathrm{HVA}}^f$ in (38) is such that $\widetilde{\mathrm{HVA}}^f(T,\cdot)=0$, $\widetilde{\mathrm{HVA}}^f(t,0)=0$ for all t and, for $u<t$ and $S\in(0,\infty)$,

$$\begin{aligned}\widetilde{\mathrm{HVA}}^f(u,S)&=\mathbb{E}\left[(f_t-f_u)+\widetilde{\mathrm{HVA}}^f(t,S_t)\,\middle|\,S_u=S\right]\\&=\mathbb{E}\left[(f_t-f_u)+\widetilde{\mathrm{HVA}}^f(t,S_t)\mathbb{1}_{\{S_t>0\}}\,\middle|\,S_u=S\right],\end{aligned}\tag{73}$$

for f as per (37). Accordingly, we approximate on $(0,\infty)$ the functions $\widetilde{\mathrm{HVA}}^f(t_i,\cdot)$ for $t_i:=i\frac{T}{10}$ along the lines of IV.§2, as follows. Set $\widehat{\mathrm{HVA}}^f(t_{10},\cdot)=0$ and assume that we have already trained neural networks $\widehat{\mathrm{HVA}}^f(t_k,\cdot)$, $i+1\leq k<10$. Based on sampled data

$$\begin{aligned}&(feature^m_{hva^f_i},label^m_{hva^f_i})=\\&\quad\left(\widetilde{S}^m_{t_i},(f_{t_{i+1}}-f_{t_i})^m+\widehat{\mathrm{HVA}}^f(t_{i+1},S^m_{t_{i+1}})\mathbb{1}_{\{S^m_{t_{i+1}}>0\}}\right),\ 1\leq m\leq M\end{aligned}$$

where each $S^m_{t_{i+1}}$ is obtained from (22) with the initial condition $S^m_{t_i}=\widetilde{S}^m_{t_i}>0$ simulated from (45), in view of (73) and of the least-squares characterization of conditional expectation (in square integrable cases), we seek for $\widehat{\mathrm{HVA}}^f(t_i,\cdot)$ in

$$\underset{\Phi\in\mathcal{NN}}{\operatorname{argmin}}\sum_{m=1}^M\left(label^m_{hva^f_i}-\Phi(feature^m_{hva^f_i})\right)^2,\tag{74}$$

where $\mathcal{NN}$ denotes the set of feedforward neural networks with three hidden layers of 10 neurons each and ReLU activation functions.

We then obtain $\widehat{\mathrm{HVA}}^f(0,S_0)=0.04613$ from $f_{t_1}+\widehat{\mathrm{HVA}}^f(t_1,S_{t_1})$ as a sample mean. The corresponding standard deviation, 95% confidence interval and relative error at 95% are $\hat{\sigma}^f\simeq 6\times10^{-3}$, $[0.04601,0.04624]$ and $\frac{1.96\hat{\sigma}^f}{\widehat{\mathrm{HVA}}^f_0\sqrt{M}}\simeq 0.25\%$, where $\hat{\sigma}^f$ denotes the empirical standard deviation of $f_1+\widehat{\mathrm{HVA}}^f(t_1,S_{t_1})$.

B Economic Capital Computations

Next we approximate $\widetilde{\mathrm{EC}}(t,\cdot)$ on $(0,\infty)$ by the two-stage scheme of IV.§3, for each $t=t_i, 1\leq i<10$. We first train a neural network $\widehat{\mathbb{V}a\mathbb{R}}(t,\cdot)$ approximating $\widetilde{\mathbb{V}a\mathbb{R}}(t,\cdot)$ based on sampled data

$$(feature^m_{var_i}, label^m_{var_i}) = \left(\widetilde{S}^m_t, (\mathcal{L}_{t'}-\mathcal{L}_t)^m\right),\ 1\leq m\leq M$$

and on the pinball loss function, seeking for $\widehat{\mathbb{V}a\mathbb{R}}(t,\cdot)$ in

$$\operatorname*{argmin}_{\Phi\in\mathcal{NN}} \frac{1}{M}\sum_{m=1}^{M}\left((label^m_{var_i}-\Phi(feature^m_{var_i}))^+ + (1-\alpha)\Phi(feature^m_{var_i})\right).$$

Note from (20) that, for $t=t_i$, sampling $\mathcal{L}_{t'}-\mathcal{L}_t$ uses the already trained neural network $\widehat{\mathrm{HVA}}^f(t_{i+1},\cdot)$. For $t=1yr$ (where the approximation should be the worst due to error accumulated on $\widetilde{\mathrm{HVA}}^f$ from dynamic programming, the twin Monte Carlo estimate that stems from IV.(8) for the distance in p-values between the estimate $\widehat{\mathbb{V}a\mathbb{R}}(t,S_t)$ and the targeted (unknown) $\mathbb{V}a\mathbb{R}_t\left(\mathcal{L}_{t'}-\mathcal{L}_t\right)$ is less than $3.6\times10^{-3}\leq 1-\alpha=10^{-2}$ with 95% probability.

We then train neural networks $\widehat{\mathrm{EC}}(t,\cdot)$ approximating $\widetilde{\mathrm{EC}}(t,\cdot)$ on $(0,\infty)$ at times $t=t_i$ based on sampled data

$$(feature^m_{ec_i}, label^m_{ec_i}) =$$
$$\left(\widetilde{S}^m_t, (1-\alpha)^{-1}\left((\mathcal{L}_{t'}-\mathcal{L}_t)^m-\widehat{\mathbb{V}a\mathbb{R}}(t,\widetilde{S}^m_t)\right)^+ + \widehat{\mathbb{V}a\mathbb{R}}(t,\widetilde{S}^m_t)\right),\ 1\leq m\leq M$$

and a quadratic loss function, i.e. we seek for $\widehat{\mathrm{EC}}(t_i,\cdot)$ in

$$\operatorname*{argmin}_{\Phi\in\mathcal{NN}} \frac{1}{M}\sum_{m=1}^{M}\left(label^m_{ec_i}-\Phi(feature^m_{ec_i})\right)^2.$$

For $t=1yr$, the twin Monte Carlo estimate of Barrera et al. (2024, Proposition 3.4(ii)) (cf. Proposition IV.2(ii)) for the L_2 norm of the difference between the estimate $\widehat{\mathrm{EC}}(t,\widetilde{S}_t)$ and the targeted (unknown) $\widetilde{\mathrm{EC}}(t,\widetilde{S}_t)$ is smaller than 0.067 (itself significantly less than the orders of magnitude of EC visible on the lower panels of Figure 69) with probability 95%.

We also compute $\widehat{\mathbb{V}a\mathbb{R}}(0,S_0)=0.0120$ (which is needed for $\widehat{\mathrm{EC}}(0,S_0)$ below) as an empirical (unconditional) value-at-risk. The corresponding 95% confidence interval and relative error at 95% are $[0.0117, 0.0123]$ and $\frac{1.96}{\widehat{\mathbb{V}a\mathbb{R}}(0,S_0)\hat{d}(\widehat{\mathbb{V}a\mathbb{R}}(0,S_0))}\sqrt{\frac{\alpha(1-\alpha)}{M}}\simeq 2.3\%$, where $\hat{d}$ denotes the empirical density of $\mathcal{L}_{t_1=1}-\mathcal{L}_{t_0=0}$. Finally we compute $\widehat{\mathrm{EC}}(0,S_0)=0.493$ using the recursive algorithm of Costa and Gadat (2021, Eqn. (4)). Using the central limit theorem for expected shortfalls derived in Costa and Gadat (2021, Theorem 1.3), a 95% confidence interval is $[0.451, 0.534]$, and the relative error at 95% is $\sqrt{\frac{b_M}{2}}\frac{1.96\hat{\sigma}^s}{(1-\alpha)\widehat{\mathrm{EC}}(0,S_0)} \simeq 0.08$, where $\hat{\sigma}^s$ denotes the empirical standard deviation of $(\mathcal{L}_1 - \mathcal{L}_0)\mathbb{1}_{\{(\mathcal{L}_1-\mathcal{L}_0)>\widehat{\mathbb{V}\mathrm{a}\mathbb{R}}(0,S_0)\}}$ and b_M is defined in Costa and Gadat (2021, Assumption H_{a_n,b_n}).

C KVA Computations

Last, we approximate $\widetilde{\mathrm{KVA}}(t,\cdot)$ at times $t = t_i$ on $(0,\infty)$, for i decreasing from 10 to 1, by neural networks $\widehat{\mathrm{KVA}}(t_i,\cdot)$, based on the following dynamic programming equation, for $0 \le i < 10$:

$$\begin{aligned}\mathrm{KVA}_{t_i} &= \mathbb{E}_{t_i}\left[\mathrm{KVA}_{t_{i+1}} + h\int_{t_i}^{t_{i+1}} (\mathrm{EC}_u - \mathrm{KVA}_u)^+\,du\right]\\ &\approx \mathbb{E}_{t_i}\left[\mathrm{KVA}_{t_{i+1}} + h(t_{i+1}-t_i)\left(\mathrm{EC}_{t_{i+1}} - \mathrm{KVA}_{t_{i+1}}\right)^+\right].\end{aligned}$$

Starting from $\widehat{\mathrm{KVA}}(t_n,\cdot) = 0$ and having already trained the $\widehat{\mathrm{KVA}}(t_j,\cdot), j > i > 0$, we train $\widehat{\mathrm{KVA}}(t_i,\cdot)$ based on sampled data

$$\begin{aligned}(feature^m_{kva_i}, label^m_{kva_i}) = \Big(&\widetilde{S}^m_{t_i},\\ &h(t_{i+1}-t_i)\left(\widehat{\mathrm{EC}}(t_{i+1},S^m_{t_{i+1}}) - \widehat{\mathrm{KVA}}(t_{i+1},S^m_{t_{i+1}})\mathbb{1}_{\{S^m_{t_{i+1}}>0\}}\right)^+ +,\\ &\widehat{\mathrm{KVA}}(t_{i+1},S^m_{t_{i+1}})\mathbb{1}_{\{S^m_{t_{i+1}}>0\}}\Big),\ 1 \le m \le M\end{aligned}$$

seeking for $\widehat{\mathrm{KVA}}(t_i,\cdot)$ in

$$\underset{\Phi\in\mathcal{NN}}{\mathrm{argmin}}\,\frac{1}{M}\sum_{m=1}^{M}\left(label^m_{kva_i} - \Phi(feature^m_{kva_i})\right)^2.$$

We then compute $\widehat{\mathrm{KVA}}(0,S_0) = 0.407$ from $h(t_1 - t_0)(\widehat{\mathrm{EC}}(t_1,S_{t_1}) - \widehat{\mathrm{KVA}}(t_1,S_{t_1})\mathbb{1}_{\{S_{t_1}>0\}}) + \widehat{\mathrm{KVA}}(t_1,S_{t_1})$ as a sample mean. The corresponding standard deviation, 95% confidence interval and relative error at 95% are $\hat{\sigma}^{kva} \simeq 6\times 10^{-2}$, $[0.4056, 0.4082]$ and $\frac{1.96\hat{\sigma}^{kva}}{\widehat{\mathrm{KVA}}_0\sqrt{M}} \simeq 0.0028$, where $\hat{\sigma}^{kva}$ denotes the empirical standard deviation of $h(t_1-t_0)(\widehat{\mathrm{EC}}(t_1,S_{t_1}) - \widehat{\mathrm{KVA}}(t_1,S_{t_1})\mathbb{1}_{\{S_{t_1}>0\}}) + \widehat{\mathrm{KVA}}(t_1,S_{t_1})$.

Bibliography

Abbas-Turki, L., S. Crépey, B. Li, and B. Saadeddine (2025). An explicit scheme for pathwise XVA computations. *Journal of Computational Finance* 28(4), 1–36, 2025.

Abbas-Turki, L., S. Crépey, and B. Saadeddine (2023). Pathwise CVA regressions with oversimulated defaults. *Mathematical Finance 33*(2), 274–307.

Abbas-Turki, L., B. Diallo, and S. Crépey (2018). XVA principles, nested Monte Carlo strategies, and GPU optimizations. *International Journal of Theoretical and Applied Finance 21*, 1850030.

Abbas-Turki, L. A., S. Vialle, B. Lapeyre, and P. Mercier (2014). Pricing derivatives on graphics processing units using monte carlo simulation. *Concurrency and Computation: Practice and Experience 26*(9), 1679–1697.

Acerbi, C. and D. Tasche (2002). On the coherence of expected shortfall. *Journal of Banking & Finance 26*(7), 1487–1503.

Albanese, C. and L. Andersen (2014). Accounting for OTC derivatives: Funding adjustments and the re-hypothecation option. ssrn.2482955.

Albanese, C., L. Andersen, and S. Iabichino (2015). FVA: Accounting and risk management. *Risk Magazine*, February 64–68. Long preprint version ssrn.2517301.

Albanese, C., Y. Armenti, and S. Crépey (2020). XVA Metrics for CCP optimisation. *Statistics & Risk Modeling with Applications in Finance and Insurance 37*(1-2), 25–53.

Albanese, C., T. Bellaj, G. Gimonet, and G. Pietronero (2011). Coherent global market simulations and securitization seasures for counterparty credit risk. *Quantitative Finance 11-1*, 1–20.

Albanese, C., D. Brigo, and F. Oertel (2013). Restructuring counterparty credit risk. *International Journal of Theoretical and Applied Finance 16*(2), 1350010 (29 pages).

Albanese, C., S. Caenazzo, and S. Crépey (2016). Capital and funding. *Risk Magazine*, May 71–76.

Albanese, C., S. Caenazzo, and S. Crépey (2017). Credit, funding, margin, and capital valuation adjustments for bilateral portfolios. *Probability, Uncertainty and Quantitative Risk 2*(7), 26 pages.

Albanese, C., S. Caenazzo, and M. Syrkin (2017). Optimising VAR and terminating Arnie-VAR. *Risk Magazine*, October.

Albanese, C., M. Chataigner, and S. Crépey (2020). Wealth transfers, indifference pricing, and XVA compression schemes. In Y. Jiao (Ed.), *From Probability to Finance—Lecture note of BICMR summer school on financial mathematics*, Mathematical Lectures from Peking University Series, pp. 219–248. Springer.

Albanese, C., S. Crépey, R. Hoskinson, and B. Saadeddine (2021). XVA analysis from the balance sheet. *Quantitative Finance 21*(1), 99–123.

Albanese, C., S. Crépey, and S. Iabichino (2021). A Darwinian theory of model risk. *Risk Magazine*, July pages 72–77.

Albanese, C., S. Crépey, and S. Iabichino (2023). Quantitative reverse stress testing, bottom-up. *Quantitative Finance 23*(5), 863–875.

Albanese, C. and M. Trovato (2008). A stochastic monetary policy interest rate model. In *Computational Methods in Financial Engineering: Essays in Honour of Manfred Gilli*, pp. 343–392. Springer.

Albanese, C. and A. Vidler (2013). A structural model for credit-equity derivatives and bespoke CDOs. *Willmott Magazine* (24 June).

Amini, H., D. Filipovic, and A. Minca (2020). *SIAM Journal on Financial Mathematics 11*(1), 60–98.

Andersen, L. and A. Dickinson (2019). Funding and credit risk with locally elliptical portfolio processes: An application to central counterparties. *Journal of Financial Market Infrastructures 7*(4), 27–70.

Andersen, L., D. Duffie, and Y. Song (2019). Funding value adjustments. *Journal of Finance 74*(1), 145–192.

Antonov, A., S. Issakov, and A. McClelland (2018). Efficient SIMM-MVA calculations for callable exotics. *Risk Magazine*, August.

Armakola, A. and J.-P. Laurent (2015). CCP resilience and clearing membership. ssrn.2625579.

Armenti, Y. and S. Crépey (2017). Central clearing valuation adjustment. *SIAM Journal on Financial Mathematics 8*(1), 274–313.

Armenti, Y., S. Crépey, and C. Zhou (2017). The sustainable Black-Scholes equations. In J. Garrido, M. Jeanblanc, and J. Londono (Eds.), *Actuarial Science and Quantitative Finance: ICASQF2016, Cartagena, Colombia, June 2016*, pp. 155–167. Springer.

Artzner, P., K.-T. Eisele, and T. Schmidt (2024). Insurance–finance arbitrage. *Mathematical Finance 34*(3), 739–773.

Avellaneda, M. and R. Cont (2013). Close-out risk evaluation (CORE): A new risk management approach for central counterparties. ssrn.2247493.

Avellaneda, M., A. Levy, and A. Paras (1995). Pricing and hedging derivative securities in markets with uncertain volatilities. *Applied Mathematical Finance 2*(2), 73–88.

Bank for International Settlements (2013). Fundamental Review of the Trading Book: a Revised Market Risk Framework. http://www.bis.org/publ/bcbs265.pdf.

Bank for International Settlements (2019a, June). A discussion paper on central counterparty default management auctions. Technical report, CPMI-IOSCO. https://www.bis.org/cpmi/publ/d185.htm.

Bank for International Settlements (2019b, February). Minimum capital requirements for market risk. https://www.bis.org/bcbs/publ/d457.htm.

Bank For International Settlements (2020, June). Central counterparty default management auctions—issues for consideration. Technical report, CPMI-IOSCO. https://www.bis.org/cpmi/publ/d192.htm.

Bank for International Settlements (2022, November). OTC derivatives statistics at end-june 2022. Technical report, Bank For International Settlements. https://www.bis.org/publ/otc_hy2211.htm.

Bank of England (2018, June). Central clearing and resolution—Learning some of the lessons of Lehmans. Technical report, Bank of England. https://www.bankofengland.co.uk/speech/2018/jon-cunliffe-speech-at-futures-industry-association-international-derivatives-expo.

Bank of International Settlements and OICV-IOSCO (2015). Public quantitative disclosure standards for central counterparties. https://www.bis.org/cpmi/publ/d125.pdf.

Banking Association, E. (2013). Capital Requirements Directive IV, Article 292. *CRD-IV 292.*

Barrera, D., S. Crépey, B. Diallo, G. Fort, E. Gobet, and U. Stazhynski (2019). Stochastic approximation schemes for economic capital and risk margin computations. *ESAIM: Proceedings and Surveys 65*, 182–218.

Barrera, D., S. Crépey, E. Gobet, H.-D. Nguyen, and B. Saadeddine (2024). Statistical learning of value-at-risk and expected shortfall. arXiv:2209.06476v2.

Barrieu, P. and G. Scandolo (2015). Assessing financial model risk. *European Journal of Operational Research 242*(2), 546–556.

Basel Committee on Banking Supervision (2005). An explanatory note on the Basel II IRB risk weight functions. http://www.bis.org/bcbs/irbriskweight.pdf.

Basel Committee on Banking Supervision (2013). Consultative document: Fundamental Review of the Trading Book: A revised market risk framework. http://www.bis.org/publ/bcbs265.pdf.

Basel Committee on Banking Supervision (2019, June). Overview of Pillar 2 supervisory review practices and approaches. https://www.bis.org/bcbs/publ/d465.pdf.

Basel Committee on Banking Supervision (January, 2012). Consultative document: Application of own credit risk adjustments to derivatives.

Bastide, D. and S. Crépey (2025). Provisions and economic capital for credit losses. *Quantitative Finance 25*(4), 617–631.

Bastide, D., S. Crépey, S. Drapeau, and M. Tadese (2023). Derivatives' risks as costs in a one-period network model. *Frontiers of Mathematical Finance 2*(3), 283–312.

Bastide, D., S. Crépey, S. Drapeau, and M. Tadese (2025). Resolving a clearing member's default: A Radner equilibrium approach. *Mathematics and Financial Economics 19*, 183–223.

BCBS (2005). The application of Basel II to trading activities and the treatment of double default effects. *Basel Committee on Bank Supervision.*

Bellini, T., M. Eichhorn, and D. Mayenberger (Eds.) (2021). *Reverse Stress Testing in Banking: A Comprehensive Guide.* De Gruyter.

Bénézet, C. and S. Crépey (2024). Handling model risk with XVAs. *Frontiers of Mathematical Finance 3*(4), 490–519.

Bénézet, C., S. Crépey, and D. Essaket (2022). Hedging valuation adjustment for callable claims. arXiv:2304.02479.

Bergstra, J. and Y. Bengio (2012). Random search for hyper-parameter optimization. *Journal of machine learning research 13*(Feb), 281–305.

Biais, B., F. Heider, and M. Hoerova (2016). Risk-sharing or risk-taking? counterparty risk, incentives, and margins. *The Journal of Finance 71*(4), 1669–1698.

Bichuch, M., A. Capponi, and S. Sturm (2018). Arbitrage-free XVA. *Mathematical Finance 28*(2), 582–620.

Bichuch, M., A. Capponi, and S. Sturm (2020). Robust XVA. *Mathematical Finance 30*(3), 738–781.

Bielecki, T. and M. Rutkowski (2002). *Credit Risk: Modeling, Valuation and Hedging.* Springer.

Bielecki, T. R. and M. Rutkowski (2015). Valuation and hedging of contracts with funding costs and collateralization. *SIAM Journal on Financial Mathematics 6*, 594–655.

Black, F. and R. Litterman (1991). Asset allocation: combining investor views with market equilibrium. *The Journal of Fixed Income 1*(2), 7–18.

Black, F. and R. Litterman (1992). Global portfolio optimization. *Financial Analysis Journal* (September), 28–43.

Board of Governors of the Federal Reserve System (2013, November). Rules and regulations, policy statement on the scenario design framework for stress testing. *Federal Register*. https://www.gpo.gov/.

Board of Governors of the Federal Reserve System (2012, May). Guidance on stress testing for banking organizations with total consolidated assets of more than $10 billion. *SR Letter 12-7*. https://www.federalreserve.gov/supervisionreg/.

Board of Governors of the Federal Reserve System (2020, October). Amendments to the regulatory capital, capital plan, and stress test rules. https://www.federalreserve.gov/newsevents/pressreleases/bcreg20200304a.htm.

Bozinovski, S. (2020). Reminder of the first paper on transfer learning in neural networks, 1976. *Informatica 44*(3).

Breuer, T., M. Jandacka, K. Rheinberger, and M. Summer (2009). How to find plausible, severe, and useful stress scenarios. *The International Journal of Central Banking 18.*

Brigo, D. and A. Capponi (2010). Bilateral counterparty risk with application to CDSs. *Risk Magazine*, March 85–90. Preprint arXiv:0812.3705.

Brigo, D., M. Morini, and A. Pallavicini (2013). *Counterparty Credit Risk, Collateral and Funding: With Pricing Cases For All Asset Classes.* Wiley.

Brigo, D. and A. Pallavicini (2014). Nonlinear consistent valuation of CCP cleared or CSA bilateral trades with initial margins under credit, funding and wrong-way risks. *Journal of Financial Engineering 1*, 1–60.

Buehler, H. (2019). Statistical hedging. ssrn.2913250.

Buehler, H., L. Gonon, J. Teichmann, and B. Wood (2019). Deep hedging. *Quantitative Finance 19*(8), 1271–1291.

Burgard, C. and M. Kjaer (2011). In the balance. *Risk Magazine*, October 72–75.

Burgard, C. and M. Kjaer (2013). Funding costs, funding strategies. *Risk Magazine*, December 82–87. Preprint ssrn.2027195.

Burgard, C. and M. Kjaer (2017). Derivatives funding, netting and accounting. *Risk Magazine*, March 100–104. Preprint ssrn.2534011.

Burnett, B. (2021). Hedging value adjustment: Fact and friction. *Risk Magazine*, February 1–6.

Burnett, B. and I. Williams (2021). The cost of hedging XVA. *Risk Magazine*, April.

Canabarro, E. and D. Duffie (2003). Measuring and marking counterparty risk. *Asset/Liability Management for Financial Institutions. Institutional Investor Books.*

Capriotti, L., Y. Jiang, and A. Macrina (2017). AAD and least-square Monte Carlo: Fast Bermudan-style options and XVA Greeks. *Algorithmic Finance 6*(1-2), 35–49.

Carmona, R. and S. Crépey (2010). Particle methods for the estimation of credit portfolio loss distributions. *International Journal of Theoretical and Applied Finance 13*(04), 577–602.

Castagna, A. (2014). Towards a theory of internal valuation and transfer pricing of products in a bank: Funding, credit risk and economic capital. ssrn.2392772.

Cesari, J., J. Aquilina, and N. Charpillon (2010). *Modelling, Pricing, and Hedging Counterparty Credit Exposure.* Springer.

Chen, J., X. Pan, R. Monga, S. Bengio, and R. Jozefowicz (2016). Revisiting distributed synchronous SGD. arXiv:1604.00981.

Cheridito, P., U. Horst, M. Kupper, and T. A. Pirvu (2015). Equilibrium pricing in incomplete markets under translation invariant preferences. *Mathematics of Operations Research 41*(1), 174–195.

Chizat, L. and F. Bach (2018). On the global convergence of gradient descent for over-parameterized models using optimal transport. *Advances in Neural Information Processing Systems 31.*

Choromanska, A., M. Henaff, M. Mathieu, G. B. Arous, and Y. LeCun (2015). The loss surfaces of multilayer networks. In *Artificial Intelligence and Statistics*, pp. 192–204. Proceedings of Machine Learning Research.

Collin-Dufresne, P., R. Goldstein, and J. Hugonnier (2004). A general formula for valuing defaultable securities. *Econometrica 72*(5), 1377–1407.

Committee of European Insurance and Occupational Pensions Supervisors (2010). QIS5 technical specifications. https://eiopa.europa.eu/Publications/QIS/QIS5-technical_specifications_20100706.pdf.

Cont, R. (2006). Model uncertainty and its impact on the pricing of derivative instruments. *Mathematical Finance 16*(3), 519–547.

Cont, R. (2017). Central clearing and risk transformation. *Financial Stability Review 21*, 141–147.

Costa, M. and S. Gadat (2021). Non asymptotic controls on a recursive superquantile approximation. *Electronic Journal of Statistics 15*(2), 4718–4769.

Crépey, S. (2013). *Financial Modeling: A Backward Stochastic Differential Equations Perspective.* Springer Finance Textbooks.

Crépey, S. (2015). Bilateral counterparty risk under funding constraints. Part II: CVA. *Mathematical Finance 25*(1), 23–50.

Crépey, S. (2022). Positive XVAs. *Frontiers of Mathematical Finance 1*(3), 425–465.

Crépey, S. (2024). Invariance times transfer properties. *Probability, Uncertainty and Quantitative Risk 9*(4), 431–452.

Crépey, S., T. R. Bielecki, and D. Brigo (2014). *Counterparty Risk and Funding: A Tale of Two Puzzles.* Taylor & Francis, New York. Chapman & Hall/CRC Financial Mathematics Series.

Crépey, S. and M. Dixon (2020). Gaussian process regression for derivative portfolio modeling and application to CVA computations. *Journal of Computational Finance 24*(1), 47–81.

Crépey, S., B. Li, H.-D. Nguyen, and B. Saadeddine (2024). CVA sensitivities, hedging and risk. *Risk Magazine*, July 1–6.

Crépey, S., W. Sabbagh, and S. Song (2020). When capital is a funding source: The anticipated backward stochastic differential equations of X-Value Adjustments. *SIAM Journal on Financial Mathematics 11*(1), 99–130.

Crépey, S. and S. Song (2015). BSDEs of counterparty risk. *Stochastic Processes and their Applications 125*(8), 3023–3052.

Crépey, S. and S. Song (2016). Counterparty risk and funding: Immersion and beyond. *Finance and Stochastics 20*(4), 901–930.

Crépey, S. and S. Song (2017a). Invariance properties in the dynamic Gaussian copula model. *ESAIM: Proceedings and Surveys 56*, 22–41.

Crépey, S. and S. Song (2017b). Invariance times. *The Annals of Probability 45*(6B), 4632–4674.

Cybenko, G. (1989). Approximation by superpositions of a sigmoidal function. *Mathematics of control, signals and systems 2*(4), 303–314.

Dalang, R. C., A. Morton, and W. Willinger (1990). Equivalent martingale measures and no-arbitrage in stochastic securities market models. *Stochastics and Stochastic Reports 29*(2), 185–201.

Default Risk Management Working Group (2016, january). Uniform CCP terminology for default management auctions. Technical report, Commodity Futures Trading Commission (CFTC).

Dellacherie, C. and P.-A. Meyer (1980). *Probabilité et Potentiel, Chapitres V à VIII.* Hermann.

Detering, N. and N. Packham (2016). Model risk of contingent claims. *Quantitative Finance 16*(9), 1357–1374.

Dordu, B., R. Edge, and D. Schwindt (2017, May). Measuring the severity of stress-test scenarios. *FEDS Notes*. Retrieved on November 15, 2021 on https://www.federalreserve.gov/econres/notes/feds-notes/.

Du, S., J. Lee, H. Li, L. Wang, and X. Zhai (2019). Gradient descent finds global minima of deep neural networks. In *International conference on machine learning*, pp. 1675–1685. Proceedings of Machine Learning Research.

Duffie, D. (2014). Resolution of failing central counterparties. Working paper no. 3256, Graduate School of Business, Stanford University.

Duffie, D. and M. Huang (1996). Swap rates and credit quality. *Journal of Finance 51*, 921–950.

Duffie, D. and W. Sharer (1986). Equilibrium and the role of the firm in incomplete market. Stanford University, Working Paper No. 915, available at https://www.gsb.stanford.edu/faculty-research/working-papers/equilibrium-role-firm-incomplete-markets.

Dybvig, P. (1992). Hedging non-traded wealth: when is there separation of hedging and investment. In S. Hodges (Ed.), *Options: Recent Advances in Theory and Practice*, Volume 2, pp. 13–24. Manchester University Press.

Weinan, E., J. Han, and A. Jentzen (2017). Deep learning-based numerical methods for high-dimensional parabolic partial differential equations and backward stochastic differential equations. *Communications in Mathematics and Statistics 5*(4), 370–398.

El Karoui, N., S. Peng, and M.-C. Quenez (1997). Backward stochastic differential equations in finance. *Mathematical Finance 7*, 1–71.

Elices, A. and E. Giménez (2013). Applying hedging strategies to estimate model risk and provision calculation. *Quantitative Finance 13*(7), 1015–1028.

Élie, R. (2006). *Contrôle stochastique et méthodes numériques en finance mathématique*. Ph. D. thesis, University Paris-Dauphine. https://pastel.archives-ouvertes.fr/tel-00122883/file/thesis.pdf.

Elouerkhaoui, Y. (2007). Pricing and hedging in a dynamic credit model. *International Journal of Theoretical and Applied Finance 10*(4), 703–731.

Elouerkhaoui, Y. (2017). *Credit Correlation: Theory and Practice*. Palgrave Macmillan.

European Bank Authority (2021, January). 2021 EU-wide stress test methodological note. https://www.eba.europa.eu/eba-launches-2021-eu-wide-stress-test-exercise.

European Bank Authority (2022, December). Risk assessment of the european banking system. Technical report, European Bank Authority.

European Central Bank (2018, November). ECB guide to the internal capital adequacy assessment process (ICAAP). *Banking Supervision*. https://www.bankingsupervision.europa.eu/home/html/index.en.html.

European Central Bank (2021). Pillar 2 guidance (P2G). https://www.bankingsupervision.europa.eu/banking/srep/html/p2g.en.html.

European Parliament (2012). Regulation (EU) no 648/2012 of the European parliament and of the council of 4 July 2012 on OTC derivatives, central counterparties and trade repositories. http://eur-lex.europa.eu/LexUriServ/LexUriServ.do?uri=OJ:L:2012:201:0001:0059:EN:PDF.

European Parliament (2013). Regulation (EU) No 575/2013 of the European Parliament and of the Council of 26 June 2013 on prudential requirements for credit institutions and investment firms and amending Regulation (EU) No 648/2012. *Official Journal of the European Union 59*(June 2013).

European Parliament (2016). Commission Delegated Regulation (EU) 2016/101 of 26 October 2015 supplementing Regulation (EU) No 575/2013 of the European Parliament and of the Council with regard to regulatory technical standards for prudent valuation under Article 105(14). *Official Journal of the European Union 59*(28 January 2016).

European Systemic Risk Board (2013). Macro-financial scenario for the 2020 EU-wide banking sector stress test. *European Banking Authority*. https://www.eba.europa.eu/eba-launches-2020-eu-wide-stress-test-exercise.

Farkas, W., F. Fringuellotti, and R. Tunaru (2020). A cost-benefit analysis of capital requirements adjusted for model risk. *Journal of Corporate Finance 65*, 101753.

Faruqui, U., W. Huang, and E. Takats (2018, December). Clearing risks in OTC derivatives markets: the CCP-bank nexus. *BIS Quarterly Review*.

Federal Register (April, 2014). Rules and regulations. *Federal Register 79*.

Ferrara, G. and X. Li (2020). Central counterparty auction design. *Journal of Financial Market Infrastructures 08*(02), 47–58.

Föllmer, H. and A. Schied (2016). *Stochastic Finance: An Introduction in Discrete Time* (4th ed.). De Gruyter Graduate, Berlin.

G20 Research Group (2009, September). G20 Leaders Statement: The Pittsburgh Summit, September 24-25, 2009, Pittsburgh. http://www.g20.utoronto.ca/2009/2009communique0925.html.

Gaunt, R. E. (2021). A simple proof of the characteristic function of Student's t-distribution. *Communications in Statistics–Theory and Methods 50*(14), 3380–3383.

Ghamami, S. (2015). Static models of central counterparty risk. *International Journal of Financial Engineering 02*, 1550011–1 to 36.

Glasserman, P. (2003). *Monte Carlo methods in financial engineering.* Springer.

Gnoatto, A., A. Picarelli, and C. Reisinger (2023). Deep xVA solver–a neural network based counterparty credit risk management framework. *SIAM Journal on Financial Mathematics 14*(1), 314–352.

Gobet, E. (2016). *Monte-Carlo methods and stochastic processes: from linear to non-linear.* Chapman and Hall/CRC.

Gobet, E., J.-P. Lemor, and X. Warin (2005). A regression-based Monte Carlo method to solve backward stochastic differential equations. *The Annals of Applied Probability 15*(3), 2172–2202.

Goodfellow, I. J., Y. Bengio, and A. Courville (2016). *Deep Learning.* MIT Press. http://www.deeplearningbook.org.

Gottardi, P. (1995). An analysis of the conditions for the validity of Modigliani-Miller Theorem with incomplete markets. *Economic Theory 5*, 191–207.

Gourinchas, P.-O., P. Martin, and T. Messer (2023, August). The economics of sovereign debt, bailouts, and the eurozone crisis. Technical report, International Monetary Fund. https://www.imf.org/-/media/Files/Publications/WP/2023/English/wpiea2023177-print-pdf.ashx.

Green, A. (2015). *XVA: Credit, Funding and Capital Valuation Adjustments.* Wiley.

Green, A., C. Kenyon, and C. Dennis (2014). KVA: capital valuation adjustment by replication. *Risk Magazine*, December 82–87.

Gregory, J. (2014). *Central Counterparties: Mandatory Central Clearing and Initial Margin Requirements for OTC Derivatives.* Wiley.

Gregory, J. (2015). *The xVA challenge: counterparty credit risk, funding, collateral and capital.* Wiley.

Grundke, P. (2011). Reverse stress tests with bottom-up approaches. *The Journal of Risk Model Validation 5*(1), 71–90.

He, S.-W., J.-G. Wang, and J.-A. Yan (1992). *Semimartingale Theory and Stochastic Calculus.* Science Press and CRC Press Inc.

Henrard, M. (2011). Adjoint algorithmic differentiation: Calibration and implicit function theorem. *OpenGamma Quantitative Research.* ssrn:1896329.

Hoeting, J. A., D. Madigan, A. E. Raftery, and C. T. Volinsky (1999). Bayesian model averaging: A tutorial. *Statistical Science 14*(4), 382–417.

Hornik, K. (1991). Approximation capabilities of multilayer feedforward networks. *Neural networks 4*(2), 251–257.

Huge, B. and A. Savine (2020). Differential machine learning: the shape of things to come. *Risk Magazine*, September.

Hull, J. and A. White (1990). Pricing interest rate derivative securities. *Review of Financial Studies 4.*

Hull, J. and A. White (2012). The FVA debate, followed by The FVA debate continued. *Risk Magazine*, July 83–85 and October 52.

Huré, C., H. Pham, and C. Warin (2020). Deep backward schemes for high-dimensional nonlinear PDEs. *Mathematics of Computation 89*(324), 1547–1579.

International Financial Reporting Standards (2013). IFRS 4 insurance contracts exposure draft.

Johnson, N. L., S. Kotz, and N. Balakrishnan (1995). *Continuous univariate distributions*, Volume 2. Wiley.

Johnson, R. A. and D. W. Wichern (2007). *Applied Multivariate Statistical Analysis* (sixth ed.). Pearson, New Jersey.

Kabanov, Y. and M. Safarian (2009). *Markets with transaction costs: Mathematical Theory.* Springer Science & Business Media.

Kalkbrener, M. (2005). An axiomatic approach to capital allocation. *Mathematical Finance 15*(3), 425–437.

Karoui, N. E., M. Jeanblanc-picqué, and S. E. Shreve (1998). Robustness of the Black and Scholes Formula. *Mathematical Finance 8*(2), 93–126.

Kidger, P. and T. Lyons (2020). Universal approximation with deep narrow networks. In *Conference on learning theory*, pp. 2306–2327. Proceedings of Machine Learning Research.

Kingma, D. P. and J. Ba (2014). Adam: A method for stochastic optimization. arXiv:1412.6980.

Kjaer, M. (2019). In the balance redux. *Risk Magazine* (November).

Kruse, T. and A. Popier (2016). BSDEs with monotone generator driven by Brownian and Poisson noises in a general filtration. *Stochastics 88*(4), 491–539.

Lamberton, D. and B. Lapeyre (1996). *Introduction to Stochastic Calculus Applied to Finance.* Chapman & Hall.

Landsman, Z. M. and E. A. Valdez (2003). Tail conditional expectations for elliptical distributions. *Scandinavian Actuarial Journal 7*(4), 55–71.

Lee, E. S. and R. N. Forthofer (2005). *Analyzing complex survey data.* Sage Publications.

Lei, Y., T. Hu, G. Li, and K. Tang (2019). Stochastic gradient descent for nonconvex learning without bounded gradient assumptions. *IEEE transactions on neural networks and learning systems 31*(10), 4394–4400.

Li, M. and F. Mercurio (2015). Jumping with default: wrong-way risk modelling for CVA. *Risk Magazine*, November.

Lipton, A. (2018). Systemic risks in central counterparty clearing house networks. In L. A. . M. Pykhtin (Ed.), *Margin in Derivatives Trading*, pp. 367–401. Risk Books.

Longstaff, F. A. and E. S. Schwartz (2001). Valuing American options by simulation: A simple least-squares approach. *The Review of Financial Studies 14*(1), 113–147.

Magill, M. and W. Shafer (1991). Incomplete markets. In W. Hildenbrand and H. Sonnenschein (Eds.), *Handbook of Mathematical Economics* (1 ed.), Volume 4, Chapter 30, pp. 1523–1614. Elsevier.

Matloff, N. (2017). *Statistical regression and classification: from linear models to machine learning.* Chapman and Hall/CRC.

McNeil, A., R. Frey, and P. Embrechts (2015). *Quantitative Risk Management: Concepts, Techniques and Tools*. Princeton University Press.

Meeker, W. Q., G. J. Hahn, and L. A. Escobar (2017). *Statistical Intervals, A Guide for Practitioners and Researchers*. Hoboken, New Jersey, United States: Wiley.

Menkveld, A. J. and G. Vuillemey (2021). The economics of central clearing. *Annual Review of Financial Economics 13*(1), 3.1–3.26.

Merton, R. (1974). On the pricing of corporate debt: the risk structure of interest rates. *The Journal of Finance 29*, 449–470.

Modigliani, F. and M. Miller (1958). The cost of capital, corporation finance and the theory of investment. *Economic Review 48*, 261–297.

Montesi, G. and G. Papiro (2018). Bank stress testing: A stochastic simulation framework to assess banks'financial fragility. *Risks 6*(3), 82.

Montesi, G., G. Papiro, M. Fazzini, and A. Ronga (2020). Stochastic optimization system for bank reverse stress testing. *Journal of Risk and Financial Management 13*(8), 174.

Murphy, D. and P. Nahai-Williamson (2014, October). Dear prudence, wont you come out to play? approaches to the analysis of central counterparty default fund adequacy. *Financial Stability Paper 30*.

Murphy, K. (2012). *Machine Learning: A Probabilistic Perspective*. The MIT Press.

Myers, S. (1977). Determinants of corporate borrowing. *Journal of Financial Economics 5*, 147–175.

NVIDIA Corporation (2020). Programming guide: Cuda toolkit documentation. https://docs.nvidia.com/cuda/cuda-c-programming-guide/index.html.

Oleschak, R. et al. (2019). Central counterparty auctions and loss allocation. Technical report, Swiss National Bank.

Pan, S. J. and Q. Yang (2009). A survey on transfer learning. *IEEE Transactions on knowledge and data engineering 22*(10), 1345–1359.

Pardoux, E. and S. Peng (1990). Adapted solution of a backward stochastic differential equation. *Systems Control Letters 14*, 55–61.

Peng, S. and Z. Yang (2009). Anticipated backward stochastic differential equations. *The Annals of Probability 37*(3), 877–902.

Piterbarg, V. (2010). Funding beyond discounting: collateral agreements and derivatives pricing. *Risk Magazine*, August 57–63.

Popper, K. R. (1934). *The Logic of Scientific Discovery.* Hutchinson.

Radner, R. (1982). Equilibrium under uncertainty. In K. J. Arrow and M. D. Intriligator (Eds.), *Handbook of Mathematical Economics*, Volume 2, Chapter 20, pp. 923–1006. Elsevier.

Rasmussen, C. and C. Williams (2006). *Gaussian Processes for Machine Learning.* Adaptive computation and machine learning. MIT Press.

Recht, B., C. Re, S. Wright, and F. Niu (2011). Hogwild!: A lock-free approach to parallelizing stochastic gradient descent. *Advances in neural information processing systems 24*.

Rockafellar, R. and S. Uryasev (2000). Optimization of conditional value-at-risk. *Journal of risk 2*, 21–42.

Rockafellar, R. T. (1970). *Convex Analysis.* Princeton Mathematical Series, No. 28. Princeton, N.J.: Princeton University Press.

Rockafellar, R. T. and R. J.-B. Wets (1998). *Variational Analysis.* Springer Berlin, Heidelberg.

Saadeddine, B. (2022). *Learning From Simulated Data in Finance: XVAs, Risk Measures and Calibration.* Ph. D. thesis, Université Paris-Saclay. https://theses.hal.science/tel-03894764v1/document.

Savine, A. (2018). From model to market risks: The implicit function theorem (IFT) demystified. *ssrn:3262571*.

Schönbucher, P. (2004). A measure of survival. *Risk Magazine 17*(8), 79–85.

Shorten, C. and T. Khoshgoftaar (2019). A survey on image data augmentation for deep learning. *Journal of Big Data 6*(1), 60.

Silotto, L., M. Scaringi, and M. Bianchetti (2021). Everything you always wanted to know about XVA model risk but were afraid to ask. *arXiv:2107.10377*.

Siu, T., H. Tong, and H. Yang (2004, 02). On Bayesian value at risk: From linear to non-linear portfolios. *Asia-Pacific Financial Markets 11*, 161–184.

Soner, H. M., N. Touzi, and J. Zhang (2012). Wellposedness of second order backward SDEs. *Probability Theory and Related Fields 153*(1-2), 149–190.

Swiss Federal Office of Private Insurance (2006). Technical document on the Swiss solvency test. https://www.finma.ch/FinmaArchiv/bpv/download/e/SST_techDok_061002_E_wo_Li_20070118.pdf.

Tan, X. and N. Touzi (2013). Optimal transportation under controlled stochastic dynamics. *The Annals of Probability 41*(5), 3201–3240.

The United States Congress (2010, July). Dodd-Frank Wall Street reform and consumer protection act. https://www.congress.gov/111/plaws/publ203/PLAW-111publ203.pdf.

Tsitsiklis, J. N. and B. Van Roy (2001). Regression methods for pricing complex american-style options. *IEEE Transactions on Neural Networks 12*(4), 694–703.

Vicente, L., F. Cerezetti, S. D. Faria, T. Iwashita, and O. Pereira (2015). Managing risk in multi-asset class, multimarket central counterparties: The CORE approach. *Journal of Banking & Finance 51*, 119–130.

Index

Acknowledgments

The book is mainly based on the following papers or variations thereof: Armenti, Crépey and Zhou (2017) (Chapter 0), Albanese, Crépey, Hoskinson and Saadeddine (2021) (Chapters I and VII), Crépey (2022) (Chapters II and III), Abbas-Turki, Crépey and Saadeddine (2023)[1] (Chapters IV and V), Crépey, Li, Nguyen and Saadeddine (2024)[2] (Chapter VI), Bastide, Crépey, Drapeau and Tadese (2023) (Chapter VIII), Bastide, Crépey, Drapeau and Tadese (2025)[3] (Chapter IX), Albanese, Crépey and Iabichino (2023) (Chapter X), Bénézet and Crépey (2024) (Chapter XI). I express my deep gratitude to all co-authors. A special word for Claudio Albanese, whose beautiful ideas are the source of the cost-of-capital XVA approach.

The research presented in this book has also greatly benefited from exchanges with the banking industry, especially in the context of interactions with various research chairs, namely (by alphabetical order): the chair Capital Markets Tomorrow: Modeling and Computational Issues, under the aegis of the Institut Europlace de Finance, a joint initiative of LPSM/UPCité and Crédit Agricole CIB; the chair Futures of Quantitative Finance, a partnership between LPSM at UPCité and CERMICS at École nationale des ponts et chaussées; the chair Stress Test, Risk Management and Financial Steering, led by the French École polytechnique and its Foundation and sponsored by BNP Paribas. Let each of these partners be warmly thanked for unique collaborative adventures.

[1]With repository https://github.com/BouazzaSE/NeuralXVA.
[2]With repository https://github.com/hoangdungnguyen/CVA-learning-and-hedging.
[3]With repository https://github.com/mekonnenta/CCP-Radner.git.

About the Author

Stéphane Crépey is Distinguished Professor in Mathematics at Université Paris Cité (UPCité), Laboratoire de Probabilités, Statistique et Modélisation (LPSM), Head of Team Mathematical Finance and Numerical Probability. His research interests are:

- counterparty credit risk and XVA analysis, central counterparties, risk analysis;
- machine learning in finance: supervised learning of prices, sensitivities and risk metrics from simulated payoffs, model calibration by neural nets or Gaussian processes regression, anomaly detection;
- model risk and uncertainty quantification;
- backward stochastic differential equations, random times modeling, enlargement of filtration.

He co-authored 80+ research papers published in journals including Annals of Probability, Annals of Applied Probability, Finance and Stochastics, Mathematical Finance, Stochastic Processes and their Applications, Electronic Journal Probability, SIAM Journal on Financial Mathematics, SIAM/ASA Journal on Uncertainty Quantification, SIAM Journal on Mathematical Analysis, Quantitative Finance, or Risk Magazine. Before this one, he wrote two books: Financial Modeling, a Backward Stochastic Differential Equations Perspective (S. Crépey, Springer Finance Textbook Series, 2013) and Counterparty Risk and Funding, a Tale of Two Puzzles (S. Crépey, T. Bielecki and D. Brigo, Chapman & Hall/CRC Financial Mathematics Series, 2014).

Stéphane Crépey graduated from ENSAE ParisTech and holds a PhD in differential games and mathematical finance from Ecole Polytechnique and INRIA Sophia Antipolis. He is currently

- UPCité head of master M2MO[1];
- scientific officer of the chair Capital Markets Tomorrow: Modeling and

[1]https://masterfinance.math.univ-paris-diderot.fr/.

Computational Issues[2], under the aegis of the Institut Europlace de Finance, a joint initiative of LPSM/UPCité and Crédit Agricole CIB;

- UPCité leader of the chair Futures of Quantitative Finance[3], a partnership between LPSM at UPCité and CERMICS at École nationale des ponts et chaussées;
- scientific collaborator of the chair Stress Test, Risk Management and Financial Steering[4], led by the French École polytechnique and its Foundation and sponsored by BNP Paribas;
- member of the scientific advisor board of the AMF[5] (French financial markets authority).

[2]https://www.institutlouisbachelier.org/recherche/programmes/banque-de-marches-de-demain-enjeux-modelisation-et-calcul.

[3]https://futures-quantitative-finance.u-paris.fr.

[4]http://www.cmap.polytechnique.fr/~stresstest/.

[5]https://www ientifique.